D0897661

THE SOCIAL ECONOMY OF CITIES

Volume 9, URBAN AFFAIRS ANNUAL REVIEWS

INTERNATIONAL EDITORIAL ADVISORY BOARD

ROBERT R. ALFORD
University of Wisconsin

HOWARD S. BECKER
Northwestern University

ASA BRIGGS
University of Sussex

JOHN W. DYCKMAN
University of California (Berkeley)

H. J. DYOS
University of Leicester

JEAN GOTTMANN
Oxford University

SCOTT GREER
University of Wisconsin (Milwaukee)

BERTRAM M. GROSS
City University of New York

ROBERT J. HAVIGHURST
University of Chicago

EIICHI ISOMURA
Tokyo Metropolitan University

WILLIAM C. LORING
U.S. Public Health Service

MARTIN MEYERSON
University of Pennsylvania

EDUARDO NEIRA-ALVA
CONDER, *Brazil*

ELINOR OSTROM
Indiana University

HARVEY S. PERLOFF
University of California (Los Angeles)

P. J. O. SELF
The London School of Economics and Political Science

KENZO TANGE
Tokyo National University

WILBUR R. THOMPSON
Wayne State University and
Northwestern University

THE SOCIAL ECONOMY OF CITIES

Edited by
GARY GAPPERT
and
HAROLD M. ROSE

Volume 9, URBAN AFFAIRS ANNUAL REVIEWS

SAGE PUBLICATIONS / BEVERLY HILLS / LONDON

DEDICATION

This volume is dedicated to our friend and colleague Kirk R. Petshek (1913-1973), who pioneered in the theory and practice of urban social economy in Philadelphia and Milwaukee.

Copyright © 1975 by Sage Publications, Inc.

All rights reserved. No part of this book may be reproduced or utilized in any form or by any means, electronic or mechanical, including photocopying, recording, or by any information storage and retrieval system, without permission in writing from the publisher.

For information address:

SAGE PUBLICATIONS, INC.
275 South Beverly Drive
Beverly Hills, California 90212

SAGE PUBLICATIONS LTD
St George's House / 44 Hatton Garden
London EC1N 8ER

Printed in the United States of America

International Standard Book Number 0-8039-0326-X

Library of Congress Catalog Card No. 73-88911

FIRST PRINTING

HT108
U7
vol. 9

CONTENTS

Introduction □ *Gary Gappert and Harold M. Rose* 7

PART I: THE CHANGING CONCERNS OF NATIONAL URBAN POLICY 37

1 □ The Political Economy of Urban America:
National Urban Policy Revisited □ *Jesse Burkhead* 49

2 □ The Large City as Fiscal Artifact □ *Seymour Sacks
with Patrick J. Sullivan* 69

PART II: URBAN HOUSING MARKETS AND THEIR INTERPRETATION 115

3 □ The Political Economy of Urbanization in
Advanced Capitalist Societies: The Case of the
United States □ *David Harvey* 119

4 □ Short-Term Housing Cycles in a
Dualistic Metropolis □ *Brian J.L. Berry* 165

5 □ Socioeconomic Transition and Housing Values:
A Comparative Analysis of Urban Neighborhoods □
Hugh O. Nourse and Donald Phares 183

PART III: ANALYZING THE DYNAMICS OF URBAN SPATIAL CHANGES 209

6 □ Emerging Patterns of Industrial Activity Within
Large Metropolitan Areas and Their Impact on the
Central City Work Force □ *Charles M. Christian* 213

7 □ The Street Gang in Its Milieu □ *David Ley* 247

8 □ Locational and Ecological Aspects of
Urban Public Sector Housing: The Israeli Case □
Amiram Gonen 275

PART IV: PERSPECTIVES ON URBAN STRESS AND POPULATION CHANGES 297

9 □ Urban Black Migration and Social Stress:
The Influence of Regional Differences in
Patterns of Socialization □ *Harold M. Rose* 301

10 □ The Periphery at the Pole: Lima, Peru □
   Milton Santos   335

11 □ Indicators of Labor Market Functioning and
   Urban Social Distress □ *Robert Mier,*
   Thomas Vietorisz, and Jean-Ellen Giblin   361

12 □ The Distribution of Coloured Immigrants in Britain □
   Ceri Peach, Stuart Winchester, and Robert Woods   395

PART V: POVERTY AND RESOURCES: ANALYSIS AND MOBILIZATION   421

13 □ Urban Work in a Changing Economy □
   Scott Greer and Ann Lennarson Greer   427

14 □ The Social Economy in Urban Working-Class
   Communities □ *Martin D. Lowenthal*   447

15 □ Rethinking Urban Problems: Inequality and the
   Grants Economy □ *John P. Blair, Gary Gappert, and*
   David C. Warner   471

16 □ Recent Trends in Corporate Urban Development
   Programs □ *Jules Cohn*   507

PART VI: THREE PERSPECTIVES ON EVALUATION   525

17 □ Social Indicators and Society:
   Some Key Dimensions □ *Michael J. Flax,*
   Michael Springer, and Jeremy B. Taylor   535

18 □ Program Evaluation and Policy Analysis of
   Community Development Corporations □
   Harvey A. Garn   561

19 □ An Evaluation of Evaluation Research:
   When Professional Dreams Meet Local Needs □
   Michael Barndt   589

AN EPILOGUE ON POLICY

20 □ Alternative Agendas for Urban Policy and
   Research in the Post-Affluent Future □
   Gary Gappert   611

THE AUTHORS   639

Introduction

GARY GAPPERT
HAROLD M. ROSE

□ THE SOCIAL ECONOMY OF CITIES is a theoretical construct. As with many of the constructs of conventional economics, its value lies perhaps more in its conceptual suggestion than in its analytical precision. The social economy of cities, simply put, is the interrelationships of the social and economic systems operating within the spatial boundaries and institutional framework of a particular urban place.

This ninth volume in the *Urban Affairs Annual Reviews* has been motivated by several related concerns associated with the social economies of cities.

Without a precise theoretical model of the urban social economy to work from, many urbanists have been forced to interject their own normative mode of analysis. This volume has been assembled to reflect the rich diversity of formulations and empirical research which can currently be said to represent the still inchoate field of urban social economy. A main concern for urban social scientists is the struggle to identify and map (cognitively as well as physically) the different and emerging patterns of socioeconomic behavior within a particular urban space, be it neighborhood or metropolis. Both traditional social and economic theories seem out of focus when it comes to much of urban phenomena.

Lacking any kind of comprehensive, systemic deductive model of the social economy of cities, the urban theorist is faced with the kind of problem posed by the philosopher, Gilbert Ryle (1954), when he wrote:

A theorist is not confronted by just one question, or even by a list of questions numbered off in serial order. He is faced by a tangle of questions, intertwined and slippery questions. Very often he has no clear idea what his questions are until he is well on the way towards answering them. He does not know, most of the time, even what is the general pattern of the theory that he is trying to construct, much less what are the precise forms and interconnections of its ingredient questions.

It is suggested that there is a continuing tension between theoretical formulations and empirical procedures. It is this tension which is often neglected by the policy analyst driven by urgent social demands upon political processes. One response to this dilemma is to attempt to provide orderly classification schemes by which urban problems can be sorted out and understood. George Rohrlich (1974), an economist at Temple University, is an advocate for such an etiological approach:

What needs to be brought together and integrated are the insights gained in the several psycho-social sciences and the applied arts and helping professions with a view to marshalling these for economic analysis and policy purposes.

He suggests a threefold process. First, available knowledge and research should be systematically scrutinized so as to provide a perspective on the predictable dysfunctional consequences of inputs in such key areas of health, education, community development and so forth. Second, the critical points of social renewal and community equilibrium by which institutions, traditions and established usages generate a degree of social coherence need to be identified. Third, the analysis of the allocative and distributive effects of given urban policies and practices must be made, along with a calculation of their costs, and these must be arrayed against the same kind of analysis for other possible courses of action. In short he is suggesting that economists and other social scientists tackle the economics of the "web of life" in all its dimensions (notably social relationships, forms of organization, and institutions) which interface and compete in the partly natural, partly man-made environment of particular urban places.

Such an attempt to build an etiological taxonomy of urban phenomena and policy process is ambitious if not premature. Kenneth Boulding (1958) has warned us that—no matter how much we improve our knowledge of the universe and society—we are extremely unlikely ever to do away with all the uncertainty about which consequences follow what action. Important questions may also remain unanswered by both empirical research and theoretical synthesis because they may relate essentially to questions of value, belief and ideology.

In the rest of this introduction we would like to raise some of these questions, examine the scope of urban economics and its correlates, and suggest some important dimensions of urban social economy.

I. THE DILEMMA OF NATIONAL ECONOMICS AND THE DEVELOPMENT OF URBAN ECONOMICS

It is ironic that at a point where the "success" of urban economics is being hailed, national (or macro) economics based on the structure of neo-Keynesian thinking has reached what partially appears to be a dead end. Nobel Laureate Kenneth Arrow has referred to the coexistence of rising, persistent inflation and growing unemployment as an "uncomfortable fact and an intellectual riddle" (*New York Times*, March 20, 1973). At the same time William Pendleton, speaking for the Ford Foundation, proclaimed the success of their "investment" in urban economics. Speaking before a meeting of the American Council on Education, he said:

Beginning in 1959, an eight year effort was begun to turn the attention of the economics profession towards the city. Administered by Resources for the Future's Committee on Urban Economics, the program spent about $1.25 million trying to pay attention to what was happening in the nation's cities. Through the eight years of its life, the Committee produced about forty Ph.D.s, helped to establish half a dozen research centers, and promoted a number of research projects in what was, at the time, virtually an unknown field.

He concluded "the money was well spent" and that "the field is here to stay" even though "we have a long way to go before we really understand the economic dynamics of the city" (Pendleton, 1974: 6-7).

Although it has been said that if you put four economists in the same room, you will get at least eight opinions on the same issue, it is perhaps possible to make a number of general observations about the state of national and urban economics with which most economists would substantially agree. From these observations the need for an explicit concern with urban social economics can be seen. (It has also been said, and this was reflected in President Ford's September 1974 "summit meeting" on the U.S. economy, that a group of economists will spend 95% of their time arguing about the 5% of an issue around which they disagree. Economists can disagree about so much simply because there is much about which they can agree.)

One general proposition is that during the last thirty years economics celebrated its great triumph. That triumph has been described as the "neo-Keynesian synthesis." The basic ingredients of the synthesis revolve around the development of a theory of national demand-management and a set of supportive policies. The strength of neo-Keynesian economics has rested upon a combination of measurement (GNP and national income accounts), theory (fiscal drag and the full employment gap), and policy prescriptions (fiscal fine-tuning). In the United States, the neo-Keynesian movement reached its greatest success with the tax cuts of the early 1960s which sparked a sustained period of economic growth and expansion. A whole generation of economists witnessed the first field test of their theory: they had split their "atom." The appearance of John Maynard Keynes' picture on the cover of *Time* magazine and the subsequent establishment of the Nobel Prize in economics culminated the celebration (Burkhead, 1971).

At the present time, however, the national economy does not face a demand-management problem, with its corollaries of unemployment and lagging growth. Instead the economy faces a problem of supply, characterized by shortages, inflation, and doomsday scenarios of the limits of growth on a finite planet. The situation is exacerbated by a structure of controls, incentives, and subsidies which enrich some and impoverish others and which have led to a further expansion of the parameters of inequality. As a result of these developments even the Republican chairman of the Council of Economic Advisors has urged that the country consider moving in the direction of "indicative planning" which has been pioneered in France and adopted in Japan (Black 1967; Coffey, 1973). Indicative planning is a system of cooperative planning undertaken by the major economic interest groups in which economic goals are disaggregated by important sectors, and a series of targets and tactics are agreed to by the major parties.

The spectre of indicative planning as a strategy for national economic planning leads, through its concern with disaggregation, to a concern with regional planning and development. (For the French experience, see Hansen, 1968.) Concern for regional planning leads to issues of planned urbanization, of selected decentralization, and other such concerns associated with the geographic distribution of economic activity. These kinds of concerns have been developed in the literature of regional economics and around the notion of a national system of cities.

Wilbur Thompson (1972) has reviewed the policy and theory issues associated with the national system of cities as an object of public policy. Wingo (1972) and Chinitz and Dusansky (1972) have also reviewed these issues. Wingo concludes that "the base of solid analysis on which decisions about national urban growth policy will have to rest is hardly strong enough to bear the weight."

Kerr and Williamson (1970) have reviewed the state of regional economics in the United States. They divide regional economics into three broad fields: the economics of location, trade multiplier theory, and interindustry (input-output) analysis. Urban economics is treated as a fourth separate field of theory and technique. Regional economics is distinguished by its concern with the spatial dimensions of the location of economic activity. Isard and others have for more than twenty years expressed concern about this dimension, which led them to promote the development of regional science. For them (see Isard and Reiner, 1966), regional science problems are those for which a spatial focus is central. They have attempted to build regional analysis into the same kind of general equilibrium framework which distinguishes neo-classical economic theory.

Another general observation might be that the further development of the field of regional economics is limited by the costly informational demands required to operationalize complex regional economic models, and by problems of social and political plausibility. Ultimately an approach to regional development is as much dependent upon political choices as it is upon analytical refinements and theoretical breakthroughs. Kerr and Williamson note the absence of behavioral formulations (and testing) within the regional economic tradition. They describe urban economics as "an area of study which is in great need of more knowledge of basic behavioral relationships and how they can be fitted into a general explanation of economic phenomena." A number of surveys of urban economics has also revealed this and related problems in the field of economics.

The field of urban economics is really the economics of land use and development, with special emphasis on the ancillary systems of housing and transportation. The economics of the growth and development of the spatial form of urbanized areas has also been the concern of numerous models built by urban economists. Based upon the logico-deductive tradition of neo-classical theory, urban economists have developed models dealing with urban growth, land values, housing discrimination, transportation congestion, and so forth. Goldstein and Moses (1973: 493) indicate that "the dominant aspect of urban economic theory is competition for land." Major difficulties with this dominant model of urban economics include: (1) the assumption of a single center of economic activity, (2) the absence of interdependencies among economic units on the choice of location, and (3) the absence of institutions including government (p. 495). Much of the work associated with the elaboration of this kind of model has also been based upon a static assumption.

More sophisticated models have been developed within what Mills and Mackinnon (1973) call the "new" urban economics. They write that "the hallmark of the new urban economics is the use of fairly sophisticated mathematics-calculus of variations, programming and control theory–to characterize some fundamental aspects of urban structure." Much of this work has been characterized as the expenditure of "large amounts of technical expertise" on "relatively minor variations of a few themes" (p. 595). A more serious reservation is expressed by Mills and Mackinnon (p. 600) when they write: "It is somewhat ironic that economists are becoming proficient at building models in which all employment is in the CBD and all housing outside it, when cities look less and less like that paradigm."

But the achievement of urban economics has indeed been considerable. The models have provided a more thorough treatment of the problems of externalities. They have developed an understanding of the problems of peak load pricing, of property rights assignments, of transaction costs, and so forth. The problems of a land tax, the housing filtration process, and vacancies chains have been identified and illuminated through the analysis of the land economy of cities.

The political economics of cities represents a more distinct concern with the economics of government activities within the city. The problems associated with various externalities generated from public and private actions in the urban land economy have been a major preoccupation of urban political economics. Environmental spillovers, the proper tax treatment of land improvements, the development of user charges, and governmental regulation have received special attention (Mackean, 1973). The literature has also emphasized the fiscal crisis of cities, problems of taxation, and the supply of, and demand for, government services (Crecine, 1970; Bish, 1971; Neenan, 1972). The value of much of this work applying the neo-classical tradition to the land economy and the political economy of cities is that the theoretical models allow us to gain insight into certain characteristics of urban areas by a rigorous treatment of a small number of relationships thought to be important to the spatial or governmental characteristic in question. Such characteristics as the relative location of different income groups, variable capital-land ratios, transportation congestion and so forth can be elegantly derived from such models.

It is harder than it may seem to describe how the economy of a city is really working. The description and analysis is difficult to get at. In many cases theory is relied upon to say that "the system is working as if . . . ."

At the same time, these models have been useful in directing some of the empirical work with the data at hand. Econometric models have been developed to determine whether or not data representing the outcomes of certain economic processes (such as housing valuations) can be shown to represent the working out of certain types of market forces. Muth's *Cities and Housing* (1969) stands out as one of the controversial classics in this area. These kinds of models are most useful in helping people think systematically about the issues at hand. They are less useful and less reliable when they are used to make growth projections over time because over time values, tastes, and institutions may change. Although evaluating and understanding urban land market forces is extremely useful, designing programs, policies and projects to intervene or to alleviate the outcome of forces becomes the urban development problem. As it turns out many solutions do not work or they work for the wrong reasons. In some cases urban projects are designed to ride with the prevailing market forces and have no chance of failure. Other projects ignore these forces and have no chance of success.

Taken on their own terms, the neo-classical models have much value. The problems with these models are the problems of traditional economics generally. First, the models reflect a concern with allocative efficiency and not with questions of equity, distributional efficiency, and so forth. Economics does not have a useful theory of income distribution. Kenneth Boulding (1973) once said that "on this subject traditional economics is dead as a dog because it has tried to steer away from interpersonal comparisons." Within conventional welfare economics, the concept of Pareto optimality has long been used to assert that we can move toward improving someone's position in society as long as we do not impair the position of others. But it is, of course, the relative position which counts in the real world, as Veblen and Henry George suggested a century ago. More recently Arartya K. Sen (1970: 22) made this assessment:

There is danger in being exclusively concerned with Pareto-optimality. An economy can be optimal in this sense even when some people are rolling in luxury and others are near starvation as long as the starvers cannot be made better off without cutting into the pleasures of the rich. . . . In short, a society or an economy can be Pareto-optimal and still be perfectly disgusting.

Second, economics has lacked a general theory of the state or of social institutions. These either have been taken as given, or they are introduced to respond to market failures, externalities and so forth (Haverman and Margolis, 1970). A theory of collective choice has been developed over the last decade. This theory is based upon the assumption that individuals pursue their own self-interest and act to maximize "utility." It then develops the conditions under which group decisions rather than individual decisions will emerge. The possibilities of utility interdependence, of a "social group will," or of an "overriding public interest" are rejected (Buchanan and Tollison, 1972).

Third, lacking a theory of institutions, economics cannot really grapple with the kinds of changes in urban conditions which are induced by the "openness" of the urban economy. It is the flows into and out of and between cities which dominate urban economic life. These forces are not contained within the walls of the city-state. The forces stem from the state of the national technology and culture. Although it is true that urban areas are characterized by the existence of various local and regional markets which can be explicated by economic theory, they are also influenced by social, cultural, and psychological forces and by political and historical circumstances. In recent times economics generally and urban economics in particular have either ignored these variables or have treated them as given. Power, status, envy, discrimination, good fellowship, and so forth are often more important determinants of economic activity than are wage rates or land prices. As Marshall (1974: 861) points out in his review of the economics of racial discrimination, "The neo-classical theory does not deal with institutional discrimination because the model assumes . . . that institutions are fixed and therefore do not interact with the basic variables in the model."

Finally, traditional economics has, for the most part, forgotten its origins as a branch of moral philosophy. Since Alfred Marshall in the late nineteenth century, economic science has developed a value-free analytical apparatus (i.e., taking values as given) and has not struggled with understanding the full range of goals associated with economic activity—at the level of the nation or the city or the community. How do social and interpersonal values relate to patterns of economic activity? How do economic development, growth, and decline effect, retard, or ignore processes of social development? What is it that people do within the city and why do they do it? They sleep, eat, drink, work, recreate, fornicate and otherwise maintain themselves. Just as economics has ignored the social person in its models derived from an "economic man-utility maximizer" construct, so have sociologists, in studying the city as a social realm, ignored the economic person. As Bookchin (1973) has noted, "Urban sociology presupposes that the city can be taken as a social isolate and examined on its own terms."

It is perhaps time to pursue the study of economic activity at the level of the city more explicitly as a series of social processes. Two perspectives on this pursuit are readily called to mind. Barbara Bergman, newly elected president of the Eastern Economic Association, has reviewed the problems of economists dealing with behavioral research. As Bergman (1974) notes:

> One habit of mind that has inhibited the accumulation of useful knowledge is the unwillingness and inability of economists to collect for themselves information about the behavior of men and women when they act as economic agents.
>
> Furthermore . . . The infatuation of economists with statistical significance has cast all empirical inquiries into the form "Does this variable affect that one?" and has caused the downgrading of investigations which have as their aim the description of practices and habits.

She urges that the economist begin to employ the research techniques of the social anthropologist and experimental psychologist in analyzing economic activity through personal and direct observation.

A second and different kind of perspective is suggested by Adolph Lowe and Robert Heilbroner. Heilbroner (1970b), in reviewing the limits of the scientific paradigm in economics, has called for the "conversion of economics into an instrument of social science whose purpose and justification was not so much the elucidation of the way society actually behaves, as the formulation of the way in which it should behave" if it wishes to achieve a postulated goal such as full employment, balanced growth, and so forth.

This reorientation of economics has been dubbed "instrumental economics" by Adolph Lowe in his classic, *On Economic Knowledge* (1965). Instrumental economics requires that the goals of social development be explicitly articulated. Once the goals (and alternatives) of social development, either national or urban, are posited, it is the role of economic analysis to begin to describe the various ways in which a social system at point A may begin to move toward point B. This is the approach that Bennett Harrison (1974) calls for when he sets the goal of "deepening" urban development.

Lowe calls this the "logic of economic goal setting." He notes that the major unknowns are (1) the terminal state and process of the macro-system, (2) the patterns of micro-behavior and micro-motivations necessary to keeping the state on that path or at that level, and (3) a state of the environment necessary to insure or permit those behaviors and motivations. Traditional economics (and social science) is employed to specify the initial state of the system including whatever laws, rules, and empirical generalizations are necessary for the understanding of the social system in question (Lowe, 1969: 17). The reintegration of economics into social science seems to be in order.

II. THE ECONOMIST AS SOCIAL SCIENTIST

Jacob Viner once suggested that economics is what economists do. Perhaps it could be suggested that social economics is what other social scientists do when they study the implications of economic activity. Certainly economists have always had a great difficulty in pursuing the analysis of economic activity beyond the realm of traditional neo-classical theory. The social outcomes of economic activity or the economic significance of social activities have not been comfortable areas of study for the economist. Vilfredo Pareto (1935) once admonished his fellow economists "to resort to other sciences and go into them thoroughly—not just incidentally—for their bearing on the given economic problem."

Harry G. Johnson has also complained about the preoccupation of the economist with his new-found mathematical models. At his Inaugural lecture at the London School of Economics, Johnson exclaimed:

Economic theory is now written in set theory, mathematical models, and regression results, with the consequence that it seems to be too abstruse and remote from reality as well as too technical, to be of any interest or use to the non-specialist concerned with understanding and improving the world in which he lives.

But at the same time the economist has ignored the tools of inquiry of the other social sciences. Johnson continues:

This impression is ... often shared by the professional economists themselves, who frequently confine the use of their scientific tools to technical and academic problems, while relying on intuition, emotion or political philosophy to form their attitudes on social questions. [Johnson, 1967: 3]

Similarly, Heilbroner (1970b) has also identified the "scientific paradigm" of economics as a model which has led the economist into a number of conceptual dead ends. He writes:

Here is where the scientific model of economics thus meets its ultimate limit. For in its unwillingness to indulge in sociopolitical guesswork out of fear of transgressing the limits of proper scientific procedure, economics is placed in a position in which its findings must be partial and incomplete and very possibly erroneous. For every economic act, from the simplest response to a change in prices to the most complex decisions of an entrepreneur, is also a social act.

It must be realized that the "economic man" abstraction of economic theory must be replaced by a broader socioeconomic definition. The economic order of the modern urban world is also a generator of power, privilege, lifestyle, and values. Economic reality in fact generates a social man concern which is reflected in the cultural and political evolution of our urban society. There is probably no easy way out of the specialization niches of the contemporary academic disciplines. No doubt some economists will attempt to build into their economic society formulations some additional concepts reflecting an abstraction of social reality. Kevin Lancaster's (1971) new theory of consumer behavior comes to mind. Other economists will seek to adopt and adapt the field method practices of sociologists and anthropologists. More significant perhaps is that other social scientists pursue their own investigations to the outcomes and patterns of economic activity, deriving their own theories as they go along.

Perhaps there are four distinct social science paradigms around which the study of the urban social economy could be pursued by economists, social scientists, or whomever. Paradigm I is the traditional, logical-deductive paradigm. This posits a number of assumptions about economic man (and the economic organization) and derives conclusions about behavior from them. This is the model provided by Bish (1971: 1-17) in his *The Political Economy of Metropolitan Areas*. To broaden the model and to extend its usefulness, a series of assumptions about sociocultural man would first have to be made. Alternatively, one might also develop a political man or a psychological man construct from which certain implications about the behavior of urban man could be logically determined. This mode of logico-deductive scholarship is popular in all the disciplines but has been most extensively developed in the neo-classical tradition of economists.

Paradigm II is the paradigm of statistical significance. This paradigm either develops or tests hypotheses by exploring the significance of data—usually secondary data derived from public sources. With cross-sectional data strong correlations are used to imply or to confirm causation. With longitudinal data causation is also often implied from correlation, and future outcomes are sometimes projected. Most modern social scientists are now facile with these techniques, but again it is the econometric tradition which is the most sophisticated manifestation of this paradigm. Muth's *Cities and Housing* is a seminal treatment of statistical correlations to infer the dynamics of the housing market in Chicago. Jay Forrester's *Urban Dynamics* represents probably the most sophisticated extension of correlation between data series to simulate a set of projected futures.

Paradigm III is the primary behavior paradigm or the paradigm of experiential learning. This paradigm is concerned with the description and interpretation of actual behavior. This paradigm contains the methods of the participant observer—the social anthropologists, the psychiatrists, and others who seek to derive explanation from exposure to actual behavior or from primary data. This is what Glaser and Strauss (1967) call "the discovery of grounded theory." It may involve the development of a survey instrument or the sorting and classification of behavior and behavioral types. Paradigm III is also concerned with the realm of planned social experimentation. The recent federal support of income maintenance and housing allowance experiments has refocused the concern of the urban social economist on the methodological difficulties of field work.

Paradigm IV is the paradigm of goal-directed analysis or the paradigm of organizational development where goals are established, and policies, structures, and programs developed to reach those goals from an initial state of the system (which also must be specified). This paradigm is present in the economic development tradition, where most economists involved in that field wound up with a concern for institution-building. In the field of urban social economy, the problems and potential of urban ghetto development represent almost a separate subfield (see Harrison, 1974a) based upon this paradigm.

III. TEN DIMENSIONS OF URBAN SOCIAL ECONOMY

Without being able to identify a complete model of the social economy of cities, a more modest approach at its definition can be achieved by listing ten major dimensions. Louis Wirth formulated the concept of urbanism around the notions of size, density, and heterogeneity. A more appropriate contemporary conception of urbanism, featuring relationships expressing underlying socio-economic realities, should probably include the notions of (1) urban population changes, (2) segmented markets, (3) the poverty and progress paradox, (4) the organization and distribution of lifestyles, (5) the problems of community or group identification, (6) the manifestations of racial conflict and cultural aversions, (7) the range of intermediate institutions and organizations, (8) innovation and diffusion, (9) the changing nature and value of centrality, and (10) the evolving uses of urban structural forms. These dimensions are treated below.

Urban population changes and the dynamic forces behind those changes are perhaps the chief manifestations of the social economic conditions of the city. The population changes reflect the push-and-pull forces of migration between and within cities. The attention of the last three decades has focused primarily upon the rural to urban population dynamic. As Niles Hansen demonstrated in *Rural Poverty and the Urban Crisis* (1970), these movements can be best understood within a framework of regional development, economic motivations, and social needs.

In the years ahead the analysis of intra- and interurban population movements will be of greater importance. As Morrison (1974) has demonstrated in his studies for The Rand Corporation, intrametropolitan movements within cities like St. Louis and Seattle, strongly influenced by federal procurement policies, represent the changing but chronic urban condition.

Associated with these population changes is the development and evolution of *segmented urban markets* for housing, labor, social services, and so forth. Earlier views of the homogeneous nature of urban economic processes have given way to more detailed analysis on the segmented nature of the supply and demand forces operating within the urban region. This is especially true of the several markets for labor which seem to exist. Both the notion and analysis of a dual labor market (Reich, Gordon and Edwards, 1973) and of an internal labor market (Piore and Doeringer, 1971) provide a much more realistic perspective on the development of urban policy.

Similar concepts of segmentation in other markets are perhaps important in understanding the *poverty and progress paradox*. Urbanists since the time of Henry George have struggled to understand the persistence of economic instability and deprivation alongside the growth of wealth and affluence. It may well be that the lower class so deplored (but accepted) by Edward Banfield and the so-called "culture of poverty" identified by Oscar Lewis are really best viewed as social adaptations to the structure of economic inequality which is so endemic to urban growth.

One of the significant spatial manifestations of the persistence of economic inequality in cities is the evolving *distribution of lifestyles* governed by socioeconomic circumstances. From Dorcester in Boston to River Hills in Milwaukee to the new town of Flower Mound in Texas, the self-selecting forces of social and economic segregation are strong, persistent, and resistant to planned change.

It may be, as David Harvey (1973) points out, that as the forces of urbanization create a more homogenous nation, the demands of lifestyle and workstyle affinity groups reestablish within the city "regional" differences. Insofar as the urbanization of the countryside has created an interrelated network of economically effective space, people attempt by all manner of social means to differentiate what the marketplace has in fact rendered homogenous. As Harvey writes: "Hence the urban space economy is replete with all manner of pseudo-hierarchical spatial orderings to reflect prestige and status in residential locations. These orderings are very important to the self-respect of people."

Out of this "created space" may come the forces of social redistribution and reciprocity. These forces are perhaps the significant forms of behavior which along with communication and the ordering of behavior make up a *sense of community* (Minar and Greer, 1969). The struggle to find or to recreate a sense of community within the competitive nexus of the urban markets remains contested throughout contemporary metropolitan America. The "privatism" of the early industrial city continues and has redefined itself in suburbia as Packard (1972) and Slater (1970) have pointed out.

It is not necessary to define the social purpose of community-seeking behavior within the spatial boundaries of the traditional urban neighborhood. This concept perhaps is more of a theoretical construct than a social reality. But the notion of neighborhood is being resurrected, either through the imperative of political decen-tralization (Kotler, 1969) or through the economic value of "neighborly behavior" (Keller, 1968). But the sense of community which seemingly has a socioeconomic logic is often torn apart by the acceleration of *racial conflict and cultural aversion*. On the one hand these emotional forces with their complex psychological roots tend to reinforce socioeconomic segregation tendencies. The social stress engendered by economic competition, instability, and the frailty of social cohesion in a changing city is exacerbated by these forces. But on the other hand it is unclear whether the further evolution of the social economy of cities can be directed so as to ameliorate these tensions. Halpern (1969), in defining racism as a symptom of incoherence between individuals and social groupings, suggests that the tension management of other kinds of relationships must be sought so as to achieve new, transformed relationships between members of different groups. Directing the social economy of the city to that kind of end may be beyond the ability of the urban political process.

Behind the myth of the partially integrated, partially pluralistic city lies the need for a concept of metropolitan citizenship which is illusive and visionary. Perhaps the preoccupation of political science with issues of metropolitan functions and structures has obscured the need to understand the historical and cultural forces present within the political process. More important to the understanding of the further evolution of the social economy of cities is the need to perceive *the range of intermediate institutions and organizations* which operate within the urban region. These intermediate insti-tutions operating at the level of the urban scale are the single most important interface between the local citizen and the national society. In spite of the dominance of the national corporate economy and the preeminence of a televised culture, it is the institutions which are operative within the 243 SMSAs that chiefly influence the pace and style of life for most Americans.

With the advent of the several forms of revenue sharing, localism in the range of urban institutions has been reinforced, if not reestablished. A more systematic analysis of the structure of local institutions is required. Warren's notion of the interorganization field in his recent *The Structure of Community Reform* (1974) is an important new construct by which interaction, coordination, innovation, and responsiveness within urban institutions can be analyzed. As Williamson (1973) has pointed out, the development of non-market forms of economic organization occurs whenever economic markets, attempting to complete a set of transactions, experience social or political frictions. In one sense urban institutions of many kinds arise from a social reaction to the workings and failures of economic markets. But at the same time, institutional or organizational failures lead to a whole series of bureaucratic pathologies which are themselves a manifestation of the "chronic urban" condition. Changes in the way in which urban institutions allocate economic resources for social needs reflect the evolving nature of a major portion of the urban social economy.

Jane Jacobs, in both her *The Death and Life of Great American Cities* (1961) and *The Economy of Cities* (1969), has exalted the tumultuous qualities of urban life in fostering creativity. The processes of *innovation and diffusion* are not well understood, but both theory and experience suggest that the supply of deviance and the demand for innovations often achieve a synergistic equilibrium in the hurly-burly of urban systems. It is unclear, however, if Jacob's model, primarily drawn from New York City in which many things have been invented ranging from brassieres to neighborhood school councils, is equally applicable to Dayton, Milwaukee, and Omaha. A city perhaps has a period of innovation and a period of diffusion, and its survival may depend upon prolonging those periods. The socioeconomic correlates of innovation in both the public and private sectors remain undetected and untested.

Associated with the processes of innovation and diffusion is the *changing nature and value of centrality*. The centrifugal forces manifested in suburbanization and the emergence of "spread city" have obscured the fact that new centers of centripetal attraction have appeared throughout the metropolis. The concern of the traditional civic elites with the reduced significance of the central business district has lead to a lack of attention to the social and economic processes which have sprung up around the airports, the office parks, the shopping centers, and so on—all of which represent the polycentric city. This neglect has been compounded by the work of the architect-planners who have designed urban forms with rigorously separated specialized functions, thereby making it more difficult to reestablish the creative congestion of traditional urban centers.

The evolving uses of urban structural forms become, in a period of rediscovered economic scarcity, of central importance in the attempt to use "natural" centripetal forces to offset the resource-wasting forces of subsidized suburbanization. Too often a wasteful "physical fix" is sought as a response to urban social problems. From the "super suburbs" called New Towns to the "defensible space" fantasies of public housing planners, the construction of new designs has been sought as some kind of solution to social malfunctions. An alternative perspective would seek to organize and motivate social and economic behavior toward the better use and adaptation of existing structural forms. The organization of low-income tenants into public safety committees is as important to the evolving use of public housing as is the formation of development districts by private businessmen to the preservation of retail shopping areas.

IV. TOWARD THE STUDY OF
URBAN SOCIAL ECONOMY

From the perspective suggested by the ten dimensions of urban social economy, a number of research orientations seem important.

First, many more literary expositions of analytically formulated urban models need to be attempted. The imaginative work begun in 1965 by Wilbur Thompson needs to be continued, but with the special inclusion of social, institutional, and organizational variables. Perhaps the work of David Harvey (1973) points to the continuation and elaboration of that tradition.

Second, alternative goal structures for urban development need to be more sharply articulated and discussed. If we are to have better models of the urban social economy so that planned interventions can be made more effective, the objective function of urban development policy needs to be more precisely formulated. As a Ford Foundation executive has plaintively wailed:

The charge to the urban agent was ambiguous and complex—somehow to tackle the problems of the city. It was never clear what the target was . . . and the indicators of success were not even specified, let alone measured. [Pendleton, 1974: 7]

Certainly the series of HUD secretaries with their emphasis on "housing unit" goals have not provided an adequate agenda for policy debate. The proper objective function for urban social policy might be to modulate income and wealth inequality, or to compensate for such inequality by providing significant support for social and technological innovations for the urban poor (for a more extended discussion, see Gappert, 1974: 49-51).

It seems explicit that the study of socioeconomic systems must directly deal with the analysis of complementary and competitive goal structures implicit in the social process being studied. The objective functions of urban policy need, at the very least, to be assessed in the light of alternative formulations of social purpose. A research director in New York City, in formulating a social welfare model, demonstrates that simple income increase is insufficient to improve the well-being of the urban poor (Leverson, 1970). If we accept that the economic man construct is an inadequate model for policy formulation, it becomes even more important to provide for research into the goal-setting and goal-seeking processes of urban social systems.

Third, it may well be that the social economies of cities vary significantly according to the age, size, locations, and functions of urban regions. As we develop more rigorous models of city types (Sutton et al., 1974), the formulation and testing of hypotheses about socioeconomic behavior within and between classes of city types will be desirable. This will require a much more rigorous approach to comparative research (Hanna, 1974).

Fourth, at the same time, at this stage of our knowledge of urban social systems, the community case study approach to urban social economy is important. The development of socioeconomic case study models, procedures, and methods is long overdue. Perhaps the work of economic geographers (McCarthy and Lindberg, 1966) provides an important threshold methodology for such work. The spatial dimensions of socioeconomic change and activity cannot be ignored in understanding the evolution and development of the urban social economy. Social space and social distance are important factors in the dynamics of local markets and need to be more thoroughly conceptualized.

As would-be urbanists attempt to discover, formulate, and understand the workings of the urban social economy, they will of course need to draw from the broadest array of social science methodology and analytical tools. On the one hand, they will continue to rely upon deductive modeling and the empirical testing of hypotheses. But on the other hand, the design of field experiments for new public programs, the development of predictive hypotheses from evolutionary models, and the synthesis of experiential processes will call forth the flourishing of the sociological imagination. Glaser and Strauss (1967) have contrasted logico-deductive theory with what they dub "grounded" theory: this is the discovery of theory from data drawn from experience. As they write:

> Generating a theory from data means that most hypotheses and concepts not only come from the data but are systematically worked out in relation to the data during the course of the research. Generating a theory involves a process of research. [Glaser and Strauss, 1967: 6]

It may well be that the only adequate theories about the social economy of cities will have to be generated from new processes of socioeconomic research.

In all these inquiries, the new breed of urban social scientist will need to grapple more successfully with three considerations. Two of these are (1) the need to employ a system perspective for more complete formulations of the social economic systems in which he is interested; and (2) the need to deal with questions of social purpose and organizational development so that the "created" nature of urban environments is better understood. As Mukerjee once wrote: "It is the notion of a special regional or historic economic system, based on the theory of institutions, which supplies the thread that holds together the beads of scattered economic data" (quoted in Gottlieb, 1971: 44). The third perspective is suggested by the economist Edgar Dunn (1971), who has built a model of social development in which the process of "social learning" has a major role. Suggesting that human social processes involve behavior-evaluating behavior, behavior-organizing behavior, and behavior-changing behavior, Dunn focuses upon the problems of revising social goals and reorganizing social organization and control systems. His concern is that social scientists should contribute to the process of making social learning more orderly and rational:

In social systems problem solving and hypothesis testing take on a different character. The basic point of departure is the fact that the social system experimenter is not exogenous to the system. He exists as an endogenous component of the system he is attempting to understand and transform. He is not dealing with the understanding and design of fully deterministic systems. He is immersed in the act of social system self-analysis, and self-transformation. *He is the agent of social learning*—a purposive, self-actuating, but not fully deterministic process. He is not interested to the same degree in establishing universals because the social system which engages his activity is phenomenologically unique and both its structure and function are temporary in character. He is engaged, rather, in formulating and testing developmental hypotheses. *The developmental hypothesis is a presupposition that, if the organization and behavior of the social system were to be modified in a certain way, the goals of the system would be more adequately realized.* This developmental hypothesis is not tested repeatedly under nearly identical or controlled conditions. Rather, it is tested by the degree to which goal convergence is realized as a result of the experimental design. Problem solving–hypothesis formulation and testing—is an iterative, sequential series of adaptations of an adaptable, goal-seeking, self-activating system. It can be characterized as evolutionary experimentation. [Dunn, 1971: 241; italics added]

This point of view returns us to the perspective of "instrumental economics." This is the role suggested by Bennett Harrison (1974: 192) in his analysis and advocacy of a new kind of urban development. He sees economists and social scientists as instruments of social purpose. As he writes, "We must help governments to design, implement and evaluate planned experiments in the management and financing of this new form of urban development."

These remarks may suggest a too audacious role model for the urban social scientist, but, willy-nilly, the research of the urban social economist cannot escape a strong commitment to the actual utilization of knowledge.

The contributions assembled for this volume are not intended to fulfill or portray the kind of social inquiry paradigms suggested above. Instead they are intended to only lay out the potential array of problems and prospects for a more intense consideration of the problems of urban social economies. In the six major sections of this book, we have attempted to provide a diverse representation of both the topics and methodologies which represent the emerging and overlapping concerns of the geographers, economists, and sociologists who are now taking a more comprehensive approach to their study of the urban social economy.

APPENDIX

A NOTE ON THE DEVELOPMENT OF

SOCIAL ECONOMY

Although the term "political economy" has had extensive and varied usage for a century or more, the term "social economy" is a less familiar one. One first notable use of the term occurs with Friedrich von Wieser's *Social Economics* (1927), which lays out the thoughts and methods of the so-called Austrian School. Gustav Cassel had earlier used the term in 1923. In 1936 John Maurice Clark first published his collected essays in *Preface to Social Economics*. The development of the area of social economy is still inchoate and is related to a number of other developments in the field of economic analysis, and a sharp urban focus has yet to emerge. However, these developments are worth reviewing.

First, there is an Association for Social Economics, which publishes the *Review of Social Economy*. This review was originated in 1942 by the Catholic Economic Association, which changed its name to the above in 1970. (They also considered renaming themselves the Association of Normative Economics, but that suggestion was rejected by the membership.) The Association is currently headed by Robert Danner, chairman of the economics department of Marquette University. Approximately half of its board comes from Catholic universities. There is also a new journal published at Hull University entitled, *International Journal of Social Economics*.

Second, the term social economics is also being used by the Brookings Institution, which has published at least eleven titles in a series called "Studies in Social Economics." These range from *The Doctor Shortage: An Economic Diagnosis* (Fein, 1967) to *Reforming School Finance* (Reischauer, 1973). The series features empirical and partially theoretical studies of selected problems in the fields of health, education, social security, and welfare. The Brookings work represents the fact that at a national policy level there is a growing desire to make economic and social policies converge. George Rohrlich, who founded the Institute for Social Economics at Temple University, has reflected upon these issues in his *Social Economics for the 1970s* (Rohrlich, 1970). He uses a framework with four dimensions to organize the concerns of social economics. These include:

(1) applied economics: the application of economic theory to social problems;

(2) applied statistics: measurement and delineation of social problems;

(3) economic ecology: the social causes (or correlates) of economic behavior;

(4) welfare economics: the social consequences of economic behavior.

This list follows the framework of a book published in Britain called *Social Economics* (Hagenbuch, 1958).

Third, at the societal level, the works of John Kenneth Galbraith (1967) and Robert Heilbroner (1970a) represent attempts to define the national social economy as a subject for special inquiry. *The Coming of Post-Industrial Society* (Bell, 1973) represents a similar attempt by a sociologist. It is unfortunate to note, however, that the new breed of "macrosociologists" neglects the concerns of national economic structures. In *Macrosociology: Research and Theory* (Coleman, Etzioni, and Porter, 1970), neither Galbraith nor Heilbroner is cited. The work of Parsons and Smelser (1965) remains a lonely seminal work in the analysis of economic concerns by sociologists. In it, they write: *

Economic theory should . . . be regarded as the theory of typical processes in the "economy," which is a subsystem differentiated from other subsystems in the society. The specifically economic aspects of the theory of social systems therefore is a special case of the general theory of the social system.

Two recent books, *The Social Economy of West Germany* (Hallett, 1974) and *The Social Economy of France* (Coffey, 1973), are simple but useful attempts to define the socioeconomic systems of two European countries in relation to questions of national policy development in the post-World War II period. As Hallett (1974: ix) writes: "Academic economists have tended to apply sophisticated techniques to an extremely limited field and to ignore the institutional, social and political factors which are so important to economic life."

Interesting methodological questions pursuant to the study of national economy from a social and behavioral perspective are raised by Adolph Lowe (1965). Lowe's notions are more thoroughly explored and critiqued in *Social Ends and Economic Means* edited by Heilbroner (1969); this is a collection of essays by nine economists. Similar kinds of questions are explored in Krupp (1966). In his volume a slightly more traditional group of economists also explore "normative" issues in economic analysis and suggest the kinds of problems associated with a neglect of behavioral and value orientations in the study of economic activity.

Fourth, in a different sense it might be suggested that the field of social economy is a reincarnation of the field of institutional economics which flourished at the University of Wisconsin at the beginning of the twentieth century around Richard Ely (whose *Land Economics* is still a classic) and John Rogers Commons. Commons, who was hounded out of Syracuse University in the late nineteenth century for advocating Sunday baseball as a recreation outlet for factory workers, synthesized much of the significance of that school of empirical work in his two-volume *Institutional Economics*. This field of work has recently been reviewed by Allen G. Gruchy (1973). Institutionalists were concerned with how economies change over time, altering the centers of power, class relationships, and the structure of industry and government activity.[1] Unlike traditional economics, institutions and technology were not taken as givens. Instead institutional economics was concerned with questions of the change and control of these institutions so that the evolution of economic life could be shaped to social purposes or goals. This tradition continues in a somewhat transformed fashion at the Madison campus of the University of Wisconsin by the Institute for Research on Poverty, headed by Robert Haveman. The Institute's conceptual and analytical work has led participants into the management of actual field experiments to determine the effects of income maintenance programs. Some of this recent work has been published in the *Journal of Human Resources*, edited at that campus.

The practice and pursuit of institutional economics seem to wax and wane with the ups and downs of the cycle of social and economic welfare. As economic conditions worsen, the more people turn to economists for solutions; and the more that society turns to economists to solve unsolved problems, the greater is the sense of dissatisfaction within the economics profession as to the validity of traditional mainstream. State governments in American society have turned to their state universities to different degrees for help with problems of economic and social distress. The Wisconsin tradition of institutional economics was vigorously used in the LaFollette period of progressivism. In California, Jessica Blanche Peirotto, the first woman to receive the rank of full professor at the University of California (in 1918), was chairman in the 1920s and 1930s of a state government committee on Research in Social Economics. Her special concern was the economics of the household especially as related to social policies.

Yet another approach to the field of social economy is represented by those economists who, through their work in the field of economic development abroad, have rediscovered the importance of social and behavioral inquiry. Everett Hagen's volume (1962) was a pioneering work in this area. More recently, Edgar Dunn (1971) subtitled his book, "a process of social learning."

Noneconomists with an interest in economic development have also contributed to socioeconomic synthesis. Charles McClelland (1969), a psychologist, has contributed to the understanding of economic achievement motivations. John Kunkel (1970), a behavioral sociologist, subtitled his work, "a behavioral perspective of social change." The philosopher Denis Goulet (1971) and David Apter (1971), a political scientist, have contributed to this discussion of the relationship between social and economic development.

A sixth approach to social economy can be cited with reference to the sociological interpretations of Neil Smelser (1963). This work summarizes the contributions of sociologists to the analysis of social structures stemming from industrial economy. The field of economic psychology and behavioral science must also be noted. This work, pioneered by George Katona at the Survey Research Center at the University of Michigan, helped advanced field surveys as an important component of economic forecasting and planning. A recent volume by Strumpel et al. (1972) provides a series of perspectives on the significance of this work. Anthropologists have also used the term social economy, as in Oberg (1973).

Regrettably, however, very few of these explorations into integrating social and economic analysis have had a distinct urban focus. The behavioral-institutional approach to the study of economic activity has been more national and international in its scope. Very seldom in the development of these different fields has attention been paid to the urban domain. Even less has the analysis concerned itself with differences in social economy across an interurban field. Interestingly enough, it has been the geographer, often without an explicit or implicit interest in socioeconomic theory, who has done the best work in "mapping" socioeconomic activities and differences onto a representation of urban space. For instance, in the April 1974 issue of *Economic Geography*, three articles reflect this approach: "The Office in Metropolis: An Opportunity for Shaping Metropolitan America," "Regional Fluctuations in Unemployment within the U.S. Urban Economic System," and "Personal Preference Patterns and the Changing Map of American Society."

Two other approaches to the study of significant social and institutional variables by economists need to be noted. First, The Union for Radical Political Economics, organized in the late 1960s, now publishes the *Review of Radical Political Economics*. Bronfenbrenner (1970), in reviewing this development, urges that "radical economics should be recognized as a legitimate field of concentration in the study and practice of economics." The field of radical economics represents the current manifestation of the tradition of political economics. It is sharply critical (both in its rhetoric and its analysis) of the political forces and institutions which have contributed to extensive economic inequality, regressive taxation, and the manipulation of international economic forces by the Western institutions of corporate finance. In its assumptions of a "power elite," it has addressed itself more to the class basis of institutional decision-making than it has to social and psychological determinants of economic processes.

Second, the field of "human resources" economics, applied to the urban domain, has had too little systematic attention. Except for a volume published in the early 1960s by Resources for the Future (Perlman, 1963), very little integrated analysis has been applied to the dynamics of human resource development. This volume conceived of the problem in terms of the characteristics of labor force and consuming households. As Perlman wrote:

> increasing emphasis will have to be put on urban centers as places to attract consuming households both in the sense of encouraging the development of the appropriate skills needed for future industry, and in the sense that the community will be able to hold its best

workers because it is a pleasant place for them to live [and hence to consume].

Unfortunately, this synthesis of work-consumption analysis never has received a great deal of additional attention.

NOTE

1. A colleague, Manuel Gottlieb, has suggested that I include mention of the German historical school at this point. Max Weber's monumental work has been relatively recently translated in part (1968); an early partial translation (1947) is still available. Some of the work of Werner Sombart (1967) has also been translated. See also the work of Joseph Schumpeter (esp. 1934). The significance of this kind of sociohistorical approach to economic analysis has been beautifully synthesized in Gottlieb's own work (1953), wherein he writes:

We should thus recognize that an economic system is a highly complex concept. It should denote an ideal-type form resolvable into a family of hybrid stages and types of systems united more or less loosely in civilizational complexes and identified individually in terms of a more or less integrated pattern of functional priorities, ways and means, and an institutional Gestalt.

Such sociohistorical concomitance to an analysis of the economic evolution of individual cities would be an important contribution.

REFERENCES

APTER, D. (1971) Choice and the Politics of Allocation. New Haven: Yale University Press.

BELL, D. (1973) The Coming of Post-industrial Society. New York: Basic Books.

BERGMAN, B. R. (1974) "Economist, poll thy people." New York Times (November 3).

BISH, R. (1971) The Public Economy of Metropolitan Areas. Chicago: Markham.

BLACK, J. (1967) "Indicative planning and stable growth." Exeter: University of Exeter, May 8.

BOOKCHIN, M. (1973) The Limits of the City. New York: Harper.

BOULDING, K. E. (1973) "Interview." Challenge (July/August): 39.

BRONFENBRENNER, M. (1970) "Radical economics in America: a 1970 survey." Journal of Economic Literature (September): 747-766.

BUCHANAN, J. and R. TOLLISON (1972) Theory of Public Choice. Ann Arbor: University of Michigan Press.

BURKHEAD, J. (1971) "Fiscal planning—conservative Keynesian." Public Administration Review (May/June): 335-345.

CASSEL, G. (1967) The Theory of Social Economy. New York: Augustus M. Kelley.

CHINITZ, B. and R. DUSANSKY (1972) "The patterns of urbanization within regions of the United States." Urban Studies 9, 3: 289-298.

CLARK, J. M. (1967) Preface to Social Economics. New York: Augustus M. Kelley.

COFFEY, P. (1973) The Social Economy of France. New York: St. Martin's Press.

COLEMAN, J., A. ETZIONI, and J. PORTER (1970) Macrosociology: Research and Theory. Boston: Allyn & Bacon.

COMMONS, J. R. (1959) Institutional Economics: Its Place in Political Economy. 2 vols. Madison: University of Wisconsin Press.

CRECINE, J. (1970) Financing the Metropolis. Beverly Hills, Calif.: Sage.

DUNN, E. (1971) Economic and Social Development: A Process of Social Learning. Baltimore: Johns Hopkins University Press.

ELY, R. (1964) Land Economics. Madison: University of Wisconsin Press.

FEIN, R. (1967) The Doctor Shortage: An Economic Diagnosis. Washington, D.C.: Brookings.

FORRESTER, J. (1969) Urban Dynamics. Cambridge, Mass.: MIT Press.

GALBRAITH, J. K. (1967) The New Industrial State. Boston: Houghton Mifflin.

GAPPERT, G. (1974) "Assessment and acceptability of urban futures." Futures 6: 42-58.

GLASER, B. G. and A. L. STRAUSS (1967) The Discovery of Grounded Theory. Chicago: Aldine.

GOLDSTEIN, G. S. and L. N. MOSES (1973) "A survey of urban economics." Journal of Economic Literature.

GOTTLEIB, M. (1971) "Mukerjee: economics become social science." Journal of Economic Issues 5, 4: 33-53.

——— (1953) "The theory of an economic system." American Economic Review 43, 2.

GOULET, D. (1971) The Cruel Choice: A New Concept in Development. New York: Atheneum.

GRUCHY, A. G. (1973) Contemporary Economic Thought: The Contribution of Neo-Institutional Economics. New York: Augustus M. Kelley.

HAGEN, E. (1962) On the Theory of Social Change: How Economic Growth Begins. Homewood, Ill.: Dorsey.

HAGENBUCH, W. (1958) Social Economics. New York: Cambridge University Press.

HALLETT, G. (1973) The Social Economy of West Germany. London: Macmillan.

HALPERN, M. (1969) Applying a New Theory of Human Relations to the Comparative Study of Racism. Denver, Colorado: Denver University Graduate School of International Studies.

HANNA, J. (1973) Comparative Urban Research. New York: City University.

HANSEN, N. (1970) Rural Poverty and the Urban Crisis: A Strategy for Regional Development. Bloomington: Indiana University Press.

——— (1968) French Regional Planning. Bloomington: Indiana University Press.

HARRISON, B. (1974) "Ghetto economic development: a study." Journal of Economic Literature 12.

HARVEY, D. (1973) Social Justice and the City. Baltimore: Johns Hopkins University Press.

HAVERMAN, R. and J. MARGOLIS (1970) Public Expenditures and Policy Analysis. Chicago: Markham.

HEILBRONER, R. (1970a) Between Socialism and Capitalism. New York: Random House.

——— (1970b) "On the limited relevance of economics." The Public Interest (Fall): 80-93.

——— (1969) Economic Means and Social Ends. Englewood Cliffs, N.J.: Prentice-Hall.

ISARD, W. and T. A. REINER (1966) "Regional science: retrospect and prospect." Regional Science Association, Papers 16: 1-16.

JACOBS, J. (1969) The Economy of Cities. New York: Random House.

——— (1961) The Death and Life of Great American Cities. New York: Random House.

JOHNSON, H. G. (1967) "The economic approach to social questions," p. 3 in Weidenfield and Nicolson, London, October, for the London School of Economics.

KELLER, S. (1968) The Urban Neighborhood: A Sociological Perspective. New York: Random House.

KERR, A. and M. WILLIAMSON (1970) "Regional economics in the United States." Growth and Change (January): 5-19.

KOTLER, M. (1969) Neighborhood Government: The Local Foundations of Political Life. Indianapolis: Bobbs-Merrill.

KRUPP, S. R. [ed.] (1966) The Structure of Economic Science. Englewood Cliffs, N.J.: Prentice-Hall.

KUNKEL, J. (1970) Society and Economic Growth: A Behavioral Perspective of Social Change. New York: Oxford University Press.

LANCASTER, K. (1971) "A new approach to consumer theory." Journal of Political Economy 74.

LEVERSON, I. (1970) "Strategies against urban poverty," in K. Boulding and M. Pfaff (eds.) Redistribution to the Rich and the Poor. Belmont, Calif.: Wadsworth.

LOWE, A. (1969) "Toward a science of political economics," in R. Heilbroner (ed.) Economic Means and Social Ends. Englewood Cliffs, N.J.: Prentice-Hall.

——— (1965) On Economic Knowledge. New York: Harper & Row.

McCARTHY, M. L. and J. LINDBERG (1966) A Preface to Economic Geography. Englewood Cliffs, N.J.: Prentice-Hall.

McCLELLAND, C. (1969) Motivating Economic Achievements. New York: Free Press.

MACKEAN, R. N. (1973) "An outsider looks at urban economics." Urban Studies 10, 1 (February): 19-32.

MARSHALL, R. (1974) "The economics of racial discrimination: a survey." Journal of Economic Literature 12: 849-871.

MILLS, E. S. and J. MACKINNON (1973) "Notes on the new urban economics." Bell Journal of Economics and Management Science (Autumn): 593-651.

MINAR, D. and S. GREER (1969) The Concept of Community. Chicago: Aldine.

MORRISON, P. A. (1974) "Guiding Urban Growth: Policy Issues and Demographic Constraints." Santa Monica, Calif.: RAND Corporation.

MUKERJEE, R. (1942) The Institutional Theory of Economics. London: Macmillan.

MUTH, R. (1969) Cities and Housing. Chicago: University of Chicago Press.

NEENAN, W. (1972) Political Economy of Urban Areas. Chicago: Markham.

OBERG, K. (1973) The Social Economy of the Tlingit Indians. Seattle, Wash.: University of Washington Press.

PACKARD, V. (1972) A Nation of Strangers. New York: McKay.

PARETO, V. (1935) Mind and Society 4: 1413.

PARSONS, T. and N. SMELSER (1965) Economy and Society. New York: Free Press.

PEIROTTO, J. B. (1967) Essays in Social Economics. Freeport, N.Y.: Books for Libraries Press.

PERLMAN, M. (1963) Human Resources in the Urban Economy. Washington, D.C.: Resources for the Future.

PENDLETON, W. (1974) Urban Studies and the University: The Ford Foundation Experience. New York: Ford Foundation.

PIORE, M. and P. DOERINGER (1971) Internal Labor Markets and Manpower Analysis. Lexington, Mass.: D. C. Heath.

REICH, G. and M. EDWARDS (1973) "A theory of labor market segmentation." American Economic Review (May).

REISCHAUER, R. D. (1973) Reforming School Finance. Washington, D.C.: Brookings Institution.

ROHRLICH, G. (1974) "The potential of social ecology for economic science." Review of Social Economy 31, 1.

RYLE, G. (1954) Dilemmas. New York: Cambridge University Press.

SCHUMPETER, J. (1934) The Theory of Economic Development: An Inquiry into Profits, Capital Credit, Interest and the Business Cycle. Cambridge: Harvard University Press.

SEN, A. (1970) Collective Choice and Social Welfare. San Francisco: Holden-Day.

SLATER, P. (1970) The Pursuit of Loneliness: American Culture at the Breaking Point. Boston: Beacon Press.

SMELSER, N. (1963) The Sociology of Economic Life. Englewood Cliffs, N.J.: Prentice-Hall.

SOMBART, W. (1967) Luxury and Capitalism. Ann Arbor: University of Michigan Press.

STRUMPEL, B., J. MORGAN, and E. ZAHN [eds.] (1972) Human Behavior in Economic Affairs. San Francisco: Jossey-Bass.

SUTTON, R. J., J. KOREY, S. BRYANT, and R. DODSON (1974) American city types: towards a more systematic urban study." Urban Affairs Quarterly 9, 3 (March): 369-401.

THOMPSON, W. (1972) "The national system of cities as an object of public policy." Urban Studies 9: 99-116.

——— (1965) Preface to Urban Economics. Baltimore: Johns Hopkins University Press.

WARREN, R. (1974) The Structure of Community Reform. Lexington, Mass.: D. C. Heath.

WEBER, M. (1968) Economy and Society: An Outline of Interpretive Sociology. New York: Bedminster.

——— (1947) Theory of Social and Economic Organization. New York: Free Press.

WIESER, F. von (1927) Social Economics. London: George Allen & Unwin.

WILLIAMSON, O. E. (1973) "Markets and hierarchies: some elementary considerations." American Economic Review 63.

WINGO, L. (1972) "Issues in a national urban development strategy for the U.S." Urban Studies 9: 3-28.

Part I

THE CHANGING CONCERNS OF NATIONAL URBAN POLICY

Introduction

□ IT IS UNCLEAR whether the concept of a national urban policy is viable. Moynihan, in his quest to define the policy space involved (1970), suggested that a national or macro approach to urban policy lacked a general theory which could guide federal policy makers toward a series of actions with reliable, predictable outcomes. It is appropriate to note that the test of Keynesian economics with the tax cut of the early 1960s occurred only several decades after the publication of Keynes' *General Theory of Employment, Interest and Money* (1936) and the passage of the Full Employment Act in 1946 with its provisions for a council of economic advisors, an annual economic report, and related technical analysis. The passage of the Housing and Urban Development Act of 1970, with its provision for a biannual national growth report, lacked not only a theoretical and analytical rationale but was also unclear as to what was the object of policy beyond some ill-defined concerns with land use, population density, and a projected housing shortage. In this introduction to Part I, it may be useful to explore some of the problems and prospects of national urban policy. First, the problem of an objective function for urban policy is examined. A number of other questions are raised. The federal bias in national resources allocation is also examined, and the problems of determining the directions and impact of a multidimensional federal urban policy are discussed.

THE PROBLEM OF AN OBJECTIVE FUNCTION

FOR URBAN POLICY

There is a great deal of confusion currently in the fields of urban and social policy. On the one hand, there is an inchoate social development paradigm behind such notions as "quality of life," "social accounting," "childhood development," and the like. On the other hand, the new breed of social scientist scurries around executing survey instruments, developing data series, and collecting correlation coefficients from multiple regression analysis. Unfortunately the new investment in urban data collection does not correspond to any useful or adequate policy paradigm. An urban policy paradigm can be successful only if it has the ability to extend the boundaries of understanding through the interpretation of new elaboration of urban phenomena. As Kuhn (1970) has stated:

The existence of the paradigm sets the problem to be solved; often the paradigm theory is implicated directly in the design of apparatus able to solve the problem. Without the *Principia*, for example, measurements made with the Atwood machine would have meant nothing at all.

It was with the apparent failure of many Great Society programs in their early development that a concern for a restatement of social and urban policy purpose became manifest. But a comprehensive national social model is still missing. Perhaps that is best. Maybe a national approach to urban and social policy would only mask regional and local needs. More significantly required is a model which provides for the interrelationships of behavior, values, and feelings of subjective welfare at the level of the urban community.

The provision of an objective function for urban social policy to national decision-makers at this time is perhaps inappropriate. There is still inadequate classification of the nature of the diversity of social phenomena within and between American cities. It was the hypothesis of the Rand Urban Study Group investigating growth in San Jose, Seattle, and St. Louis that cities in the United States are so diverse in their structures, problems, and other policy characteristics that any national urban policy will necessarily affect them differentially. To the extent that this is the case, any national policy discriminates with respect to its application to specific urban areas.

The kindest thing which can be said about existing national policies is that their effect on different urban areas is unintended. The most commonly expressed objective function for national urban policy concerns the reduction of population density. Documents of both the legislative and executive branches of government deplore the concentration of too much of the population on too little of the land. This view is based upon the notion that the urban crisis is uniquely tied to the density of cities. Others, of course, like Jane Jacobs and William Whyte, suggest that "concentration," as such is the saving grace of cities: achieve more attractive higher densities and you will achieve an acceptable urban culture.

Another objective function has been the expansion of publicly sponsored housing according to certain "minimum adequate standards." These standards, of course, have been based upon physical dimensions and not upon any systems view of the "use" of housing (and its immediate environment) by members of particular social sub-groups.

Other programs, such as the FHA and VA mortgage programs and the expressway and mass transit programs, have as their objective functions the provision of convenience to the mobile middle classes. In the words of Banfield (1970), "they aim at problems of comfort, convenience, amenity and business advantage, not at ones involving the essential welfare of individuals or the good health of society."

Newer, still emerging objective functions have to do with the provision of (a) a guaranteed income floor (but the Moynihan-Nixon bill would have had its greatest impact in the rural South); (b) a surrogate guaranteed employment program through a subsidy for public employment in selected areas; and (c) the provision of quasi-market power over the provision of social services to the poor through voucher systems of one kind or another.

On the other hand, some of the programs developed under the Nixon Administration—such as revenue sharing, an emphasis on career education, aid to minority businesses, and so on—also have an emphasis on providing more or better means by which certain traditional values can be enhanced or diffused to different groups. These policies aim for the restoration of the local government ethic, the work ethic, the business ethic, and the like.

At different levels some of these latter programs and proposals represent a struggle to express or forge a social development paradigm from which more attainable objective functions for urban places can be derived. In fact, however, it is better postulated that the urban problem is foremostly a problem of the distribution of income and economic activity. Although it is now clear that we have during the last generation transferred our rural poor to the cities, it is less clear perhaps that the cities and their metropolitan regions face not only a future as the "poor houses" of the nation, but also continuing and intensifying conflict over the resources allocations of all kinds. The energy crisis will be just one resource allocation problem which will affect cities and urban populations differentially.

On the one hand, this is a reflection that national market forces will continue to operate in differential ways. But on the other hand, market forces within cities and their regions will also continue to operate differentially. Miller and Bloomberg (1968), in calling for a multidimensional approach to urban poverty, suggest a deeper conception of what is meant by "impoverished" and "deprivation." They suggest that as we seek to extend the conditions of well-being to the whole of the population, we must come to grips with the notion of a "left-out" segment of the population. It may well be that cities will always be the home of those that are "left out." If this is true, perhaps the conceptual problem for the analysis of urban and social policies is to identify the nature of inequality, "left-out-ness," and scarcity in our postindustrial future. The decent objective function for urban social policy might be to modulate income and wealth inequality, or to compensate for it by providing significant support for social and technological innovations for the poor. It may well be said that the only thing homogenous about urban policy space is that it contains a growing proportion of the population denied access to affluence.

OTHER QUESTIONS FOR URBAN POLICY

Besides suggesting that a more explicit objective function for national urban policy needs to be developed, a number of other questions can also be posed. Whose policy should it be: the President's, the Congress's, the mayors'? How should it be formulated? How should it be developed? What are the forces behind the implicit national policy toward cities which already exist? How can federal inputs of money and laws and coordination be translated into reliable outcomes? What are the strategies and instruments of intervention which can connect the national or federal policy purpose (if it were to exist) to the areas of greatest need (if it could be measured) within the national system of cities?

Two midwestern foundations, in pursuing an analysis of urban policy, developed nine categories of organizations or agencies which contribute to a greater or lesser degree to the formulation or advocacy of urban and social policy. These are shown in Table 1. In polling elected and appointed officials the organizations' most consulted on policy issues were identified as the Brookings Institution, U.S. Chamber of Commerce, National Governors Conference, AFL-CIO, National League of Cities, and the NAACP. Other

TABLE 1

MACRO-ORGANIZATIONS CONCERNED WITH URBAN RELATED ISSUES

Urban Policy Research Organizations
The Brookings Institution
Center for Community Change
National Academy of Sciences

Government Officials' Organizations
Conference of Mayors of the U.S.
National League of Cities
Council of State Governments

Urban Professionals' Organizations
American Institute of Architects
Institute of Public Administration
American Institute of Planners

Other Urban-Related Professional Organizations
American Bar Association
American Medical Association
National Association of Social Workers

Population Sector Representatives
American Association of Retired Persons
Americans for Indian Opportunity
National Association for the Advancement of Colored People

General Human Rights Organizations
American Civil Liberties Union
B'nai B'rith
League of Women Voters

Educational and Cultural Agencies
American Association of Museums
National Education Association
American Association for the Advancement of Science

Employment-Oriented Organizations
National Alliance of Businessmen
United Automobile Workers
Aeronautical and Agricultural Implement Workers of America
Opportunities Industrialization Centers of America

Development and Other Trade Associations
Chamber of Commerce of the U.S.
Institute for Rapid Transit
Urban Land Institute

organizations felt to have an impact on urban legislation were identified as the League of Women Voters, American Civil Liberties Union, Common Cause, American Institute of Planners as well as AFL-CIO and NAACP. Irrespective of the merits of these groups serving ad hoc representatives roles in the struggle to define policy issues, it is clear that a series of macro-level organizations are significantly involved in formulating national urban and social policy. The problem, of course, is that their perspective will differ appreciably from a local level perspective. The national policy agenda from any perspective will naturally be concerned with broad questions of national constraints, regulations, and resource allocation which do not have a direct interface with the immediate concerns of the local urban system of governance and management.

Table 2 provides a ranking of major urban problems by a group of mayors and councilmen polled by the National League of Cities. It is

TABLE 2

MAJOR URBAN PROBLEMS PERCEIVED BY LOCAL OFFICIALS

Problems	Ranking by Mayors	Total Mentions
Refuse and solid waste	1	675
Law enforcement	3	600
Streets and highways	2	573
Relations with county	4	543
Fiscal/tax policies	5	503
Downtown development	6	490
Planning and zoning	8	486
Citizen participation	7	444
Public transit	12	420
Use of general revenue sharing	11	393
Parks and recreation	10	385
Energy shortage	9	380
Housing	13	310
Relations with state	15	309
Upgrading city staff	16	309
Economic development	17	286
Water quality	14	269
Relations with federal government	16	260
Collective bargaining	19	240
Manpower development	18	233
Relations with region	21	228
City government reorganization	22	223
Health care	20	213
Education	23	191
Social services	24	189
Race relations	26	163

SOURCE: National League of Cities, *America's Mayors and Councilmen: Their Problems and Frustrations* (Washington, D.C., 1974).

perhaps obvious that their interest in "refuse and solid waste" and "relations with county" are not exactly issues that are associated with Brookings, NAACP, the League of Women Voters, and so forth. Similarly, at an even lower level of the scale, polls of urban citizens reflect normally not a concern with the management of municipal systems, but with improper or disruptive behavior in public places (Wilson, 1968) and the equitable distribution of services.

These differences in perception of urban needs by members of different institutional and social systems mean, at the very least, that any national needs-assessment approach to defining the basis of national urban policy will be both inadequate and confusing. The issue is the need to grapple with the "multiple realities" reflected from different positions within the domain of urban policy space, and to provide for the necessary integration and compromises with the various values and interests represented. The problems of revenue sharing and the New Federalism are reflective of the inability to achieve such an integration.

BEYOND REVENUE SHARING: RESOURCE ALLOCATION WITHIN A FEDERAL SYSTEM

It is clear that one of the consequences of revenue sharing is that it did not substantially increase the flow of resources to the major and largest central cities. It did instead provide new resources to a large number of small cities and large suburbs. The fiscal outcome of the New Federalism is discussed in more detail in the chapter by Professor Jesse Burkhead in this section. This outcome is perhaps not unexpected, given the nature of the American federal system. The American federal system causes two problems to any notion of a national urban policy which would redirect major quantities of resources to cities. First, the 50 states as the member units of the nation each represent and define the urban policy space in different ways. Table 3, for instance, lists the top 20 "urban" states using three different indicators. These are density, percentage urban by census definition, and percentage of population living in cities over 25,000. The anomalies are interesting. Of the top 20 states in density the range is from almost 1,000 people per square mile in New Jersey (with a density of 44,081 per square mile in Union City) to less than

TABLE 3
THE TWENTY "MOST URBAN" STATES: 3 INDICATORS

Density Per Square Mile		Percent Urban By Census Definitions*		Percent Population (Cities over 25,000)	
New Jersey	953	California	90.9	California	63
Rhode Island	905	New Jersey	88.9	Arizona	63
Massachusetts	727	Rhode Island	87.0	New York	57
Connecticut	624	New York	85.6	Rhode Island	57
Maryland	397	Massachusetts	84.6	Texas	55
New York	381	Illinois	83.0	Colorado	55
Delaware	277	Colorado	83.0	Illinois	54
Pennsylvania	262	Hawaii	83.0	Hawaii	53
Ohio	260	Nevada	80.9	Massachusetts	50
Illinois	199	Utah	80.9	Michigan	49
Michigan	156	Florida	80.5	Nevada	48
Indiana	144	Michigan	79.8	Ohio	43
California	128	Texas	79.5	Oklahoma	42
Florida	126	Arizona	79.5	Wisconsin	42
Hawaii	120	Colorado	78.7	Connecticut	41
Virginia	117	Connecticut	77.3	Minnesota	40
North Carolina	104	Maryland	76.6	Indiana	40
Tennessee	95	Ohio	75.3	New Mexico	40
South Carolina	86	Michigan	73.9	Florida	39
New Hampshire	82	Washington	72.6	Missouri	39
		Delaware	72.1		
		Pennsylvania	71.5		

*Places over 2,500 population.

82 per square mile in New Hampshire (with a density of 2,734 in Manchester). Using percentage urban (population living in places of over 2,500) 6 other states join the domain of urban policy space including Texas and Nevada. Using the third criterion, the percentage living in cities over 25,000 population, 5 other states including Wisconsin and Oklahoma appear and 7 of the original list drop out.

The problem of trying to establish a rather homogeneous segment of the urban problem space consisting of at least 20 states is related to the second problem which is that the 20 "least urban" states control 40% of the votes in the U.S. Senate. This problem can be exaggerated, but the realities of the federalized distribution of power within the U.S. Senate mean that 20 states acting in unison can essentially block any policy of resource allocation which does not consider their interests. In 1970, for instance, out of $24 billion in federal aid to state and local governments, about $5.8 billion (26%) went to the 20 states with the least amount of population in cities over 25,000, while the 20 states with the most population in cities over 25,000 (from California to Missouri) received $13.4 billion or not quite 60% of total federal aids. These 20 states contain about 60% of the population of the country. The 20 least urban (with only

20% of the national population) received about a 5% federal aid premium for being "underpopulated."

The federal dilemma for national urban policy is that even if the 20 most urban states could achieve a common definition of urban needs and policy, the 20 least urban states are unlikely to agree with the provision of resources to the governments with the greatest proportion of people living within cities. Indeed, as Burkhead points out in his revisitation of national urban policy, the federal government with its direct outlays (for defense, and so on) winds up with more of an effect on urban populations through its direct influence on the location of federal procurements. Professor Seymour Sacks in his chapter goes on to demonstrate that the city is a fiscal artifact and should be approached with caution when generalizations about its economic conditions are used as a guide to policy. It perhaps should be apparent that major new aid to the state and city governments within the national urban policy space is unlikely to be a key element of national urban policy.

What then of the city and a national policy directed to the real and apparent needs of large city areas? If the heterogeneous nature of the urban policy space remains misunderstood, and if the political rules of the federal system prevent any significant agreement on resource reallocation, what can be expected to unfold as national urban policy in the decade ahead?

First, the problem needs to be redefined. Some kind of multi-dimensional goal structure for populations living within the metropolitan areas needs to be formulated. The goal structure should relate more to the socioeconomic conditions and circumstances and less to the statistical aggregates of arbitrarily defined political boundaries. As shown in Table 4, statistical categories represented as "richest" and "fastest growing" counties and as "central cities" have experienced more concentrated flows of federal expenditures. But it is unclear, of course, what the distributional impact of those expenditures was among the population within those units so categorized.

Second, a further redefinition of the problem should also concern itself more explicitly to defining "metropolitan" policy space and the inequalities which obtain within those areas. Added to the notion of standard metropolitan statistical area (SMSA) should be the notion of functional economic area (FEA). Burton and Garn (1972) employ similar ideas in sketching out the dimensions of a national "metropolitan" policy which incorporates the concept of "industrial manpower communities" as a device to achieve progress in social and

TABLE 4
LOCATIONAL IMPACT OF FEDERAL EXPENDITURES—
FISCAL YEAR 1969

Area Type	Population % of U.S.	Program % of U.S.	Concentration Ratio
Poorest counties	10.0	6.1	0.61
Richest counties	10.1	13.2	1.31
Slowest-growing counties	10.2	11.7	1.15
Fastest-growing counties	10.0	12.4	1.24
SMSAs >1 million in 1966	37.3	42.7	1.14
SMSAs < 1 million in 1966	29.6	30.5	1.03
Non-SMSA urban counties	11.0	9.9	0.90
Rural counties	22.1	16.9	0.76
Central cities	12.5	18.7	1.50
Suburbs	8.4	8.2	0.98
EDA counties	38.3	39.4	1.03

SOURCE: Report of the Joint Locational Analysis Project Office of Management and Budget/Economic Development Administration, September 1, 1970.

economic integration. Additional reconceptualization of the policy space to emphasize socioeconomic differences is needed.

Third, perhaps it is also appropriate to indicate that the real concern of national urban policy must be to focus on those forces which determine the national location of economic activities. Companion to that concern is a concern with the problems of social stress associated with economic instability and dislocation. At some future point, when national economic policy moves in the direction of indicative planning, perhaps a consensus on federal support for at least minimum levels of employment in declining areas (both urban and rural) can be achieved.

Fourth, at the very least, the development of an urban impact statement approach to federal policies and activities is long overdue. Figure 1, drawn from a report by Hartley and Patton to HUD (1973), is an initial attempt to begin to understand the locational impact of different federal policies. Additional work is being done by the Rand Corporation in this area. This essentially leaves us with four sets of questions:

(1) How should we think about the goals of national urban policy? Is the evolutionary perspective of the Burkhead paper an accurate portrayal of the problems of formulating a national policy? Will national urban policy ever be precisely and explicitly worked out? If it is, what policy conditions will have to change?

(2) How should we remember and analyze the decade of the 1960s with relation to urban problem-solving? Was the "benign neglect" of

Components of Development	Federal Instruments for Meeting National Needs					
	Capital Investment (Physical Infrastructure)		Tax Structure	Procurement of Goods and Services, and Location of Installations	Provision of Credit	Regulation of Economic Activity
	Direct Federal Public Works	Assistance to State/Local Governments				
HOUSING						
Regional Development	Negligible	Negligible	Negligible	Negligible	Negligible	Negligible
Urban Development	Negligible	Slight	Strong	Slight	Strong	Negligible
TRANSPORTATION						
Regional Development	Strong	Strong	Negligible	Slight	Slight	Strong
Urban Development	Strong	Strong	Negligible	Slight	Slight	Strong
SOCIAL SERVICES						
Regional Development	Negligible	Negligible	Negligible	Negligible	Slight	Negligible
Urban Development	Slight	Strong	Negligible	Negligible	Negligible	Negligible
ENVIRONMENTAL PROTECTION						
Regional Development	Slight	Slight	Slight	Negligible	Slight	Slight
Urban Development	Strong	Strong	Slight	Slight	Slight	Strong
LAND USE						
Regional Development	Strong	Strong	Moderate	Moderate	Moderate	Moderate-
Urban Development	Moderate	Strong	Strong	Strong	Strong	Strong
ASSISTANCE TO LAGGING AND RURAL AREAS						
Regional Development	Strong	Moderate	Strong	Strong	Strong	Moderate
Urban Development	Strong	Slight	Strong	Slight	Slight	Strong
ENERGY						
Regional Development	Strong	Slight	Moderate	Slight	Slight	Slight
Urban Development	Slight	Slight	Slight	Slight	Slight	Slight
COMMUNICATIONS						
Regional Development	Strong	Slight	Strong	Strong	Slight	Strong
Urban Development	Slight	Slight	Slight	Slight	Slight	Strong

SOURCE: Hartley and Patton (1973).

Figure 1: EFFECT OF FEDERAL ACTIVITIES ON REGIONAL AND URBAN DEVELOPMENT

Nixon's New Federalism an appropriate reaction to the excess of expectations generated by Great Society rhetoric? What policy is likely to follow the New Federalism now that we are a metropolitan/suburban society? Will it benefit cities any more than current policy and in what ways?

(3) If we need to differentiate between different types of cities, how do we establish such a framework? As indicated in the Sacks and Sullivan paper, should we distinguish between the fiscal circumstances of different kinds of cities? Should there be more emphasis on socioeconomic differences between and within cities? How would different typologies affect the determination of national urban policy?

(4) What are the politics and dynamics of national urban policy? How can the evolutionary nature of that policy-making process be best understood, analyzed, and monitored? Are nonurban aspects of national policy more important to cities than the more explicitly urban aspects of national policy?

REFERENCES

BANFIELD, E. (1970) The Unheavenly City. Boston: Little, Brown.

BURTON, W. and H. GARN (1972) The President's Growth Report—A Critique and Alternative Formulations. Washington, D.C.: Urban Institute.

HARTLEY, D. K. and J. W. PATTON (1973) "The impact of federal activities on development." Washington, D.C.: U.S. Department of Housing and Urban Development, Growth Policy Series, November.

KEYNES, J. M. (1936) General Theory of Employment, Interest and Money. London: Macmillan.

KUHN, T. S. (1962) The Structure of Scientific Revolutions. Chicago: University of Chicago Press.

MILLER, S. M. and W. BLOOMBERG (1968) "Shall the poor always be impoverished?" in W. Bloomberg and H. J. Schmandt (eds.) Power, Poverty and Urban Policy. Beverly Hills, Calif.: Sage.

MOYNIHAN, D. P. (1970) Toward a National Urban Policy. New York: Basic Books.

WILSON, J. Q. (1968) Metropolitan Enigma. Cambridge: Harvard University Press.

1

The Political Economy or Urban America: National Urban Policy Revisited

JESSE BURKHEAD

☐ AT A CONFERENCE IN JANUARY 1967 Alan K. Campbell and the present writer assumed responsibility for surveying the status of national urban policy. The conclusion reached at that time was:

There is in fact no general agreement on any of the [se] prerequisites to an urban policy and it is, in part, this lack of agreement which has produced the variety of ad hoc approaches to urbanism and metropolitanism that today, together, constitute urban policy. [Campbell and Burkhead, 1968: 638]

We recognized that there were demands at that time for a national urban policy—we were writing at a time when some inner cities were in flames—but we argued that incrementalism would dominate and that anything resembling a national policy would not emerge.

For the next three or four years it looked as if we were quite wrong. The reports of the Douglas Commission, the Kaiser Com-

AUTHOR'S NOTE: *I am indebted to Alan K. Campbell and David Puryear for comments on an earlier draft, and to J. Michael McGuire and Shawna Grosskopf for computational assistance.*

mittee, the Kerner Commission, and *Urban and Rural America: Policies for Future Growth*, by the Advisory Commission on Intergovernmental Relations, all pointed to the possibilities for a more positive, coordinated federal role with respect to the cities.[1] The high-water mark of such proposals was reached in 1969 with Daniel P. Moynihan's ten-point program. Most of the points were concerned with an improvement in the conditions of a dependent poverty class, an improvement in federal programs that obstruct state-local ability to handle their problems, and local government reorganization.[2]

Ironically enough, if Moynihan's ten-point program and its influence on the Administration was the high-water mark, the low mark was reached rather quickly—the abandonment by the Administration and the rejection by the Congress in the fall of 1972 of the Family Assistance Plan, initially promoted by Moynihan as the federal centerpiece for national urban policy.[3]

For the past three years it has been downhill all the way. Unfortunately, the pessimism of 1967 once again seems appropriate. It is the purpose of this essay to attempt a redefinition of the issues and an assessment of immediate prospects. This is not a cheerful assignment; the outlook for an appropriate political coalition with an activist approach is not promising. Neither is it a propitious time for such an assessment; the federal "system" is in the midst of basic restructuring under the New Federalism and the policy outcomes of that restructuring are very much in doubt.

DEFINITIONAL CONFUSION

The first difficulty is that it may not even be appropriate to speak of a national urban policy in terms that suggest a distinction between "urban" and "domestic." With 68.6% of the nation's population, as of the 1970 Census, living within SMSAs (Standard Metropolitan Statistical Areas), and with that proportion certain to increase for the remainder of the century, we are now and for the foreseeable future will be a highly urbanized society. This suggests that the policy issue is one of the nature of federalism—the respective fiscal and programmatic roles that should be played by the federal government, the states, by substate regions, and by municipalities. Thus, urban policy is both substantive and structural.

The second definitional difficulty is that we have had, since World War II, a national urban policy fashioned in the great tradition of American pragmatism. This policy perpetuated the fragmentation of local government, rather than encouraging its reform. It has consisted of federal stimuli to urban sprawl by way of the highway program, FHA- and VA-insured mortgages for the middle class, tax concessions to homeowners as compared to renters, a minimum volume of public housing constructed almost entirely in central cities, and exclusionary local government land-use policies that have successfully confined the poor, the aged, and ethnic minorities to the inner city.[4] It follows that those who speak of a national urban policy should perhaps be talking about a settlement policy or a land-use policy, if it is intended to change this pattern (see Wingo, 1972: 6-10). And in land use the issue is again one of federalism—the role to be played by the national and by state and local governments.

A third difficulty, perhaps more conceptual than definitional, is describing anything relevant to urban policy in isolation from the national macroeconomy and its impact on the federal budget. The continuation of two-digit inflation accompanied by relatively high levels of unemployment, balance-of-payments disequilibria, and a credit squeeze, all on a worldwide scale, may so dominate U.S. domestic policy within the next few years that urban problems and policies will be relegated to an obscure position on the agenda.[5]

And the final difficulty, writing at "this point in time," is what Heilbroner (1974) has called "the pervasive unease of our contemporary mood." Not since the Great Depression have we been so belabored by visions of apocalypse. Overpopulation, combined with supply shortages, will overwhelm us. Air pollution will strangle us; water pollution will poison us; the seas will turn to sludge heaps. And if this is not enough to do us in, thermal pollution will descend within a century, or some mad physicist-terrorist will hold the world at ransom with a homemade atomic weapon. And in the meantime, with the demise of neo-Keynesian economics, we may have a repetition of the worldwide Great Depression.[6] Since any or all of these calamitous events may indeed come to pass, any attempt to prescribe or predict a national urban policy may be a bit like painting a fresco on a wall that is crumbling along the edges.

Nevertheless, with the caveats to one side, it must be hoped that we have learned something about urban policy and perhaps that we may even be able to profit from experience.

THE LEGACY OF THE SIXTIES

Even as our defeat in Vietnam forced a reappraisal of U.S. foreign policy, so have the domestic policy experiences of the 1960s initiated a reappraisal. The subject of "urban policy reexamined" or "what happened to the Great Society?" is already a familiar theme for academics and journalists and will continue to be such.[7] Reexamination should require an in-depth probe of social, economic, and political behavior patterns and the policy thrusts and counter-thrusts that emerged—an assignment far beyond the scope of this essay. Therefore, what will be attempted here is an impressionistic account of what happened, what did not happen, and the political-economic forces that have survived. That which happened will be organized around four general topics: (1) the economic character of metropolitan areas, (2) governmental organization in SMSAs, (3) fiscal behavior, and (4) ethnicity.

WHAT HAPPENED

(1) The trends in the out-movement of jobs and residences that were accelerated immediately after World War II were further accentuated during the prosperous years of the 1960s. As is commonly observed, the central-city portion of SMSA employment declined almost everywhere, giving rise to new patterns of out-commuting and cross-commuting, and increasing the dependence of the work force on the automobile. This dispersion and accentuated urban sprawl is often said to have contributed to a substantial mismatch of jobs and people, with the unskilled jobs available in the suburbs, and inner-city residents unable to reach them or establish residence near them. However, sweeping generalizations on this point are dangerous. In some of the largest SMSAs there appears to be an adequate supply of unskilled labor in the suburbs, and inner-city residents would benefit little by improved mass transit for reverse commuting. A careful study of the New York SMSA found, for example, that Nassau County had far more unskilled residents seeking employment than there were jobs to be filled (see Regional Plan Association, 1973).

The patterns of employment and residential distribution of the 1960s contributed to an accentuation of the traditional dichotomy between the central city (CC) and outside central city (OCC), although each SMSA would need to be examined specifically on this point.[8] Some suburbs, of course, came to look more like central cities in terms of their housing problems, their economic base, and their land-use patterns, but this is more the case with the older SMSAs in the northeast and middle west than in the southeast and southwest. The CC-OCC dichotomy continues as a pervasive feature of U.S. metropolitan areas.

It could be argued that the justification for a national urban policy is strengthened if SMSAs are coming to be more homogeneous. If such were the case a national program for housing or for income maintenance would have more uniform impact.

Although homogeneity has many dimensions, one of the most significant is median family income. Table 1 shows computations of the coefficient of variation of median family income among SMSAs with a population of more than 200,000 for the years 1959 and 1969. These are necessarily rough comparisons since the 140-plus SMSAs in this category span a wide range of population size. The coefficient of variation is the mean (of median family income)

TABLE 1

THE COEFFICIENT OF VARIATION FOR MEDIAN FAMILY INCOME FOR SMSAs OVER 200,000 POPULATION, 1959 AND 1969

Areas	1959 Median Family Income	1969 Median Family Income
All SMSAs	.137	.132
Geographic Divisions		
New England	.120	.136
Middle Atlantic	.116	.107
South Atlantic	.150	.136
East South Central	.068	.082
West South Central	.086	.099
East North Central	.062	.061
West North Central	.062	.082
Mountain	.084	.085
Pacific	.087	.109
Regions		
Northeast	.123	.129
South	.124	.118
North Central	.063	.069
West	.087	.106

SOURCE: There were 147 SMSAs with population over 200,000 in 1959 and 148 in 1969. Data are from *U.S. Census of Population: 1960*, Part I, U.S. Summary, General Social and Economic Characteristics, Tables 148, 154; ibid., *1970*, Tables 184, 188.

for each group of SMSAs divided by the standard deviation of the group. A decline in the coefficient is evidence of increased homogeneity; an increase in the coefficient is evidence of greater diversity with respect to SMSA income.

The computations for four major regions suggest that only in the South was there a slight movement toward homogeneity with respect to income—the coefficient of variation dropped from .124 to .118. There were slight movements away from homogeneity in the other three regions, with the most pronounced movement in the West —from .087 to .106. The geographic divisions show somewhat more diversity, as would be expected from further disaggregation, but again, no pronounced national trend. Homogeneity with respect to family income increased in the South Atlantic states from .150 to .136 and decreased in the Pacific states where the coefficient of variation changed from .087 to .109. It must be concluded that there continues to be substantial diversity among SMSAs in different regions when diversity is viewed in terms of median family income.

Table 2 exhibits computations of the coefficient of variation for the SMSAs in terms of per-capita income. It is possible that per-capita income behaved differently during the decade of the

TABLE 2
THE COEFFICIENT OF VARIATION FOR PER CAPITA INCOME FOR SMSAs OVER 200,000 POPULATION, 1959 AND 1969

Areas	1959 Per Capita Income	1969 Per Capita Income
All SMSAs	.144	.130
Geographic Divisions		
New England	.097	.110
Middle Atlantic	.153	.150
South Atlantic	.160	.119
East South Central	.106	.128
West South Central	.124	.102
East North Central	.087	.080
West North Central	.096	.100
Mountain	.086	.076
Pacific	.103	.128
Regions		
Northeast	.140	.143
South	.141	.117
North Central	.089	.083
West	.106	.128

SOURCE: Data were available on per capita income for 144 SMSAs in 1959 and 145 SMSAs in 1969, from Survey of Current Business, May 1973, pp. 24-31.

1960s than did median family income because of the increase in the ratio of unrelated individuals to families. However, the movements toward or away from per-capita income homogeneity, with the exception of the western region, are not markedly different than those for median family income. Again, for the geographic divisions, South Atlantic SMSAs show the greatest movement toward homogeneity—as the coefficient of variation drops from .160 to .119, and the Pacific states show the greatest movement away from homogeneity as the coefficient of variation increases from .103 to .128.

(2) Government organization in metropolitan areas altered very little, with fragmentation continuing to dominate in spite of such isolated cases as Jacksonville-Duval and Indianapolis. The number of districts organized for special purposes continued to increase. In SMSAs over 200,000 population the Bureau of the Census reported 15,371 governmental units in 1962 and 17,856 units in 1967. Only in Minneapolis-St. Paul was there a significant movement toward metropolitan government on a model that might be more widely copied than city-county consolidation. General revenue sharing, enacted in October 1972, will have the effect of freezing the existing structure of general government and tend to dampen the limited possibilities for local government reorganization.

The interesting and possibly significant developments of the decade in matters of government organization emerged in the slow strengthening of substate regionalism, spurred in part by the federal requirement that applications for federal aid be reviewed by such agencies.[9] There are now 585 substate regional councils and all but five states have such a pattern. The regional councils are about evenly divided between metropolitan and nonmetropolitan areas. In addition, 42 states have organized a pattern of substate districts for their own administrative and planning purposes.[10] These districts are usually independent of the agencies that review federal grant applications.

The second significant development is, of course, neighborhood government, as it emerged from a most complex set of forces including the civil rights movement, OEO community action programs, Model Cities, and neighborhood renewal programs.[11]

A third development, toward the end of the decade, with some promise of continuation, is the growth of state agencies with responsibility for urban problems. A number of states established offices of local government, or began slowly and haltingly with encouragement from some governors and state legislatures to "intervene" in that which had hitherto been regarded as the sacrosanct territory governed by "home rule." The land-use statutes of Florida, Delaware, and Oregon are symptomatic, as is the Urban Development Corporation of New York State.

(3) As for fiscal affairs, the decade marked some sharp departures from the years after World War II. Federal aid to metropolitan areas increased sharply, as is indicated in Table 3. Much of this increase occurred through the addition of categorical grant programs which started with Kennedy, were greatly accentuated in the Johnson Administration, and continued through the first two years of the Nixon Administration. Depending on how one counts titles and subtitles, the number of categorical grants increased from about 400 in 1964 to about 1,200 in 1970, and a very large proportion of these were directed toward urban areas.

Table 3 shows the magnitude of the dramatic increases in federal aid since 1961. From that year to 1969 federal outlays that impacted on SMSAs increased about 3.5 times. And the Office of Management and Budget projects a further increase of 2.5 times between 1969 and fiscal 1975. The largest amounts of increase came in education, medicaid, and food stamps. Public housing operating subsidies accounted for the largest increase in the category of community development and housing. The share of federal aid outlays that ends

TABLE 3
FEDERAL AID OUTLAYS IN SMSAs (in millions)

Function and Program	FY 1961	FY 1969	FY 1975 (estimate)
National defense	$ 10	$ 30	$ 38
Agriculture and rural development	155	417	440
Natural resources and environment	54	180	2,852
Commerce and transportation	1,435	2,539	3,944
Community development and housing	213	1,610	3,165
Education and manpower	558	2,963	5,097
Health	99	2,297	5,560
Income security	1,341	3,899	9,402
General government	25	129	1,082
General revenue sharing	--	--	4,322
Other	--	--	29
TOTAL	$3,890	$14,064	$35,931

SOURCE: *Special Analyses, Budget of the United States Government, Fiscal Year 1975* (Washington: Government Printing Office, 1974), p. 212.

up in SMSAs, as compared with total outlays to state and local governments, is about the same as the SMSAs' proportion of the total population—69%.

Federal aid tends to redress the fiscal perversity of state aid as between central city and outside central city. A 1970 estimate prepared for the Advisory Commission on Intergovernmental Relations shows that in the 72 largest SMSAs the central cities received 16% more per capita in state and federal aid than did their suburbs. This slight advantage was attributable to federal aid patterns, not to state aid patterns (Sacks and Callahan, 1973: 134-143).

The very large dollar increases in federal aid outlays must, of course, be looked at in relative terms. As a proportion of state-local expenditures, total federal aid outlays increased from 12.6% in 1961 to 17.4% in 1969. The peak year was fiscal 1973 at 23.5%, and 1975 is estimated to register a decline to 22.4% (Executive Office of the President, Office of Management and Budget, 1974: 210). It is apparent that neither state and local governments in the aggregate nor SMSAs can anticipate an increased share of federal funds unless the federal fiscal outlook changes dramatically.

Toward the end of the decade and continuing into the mid-1970s a significant fiscal development occurred at the state level. As an increasing number of states adopted personal income taxes and expanded sales tax bases, their revenue structures became more elastic. This, plus the one-third share of general revenue sharing that the states received, had a marked impact on their fiscal position. In 1973, for the first time since World War II, more states reduced taxes than increased them, and the trend has continued into 1974.

It is impossible to generalize about the fiscal position of local governments in metropolitan areas. As is customary, most large central cities appear to face a fiscal crisis that is as severe as ever, with recent accentuation from inflation, the demands of public employee unions, an apparent increased antipathy to the property tax, and all of this with dependence on traditional inelastic revenue sources. The central cities of the South and West have maintained more vitality than in other regions largely by capturing a part of suburban growth through annexation and consolidation.

The pattern of fiscal disparities between the central city and outside central city has not changed very much. The central cities have benefited somewhat more from federal aid than have areas outside the central city, but fiscal disparities in terms of tax rates, expenditure levels, and intergovernmental aid persist in a pattern not very different from that prevailing in 1957 (Sacks and Callahan, 1973: 91-152).

(4) The relative position of urban minorities, particularly the black minority in metropolitan areas, is not susceptible to easy generalization, as the controversies over the last two or three years indicate (see, for example, Wattenberg and Scammon, 1973; Brimmer, 1974). The educational attainment of blacks has improved and incomes of black high school and college graduates have increased. Younger blacks with a college degree appear to be closing the gap, both relatively and absolutely as compared with their white counterparts, but absolute income differentials at all educational levels remain substantial. In 1969, for all age levels, the black high school graduate was earning about the same as the white eighth-grade graduate (Brimmer, 1974: 153). In the poverty-welfare category recent experience is even less cheerful; female heads of families among blacks increased in the decade, relatively and absolutely, and their movement above the poverty line was minimal.

But the gains for urban blacks in the 1960s are not revealed wholly by looking at income profiles. The civil rights movement generated an increased political consciousness; many employment barriers were removed; middle-class black family movement to the suburbs was proportionately impressive in relative terms, although not in absolute numbers. There are far more black elected political officials today, both in the North and the South, than there were in 1960. The political gains may exceed the economic gains.

(1) The failure of the Great Society programs to enter the lists against the forces of localism and home rule was reflected not only in the lack of attention to the reform of metropolitan government structures but, and more importantly, in the consequent undisturbed pattern of land-use regulations. The continued preoccupation of urban economists with the externalities of public and private economic activity in metropolitan areas seems to have had little impact on public policy.

There are ways in which some of the major externalities could be overcome other than by the reorganization of metropolitan governments.[12] Metropolitanwide transportation planning would be helpful, and some progress has been made along these lines. But here, at least until the energy crisis in the winter of 1973-1974, little attention was paid to other than automobile transit. And it is somewhat utopian to imagine that the recently increased federal funds for mass transit will "save the cities" or alter dramatically the distribution of economic activity and residences in the metropolitan area.

The second program area where metropolitanization could be accomplished is housing. The decade recorded one moderate success in this area—the Dayton Fair Shares housing plan (Gruen and Gruen, 1972). This experience, in which a five-county area worked together to provide low and moderate income housing for the whole of the metropolitan region, has not yet been replicated elsewhere. Opening up the suburbs remains on the agenda.[13]

(2) The second major failure was in fiscal affairs. The largest metropolitan areas are surely in no stronger fiscal position now than twenty years ago, and many central cities, such as Newark and St. Louis, are in a far worse state of fiscal well-being. In the meantime, with the rapid rates of suburbanization in the 1960s, a high rate of commercial and industrial construction, and the period of sustained national economic growth from 1962 to 1970, it is evident that the private sector received very large development values—from construction and from land value increments. (Even Henry George would have been astonished.) There has not yet been devised, let alone enacted, a tax instrument to capture for community benefit the private gains of community development.

(3) The third prominent failure, closely linked with the two foregoing, was the inability to halt the process of urban decay in lower middle-income neighborhoods (see Adubato and Krickus, 1974). Central city decay is not a recent phenomenon; urban renewal programs dating from the Housing Act of 1949 were intended to deal with it. But its most recent manifestations—the "redlining" of neighborhoods by insurance companies and mortgage institutions, the subsequent abandonment of housing, and a pattern of urban blight that affects black and white ethnic neighborhoods alike—have started a particularly vicious circle which appears to be almost irreversible.

FORCES IN BEING

It is not quite clear when national policy started to move in the direction of urban problems and a concern with poverty—the two cannot be dissociated. The date might be put at Kennedy's election in 1960, or the civil rights march on Washington in 1964, or Johnson's announcement of the War on Poverty in that year. But one thing is quite clear: the mid-seventies are not the mid-sixties with respect to urban policy. With the possible exception of health insurance, we are not on the brink of major innovative policies, and, indeed, are still in the process of dismantling some of the experiments of ten years ago, as with OEO.

It may not be inappropriate to describe the course of urban policy from about 1968 to 1974 as a kind of counter-revolution. On the ideological front the publication of *The Unheavenly City* in 1970, authored by Edward C. Banfield, is typical, and his views had obvious influence in the conservative press and possibly an impact on public policy. The republication as *The Unheavenly City Revisited* in 1974, with a basic laissez-faire attitude toward urban policy and the role of the federal government therein largely unchanged, suggests that the counter-revolution is far from over.

The chronicle of events in the counter-revolution must also include the mood of political conservatism so evident in the 1972 elections. All crystal balls are cloudy when faced with forecasting the continuation of this mood.[14]

But the major factor that appears to have shaped the complex of forces that determine the course of national urban policy is fiscal. Here one must start with the national economy. The halcyon days of 1965, when it appeared that economic growth rates could be sustained indefinitely without serious inflation or balance-of-payments difficulties, disappeared by 1970. This meant the disappearance of the fiscal dividend, which in turn meant that federal funds were not available for new urban programs or for general tax reduction. The cessation of hostilities in Vietnam did not bring a peace dividend or curtail the demands of the military for larger appropriations. In consequence, the multitudinous grant-in-aid programs of the Johnson Administration could not be funded, and indeed, the Nixon Administration argued, must be sharply curtailed, on the ground that they "hadn't worked"—although many had hardly been given an adequate trial.[15] The next steps were reduced appropriation requests, impoundments, the suspension of new starts in neighborhood renewal and public housing, and the New Federalism.

It may well be that "the attack that was launched by the conservatives was if anything overdue" (Ginsberg and Solow, 1974: 124). What makes the counter-revolution so fascinating a chapter in the history of political economy is that the conservatives have been joined by many liberals and in some instances by the groups that had benefited from Great Society programs (Ginsberg and Solow, 1974: 4-13). Within the space of about two years—roughly 1970 to 1972—the national concern for the poor and for minority groups and a national policy for the eradication of poverty seemed to disappear. The form of disappearance was embodied in New Federalism. The same month—October 1972—that marked the demise of the Family Assistance Plan also marked the enactment of general revenue sharing.

It is difficult to separate the reality of New Federalism from its rhetoric and romance. The reality would appear to consist of two elements. First, given the assumed inviolability of the military budget, it was necessary to reduce the federal commitment to urban programs. Second, the system of categorical grants had indeed become most cumbersome, and state and local governments, with reason, came to resent increasingly federal "intervention" with respect to state-local decisions. The romance and rhetoric consisted of the contentions that New Federalism would return government to the people, revitalize local government, provide for greater community participation, and result in a more effective use of resources since decisions would be made locally in terms of local needs and preferences.

The two years' experience with general revenue sharing is most difficult to evaluate. It is possible to ascertain with accuracy data on the governmental units that received the funds, but it is not possible to ascertain with accuracy what was done with the funds, since they became general revenue comingled with other sources. Some highlights of the experience may, however, be noted.

(1) The total amounts involved have not been large in relation to the aggregate of state-local spending. The first year's distribution of $5.6 billion was equivalent to 5.3% of state-local noneducation spending in that year.

(2) The "insignificant" general government units, such as New England counties and midwestern townships with very limited responsibilities, received more in funds than would appear to be warranted by their program needs. Many such governments probably used the proceeds of general revenue sharing for one-time capital improvements, for debt retirement, or for tax reduction, not for current operating programs.

(3) The states' share (one-third) appears to have been devoted about 65% to tax reduction (Nathan, 1974).

(4) The large city picture is very mixed. Some, such as New York City, used the funds for current operating programs; others, such as Newark, for tax reduction (Nathan, 1974).

Since general revenue sharing expires in 1976, the Congress must, in 1975, make some decisions about its future. Although the program has been generally popular with governors and with mayors it is by no means certain that the Congress will reenact it. If it is continued it would be an appropriate occasion to tidy up the statute in the interests of large central cities. This could be done, political processes permitting, by reducing the states' share from one-third to perhaps one-quarter, and the channeling of additional monies to large cities. The 7.7% under-count of the black urban population is most evidently in need of correction, as are the ceiling provisions applicable to local governments.[16] To improve the elasticity of the shared revenues in an inflationary situation it would also be useful to tie the appropriation to the personal income tax base, as was originally proposed in the Heller-Pechman revenue sharing scheme.

The fiscal future of national urban policy is murky not alone because of the uncertainties that surround general revenue sharing, but in particular because of the uncertainties surrounding special revenue sharing.

The Administration's proposals for special revenue sharing were first enunciated in January 1971 and have been repeated and strengthened in succeeding budget messages. The fiscal concept is that of grant consolidation under a formula that would give states or local governments broader discretion in the allocation of resources within the consolidated grant. Little or no new money is to be added. As of the summer of 1974, the Administration had secured this kind of legislation for law enforcement, rural development, and manpower training. The Congress is considering special revenue sharing legislation for education, community development, and transportation. In addition, the Administration has proposed a Responsive Governments Act to extend modest sums for technical assistance in planning and management to states and local governments. It is by no means certain that any of this legislation will be adopted in the form submitted. Although it seems likely that there will be some grant consolidation, the form, content, and funding level is very much to be determined. It is highly probable that a large number of categorical grants will remain; and those who are concerned about the future of inner-city programs should favor the retention of a number of categoricals, particularly for housing and education in poverty areas.

Special revenue sharing, if adopted for community development and education, will change the rules of the game of grantsmanship. Funds will come to the cities on formula, to be allocated by means of traditional city budgetary processes. Mayors and city councils will, understandably, allot such resources in accordance with conventional practice, which may or may not extend to a concern for inner-city residents. Redistributional programs that are not popular with the middle-class voter may get short shrift. Only federal categorical grants can save the well-meaning mayor in this situation.

This point was recognized in a resolution of the International City Management Association at its meetings in October 1973:

Categorical grants are desirable when national priorities are at stake and state, local and private funding is scarce or unavailable. Or when the problems or matters being addressed occur only in a relatively small number of communities or *when the political risks are too high for more responsive local or state governments to bear* [italics added].

The game of grantsmanship will be changed in other ways if special revenue sharing is adopted for education and community development. With the categoricals many large and medium-sized cities developed competent staffs of "federal aid coordinators," sometimes with offices in Washington. Cities that staffed-up for grants and mastered the routines have tended to secure a "disproportionate" share. Such cities, and they are very often the most impecunious, will suffer, relatively, from special revenue sharing.

Although fiscal considerations have dominated urban policy in the last few years, there are other forces in being that will have an impact. It would now appear that both the Congress and the Administration are at last moving toward a major improvement in the delivery and financing of health services. This legislation could make a major contribution to the real welfare of lower- and middle-income groups in the cities. It also appears that neighborhood legal services, once an important part of OEO programs, will continue in a form that will be particularly helpful to inner-city residents.[17]

The counter-revolution against the poor and minority groups has not yet won all the recent engagements. But, to illustrate the complexity of current trends at a time when neighborhood legal services are apparently to be securely established, the Supreme Court has recently moved to reinforce home rule in land use and to narrow the scope of class action suits.[18] This may hamper the immediate possibilities for relief from exclusionary suburban zoning and housing by way of suits under the equal protection clause of the 14th Amendment.

THE IMMEDIATE FUTURE

This survey of recent developments in national urban policy and in programs that are significant for urban areas does not suggest that we are on the threshold of a ten-point program that will chart the future. In an inflationary economy, with substantial unemployment and low rates of real growth, the fisc will continue to dominate policy and program.

The cliche has it that everything depends on everything else in an urbanized society. Nonetheless, some things are more important than others, and some aspects of urban policy are more important than others. If the economic outlook brightened, and if there were more resources available for the cities, and if the national mood changed, what should be done? It is cautiously suggested here that three kinds of things should then be put on the agenda.

The first is informational. Social indicators should be developed on an SMSA basis. This means a disaggregation of the excellent contribution of OMB to nationwide indicators including time series data, on health, public safety, education, employment, income, housing, leisure and recreation, and population.[19] Social indicators at the SMSA level have their severe limitations. They do not provide information about one's neighborhood—whether streets are safer or housing has improved over time—the consequences of urban policy that interest the citizen-voter. But they are, even on an SMSA level, a valuable index of trends and are basic data for an appraisal of progress or the lack of it.

The second item that should be on the agenda is programmatic. The core of programs for a national urban policy should consist, ideally, of income maintenance, housing, and environmental protection.

Apart from middle- and upper-income antagonisms toward income maintenance, even when it is described as "welfare reform," the preparation of reasonably equitable legislation becomes more difficult each year. Since the fall of 1972 we have continued down the path of categorical assistance programs. Food stamps have been expanded; the Supplemental Security Income Program went into effect in January 1974; and if health insurance is provided with special attention to the medically indigent, another significant piece of categorical welfare aid will have been added. Thus, unless all of most of the categorical welfare aids are eliminated, which is an unlikely prospect, the task of integrating a national income maintenance program with the range of categorical poverty assistance becomes most formidable. This was the "technical" reason for the downfall of the Family Assistance Plan. Nevertheless, the task remains; it will be a complicated piece of legislation.

Although data from the 1970 Census suggest that there was substantial improvement in the housing stock as compared with 1960, major new construction, rehabilitation, and some relocation remain on the agenda, if the goals of the 1968 Housing Act are to be achieved. The proposals of the Administration for housing allowances have not yet been taken seriously by the Congress. They may or may not be the wave of the future in this area.[20] In the meantime, 1974 and 1975 promise to be very bad years for residential construction, both public and private. A highly restrictive monetary policy, thought to be necessary for the control of inflation, has been a serious setback. Unless domestic inflation is brought under control there will be little improvement in the quantity and quality of the housing stock in the near future.

Those aspects of environmental protection that are most significant for metropolitan areas are, of course, directed toward the reduction of air and water pollution and the provision of open space for recreation. Surprisingly enough, air and water pollution standards seem to have withstood very well the onslaught of attack from the oil companies, automobile manufacturers, and utilities that characterized the energy crisis in the early months of 1974. Environmental protection has been a rallying point for what would appear to be a most fragile consortium of middle-class interest groups. Nonetheless, this consortium has held together very well.

Governmental organization in metropolitan areas, described above as one of the major unfinished tasks of the last ten years, is likely to remain unfinished for the next ten. Structural reform could broaden the fiscal base of governments in metropolitan areas and contribute importantly to greater equality in the distribution of metropolitan tax burdens. Structural reform is also of greatest importance for land use. It is very difficult to see how existing patterns of urban sprawl can be reversed, or how the suburbs can be opened up for low- and middle-income housing, or how metropolitan transportation can move toward improved mass transit without the transfer of zoning and planning powers to the SMSA level and, in some instances, to even larger geographic regions. Substate regionalism is a hopeful development, but substate regionalism will accomplish very little without a sharp reduction in the strength of the localism that now controls land use.

Since the prospects for structural reform now appear so dim, if progress is to be made toward an increased viability for urban areas such progress will very likely take the form of strengthening domestic programs that impact heavily on such areas. As noted, these are income maintenance, housing, and environmental protection. Localism in land-use controls is as much structural as programmatic; unfortunately, it cuts across the programmatic lines.

It must be concluded that the prospects for the kinds of organizational and programmatic changes that would underpin a national urban policy are not very bright. But the directions toward which policy should move are much clearer than ten years ago.

NOTES

1. For an excellent review of these developments, see Wingo (1972).

2. The ten-point program emerged first in a speech at Syracuse University in May 1969 and later, with modification, in Moynihan (1970: 3-25).

3. The demise is very well described in Moynihan (1973).

4. See the comments of Ada Louise Huxtable on a speech by Mayor Tom Bradley in the *New York Times* (June 9, 1974), p. D19.

5. The Brookings volumes on priorities are significant harbingers of trends in national policy concerns. About three-quarters of the volume published in 1973 was devoted to domestic issues that were at least generally related to urban problems. In the 1974 volume only about one-third is devoted to such issues. See Fried et al. (1973) and Blechman et al. (1974).

6. On the latter theme, see Barraclough (1974).

7. See, for example, the special issue devoted to "The Great Society: Lessons for the Future," *Public Interest* (Winter 1974).

8. Additional counties were added to some SMSAs between 1959 and 1969 which makes homogeneity dimensions somewhat noncomparable. There appears to be no way to overcome this difficulty.

9. These requirements and the resulting stimulus were initiated in 1966 with the Demonstration Cities and Metropolitan Development Act and later formalized in Office of Management and Budget Circular A-95.

10. The Advisory Commission on Intergovernmental Relations has been sponsoring the creation of Umbrella Multi-Jurisdicational Organizations (UMJOs). For an exhaustive treatment of substate regional patterns, see ACIR (1973).

11. The literature is voluminous. An excellent early contribution is Kotler (1969).

12. The report of the Committee for Economic Development (1970) for a two-tier government for metropolitan areas provided an occasion for useful academic discussion. At present writing the National Academy of Public Administration is exploring its potential for application in Rochester-Monroe and Tampa-St. Petersburg.

13. The difficulties and possibilities that will be encountered are well explored in Downs (1973).

14. The semantics of the mood are interesting and possibly significant. When welfare mothers occupy the district welfare office to call attention to their grievances, this action may be described as the politics of violence (illegitimate) or the politics of confrontation (legitimate).

15. See, for example, the examination of OEO programs in Kershaw (1970).

16. Consolidated city-county governments—such as New York City, Baltimore, and Philadelphia—appear to have received inadequate consideration in the 1972 formula.

17. For perceptive observations on "using the system to change the system," see Liebman (1974).

18. Village of Belle Terre v. Boraas 94 S. Ct. 1536 (1974); Eisen v. Carlisle & Jacquelin, 94 S. Ct. 2140 (1974).

19. Executive Office of the President, Office of Management and Budget (1973).

20. Significant social experiments with housing allowances are now underway, sponsored by HUD, with participation by the RAND Corporation. The outcomes of the experiments may be helpful in drafting future legislation to increase and improve the stock of housing. (See U.S. Department of Housing and Urban Development, 1973.)

REFERENCES

ADUBATO, S. N. and R. J. KRICKUS (1974) "A strategy for the cities." Nation (May 18): 623-628.

Advisory Commission on Intergovernmental Relations (1973) Regional Decision-Making: New Strategies for Substate Districts and Regional Governance—Promise and Performance. Washington, D.C.: Advisory Commission on Intergovernmental Relations.

BANFIELD, E. C. (1970) The Unheavenly City. Boston: Little, Brown.

BARRACLOUGH, G. (1974) "The end of an era." New York Review (June 27): 14-20.

BLECHMAN, B. M. et al. [eds.] (1974) Setting National Priorities, the 1975 Budget. Washington, D.C.: Brookings.

BRIMMER, A. F. (1974) "Economic developments in the black community." Public Interest (Winter): 146-163.

CAMPBELL, A. K. and J. BURKHEAD (1968) "Public policy for urban America," in H. S. Perloff and L. Wingo, Jr. (eds.) Issues in Urban Economics. Baltimore: Johns Hopkins University Press.

Committee for Economic Development (1970) Reshaping Government in Metropolitan Areas. New York: Committee for Economic Development.

DOWNS, A. (1973) Opening Up the Suburbs. New Haven, Conn.: Yale University Press.

Executive Office of the President, Office of Management and Budget (1974) Special Analyses, Budget of the United States Government, Fiscal Year 1975. Washington, D.C.: Government Printing Office.

——— (1973) Social Indicators, 1973. Washington, D.C.: Government Printing Office.

FLAX, M. J. (1972) A Study in Comparative Urban Indicators. Washington, D.C.: Urban Institute.

FRIED, E. R. et al. [eds.] (1973) Setting National Priorities, the 1974 Budget. Washington, D.C.: Brookings.

GINSBERG, E. and R. M. SOLOW (1974) "Introduction." Public Interest (Winter).

GRUEN, N. J. and C. GRUEN (1972) Low and Moderate Income Housing in the Suburbs. New York: Praeger.

HEILBRONER, R. L. (1974) An Inquiry into the Human Prospect. New York: Norton.

KERSHAW, J. A. (1970) Government Against Poverty. Washington, D.C.: Brookings.

KOTLER, M. (1969) Neighborhood Government. Indianapolis: Bobbs-Merrill.

LIEBMAN, L. (1974) "Social intervention in a democracy." Public Interest (Winter): 14-29.

MOYNIHAN, D. P. (1973) The Politics of a Guaranteed Income: The Nixon Administration and the Family Assistance Plan. New York: Random House.

——— (1970) Toward a National Urban Policy. New York: Basic Books.

NATHAN, R. P. (1974) "Whither revenue sharing?" New York Times (June 23): E3.

Regional Plan Association (1973) Transportation and Economic Opportunity: Report to the Transportation Administration of the City of New York. New York: Regional Plan Association.

SACKS, S. and J. CALLAHAN (1973) "Central city suburban fiscal disparity," in City Financial Emergencies: The Intergovernmental Dimension. Washington, D.C.: Advisory Commission on Intergovernmental Relations.

U.S. Department of Housing and Urban Development (1973) First Annual Report of the Experimental Housing Allowance Program. Washington, D.C.: Government Printing Office.

WATTENBERG, B. J. and R. M. SCAMMON (1973) "Black progress and liberal rhetoric." Commentary (April): 35-44.

WINGO, L., Jr. (1972) "Issues in a national urban development strategy for the United States." Urban Studies (February): 3-27.

2

The Large City as Fiscal Artifact

SEYMOUR SACKS with
PATRICK J. SULLIVAN

What also seems to be beyond debate is that the fiscal condition of the local governments of these cities is as bad as or worse than it has ever been. The diagnosis is quite simple. The problems put pressure on the demand side: more money for welfare payments; more money for education; more money for police and fire protection, sanitation, and almost everything else on the list of locally provided goods and services. [Chinitz, 1972: 79]

It seems reasonable to expect that a community with a high income level and federal and state aid can spend more than a community which is deprived of such affluence. The more money a city raises, the more it can spend. Only if the richer cities spent less and those with lower income spent more would the phenomenon be sufficiently surprising to deserve serious explanation. [Meltsner and Wildavsky, 1970: 314]

☐ LARGE CITIES are complex phenomena. They vary greatly, both between and within states. They represent elements of the federal-state-local system of governments wherein no individual part can be analyzed independently of the other parts. This apparent truism is nowhere so important as in fiscal matters, yet a continuing approach views cities as singular independent entities, without regard to whether their fiscal or, for that matter, their social or economic dimensions are being considered. General statements are made in the literature concerning cities, while the courts, legislatures, and administrations on all levels act as though widely varied situations can be accurately described and policy measures enacted by a single approach which encompasses all the issues and solves all the problems.

One way of considering large cities is to view them as fiscal artifacts. Other municipalities and areas may be analyzed in similar fiscal terms, for the approach is not exclusively related to large cities. Large cities, however, have been diagnosed as having inordinate fiscal as well as social and economic problems which differentiate them as a class from other areas.[1]

The complexity asserted at the outset involves not only differences in the underlying character of areas studied, but rather important formal or definitional considerations which have substantive implications which are bypassed in most studies of large cities. Perhaps the most important formal distinction, and the one which entails the most distortion, separates the city government from other *local* governments serving the city area. Only in the rarest instance is the city serviced by a single, general-purpose municipal government. This means that in order to understand all of the problems involved, it is necessary to know something not only about the municipality, but also about the other local governments servicing the area.

Cities and other local governments also differ in the extent to which they are assigned functional responsibilities by their overlying state governments and the extent to which they attempt to meet these responsibilities. To cite the most important case, aggregate data indicate that the school district, not the municipality, is the most important class of local government in terms of total revenues and expenditures. But what may be true in the aggregate is not necessarily true in detail, especially when the detail includes the large cities.

The purpose of this paper is to examine the large cities in a context which recognizes and explains the existence of large city governments and large city areas as fiscal artifacts. The first step is to indicate the differences between the large city (municipal) governments and the local governmental systems servicing large city areas. The traditional analysis, which attempts to compare governments in terms of numbers (often the irrelevant count of suburban governmental structure in an SMSA), will be supplemented by measures which also indicate the relative importance of different classes of governments in large city areas. The second step examines the evolution of the city. In this section the paper explains the change in the city's perceived position. The large city—once viewed as fiscally advantaged (at a time when there were few alternatives to the municipal government apart from the school district)—has recently emerged in the literature and in policy as the major problem area and the major area of exploitation within the entire society—that is, the fiscally disadvantaged area (see, for example, Neenan [1972]). Reasons for this will be considered in some detail with reference to the evolving fiscal area.

The third step in our analysis examines the fiscal characteristics of the large city and large city areas in greater detail than usual, not only with respect to total expenditures and revenues, but to those of the component governments. The fourth step of the paper inspects the problem of the large city compared to its surrounding (outside central city) area in a context which recognizes differences in functional assignment of expenditures and revenue responsibilities and explicitly includes the interactive metropolitan context ot large cities. In this analysis, cities are compared to their surrounding areas rather than to other cities.

1. GOVERNMENTAL STRUCTURE:
THE CITY AS A MUNICIPALITY AND THE
CITY AS A GEOGRAPHIC AREA

The literature has generally tended to mingle the concepts of the city as a municipality and the city as a social or geographic area.[2] The organization of the municipality has been viewed as the governmental organization of the city area, and the complex governmental organization of the metropolitan area has therefore been treated as though it indicated significant problems of the city area. One alternative, of course, is to look at the metropolitan area as an urban area, the "true" city area, but this perspective is not usually adopted rigorously by those who use the concept. This portion of the paper examines the governmental organization of the large city area in a systematic fashion.

Excluded from discussion are the direct activities of overlying higher levels of government, the state and the federal government. In all cases, these activities have indirect consequences. Very direct consequences of state government activity appear in some cases. Included in this discussion are those other local governments which, whether or not they are coterminous, provide governmental services and raise revenues from the residents and users of the large city area.

As already indicated, the fiscal nature of the large city area is rarely coextensive or coterminous with fiscal boundaries of the municipality. In most cases in the United States, the activities of the municipal government, i.e., the city, represent only a fraction of the fiscal activities of local governments within the boundaries of the city. The situation is probably even more extreme outside the boundaries of the large cities, but at this point we consider only the relationships existing in the central cities.

The most general statement that can be made is that city (municipal) governments represent an important but varying segment of the local fiscal process. The importance of each municipality is modified by the existence of overlapping governments and the extent to which cities are assigned, have inherited, or have chosen to undertake functional responsibilities. Even where cities are not of overwhelming importance, they, like other minor civil divisions and counties, are important because they are the units for which information is available and as such they are objects of policy decisions. It is possible to construct a governmental decision-making system in terms of the location of school districts which would be very different from a system based on municipalities.[3]

The major source of information on the structure of local government in the United States is contained in Volume I of the Census of Governments, *Governmental Organization*. The information is detailed for metropolitan areas and more cursory for county and nonmetropolitan areas and for those places without county governments or their equivalents. Thus, very little information is provided on a city basis or indeed for other smaller units; consequently, it is often assumed that what is true for an entire area is true for the city. As shown in Table 1, the greatest number of governments within metropolitan areas are in outside central city areas. While few city areas show a considerable number of governments, this phenomenon is much more important than the standard "numbers game" used without detailed explanation in analyzing metropolitan areas would indicate. The principal set of fiscal questions in the case of the large city involves the existence of an operative county, independent school districts, and independent special districts. This, in turn, leads to the question of what constitutes a government, which will be addressed in the discussion of special districts below.

Perhaps the most important single difference among city governments is whether they have the governmental responsibility for elementary and secondary education—the case in which education is a dependent agency of the municipality and where there is no independent school district financing education. The scale of educational involvement has become important in recent years since the educational portion of municipal taxes must be subtracted out to get the "adjusted taxes" used in the local distribution of general revenue sharing monies (see U.S. Department of the Treasury, Office of General Revenue Sharing, 1973: ii-ix). The dependent school district is characteristic of the northeastern portion of the nation, especially New England, New Jersey, and cities with populations over 125,000 in New York, Tennessee, and Virginia. As can be seen from Table 1, the Northeast had 13 of the 19 dependent school districts in this sample. In other states, education is provided by independent school districts of a city or county nature except in Honolulu, where it is provided by the state. A very great difference exists, however, in the geographic boundaries of the independent school systems servicing large cities. They may be coterminous with the large cities, or larger than the cities, or smaller than the cities, or they may cut across city boundaries.

This complexity is characteristic of a few cities of extensive size and low density in the Midwest, i.e., Indianapolis and Kansas City, Missouri. It is also characteristic of Oklahoma, Texas, California, Arizona, and Oregon. No national generalization seems possible, only state typologies appear to be meaningful.[4]

In addition to the question of where school district responsibility is located, two general questions draw attention. The first asks whether an overlying county is present. Counties are important because they provide a number of functions which pass beyond city boundaries, where the county extends past city boundaries. Counties are assigned a number of important functions, most notably public welfare when it is a local rather than a state function, and health and hospital, and highway responsibilities. The absence of a county generally means that the responsibility devolves on the city or occasionally on a special district. The absence of county governments is characteristic of New England, Virginia, and a few city-county consolidations which are enormously important in the analysis of large city governments because they cover such cities as New York,

TABLE 1
LARGE CITIES: GOVERNMENTAL CHARACTERISTICS

Region	Local Governments in SMSA	City Area[a] Number of Governments in 1972		
		County	School District(s)	Special District(s)
Northeast				
Hartford	68	0	0	2
Wilmington	66	1	1	1
Washington, D.C.	90	0	0	2
Baltimore	29	0	0	0
Boston	147	0	0	4
Springfield	48	0	0	2
Jersey City	33	0	0	5
Newark	207	1	1[b]	3
Paterson	205	1	0	5
Albany	216	1	0	0
Buffalo	142	1	0	0
New York City	538	0	0	2
Rochester	198	1	0	1[c]
Syracuse	182	1	0	1
Allentown	205	1	2[b]	4
Harrisburg	202	1	1	5
Philadelphia	852	0	2[b]	4
Pittsburgh	698	1	2[b]	8
Providence	84	0	0	2
Midwest				
Chicago	1172	1	1	7
Gary	129	1	1	2
Indianapolis	296	0	9	14
Wichita	139	1	2	2
Detroit	241	1	1	3
Flint	95	1	1	1
Grand Rapids	93	1	2	1
Minneapolis	218	1	1	3
Kansas City, Mo.	256	3	16[b]	2
St. Louis	483	0	2	3
Omaha	234	1	4[b]	5
Akron	98	1	1	4
Cincinnati	260	1	3[b]	5
Cleveland	210	1	3	4
Columbus	128	1	2[b]	4
Dayton	161	1	2[b]	5
Toledo	137	1	3	7
Youngstown	107	1	1	5
Milwaukee	149	1	2[b]	1

TABLE 1 (continued)

Region	Local Governments in SMSA	City Area[a] Number of Governments in 1972		
		County	School District(s)	Special District(s)
South				
Birmingham	92	1	1	3
Mobile	35	1	County	5
Jacksonville	9	0	County	2
Miami	33	1	County	3[c]
Tampa	45	1	County	3
Atlanta	86	2	2[b]	3
Louisville	181	1	1	2
New Orleans	42	0	1	2
Greensboro	39	1[d]	0	3
Oklahoma City	77	3	14	2
Tulsa	115	2	3	3
Knoxville	33	1	0	2
Memphis	51	1	0	3
Nashville-Davidson	38	0	0	4
Dallas	201	1	7[b]	3
Fort Worth	87	1	10	5
Houston	304	1	11[b]	7
San Antonio	69	1	13[b]	7
Norfolk	10	0	0	2
Richmond	7	0	0	0
West				
Phoenix	112	1	24[b]	10
Anaheim	111	1	6[b]	7
Fresno	200	1	3	24
Los Angeles	232	1	8	7
Sacramento	210	1	7[b]	8
San Bernardino	233	1	2	6
San Diego	151	1	10[b]	3
San Francisco	302	0	2[b]	4
San Jose	75	1	20[b]	6
Denver	272	0	1	7
Honolulu	4	0	State	2
Portland	298	1	5	5
Salt Lake City	66	1	1	4
Seattle	269	1	1	6
Also				
St. Paul	See Minneapolis	1	1	6
Long Beach	See Los Angeles	1	6	5
Oakland	See San Francisco	1	1	7

a. Excludes city government
b. Includes higher education
c. Special Districts in existence in 1970
d. County Dependent School District
SOURCE: U.S. Bureau of the Census (1973) Vol. 1, *Government Organization.*

TABLE 2
DISTRIBUTION OF EXPENDITURE RESPONSIBILITIES
IN LARGE CITIES, 1970 (in percentages)

Region	Per Capita Amount	City	County	School District(s)	Special District(s)
Northeast					
Regional Average	$ 551	73.4	12.1	9.1	5.4
Hartford	501	89.7	0.0	0.0	10.3
Wilmington	679	79.4	4.9	0.0	15.5
Washington, D.C.	1006	100.0	0.0	0.0	0.0
Baltimore	638	100.0	0.0	0.0	0.0
Boston	530	95.3	0.0	0.0	4.7
Springfield	393	98.7	0.0	0.0	1.3
Jersey City	454	72.0	19.7	0.0	8.3
Newark	734	55.1	24.1	0.0	20.8
Paterson	380	66.8	26.2	0.0	7.0
Albany	476	65.9	34.1	0.0	0.0
Buffalo	528	69.4	30.6	0.0	0.0
New York City	894	92.6	0.0	0.0	7.4
Rochester	697	75.8	24.1	0.0	0.1
Syracuse	561	64.9	35.1	0.0	0.0
Allentown-Bethlehem-Easton	312	33.9	8.6	44.9	12.6
Harrisburg	359	33.7	7.7	52.9	5.6
Philadelphia	494	60.3	0.0	39.7	0.0
Pittsburgh	449	42.0	15.2	34.7	8.1
Providence	392	100.0	0.0	0.0	0.0
Midwest					
Regional Average	480	43.2	15.3	36.5	5.0
Chicago	478	42.5	10.9	33.4	13.2
Gary-Hammond-East Chicago	465	31.7	13.2	52.7	2.4
Indianapolis	355	35.4	13.9	40.6	10.1
Wichita	474	46.9	18.3	34.6	0.0
Detroit	371	47.6	14.2	37.9	0.3
Flint	747	39.9	21.4	38.7	0.0
Grand Rapids	434	37.5	16.7	45.8	0.0
Minneapolis-St. Paul	539	32.1	24.9	29.1	13.9
Kansas City, Mo.	484	50.4	12.6	35.9	1.1
St. Louis	463	48.6	0.0	38.2	13.2
Omaha	335	30.0	25.0	35.9	9.4
Akron	412	48.3	13.6	36.0	2.2
Cincinnati	760	65.8	9.5	21.6	3.1
Cleveland	512	38.6	15.6	40.0	5.7
Columbus	397	48.1	13.6	34.2	4.1
Dayton	456	49.0	12.4	35.0	3.5
Toledo	444	46.1	15.3	35.2	3.3
Youngstown-Warren	335	42.2	14.9	39.9	2.9
Milwaukee	561	37.6	27.2	29.2	6.1

TABLE 2 (continued)

Region	Per Capita Amount	City	County	School District(s)	Special District(s)
South					
Regional Average	$ 383	53.3	10.5	30.0	6.2
Birmingham	334	39.6	11.4	33.2	15.8
Mobile	333	30.7	6.8	45.5	17.0
Jacksonville	308	43.5	0.0	56.5	0.0
Miami	481	27.9	24.5	41.9	5.6
Tampa-St. Petersburg	372	36.2	17.2	44.8	1.7
Atlanta	554	36.7	10.2	39.4	13.7
Louisville	508	63.0	14.0	22.6	0.1
New Orleans	333	54.4	0.0	37.9	7.8
Greensboro-Winston-Salem-Highpoint	433	46.1	51.3	0.0	2.3
Oklahoma City	296	51.7	8.2	40.0	0.0
Tulsa	306	38.9	9.9	51.2	0.0
Knoxville	372	77.0	15.9	0.0	7.8
Memphis	370	78.4	12.5	0.0	10.8
Nashville-Davidson	378	90.5	0.0	0.0	9.5
Dallas	352	44.2	6.3	42.0	7.4
Fort Worth	315	37.5	9.2	47.2	6.1
Houston	305	35.8	8.5	48.3	7.4
San Antonio	252	34.5	4.7	50.1	10.7
Norfolk-Portsmouth	453	100.0	0.0	0.0	0.0
Richmond	529	100.0	0.0	0.0	0.0
West					
Regional Average	541	41.8	19.8	34.8	3.6
Phoenix	375	34.5	15.7	49.8	0.0
Santa Ana-Garden Grove, Anaheim	410	24.0	25.8	46.5	3.7
Fresno	686	30.4	35.9	31.1	2.5
Los Angeles-Long Beach	623	28.8	38.9	31.6	0.8
Sacramento	683	30.5	37.6	31.3	0.6
San Bernardino-Riverside-Ontario	635	24.0	33.7	42.0	0.2
San Diego	484	27.8	31.2	38.6	2.4
San Francisco	779	71.1	0.0	24.5	4.4
Oakland	747	31.0	26.0	32.0	11.0
San Jose	553	27.4	30.0	42.3	0.4
Denver	501	60.9	0.0	36.1	3.0
Honolulu	198	100.0	0.0	0.0	0.0
Portland	485	39.7	13.0	40.4	6.9
Salt Lake City	305	34.9	18.4	46.4	0.3
Seattle-Everett	524	37.1	13.7	29.4	19.8
Overall Average	478	53.6	14.1	27.2	5.1

SOURCE: Unpublished documents from the Government's Division used in "Central City Suburban Fiscal Disparity," Appendix B of Advisory Commission on Intergovernmental Relations (1973a) *City Financial Emergencies: The Intergovernmental Dimension.*

TABLE 3

DISTRIBUTION OF TAX RESPONSIBILITIES
IN LARGE CITIES, 1970 (in percentages)

Region	Per Capita Amount	City	County	School District(s)	Special District(s)
Northeast					
Regional Average	$ 272	76.6	14.1	9.2	0.1
Hartford	354	98.7	0.0	0.0	1.3
Wilmington	213	90.7	9.3	0.0	0.0
Washington, D.C.	516	100.0	0.0	0.0	0.0
Baltimore	221	100.0	0.0	0.0	0.0
Boston	369	100.0	0.0	0.0	0.0
Springfield	210	100.0	0.0	0.0	0.0
Jersey City	240	75.6	24.4	0.0	0.0
Newark	352	79.6	20.4	0.0	0.0
Paterson-Clifton Park-Passaic	221	79.7	20.3	0.0	0.0
Albany	219	75.3	24.7	0.0	0.0
Buffalo	236	60.6	39.4	0.0	0.0
New York City	384	100.0	0.0	0.0	0.0
Rochester	272	62.0	38.0	0.0	0.0
Syracuse	272	48.9	51.1	0.0	0.0
Allentown-Bethlehem-Easton	176	35.9	12.8	51.3	0.1
Harrisburg	180	43.6	12.3	44.5	0.0
Philadelphia	250	57.2	0.0	42.8	0.0
Pittsburgh	294	47.8	16.1	36.0	0.0
Providence	196	100.0	0.0	0.0	0.0
Midwest					
Regional Average	240	40.9	13.1	43.7	2.3
Chicago	244	46.1	7.6	36.8	9.4
Gary-Hammond-East Chicago	266	38.5	17.4	41.3	2.8
Indianapolis	226	41.0	16.1	36.2	6.6
Wichita	209	34.1	24.6	41.3	0.0
Detroit	255	57.8	11.0	30.8	0.4
Flint	262	36.3	9.3	54.4	0.0
Grand Rapids	177	42.3	12.4	45.3	0.0
Minneapolis-St. Paul	227	38.8	18.2	42.2	0.7
Kansas City, Mo.	253	43.8	14.4	41.8	0.0
St. Louis	267	60.8	0.0	30.0	9.2
Omaha	195	33.9	15.8	49.6	0.7
Akron	225	34.1	9.7	56.3	0.7
Cincinnati	251	45.3	12.9	39.7	2.1
Cleveland	296	42.9	13.3	43.0	0.8
Columbus	198	29.1	12.9	57.5	0.6
Dayton	264	43.5	8.3	46.0	2.1
Toledo	228	39.0	9.2	50.9	0.9
Youngstown-Warren	203	38.7	9.6	50.9	0.8
Milwaukee	306	29.6	26.6	36.4	6.5

TABLE 3 (continued)

Region	Per Capita Amount	City	County	School District(s)	Special District(s)
South					
Regional Average	$ 164	61.0	14.7	22.6	1.7
Birmingham	125	54.7	34.4	10.9	0.0
Mobile	124	55.9	15.8	25.1	3.2
Jacksonville	111	64.0	0.0	36.0	0.0
Miami	221	47.7	25.8	26.5	0.0
Tampa-St. Petersburg	170	52.2	18.9	28.9	0.0
Atlanta	252	36.6	24.8	36.7	1.9
Louisville	181	52.4	20.3	27.3	0.0
New Orleans	148	68.9	0.0	31.1	0.0
Greensboro-Winston-Salem-Highpoint	155	54.3	45.7	0.0	0.0
Oklahoma City	152	46.1	14.0	39.9	0.0
Tulsa	160	32.3	18.7	49.0	0.0
Knoxville	163	81.4	18.6	0.0	0.0
Memphis	161	82.5	17.5	0.0	0.0
Nashville-Davidson	163	100.0	0.0	0.0	0.0
Dallas	211	49.9	7.9	36.4	5.7
Fort Worth	157	45.8	8.8	38.8	6.6
Houston	181	45.4	11.6	37.4	5.6
San Antonio	102	48.9	10.4	28.8	11.9
Norfolk-Portsmouth	138	100.0	0.0	0.0	0.0
Richmond	209	100.0	0.0	0.0	0.0
West					
Regional Average	249	41.4	19.0	37.8	1.8
Phoenix	172	41.9	20.4	37.7	0.0
Anaheim-Santa Ana-Garden Grove	235	25.0	26.1	42.6	6.3
Fresno	284	31.3	30.0	36.5	2.2
Los Angeles-Long Beach	329	32.2	27.8	38.0	1.9
Sacramento	284	35.5	30.0	34.1	0.4
San Bernardino-Riverside-Ontario	261	23.0	31.3	45.0	0.6
San Diego	206	31.1	25.6	43.8	0.2
San Francisco	485	61.9	0.0	38.1	0.0
Oakland	338	36.2	20.1	37.2	6.6
San Jose	250	26.6	23.6	49.4	0.1
Denver	272	62.7	0.0	37.1	0.2
Honolulu	135	100.0	0.0	0.0	0.0
Portland	260	36.5	13.6	47.9	2.0
Salt Lake City	207	38.8	19.2	38.9	3.0
Seattle-Everett	203	37.0	18.0	40.8	4.2
Overall Average	235	55.8	15.0	27.7	1.5

SOURCE: Unpublished documents from the Government's Division used in "Central City Suburban Fiscal Disparity," Appendix B of Advisory Commission on Intergovernmental Relations (1973a) City Financial Emergencies: The Intergovernmental Dimension.

Baltimore, Denver, Philadelphia, St. Louis, New Orleans, San Francisco, and Honolulu City-County, as well as the more recent consolidations in Nashville-Davidson, Indianapolis (Marion County), Jacksonville (Duval County), and Baton Rouge. The existence of a city-county with school and public welfare responsibilities is of great importance in evaluating any set of local fiscal requirements, for it represents a concentration of fiscal activities which is not comparable to most large cities.

The second question to be asked of local government organization is whether independent special districts affect the fiscal behavior of an area under study. In a number of states, special districts are very important in the provision of suburban services. Special districts also provide a number of very large-scale services—especially mass transit, electricity, and so on—where they provide services in large city areas. They seem to be a far more common element in local government than are independent school districts. Only a few cities entirely lack independent special districts.

Special districts appear most prevalent in the West, but they are also common in the Midwest and East as well. The count of independent special districts we have used is that of the Bureau of the Census and does not measure fiscal importance. Far more important is the great variety of subordinate agencies and taxing areas which do not have all the characteristics of an independent government, but which have responsibilities to be recognized in evaluating the fiscal structure of local governments. This is most notably true of housing and urban renewal agencies which are dealt with as a single class by financial services, regardless of how they are classified by the Census Bureau. Where housing and urban renewal agencies receive very large federal subsidies and grants, the observed pattern of finances of individual city governments may be distorted. Special districts in cities serve other major purposes, but they are often the counterparts of agencies which, in other cities, are parts of the general municipal purpose governments.

Tables 2 and 3 contain an analysis of the distribution of expenditure and tax responsibilities, respectively, of the city governments and other local governments providing services to the city. These tables represent the first major attempt to allocate the city's share of fiscal activity when the governments involved go beyond the city boundaries, as well as the simpler case in which all activities of the special district and school district are contained within the city under consideration. The fiscal counterparts of the different governmental arrangements are shown on a regional and state basis.

It should also be recognized that relationships exist between the city and surrounding areas when they are part of the same government, for example, county, school district, and special district.[5] These areas also interact in their metropolitan context, a context partly established by the overall assignment of functional responsibilities by the state government and the subsequent modification of this assignment on a local level. One must remember also the existence of state and federal aid, which is influenced by the assignment of functional responsibility; the fact of assignment explains why municipalities receive such a small portion of total federal aid, and even why their share of direct aid prior to general revenue sharing was and is so small compared to their presumed behavior and "needs."

2. THE NATURE OF THE CITY AREA: 1900 TO 1970

The large city is defined by its municipal boundaries. The expansion of municipal boundaries since the turn of the century has been characteristic of most large cities, although a few such cities reached maturity before the beginning of the twentieth century. The expansion of the city boundaries had important consequences for governmental organization, although here, as in other aspects of large city fiscal behavior, no universal statement is possible.

Table 4 shows the changes in the area of the large cities since the turn of the century. No information on area is provided prior to the time when the city had a population of at least 30,000. It is assumed that most cities were small when they had populations of less than 30,000 and, if they had extensive areas, that they were very definitely overbounded. Further, a population of less than 30,000 also indicates the relative immaturity of a city's experience in providing local public services. Table 4 differs from the typical approach wherein population growth serves as a single measure of city growth. Instead, the data provided indicate the differences among large cities especially in the Northeast as compared to other parts of the nation. Contrasts would appear even more dramatic if smaller cities with fixed boundaries had been included. Of the large cities with the same boundaries in 1900 and 1970, 12 of the 16 were

TABLE 4

LARGE CITIES: AREA AND DENSITY CHARACTERISTICS

Region	Area (Thousands of Acres)			Persons Per Acre		
	1970	1930	1900	1970	1930	1900
Northeast						
Hartford	11.14	10.16	10.96	14.19	16.15	7.29
Wilmington	8.26	4.60	4.03	9.74	23.17	18.98
Washington, D.C.	39.30	39.68	38.41	19.25	12.60	7.26
Baltimore	50.11	50.38	20.26	18.08	15.98	25.12
Boston	29.44	28.10	24.68	21.78	27.82	22.73
Springfield	20.29	20.29	20.29	8.08	7.39	2.60
Jersey City	9.66	8.32	8.32	26.97	38.07	24.81
Newark	15.04	15.09	13.06	25.42	29.31	18.84
Paterson	5.38	5.16	5.16	27.96	26.84	20.38
Albany	13.38	12.08	6.91	8.66	10.55	13.63
Buffalo	26.43	24.89	24.79	17.51	23.02	14.21
New York City	191.81	191.36	183.56	41.16	36.22	18.73
Rochester	23.49	21.91	10.19	12.61	14.98	15.96
Syracuse	16.51	16.22	10.84	11.96	12.91	10.00
Allentown	11.39	7.30	1.78	9.61	12.68	19.90
Harrisburg	4.86	3.96	2.37	13.81	20.28	17.48
Philadelphia	82.24	81.92	83.34	23.69	23.82	15.52
Pittsburgh	35.33	32.84	18.10	14.73	20.40	17.77
Providence	11.58	11.41	11.35	15.48	22.17	15.47
Midwest						
Chicago	142.46	129.22	117.19	23.61	26.13	14.49
Gary	26.88	25.81	a	6.53	3.89	a
Indianapolis	242.99	34.66	18.18	3.06	10.51	9.33
Wichita	55.36	13.26	a	5.00	8.38	a
Detroit	88.32	88.26	18.14	17.11	17.77	15.75
Flint	20.99	18.99	a	9.22	8.24	a
Grand Rapids	28.74	14.74	10.73	6.88	11.44	8.16
Minneapolis	35.26	35.45	32.07	12.32	13.09	6.32
Kansas City, Mo.	203.07	37.47	16.70	2.50	10.67	9.81
St. Louis	39.17	39.28	39.28	15.84	20.93	14.64
Omaha	49.02	25.03	15.40	7.09	8.55	6.66
Akron	34.69	24.06	7.47	7.94	10.60	5.72
Cincinnati	49.98	45.88	22.54	9.03	9.83	14.96
Cleveland	48.58	45.29	22.58	15.46	19.88	16.91
Columbus	86.14	24.68	10.21	6.26	11.77	12.29
Dayton	24.51	14.82	6.47	9.93	13.56	13.19
Toledo	51.97	21.79	16.03	7.39	13.34	8.22
Youngstown	21.50	21.60	6.14	6.55	7.87	7.31
Milwaukee	60.80	26.34	13.06	11.79	21.95	21.85

TABLE 4 (continued)

Region	Area (Thousands of Acres)			Persons Per Acre		
	1970	1930	1900	1970	1930	1900
South						
Birmingham	50.88	32.17	4.15	5.91	8.07	9.26
Mobile	74.62	9.53	3.64	2.55	7.16	10.57
Jacksonville	490.24	16.88	5.92	1.02	7.67	4.80
Miami	21.95	27.53	a	15.26	4.02	a
Tampa	54.08	12.16	a	5.14	8.32	a
Atlanta	84.16	22.27	10.56	5.91	12.14	8.51
Louisville	38.40	23.02	12.73	9.43	13.37	16.09
New Orleans	126.14	125.16	125.60	4.70	3.65	2.29
Greensboro	34.82	11.54	a	4.14	4.64	a
Oklahoma City	406.85	19.42	a	.83	9.55	a
Tulsa	110.02	13.84	a	3.02	10.21	a
Knoxville	49.28	16.90	2.54	3.54	6.26	12.85
Memphis	139.14	29.23	9.77	4.49	8.86	10.47
Nashville-Davidson	324.99	16.62	6.30	1.31	9.26	12.84
Dallas	169.98	26.74	5.33	4.97	9.74	8.00
Fort Worth	131.20	29.70	8.29	3.00	5.51	3.22
Houston	277.70	45.95	5.74	4.43	6.38	7.78
San Antonio	117.76	22.86	22.91	5.56	10.13	2.33
Norfolk	33.66	17.92	3.66	9.14	7.23	12.74
Richmond	38.59	15.36	2.88	7.42	11.91	29.53
West						
Phoenix	158.66	4.11	a	3.66	11.70	a
Anaheim	21.31	a	a	7.80	a	a
Fresno	26.75	5.50	a	6.20	9.55	a
Los Angeles	296.77	281.22	27.40	9.49	4.41	3.74
Sacramento	60.03	8.77	2.89	4.23	10.19	10.13
San Bernardino	28.42	11.45	a	3.67	a	a
San Diego	202.82	59.93	a	3.43	2.47	a
San Francisco	29.06	26.88	29.76	24.63	23.60	11.52
San Jose	87.17	4.96	a	5.13	11.62	a
Denver	60.93	37.09	36.70	8.44	7.76	3.65
Honolulu	53.70	53.70	53.70	6.05	2.56	.73
Portland	57.02	40.61	22.17	6.69	7.43	4.05
Salt Lake City	37.95	33.31	26.73	4.63	4.24	2.00
Seattle	53.50	43.84	17.46	9.92	8.33	4.62
Also						
St. Paul	33.41	33.39	33.39	9.21	8.13	4.88
Long Beach	31.17	18.21	a	11.51	7.80	a
Oakland	34.18	34.02	5.76	10.58	8.35	11.63

a. Population less than 30,000.
SOURCE: U.S. Bureau of the Census (1902-1971/1972) *Financial Statistics of Cities*, 1909 and 1930.

in the Northeast. On the other hand, only two cities in the South and four in the West showed any stability in their boundaries in the shorter period from 1930 to 1970. Areal change was the principal characteristic of most large cities in the nation, both in the long term and the short term.

The areal changes that characterized cities in the Northeast and the few in other parts of the nation had taken place during the nineteenth century. In a number of instances the stability from 1900 was misleading, since only part of the city area was settled. This was most dramatically true even of New York City which, at the turn of the century, was of relatively modest density, 18.73 persons per acre, a figure exceeded by a number of cities. Manhattan itself peaked in 1910 at 166.56 persons per acre, and this peaking took place during a period when Manhattan was not fully developed. American city densities, even those at the high end of the spectrum, are modest by international standards. Hong Kong-Kowloon has a density of approximately 1,000 persons per acre, and Calcutta, with a smaller area than Rochester, has over ten times as many people.[6]

New York is often cited either explicitly or implicitly used as the stereotype of American cities. The picture drawn from Table 4 indicates that this is far from the case. Unlike New York City, there is an absence of very high density in large cities in general and a relative decline in the densities of most large cities as a result of annexation, or because of a reduction in the housing densities (persons per occupied housing unit), or, most usually, both developments. The latter is brought out by the figures showing the declines in old boundaries of cities which otherwise showed prodigious change. While New York City has remained relatively stable in area and has increased dramatically in population density since 1900, most other large cities have increased in size and have remained relatively stable in density over this same period. The general pattern indicates the capturing of a considerable portion of suburban development by large cities in the United States.

The fiscal implications of these developments are of two kinds. The first has to do with the patterns of governmental organization which accompanied the massive annexations in many parts of the nation; the second involves the interpretation of the changes in the resource base if one takes into account the preceding restricted character of city areas (see Harrison and Kain, 1970).

One interpretation that lacks universal validity, but is usefully applied to a number of areas, involves the creation of a variety of governments other than the municipal governments to carry on governmental functions. Stable boundaries are accompanied by rather simple governmental relationships as are evident in the Northeast. There is some modification, but the pattern is relatively simple: a few special districts are added to an existing city, school district, and county. The situation is very different where massive annexations and consolidations occurred. In some cases, as in the South, the annexations and consolidations took place because the school districts were already countywide; annexation or consolidation did not alter the pattern of school district organization. In other portions of the nation the annexations involved municipal areas, but were not accompanied by any change in school district organization—a situation characteristic of the Southwest and California but also existing elsewhere.

These characteristics are evident from the numbers of governments involved, but also from the limited role played by the municipal government in these annexing areas. This is apparent not only on a percentage basis, but also in per-capita terms as shown in Tables 5 and 6, respectively, and is particularly true if one considers the different styles extant in the South. States and the federal government had never assumed such responsibilities prior to the New Deal.

The earlier development of cities in the Northeast and residual developments since the turn of the century in many areas have had major consequences for both the range of public expenditures and the resource base. First, the cities were the principal governments in the nation, not only compared to other local governments but compared to all governments including state and federal governments.[7] If governmental services were provided they were made available by city governments. The situation which initially led to the provision of services has (been) altered, but the provision of these services by cities was not reversed. This is partly because some costs had become fixed, such as pensions and interest payments, and partly because of the range of services and the size of public employment which, once established, developed very considerable vested interests. At the beginning of the period large cities alone were thought to have the resource base to carry major public programs, especially since states and the federal government severely limited the scope of their activities.

The problems of the large city today are thus in part a function of their advantaged position in the past as well as of their current

TABLE 5

PER CAPITA GENERAL EXPENDITURES IN LARGE CITIES, 1970

(in dollars)

Region	Amount	City	County	School District(s)	Special District(s)
Northeast					
Regional Average	551	404	67	50	30
Hartford	501	449(W)	0	0	52
Wilmington	679	539	33	0	105
Washington, D.C.	1006	1006(H/W)	0	0	0
Baltimore	638	638(W)	0	0	0
Boston	530	505	0	0	25
Springfield	393	388	0	0	5
Jersey City	454	327	89(W)	0	38
Newark	734	404	177(W)	0	153
Paterson-Clifton Park-Passaic	380	254	100(W)	0	27
Albany	476	314	162(W)	0	0
Buffalo	528	366	162(W)	0	0
New York City	894	828(H/W)	0	0	66
Rochester	697	528	168(W)	0	1
Syracuse	561	364	197(W)	0	0
Allentown-Bethlehem-Easton	312	106	27	140	39
Harrisburg	359	121	28	190	20
Philadelphia	494	298	0	196	0
Pittsburgh	449	189	68	156	36
Providence	392	392	0	0	0
Midwest					
Regional Average	480	207	73	175	24
Chicago	478	203	52	160	63
Gary-Hammond-East Chicago	405	147	61	245	11
Indianapolis	355	126	49	144	36
Wichita	474	222	87	164	0
Detroit	471	224	67	179	1
Flint	747	298	160	298	0
Grand Rapids	434	162	72	199	0
Minneapolis-St. Paul	539	173	134(W)	157	75
Kansas City, Mo.	484	244	61	174	5
St. Louis	463	225	0	177	61
Omaha	335	101	84	120	31
Akron	412	199	56	148	9
Cincinnati	760	500(H)	72	164	24
Cleveland	512	198	80	205	29
Columbus	397	291	54	136	16
Dayton	456	223	57	160	16
Toledo	444	205	68	156	15
Youngstown-Warren	335	141	50	134	10
Milwaukee	561	211	153(W)	164	34

TABLE 5 (continued)

Region	Amount	City	County	School District(s)	Special District(s)
South					
Regional Average	383	204	40	115	24
Birmingham	334	132	38	111	53
Mobile	333	102	23	152	57
Jacksonville	308	134	0	174	0
Miami	481	134	118	202	27
Tampa-St. Petersburg	372	135	64	167	6
Atlanta	554	203	56	218	76
Louisville	508	320(H)	71	115	1
New Orleans	333	181	0	126	26
Greensboro-Winston-Salem-Highpoint	433	200	222	0	10
Oklahoma City	296	153	24	118	0
Tulsa	306	119	30	157	0
Knoxville	372	342	59	0	29
Memphis	370	156	46	0	40
Nashville-Davidson	373	342	0	0	36
Dallas	352	156	22	148	26
Fort Worth	315	118	29	149	19
Houston	305	109	26	147	23
San Antonio	252	87	12	126	27
Norfolk-Portsmouth	453	453(W)	0	0	0
Richmond	529	529(W)	0	0	0
West					
Regional Average	541	226	107	188	19
Phoenix	375	129	59	187	0
Anaheim-Santa Ana-Garden Grove	410	98	106(W)	191	15
Fresno	686	209	246(W)	213	17
Los Angeles-Long Beach	623	179	242(W)	197	5
Sacramento	683	208	257(W)	214	4
San Bernardino-Riverside-Ontario	635	152	214(W)	267	1
San Diego	484	134	151(W)	187	12
San Francisco	779	554(W)	0	191	34
Oakland	747	232	194(W)	239	82
San Jose	553	152	166(W)	234	2
Denver	501	305(W)	0	181	15
Honolulu	198	198	0	0	0
Portland	485	193	63	196	33
Salt Lake City	305	106	56	142	1
Seattle-Everett	524	194	72	154	104
Overall Average	478	256	67	130	24

(H) Major Higher Educational Expenditure; (W) Major Public Welfare Expenditure
SOURCE:: Unpublished documents from the Government's Division used in "Central City Suburban Fiscal Disparity," Appendix B of Advisory Commission on Intergovernmental Relations (1973a) *City Financial Emergencies: The Intergovernmental Dimension.*

TABLE 6
PER CAPITA TAXES IN LARGE CITIES, 1970
(in dollars)

Region	Amount	City	County	School District(s)	Special District(s)
Northeast					
Regional Average	272	208	38	25	1
Hartford	354	349	0	0	5
Wilmington	213	195	20	0	0
Washington, D.C.	516	516	0	0	0
Baltimore	221	221	0	0	0
Boston	369	369	0	0	0
Springfield	210	210	0	0	0
Jersey City	240	181	59	0	0
Newark	352	281	71	0	0
Paterson-Clifton Park-Passaic	221	176	45	0	0
Albany	219	165	54	0	0
Harrisburg	180	78	22	80	0
Buffalo	236	143	93	0	0
Philadelphia	250	143	0	107	0
New York City	384	384	0	0	0
Pittsburgh	294	141	47	106	0
Rochester	272	167	103	0	0
Syracuse	272	133	139	0	0
Allentown-Bethlehem-Easton	176	63	23	90	1
Providence	196	196	0	0	0
Midwest					
Regional Average	240	98	38	105	6
Chicago	244	112	19	90	23
Gary-Hammond-East Chicago	266	102	46	110	7
Indianapolis	226	93	36	82	15
Wichita	209	71	51	86	0
Detroit	255	147	28	79	1
Flint	262	95	24	143	0
Grand Rapids	177	75	22	80	0
Minneapolis-St. Paul	227	88	41	96	2
Kansas City, Mo.	253	111	36	106	0
St. Louis	267	162	0	80	25
Omaha	195	66	31	97	1
Akron	225	77	22	127	0
Cincinnati	251	114	32	100	5
Cleveland	296	127	39	127	2
Columbus	198	58	26	114	1
Dayton	264	115	22	121	6
Toledo	228	89	21	116	2
Youngstown-Warren	203	79	19	103	2
Milwaukee	306	91	81	111	20

TABLE 6 (continued)

Region	Amount	City	County	School District(s)	Special District(s)
South					
Regional Average	164	99	25	38	3
Birmingham	125	68	43	14	0
Mobile	124	69	20	31	4
Jacksonville	111	71	0	40	0
Miami	221	105	57	59	0
Tampa-St. Petersburg	170	89	32	49	0
Atlanta	252	92	62	92	5
Louisville	181	95	37	49	0
New Orleans	148	102	0	46	0
Greensboro-Winston-Salem-Highpoint	155	84	71	0	0
Oklahoma City	152	70	21	61	0
Tulsa	160	52	30	78	0
Knoxville	163	133	30	0	0
Memphis	161	133	28	0	0
Nashville-Davidson	163	163	0	0	0
Dallas	211	105	17	77	12
Fort Worth	157	72	14	61	10
Houston	181	82	21	68	10
San Antonio	102	50	11	29	12
Norfolk-Portsmouth	138	138	0	0	0
Richmond	209	209	0	0	0
West					
Regional Average	249	103	47	94	4
Phoenix	172	72	35	65	0
Anaheim-Santa Ana-Garden Grove	235	59	61	100	15
Fresno	284	89	85	104	6
Los Angeles-Long Beach	329	106	91	125	6
Sacramento	284	101	85	97	1
San Bernardino-Riverside-Ontario	261	60	82	117	2
San Diego	206	64	53	90	1
San Francisco	485	300	0	185	0
Oakland	338	122	68	126	22
San Jose	250	67	59	124	1
Denver	272	171	0	101	1
Honolulu	135	135	0	0	0
Portland	261	95	35	125	5
Salt Lake City	207	80	40	81	6
Seattle-Everett	203	75	37	83	9
Overall Average	235	131	35	65	4

SOURCE: Unpublished documents from the Government's Division used in "Central City Suburban Fiscal Disparity," Appendix B of Advisory Commission on Intergovernmental Relations (1973a) *City Financial Emergencies: The Intergovernmental Dimension.*

position, which, if not absolutely disadvantaged, is disadvantaged compared to their earlier standing (see Sacks, Ranney, and Andrew, 1972: 46-58). The economic advantages accruing to the large cities were also fiscal advantages. If the cities had much greater expenditure responsibilities than any other class of government, they also had the greatest concentrations of fiscal resources. In this earlier period, intergovernmental flows played an even less important role for municipalities than they do today (see Table 7).

The principal advantages of the large cities, which were converted into fiscal resources via their effect on the property tax base, were three in number. The first was the enormous concentration of retailing and other types of economic activity in the Central Business District (CBD); the second was the concentration of manufacturing in large cities; the third aspect combined the growth of city areas with the holding of a broad spectrum of the population, including the wealthy, in the cities, particularly in the downtown areas at the beginning of the century and through World War II. Each of these factors augmented the fiscal advantages of the cities compared to other locations within the state, and within their own metropolitan areas in general, if not to all other places in particular (see Cubberley, 1905).

Dominating the large central city was, of course, the Central Business District. Major variations existed, particularly in the Los Angeles type of area, but in the Central Business Districts where modes of transportation were fixed by subways and streetcars, business thrived enormously. The result in these areas was very high land values and comparable assessed values of properties. Using a sample which is slightly different from the earlier sample cited, the change in the importance of CBDs during the period 1948-1967 was enormous (unpublished data of the Urban and Regional Workshop of the Economics Department of Syracuse University). In 1948, the CBDs of those large city areas for which data were available averaged some 30.8% of all retail sales, using the 1970 definition of SMSAs. By 1967 this share had fallen to 11.4% as new shopping centers and the outward movement of population reduced the attraction of the downtowns. Several CBDs registered dollar volumes of sales in 1967 that were lower in absolute dollars than in 1948.

The decline in the importance of the Central Business District was associated with a substantial loss in the most valuable portion of the real property tax base. Not only was there a decline in assessments associated with the decline in activity, but properties were removed from the tax roles and where they were replaced it was often by governmental (tax exempt) or urban renewal (tax write-down) properties. The basic strength of many, if not most, large cities came from the combined interaction of high land values and commercial activity which was no longer operative. The cause of this very high level of values was reinforced by a fixed transportation network focused on the central city.

The transportation network also contributed substantially to the fiscal strength of the central city insofar as it led to a concentration of manufacturing activity. The low-value, labor-intensive activities that surrounded or were located within the Central Business District often contributed more to jobs and social dislocations, but they did contribute to the tax base of cities (Hoover and Vernon, 1959). The concentration of transportation, particularly but not exclusively railroading, enhanced the fiscal position of the city. Not only was manufacturing more concentrated in the city portions of SMSAs, but manufacturing itself was more important relative to other aspects of economic activity during the period through World War II than it has been since then. The simultaneous decline in the relative importance of manufacturing activity and the redistribution of such activity to the outer portions of the SMSAs and to other regions in the country reduced manufacturing's relative share of the local property tax base. At least one other factor also contributed to this decline, namely, the reduction of the general property tax base to a tax base composed of real estate and land. The substitution of machinery and equipment for real estate also made the property tax base less responsive to economic growth (Goldsmith, 1962). Other elements that were responsive, such as inventories, were increasingly withdrawn from the tax base as a result of legislation or judicial review.

The ultimate straw that broke the back of cities with fixed boundaries, and the downtown areas of those cities which were able to annex, was the outward movement of many rich but mainly middle-income families, as well as the cities' failure to maintain the existing housing value/income structure vis-à-vis their suburban areas. Some of these areas were taken over by other uses, primarily commercial and service, but transitional areas were not rebuilt. These areas, together with the Central Business District, had played a significant role in giving a fiscal strength to large cities far greater in relative terms than they could ever again play in any city.

The new Central Business District, the substitutes, if any, for manufacturing, and the age distribution of the older residents and the socioeconomic characteristics of the newer inhabitants of the large cities were incapable of sustaining a level of fiscal strength anywhere near their earlier levels. The real property base which had encountered traumatic problems during the recession had a euphoric growth in the immediate post-World War II period which belied its long-term trend. Local officials were unable to recognize the fundamental magnitude of the changes that were operating against them and were sanguine in this period. When the changes were finally recognized, the substitutes for departing manufacturing no longer had a built-in dynamic operating in their favor, and they were much weaker from a fiscal point of view.

The outward movement of retailing and manufacturing was associated with substitution of new service activities in the Central Business District and elsewhere, often governmental or tax exempt because they involved educational, hospital, or similar types of activity. Urban renewal was associated with writedowns of various sorts, and in whole areas no substitute for departing manufacturing was found; in some instances it was not even sought—even before ecological concerns became manifest.

The result was a basic shift in the relative strength of many large cities which were unable to annex (internalize) metropolitan economic developments. The cities as municipalities grew less important compared to other governments servicing their areas, particularly in those cases where independent school districts and overlying counties with important functional responsibilities were present. But the decline in the fiscal importance of cities was not universal even where their tax bases were eroded. Some cities had school and county functions, and even if they did not, the position of the municipality was not relegated to an inferior status from a fiscal point of view.

The realities of the situation did not escape the notice of city (municipal) governments. If the property tax base was debilitated and there were other claims upon it, then substitute revenue sources were to be found. Insofar as large city governments were concerned, the use of the property tax for city purposes was severely curtailed except when legal limitations interfered and nonproperty taxes could not be substituted. Using national aggregates which understate the large city use of nonproperty taxes, property tax revenues fell from 95.9% of total city taxes in 1932 to 64.4% of the total in 1972. In the meantime, no such change occurred for other local governments. Property taxes, which comprised 98.6% of other local government tax revenues in 1932, still comprised 93.8% of total revenues in 1972.

SEYMOUR SACKS with PATRICK J. SULLIVAN [93]

The result of all these factors was change expressed in unsystematic ways so far as local governments in city areas were concerned. Due to their annexation characteristics, some cities resembled metropolitan areas more than they resembled other cities. But far more fundamental causes of intercity variations were the redistribution of responsibilities between governments operating on the local level and the interactions between the governments and higher levels, specifically the state and national governments.

3. THE FISCAL CHARACTERISTICS OF LARGE CITY MUNICIPALITIES AND LARGE CITY AREAS

The literature on large city fiscal behavior has been preoccupied with the activities of the expenditure patterns of large city governments.[8] An alternative literature on metropolitan fiscal disparities covers the systems of government providing services in large city areas.[9] A third literature deals with the exploitation of the central city.[10] In this section, the relationship between the city and governments serving the central city will be considered in somewhat greater detail in order to indicate the observed differences between and within states. Not only will expenditure patterns be considered, but also the sources of revenues which are so important in evaluating large city fiscal problems.

As has already been shown, city areas are serviced by a variety of governments (Table 1) which have a variety of distributional characteristics among cities. Our approach was mainly independent of the absolute amounts, emphasizing the proportionate shares of tax and expenditure revenues handled by the governments involved. Tables 5 and 6 show the patterns of the absolute values involved in expenditure and tax revenues, respectively.

The levels of general expenditure[11] are shown in Table 5 to be very different among cities, ranging from $1,006 per capita in Washington, D.C., to $87 in San Antonio. In general, city expenditures are low when there are independent school districts and overlying county governments servicing the city area. City expenditures averaged $422 per capita in the northeastern portion of the nation, while they averaged $210 in the Midwest, $205 in the South, and $203 in the West. This surprising finding is altered when account is taken of the overlying governments servicing the city area, however. The Northeast and the West clearly have higher average levels of total expenditures within the city, having $551 and $541 per capita, respectively. The Midwest had an average per-capita total expenditure of $480, and the South, as expected, had the lowest average at $383. City expenditures are further modified by the extent to which state governments provide services which in other states are provided locally. The number of categories into which these cities can be divided indicates that large cities differ enormously among themselves. Special circumstances such as the existence of a major institution of higher education as part of the city (as in Cincinnati with $180 per capita and Louisville with $133 per capita) may distort the totals considerably. What is clear from Tables 5 and 6 is that the most important single influence upon the pattern is the assignment of school responsibilities as reflected in the governmental distributions.

The range of variation is reduced when the overlying governments are taken into consideration. The highest level of expenditure by all local governments combined is still in Washington, D.C.; the lowest level of expenditure is in the wealthiest large city in the nation, Honolulu. The reason for this situation and its existence in other parts of the nation may be found in the distribution of functional responsibilities. Honolulu is the only major city in the nation where local elementary and secondary education is supplied by the state government. Further, welfare is a responsibility of the state government, as is the provision of hospital services. The absence of these three elements pushes the Honolulu area down to the lowest expenditure level of any large city area in the United States. The situation in Washington, D.C., is similarly explicable, since the city itself represents a combination of state, municipal, school district, and special district functions. In between, the results are ambiguous. In some cities, such as New York and San Francisco, high expenditures are associated with great wealth and functional responsibilities; in other areas, such as Newark and Baltimore, high expenditures are associated with limited wealth. The position asserted by Meltsner and Wildavsky cited at the beginning of this paper, except as a tautology, does not operate in the United States; large city expenditures are not a function of wealth.

As in the case of expenditures, the situation on the revenue side is far from simple. Taxes are but one source of revenue. Not only do other local sources exist—fees and charges and other miscellaneous revenues—but aids come from the federal and state governments. Further, taxes are not only raised to pay for current expenditures (including interest) but also to pay the principal on previously incurred debt. While the levels of expenditure and the levels of revenue may be similar, the connecting link between them is anything but simple. In individual areas, borrowing may be responsible for major differences between the levels of general expenditure and the levels of general revenue as they are defined for most governmental purposes. A similar situation exists in the case of grants. While it is sometimes asserted that grants are parts of the observed expenditure patterns, this is true only to a limited extent. Grants may be of a reimbursing nature, that is, they may pay for costs previously incurred; they may be used in the future and kept in the form of cash balances; or they may be used to pay off debt previously incurred in the period in which they were received. The point is not to assume away the problem, but to face it head on.

Tax assignments also vary among governments. The considerations mentioned regarding expenditures operate too for taxes. Although it is no longer true that Honolulu, the wealthiest city, has the lowest absolute level of taxes, it has the fourth lowest tax level. There is no clear-cut relationship between per-capita levels of taxes and either city taxes or city area taxes. Measured either way, enormous differences emerge which cannot be pushed aside unless the analysis includes the assignment of functional responsibilities and recognizes the levels of state and federal aid.

City taxes vary on a regional basis, but there is also considerable variation within regions. In the Northeast, large city government taxes averaged $216 per capita, whereas in the Midwest, South, and West, per-capita taxes averaged $99, $99, and $106, respectively. As was the case with expenditure levels, the major portion of the difference could be accounted for by the differential levels of responsibility. When account is taken of all local governments servicing cities, then the Northeast, Midwest, and West with per-capita taxes averaging $272, $240, and $249, respectively, fall into a roughly similar group distinguished from the South with average per-capita taxes of $164. Regional patterns are clearly state-dominated. Counties and school districts participate far more heavily in intergovernmental flows of funds as actors in behalf of state governments than do cities. Cities receive such funds when they undertake the same functions as to counties and school districts, with public welfare and schools clearly dominant. Generalizations developed from aggregate data are sometimes misleading and, in the case of large cities, may overlook the more important aspects of their fiscal nature. Some of these, such as the effects of dependent school districts, the absence of overlying counties and differences in state-local assignment systems, have already been considered, but other aspects remain to be considered. While a number of these factors are recognized in the volumes by Brazer (1959) and Bahl (1968b), which specifically focused on city governments, their studies tended to focus on expenditures, especially common expenditures. The extent to which sources of revenue and assignment systems are related to the levels of expenditure are not dealt with explicitly. The two problems of expenditure responsibilities and revenue sources cannot be separated without doing violence to an understanding of the issues.

The aggregate picture of the character of municipal government reflects the inclusion of many small jurisdictions whose character is very unlike the larger jurisdictions in many crucial fiscal aspects. Similarly, including New York City in the national totals thoroughly biases the conclusions concerning cities in general, particularly large cities. There are more or less "common functions"; other functions differences in noncommon functions, and because the noncommon functions also influence the level of some common functions, i.e., financial administration, general control, general public buildings, and interest (unpublished data of the Urban and Regional Economics Workshop of the Economics Department of Syracuse University). However, levels of aggregate expenditure differ primarily because of —notably schools, higher education, public welfare, hospitals, housing and urban renewal—and a number of very special considerations, such as airports and seaports (water transport), are excluded.

In addition to local school expenditures, which are almost exclusively a local responsibility (hence the name), the cities differ in the extent to which they have functions which are variable parts of the state-local system, primarily welfare and hospitals. The overall pattern which emerges, after excluding New York City and considering all other cities with populations over 200,000, is very different from what is commonly assumed. This new picture highlights aspects of large city fiscal behavior that have not been appreciated because of the use of aggregate data. These can be seen on the revenue side, but they can only be explained from the expenditure side for it is functional responsibility that ultimately determines revenue needs.

Using the 1970 data and comparing New York City to both the national data for *all* local governments (municipalities, counties, school districts, townships and special districts) and the data for all large cities, it is quite apparent that large cities are not major recipients of state aid or the federal aid that is channeled to cities via state aid. In 1970, New York City received twice as much "state aid" as all other large cities over 200,000 put together.[12] New York City received 10.7% of "state aid" received by all local governments while other large cities received only 5.2%. The situation with respect to direct federal aid to large city governments was very different. Large cities, exclusive of New York, received 28.9% of all direct local federal aid, and it is probably true that some independent housing authorities and other special districts located within their borders received additional sums. New York City received only 6.1% of direct federal aid, receiving the bulk of its federal aid indirectly via the route of state aid. The much greater relative dependence of large cities on federal aid is not the result of statistical aberration. In the past, the cities got the major share of a small pot in the case of federal aid, while they got a small portion of a much larger amount in the case of state aid.[13] The total amount of direct federal aid to large cities in 1972, the most recent year for which information is available, was about equal to the total amount of state aid inclusive of the federal aid portion of the state aid totals (see Martin, 1965).

The result is very surprising insofar as state-city relationships are concerned. It clearly reinforces the point made earlier—"city governments" as such are outside the general state-local pattern of intergovernmental aid (see Campbell, 1970). This is even more surprising when one recognizes that a number of cities have dependent school districts, welfare responsibilities, or both, as is true of Baltimore. The federal figure is slightly overstated because Washington, D.C., is included; there, all aid is assumed to go to the municipality, but only some classes of aid are of a municipal kind.

The specific reasons for the absence of notable state aid is, of course, that the functions for which localities receive aid are not the functions which municipalities undertake. This underscores the inappropriateness of the argument which ranks public welfare as the foremost fiscal problem of large cities (Chinitz, 1972). New York City alone shows 25.1% of all local expenditures on public welfare and 73.5% of all city expenditures on public welfare. The single fact that New York City had 11.1% of all state and local welfare expenditures with 3.9% of the national population is enormously

TABLE 7
FINANCIAL CHARACTERISTICS, 1969-70 (in thousand dollars)

	State-Local	Local	NYC	All Other Cities Over 200,000	NYC as % of All Local	All Other Cities as % of Local	NYC as % of All State-Local
General Revenue	130,756	80,916	6,622	8,819	8.2	10.9	5.1
Federal Aid	21,857	2,605	158	753	6.1	28.9	.7
State Aid	—	26,920	2,867	1,397	10.6	5.2	10.6
Taxes	86,795	38,833	3,023	4,691	7.8	12.1	5.4
Property	34,054	32,963	1,848	2,815	5.6	8.5	5.4
Non-Property	52,741	5,870	1,175	1,876	20.0	32.0	2.2
General Expenditure	131,332	82,582	6,619	9,251	8.0	11.2	5.0
Education	52,718	38,938	1,699	1,799	4.4	3.1	3.2
Highways	16,427	5,383	165	758	3.1	14.1	1.0
Public Welfare	14,680	6,477	1,628	437	25.1	6.7	11.1
Health-Hospitals	9,668	4,880	807	612	16.5	12.5	8.3
Police	4,494	3,806	487	1,169	12.8	30.7	10.8
Local Fire Protection	2,024	2,024	206	654	10.2	32.3	10.2
Sewerage	2,167	2,167	72	476	3.3	22.0	3.3
Sanitation	1,246	1,246	163	418	13.1	33.5	13.1
Local Parks/Recreation	1,888	1,888	104	560	5.5	29.7	5.5
Housing-Urban Renewal	2,138	2,115	327	447	15.5	21.1	15.3
All Other	23,912	13,658	961	2,528	7.1	18.5	4.0
Percent Population					3.9	18.9	3.9

SOURCE: U.S. Bureau of the Census (1962-1971/1972) Government Finances in 1969-1970. City Government Finances in 1969-70.

impressive. The larger figures indicate that, in most large cities, other governments held the public welfare responsibilities. A similar but less spectacular situation existed in the case of hospitals in New York, but other cities did hold hospital responsibilities. Table 7 shows the revenue and detailed expenditure patterns for New York and other large cities as a proportion of all local aggregates and of state and local aggregates.

The major conclusion of this analysis is that some currently generalized assumptions about the reasons city governments face fiscal difficulties are, instead, highly specific to certain individual governments.[14] This is particularly true of the fiscal problems of schools, public welfare, and hospitals. Further, city governments as such, apart from those in a few states, have not participated in the state aid system and in that portion of federal aid which arrives via state aid. This is a situation whose nature has not been fully appreciated.

THE PROPERTY TAX

The financial problems of large cities as well as other governments have been uniquely tied to the operation of the property tax. Two developments, however, have contributed to a reevaluation of the problems involved. The first is a major expansion in the use of nonproperty taxes by city governments and by governments in large city areas.[15] The second involves the recent reexamination of the regressivity of the property tax. For most city governments the property tax has been supplemented and, in some cases, surpassed in importance by nonproperty taxes. Not only does New York City, the historic pioneer in this area, use sales, income, and a host of other nonproperty taxes, but such taxes are characteristic of most cities where the property tax does not constitute the only tax base legally available. As shown in Table 7, New York City collected 20.0% of all local nonproperty taxes and large cities other than New York collected 32.0% of all local nonproperty taxes. These figures contrast very sharply with the fact that New York City collected only 5.6% and other large cities collected only 8.5% of all local property taxes in 1970. The latter figures are due to the fact that counties and school districts depend almost exclusively on the local property tax base (see Sacks, Andrew, O'Farrell, and Wade, 1974).

Dominating the discussion of large city fiscal problems on the revenue side has been the property tax. Ever since Henry George, an enormous literature on the property tax has been very much centered on the problems of large cities.[16] This tradition has been carried through to the very present, as is illustrated by the recent debates on property tax reform (see Peterson [1973a] and Advisory Commission on Intergovernmental Relations [1973b] for two major discussions of the issues).

The property tax has been attacked for a number of reasons ranging from considerations which are especially related to large cities to issues which are related to the theoretical limitations of the property tax as an institution. Property tax rates in cities vary enormously whether measured in terms of city governments or of all governments servicing large city areas (Bureau of the Census, 1973: 110-143, and 1971). While it is certainly true that some cities, notably Newark, Boston, and Milwaukee, have extraordinarily high tax rates and have been subject to a great deal of analysis, many other cities have very nominal rates (see Advisory Commission on Intergovernmental Relations, 1967). The effects of property taxes on the provision and maintenance of urban housing is another major issue under consideration (see Peterson, 1973b: 107-124, as well as Netzer, 1966: 67-86). The importance of the property tax as the source of large city fiscal problems has, in part, been overstated (see Sacks, Andrew, O'Farrell, and Wade, 1974: 150-152).

The area where the actual pattern of local taxes interacts with the theoretical limitations of the property tax is in its so-called "regressivity" or "antipoor bias." This has become especially important in the wake of the demands for property tax relief and reform (National Commission on Urban Problems, 1968: 358-361). At the same time as these demands have been undertaken there has been a revived interest in the theoretical underpinning of the property tax, especially in what is called its differential incidence. As will be indicated, the results of this new analysis are much more consistent with a refined set of data on the observed patterns of property taxes on residential housing in large cities as well as elsewhere (Bureau of the Census, 1973 and 1971).

As Henry Aaron has indicated recently, "Nearly all popular discussions of property taxation take its regressivity for granted . . . such studies assume that homeowners bear property tax burdens directly in their capacity as occupants and are unable to shift the tax to anyone else; (2) owners of rental properties shift the tax substantially to renters who bear the tax in proportion to rents paid."[17] Most of the analyses have been based on national aggregates, some have been based on regional analyses, but the results have been assumed to be applicable to all areas including large cities. The theoretical models of these empirical analyses have been questioned in recent years and as a result, at least among economic researchers, the assumption of regressivity has been questioned. The implications of the newer theoretical models and much more carefully collected and analyzed data indicate that the presumed regressivity has to be reevaluated not only between taxing jurisdictions, but within taxing jurisdictions, such as large cities, as well.

Specifically, the underlying assumption for the regressivity of the property tax on housing is that "housing takes a larger part of the income of the poor than of prosperous households" (National Commission on Urban Problems, 1968: 359) and that the taxes paid by all properties in a given jurisdiction are a good measure of the taxes on housing. Taking the two points in reverse order, it should be noted that property taxes are not only levied on housing, but on commercial, industrial, and utility real properties and often on tangible personal property (inventories and fixtures and equipment). These taxes are often "exported" outside the area where the property is located. This class of property is very important in most cities, and is very important in most cities compared to their outside areas.[18] This is not a tax on housing, although it may be a burden on the local residents. The residential tax, instead of being a single tax on housing, operates differently on different classes of housing. Not only is there the difference between rental and owner-occupied housing, but rental housing may be taxed or may be tax exempt as in the case of low-income public housing.

While there is considerable evidence that "housing takes a larger part of the income of the poor than of prosperous households," the property tax is not a tax on housing but on assessment of the capital value of the housing property. The patterns of capital values relative to income are far more consistent with the newer theoretical models concerning the incidence of the property tax than are the older assumptions. Because the factors involved are so important in the case of large cities and elsewhere, an attempt will be made to indicate how they emerge from the 1970 Census of Population and Housing.[19] The Census volumes not only contain information on national aggregates of central cities but on most individual large cities as well.

First, the Census gives detailed information on value-income relationships as well as on rent-income relationships. These relationships may be looked at in terms of changes in income or changes in value. When looked at in income terms, the problem is very confused. It shows that at the lowest income levels as income increases there is very little change in the value of housing; that the very poor have housing not unlike those with middle incomes. This is due to an age-household composition factor for which the Census now provides detailed information. In essence, the high value-income ratios are associated with households with heads over 65 and with one-person households. The value-income ratio rises as incomes rise in the case of the owner-occupied housing with heads less than 65. The apparent inverse relationship between the level of income and the value-income ratio is due to the dominance of the elderly in the low-income categories.

The reasons for the high value-income ratios in the case of the elderly in general and for isolated other individuals have been attributed to factors which are more important to low-income households than to higher-income households. The first is the use of current income rather than some measure of permanent income. The incomes on which the decision to purchase a house were based were quite different than the observed current income of the elderly. Further, in the case of the elderly, they are far more likely to have nonmortgaged property than the nonelderly, with the result that taxes make up the predominant recurrent cost element in the case of elderly, but not in the case at other age groups. The second reason is that the imputed value of owner-occupied housing is excluded from the calculation of income. This would lead to a much greater proportional change in the case of the high value-low income households, i.e., the elderly and the "temporarily" poor, than in those with higher income, even though they may have much higher value housing.

The tax on owner-occupied housing and on rental housing is, however, on the capital value of the property as it appears on the assessment rolls. Viewed in terms of capital values, the pattern is very different. It indicates that the poor have very low-value housing and the rich have high-value housing. Based on housing values, the distorting effect of the elderly is spread among housing value classes. Because the pattern for all central cities provides a good representation of the pattern for the individual large cities, the relationships between housing values and income as housing values change is shown in Table 8 (data derived from U.S. Census of Housing, 1970).

TABLE 8

HOUSING VALUE-INCOME RATIOS

Value of Housing	Median Income	Value-Income Ratio
Less than $ 5,000	$ 4,100	Less than 1.22
$ 5,000 to 7,499	5,800	1.01
7,500 to 9,999	7,200	1.22
10,000 to 12,499	8,400	1.34
12,500 to 14,999	9,400	1.46
15,000 to 19,999	10,800	1.62
20,000 to 24,999	12,500	1.80
25,000 to 34,999	14,500	2.07
35,000 to 49,999	18,800	2.26
More than 50,000	27,600	More than 1.81

The value-income ratio increases as housing values increase in part because of the increased availability of mortgage money to higher-income households as compared to the lower-income households. The results are consistent with the findings of Peterson (1972), who found that the market value to rent increases with rent.

In fact, the situation in the case of renters, especially in large cities, is far more complex than is usually depicted. Insofar as they live in tax exempt, low-income public housing and are not receiving rent supplements designed in part to cover tax costs, the analysis is in principle inappropriate. Since these households fall preponderantly in the low-income brackets, a new income distribution appropriate to those living in nontax exempt nonsubsidized housing must be determined. In a city such as New York this would be of major importance (see Bahl and Vogt, 1974).

These empirical findings are consistent with the newer theoretical model developed by Arnold Harberger and Peter Mieszkowski, which has produced a coherent theoretical base for differential analysis of property tax incidence (see Mieszkowski, 1969 and 1972). In the case of owner-occupied housing, it indicates that there is a capital element in the ownership of housing in addition to the conventional desire for housing services. This element is important when the housing taxes relative to income in the central city are compared to the housing taxes relative to income in their surrounding areas. The conventional notion of the higher rates of taxation in central cities relative to income will have to be reevaluated in the light of the more appropriate theoretical models and empirical data now available.

The result of all these findings is that the conventional litany of fiscal ideas concerning property taxes in city areas must be questioned. There is no uniformity to cities as fiscal artifacts, and presumed behavior patterns, especially insofar as they involve the property tax, have to be reevaluated. This does not mean that emotionally and politically these elements are not important; quite the opposite may be the case. However, if policy is made on the basis of theoretically inappropriate models and empirically inaccurate data, the result may be contrary to the interests of the large cities.

4. CENTRAL CITY AND OUTSIDE CENTRAL CITY: THE CITY PROBLEM IN ITS METROPOLITAN CONTEXT

The traditional analyses, which examine cities in terms of other cities, or cities in terms of their metropolitan nature, omit the potentially crucial element in understanding the problems facing the large cities—the fiscal interactions between the cities and their own outside areas.[20] In conjunction with the assignment of responsibilities and the historical evolution of large cities, the relationship between the cities and their outside areas is essential to a comprehension of the city as a fiscal artifact, for both population and industry have moved largely to the immediate outside areas. The city is not only part of the metropolitan community in substance, it is also part of the metropolitan community because it shares common governments with surrounding areas. Insofar as cities are able to annex or capture the growth around them, they may alter their fiscal position, but that is only one of a number of elements now entering into the decisions concerning annexation.

The situation of the city relative to its outside areas has been depicted elsewhere in considerable detail, but a summary here is valuable.[21] The picture that emerges indicates initially the enormous variety characteristic of the nation.

As shown in Table 9, the overall pattern for the large cities in 1970 was that on a per-capita basis their expenditures vastly exceed those of their own suburban areas, regardless of how they compare to other large cities. Only in 9 of the 70 suburban areas reported were expenditure levels higher than those of their central cities. This was due to the fact that, unlike interstate comparisons, intrametropolitan comparisons hold constant most assignments in functional responsibilities except where the metropolitan area crosses state boundaries. The overall differentials are great.

The nature of the city as a fiscal artifact can be seen by the separation of total expenditures into local school and other (nonlocal school) expenditures. The overall difference is a function of two very different patterns. Of the 70 areas, per-capita local school expenditures were lower in the large-city areas than in their outside central city areas in all but 9 cases, while in only one case, San Jose, were nonlocal school expenditures higher in the outside area than in the city. This result emerges from the fact that San Jose itself may be interpreted as a suburb, its enormous amount of annexation in the last twenty years being rather unrelated to a truly large city set of problems. The total is further modified by a considerable junior college expenditure and higher education expenditures thrown into the total, which affects outside areas to a greater extent than cities.

The overall pattern of large cities requires the analyst to recognize not only the levels of behavior, but the expenditure distributions which differ considerably between school expenditures and other local expenditures.

SUMMARY

While they are indissolubly related, large city governments (municipalities) and large city areas differ greatly from a fiscal point of view. Further, enormous fiscal differences are found among cities; these differences, while not fully appreciated, must be taken into account to achieve meaningful understanding of the financial nature of large-city areas. Toward that end, large city areas have been considered in greater detail than usual.

The first step of the analysis distinguished between the municipality and the local governmental systems which service large city areas. The purpose of this step was not only to indicate the existence of competing governments in large city areas, but to suggest the complexity of the process whereby local governments mobilize resources in large city areas. The current emphasis upon the city governments is shown to be fiscally incomplete.

The second step examined the city from an historical perspective. This helps one to understand the underlying governmental organi-

TABLE 9
PER CAPITA EXPENDITURES, 72 LARGEST SMSAs: 1970
(in dollars)

Region	Total Expenditure		Local School Expenditure		Non-Local School[a] Expenditure	
	CC	OCC	CC	OCC	CC	OCC
Northeast						
Hartford	501	399	208	214	293	185
Wilmington	679	285	251	210	428	75
Washington, D.C.	1006	426	211	236	795	190
Baltimore	638	348	216	205	422	143
Boston	530	365	139	177	391	188
Springfield-Chicopee-Holyoke	393	310	155	164	238	146
Jersey City	454	357	128	108	326	249
Newark	734	441	209	196	525	245
Paterson-Clifton Park-Passaic	380	418	131	197	249	221
Albany-Schenectady-Troy	473	495	150	282	323	213
Buffalo	528	525	165	255	363	270
New York City	894	644	182	324	712	320
Rochester	697	548	215	317	482	230
Syracuse	561	586	148	311	413	275
Allentown-Bethlehem-Easton	312	347	135	224	176	123
Harrisburg	359	298	189	208	170	90
Philadelphia	494	325	171	197	323	128
Pittsburgh	449	309	154	173	295	136
Providence	392	265	139	146	263	119
Regional Average	551	405	173	218	378	187
Midwest						
Chicago	478	346	143	198	335	148
Gary-Hammond-East Chicago	465	310	244	184	221	126
Indianapolis	355	305	144	194	211	111
Wichita	474	360	164	213	310	147
Detroit	471	464	177	242	294	222
Flint	747	449	264	237	483	212
Grand Rapids	434	364	184	186	250	178
Minneapolis	539	520	154	284	385	236
Kansas City, Mo.	484	346	158	186	326	160
St. Louis	463	291	138	184	325	107
Omaha	335	338	135	238	200	100
Akron	412	311	149	186	264	125
Cincinnati	760	262	174	131	586	131
Cleveland	512	368	201	185	302	183
Columbus	397	289	133	177	264	112
Dayton	456	291	165	168	291	123
Toledo	444	294	146	148	298	146
Youngstown-Warren	335	236	135	144	200	92
Milwaukee	561	486	164	247	397	239
Regional Average	480	348	166	196	314	152

TABLE 9 (continued)

Region	Total Expenditure CC	Total Expenditure OCC	Local School Expenditure CC	Local School Expenditure OCC	Non-Local School[a] Expenditure CC	Non-Local School[a] Expenditure OCC
South						
Birmingham	334	244	111	153	223	91
Mobile	333	188	109	106	224	82
Jacksonville	308	b	134	b	173	b
Miami	481	390	176	176	305	214
Tampa-St. Petersburg	372	289	151	151	221	138
Atlanta	554	315	218	187	336	128
Louisville	508	302	116	212	392	90
New Orleans	334	325	121	123	213	202
Greensboro-Winston-Salem-Highpoint	453	244	151	145	282	99
Oklahoma City	297	264	118	157	178	107
Tulsa	306	202	146	121	160	81
Knoxville	372	228	134	157	238	71
Memphis	370	240	135	183	235	57
Nashville	378	172	168	115	210	57
Dallas	352	279	139	153	213	126
Fort Worth	315	286	142	161	173	125
Houston	305	307	137	181	168	126
San Antonio	252	257	106	198	146	59
Norfolk-Portsmouth	453	293	160	174	293	119
Richmond	529	305	162	209	367	96
Regional Average	386	270	142	161	244	109
West						
Phoenix	375	387	164	223	211	164
Anaheim-Santa Ana-Garden Grove	410	373	169	233	241	140
Fresno	686	642	207	262	479	380
Los Angeles-Long Beach	623	560	174	181	449	379
Sacramento	683	566	167	230	516	336
San Bernardino-Riverside-Ontario	635	522	234	206	401	316
San Diego	484	472	160	206	324	266
San Francisco-Oakland	768	591	178	234	597	357
San Jose	554	612	222	259	331	353
Denver	501	305	170	188	331	117
Honolulu[c]	198	b	0	b	198	b
Portland	485	328	160	210	325	117
Salt Lake City	305	283	136	208	169	75
Seattle	524	471	150	219	374	252
Regional Average	541	470	176	204	365	266

a. Includes higher education.
b. No outside central city area.
c. Excluded from computation because there is no outside central city area.
SOURCE: Unpublished documents from the Government's Division used in "Central City Suburban Fiscal Disparity," Appendix B of ACIR (1973a).

zation and also distinguishes among cities. Areal change has been a fundamental attribute of most large cities, but such changes have taken place outside the Northeast. There have also been major reductions in the densities of most large cities (except those which contained suburban developments within their own boundaries). Much of the suburban development in the United States has taken place within city boundaries. This was true even of New York City viewed from a 1900 perspective. Such cities differ very considerably from those which are primarily of an "inner-city" nature. In this historical section, the former economically advantaged position of the cities is considered in some detail insofar as this is an explanation of their current disadvantages.

The fiscal characteristics of large city areas are analyzed in some detail in the third section. The position of the large cities is examined, especially in terms of the characteristic problems usually attributed to them. Some problems are shown to be very special case problems, notably public welfare and hospitals. Welfare is not a city problem; furthermore, apart from a number of states—New York, California, Minnesota, Wisconsin, and New Jersey—it is not even a local government problem. The most critical problems involve the large cities (municipalities) and the assignment of functional responsibilities. Large city governments seem to be outside the standard state-local set of intergovernmental flows. This was historically due to the fiscal strength of large city areas, which is no longer omnipresent. It appears that the connection between the cities and the federal government is strong not only in relative terms, but in absolute terms as well, while the levels of state aid received by the municipal governments of large city areas is extraordinarily small.

Much of the literature's standard credo concerning large city fiscal behavior must be considered at least questionable. Statements that must be doubted include the dependence of cities on the property tax and the regressive nature of the property tax base. The latter is particularly important if account is taken of the suburban nature of some cities and the income characteristics of those who own their own housing.

The final portion of this analysis indicates that the overall patterns of large city areas draw general meaning from their metropolitan context. Not only is the relative position of cities compared to other cities of interest, but perhaps most meaningful is the relationship between the city area and its outside area.

The complexity of large city areas and the extreme caution with

SEYMOUR SACKS with PATRICK J. SULLIVAN [109]

which generalizations about them must be approached should be underscored. The complexity of the large city as a fiscal artifact must be recognized if meaningful public policy is to be created. The extent to which these fiscal considerations, in turn, influence the social, economic, and fiscal behavior of large city areas is the subject of another analysis, which must at a minimum recognize the factors discussed in this paper. The specific context varies, but national, state, and even metropolitan fiscal policies aimed at large city areas require a much more careful statement of the issues involved than is generally characteristic of either the literature or policy.

NOTES

1. Perhaps nowhere has this been demonstrated as clearly as in the work completed by the National Commission on Urban Problems in 1968 and in the debate over the large cities' portion of General Revenue Sharing monies which has been carried on since its enactment prior to the 1972 elections.

2. See Berry (1972) for an indication of the range of approaches used in examining the city. This is also true of many recent studies which analyze the city in terms of its urban core, inner city, and central functions. For an important study which recognizes the specific nature of these problems on a comparative basis, see University of Amsterdam Socio-graphical Department (1967).

3. This has been true in several of the analyses of large city school districts. For a survey of the literature, see the bibliography in Sacks, Ranney, and Andrew (1972: 179-194).

4. Detailed information is available for the extent of dependence and coterminality of school districts on a national, regional, state, and metropolitan basis in U.S. Bureau of the Census (1972).

5. The basis of the location between the city and outside central city governments is population, except for schools, where the basis is enrollments. Other allocational devices have been used, such as valuations, but these do not seem to be as useful as population and enrollment measures.

6. For an examination of the development of the city in the United States, as well as elsewhere, see the remarkable study originally published in 1899 by Weber (1967).

7. See *Financial Statistics of Cities*—more recently, *City Government Finances*—published annually, with minor exceptions, since 1902 by the Bureau of the Census.

8. See, for example: Bahl (1968a and 1968b); Brazer (1959); Campbell and Sacks (1967); Elliot (1905); Fabricant (1952); Fisher (1961 and 1964); Hirsch (1957); Thomas, Kelly, and Garms (1966); Morss (1966); Sacks (1963); Sacks and Harris (1964); and Sacks and Hellmuth (1961).

9. See, for example: Advisory Commission on Intergovernmental Relations (1973a, 1967, and 1965); Brazer (1962 and 1958); Kee (1968); and Margolis (1957).

10. See, for example: Advisory Commission on Intergovernmental Relations (1971 and 1963); Davies (1965); Hawley (1959); Mace (1961); Pettengill and Uppal (1974); and Reischauer (1973).

11. General expenditures are defined as all expenditures other than the specifically enumerated kinds of expenditures classified as utility expenditures, liquor store expenditures, and employee retirement or other insurance trust expenditures.

12. The Census Bureau includes federal aid which is channeled through state grants as "state aid." Detailed information on all federal aid given to large cities is available only where individual states and cities have collected the information themselves.

13. The dollar amounts of direct federal aid will change dramatically relative to state aid when General Revenue Sharing appears in the governmental accounts.

14. For a detailed analysis of the problems involved, see Advisory Commission on Intergovernmental Relations (1973a: 1-57).

15. See Advisory Commission on Intergovernmental Relations (1974); sales taxes are examined on pp. 252-256, while income taxes are on pp. 291-294.

16. The committee on Taxation, Resources and Economic Development—as a direct descendent of Henry George—has produced an outstanding set of volumes dealing with the property tax. For instance, see Holland (1970); but the literature itself is overwhelming. The last attempt at a synthesis, but coming prior to the recent revolutionary developments, is Netzer (1966).

17. Aaron (1973: 212). See also Netzer (1973). For the assumptions of noneconomists on the regressivity of the property tax in urban areas, see Lineberry and Sharkansky (1971: 211).

18. A comparison between large cities and their outside counties is shown in Table A-17 of Advisory Commission on Intergovernmental Relations, 1967: 110.

19. See the individual volumes in the Metropolitan Housing Characteristics HC(2) series. The tables involved end with the number 7.

20. This material is an adaptation of Appendix B of Advisory Commission on Intergovernmental Relations (1973a).

21. See note 9, above, on metropolitan fiscal disparities.

REFERENCES

AARON, H. (1973) "A new view of property tax incidence." American Economic Association Papers and Proceedings: 212-221.

Advisory Commission on Intergovernmental Relations (1974) Federal-State-Local Finance: Significant Features of Fiscal Federalism. Washington, D.C.: ACIR.

——— (1973a) City Financial Emergencies: The Intergovernmental Dimension. Washington, D.C.: ACIR.

——— (1973b) Financing Schools and Property Tax Relief-A State Responsibility. Washington, D.C.: ACIR.

——— (1971) Measuring the Fiscal Capacity and Effort of State and Local Areas: Washington, D.C.: ACIR.

——— (1967) Fiscal Balance in the American Federal System. Vol. 2: Metropolitan Fiscal Disparities. Washington, D.C.: ACIR.

——— (1965) Metropolitan Social and Economic Disparities: Implications for Intergovernmental Relations in Central Cities and Suburbs. Washington, D.C.: ACIR.

——— (1963) Performance of Urban Functions: Local and Areawide. Washington, D.C.: ACIR.

BAHL, R. W. (1968a) "Studies on determinants of expenditures: a review," in S. Mushkin and J. Cotten (eds.) Functional Federalism: Grants in Aid and PPB Systems. Washington, D.C.: George Washington University Press.

——— (1968b) Metropolitan City Expenditures: A Comparative Analysis. Louisville: University of Kentucky Press.

——— and W. VOGT (1974) "State assumption of welfare and education financing: income distribution consequences." Syracuse, N.Y.: Syracuse University, Metropolitan Studies Program. (unpublished)

BECK, M. (1963) Property Taxation and Urban Land Use in Northeastern New Jersey. Washington, D.C.: Urban Land Institute.

BERRY, B. [ed.] (1972) City Classification Handbook. New York: Wiley-Interscience.

BRAZER, H. E. (1962) "Some fiscal implications of metropolitanism," in G. S. Burkhead (ed.) Metropolitan Issues: Social, Governmental, Fiscal. Syracuse, N.Y.: Maxwell Graduate School of Citizenship and Public Affairs.

——— (1959) City Expenditures in the United States. New York: National Bureau of Economic Research.

——— (1958) "The major metropolitan centers in state and local finance." American Economic Review 47, 2 (May): 305-316.

BREAK, G. E. (1967) Intergovernmental Fiscal Relations in the United States. Washington, D.C.: Brookings.

CAMPBELL, A. [ed.] (1970) The States and the Urban Crisis. New York: Columbia University Press.

——— and S. SACKS (1967) Metropolitan America: Fiscal Systems and Governmental Patterns. New York: Free Press.

CHINITZ, B. (1972) "Economy of the center city: an appraisal," pp. 79-95 in M. Perlman, C. J. Leven, and B. Chinitz (eds.) Spatial, Regional and Population Economics. New York: Gordon & Breach.

CRECINE, J. P. [ed.] (1970) Financing the Metropolis: Public Policy in Urban Economics. Beverly Hills, Calif.: Sage.

CUBBERLEY, E. P. (1905) School Finances and Their Apportionment. New York: Columbia University Teachers College.

DAVIES, D. (1965) "Financing urban functions and services." Law & Contemporary Problems 30, 1 (Winter).

ELLIOT, E. C. (1905) Some Fiscal Aspects of Public Education. New York: Columbia University Teachers College.

FABRICANT, S. (1952) The Trend in Government Activity Since 1900. New York: National Bureau of Economic Research.

FISHER, G. W. (1964) "Interstate variation in state and local government expenditures." National Tax Journal 17 (March): 55-74.

——— (1961) "Determinants of state and local government expenditures: a preliminary analysis." National Tax Journal 14 (December): 349-355.

GAFFNEY, M. (1971) "The property tax is a progressive tax." Proceedings of the National Tax Association: 408-426.

GOLDSMITH, R. (1962) National Wealth of the United States in the Postwar Period. Princeton, N.J.: Princeton University Press.

HARRISON, D., Jr. and J. F. KAIN (1970) "An historical model of urban form," pp. 146 in Proceedings of the Second Research Conference of the Inter-University Committee on Urban Economics. Chicago: University of Chicago, Center for Continuing Education.

HAWLEY, A. (1959) "Metropolitan population and municipal government expenditures," in P. J. Hatt and A. J. Reiss (eds.) Cities and Society. New York: Free Press.

HEILBRUN, J. (1974) Urban Economics and Public Policy. New York: St. Martin's.

——— (1966) Real Estate Taxes and Urban Housing. New York: Columbia University Press.

HIRSCH, W. (1970) The Economics of State and Local Government. New York: McGraw-Hill.

——— (1968) "The supply of urban services," pp. 477-525 in H. Perloff and L. Wingo, Jr. (eds.) Issues in Urban Economics. Washington, D.C.: Resources for the Future.

——— (1957) "Measuring factors affecting expenditure levels for local government services." St. Louis: Metropolitan St. Louis Survey. (mimeo)

HOLLAND, D. [ed.] (1970) The Assessment of Land Values. Madison: University of Wisconsin Press.

HOOVER, E. M. and R. VERNON (1959) Anatomy of a Metropolis. Cambridge, Mass.: Harvard University Press.

KEE, W. S. (1968) "City-suburban differentials in local government fiscal effort." National Tax Journal (June): 183-189.

LINEBERRY, R. L. and I. SHARKANSKY (1971) Urban Politics and Public Policy. New York: Harper & Row.

MACE, R. L. (1961) Municipal Cost-Revenue Research in the United States. Chapel Hill: University of North Carolina, Institute of Government.

MARGOLIS, J. (1957) "Municipal fiscal structure in a metropolitan region." Journal of Political Economy 65 (June): 225-236.

MARTIN, R. (1965) The Cities and the Federal System. New York: Atherton.

MAXWELL, J. (1965) Financing State and Local Governments. Washington, D.C.: Brookings.

MELTSNER, A. and A. WILDAVSKY (1970) "Leave city budgeting alone! A survey, case study and recommendations for reform," in J. Crecine (ed.) Financing the Metropolis. Beverly Hills, Calif.: Sage.

MIESZKOWSKI, P. (1972) "The property tax: an excise tax or a profits tax?" Journal of Public Economics (April): 73-96.

——— (1969) "Tax incidence theory: the effects of taxes on the distribution of income." Journal of Economic Literature (December): 1103-1122.

MORSS, E. R. (1966) "Some thoughts on the determinants of state and local expenditures." National Tax Journal 19 (March): 95-103.

National Commission on Urban Problems (1968) Building the American City.

NEENAN, W. B. (1972) Political Economy of Urban Areas. Chicago: Markham.

NETZER, D. (1973) The Incidence of the Property Tax Revisited. National Tax Journal (December): 515-535.

——— (1968) Impact of the Property Tax: Its Economic Implications for Urban Problems. Washington, D.C.: Joint Economic Committee, U.S. Congress.

——— (1966) Economics of the Property Tax. Washington, D.C.: Brookings.

PERLOFF, H. and N. RICHARD [eds.] (1968) Revenue Sharing and the City. Baltimore: Johns Hopkins Press.

PETERSON, G. E. (1973a) Property Tax Reform. Washington, D.C.: Urban Institute.

——— (1973b) "The property tax and low income housing markets," in G. E. Peterson (ed.) Property Tax Reform. Washington, D.C.: Urban Institute.

——— (1972) "The regressivity of the residential property tax." Washington, D.C.: Urban Institute Work Paper 1207-10, November.

PETENGILL, R. B. and J. S. UPPAL (1974) Can Cities Survive? New York: St. Martin's.

REISCHAUER, R. D. (1973) "Fiscal problems of cities," pp. 291-317 in Setting the National Priorities: The 1973 Budget. Washington, D.C.: Brookings.

SACKS, S. (1963) "Spatial and locational aspects of local government expenditures," in Public Expenditure Decisions in the Urban Community. Baltimore: Johns Hopkins University Press.

——— and R. HARRIS (1964) "The determinants of state and local government expenditures and intergovernmental flow of funds." National Tax Journal 17, 1 (March).

SACKS, S. and W. F. HELLMUTH (1961) Financing Government in a Metropolitan Area: The Cleveland Experience. Glencoe, Ill.: Free Press.

SACKS, S., R. ANDREW, P. O'FARRELL, and J. WADE (1974) "Competition between local school and non-school functions for the property tax base," pp. 147-161 in C. Lindholm (ed.) Property and the Finance of Education. Madison: University of Wisconsin Press.

SEYMOUR SACKS with PATRICK J. SULLIVAN [113]

SACKS, S., A. RANNEY and R. ANDREW (1972) City Schools–Suburban Schools: A History of Fiscal Conflict. Syracuse, N.Y.: Syracuse University Press.

STUDENSKI, P. (1930) The Government of Metropolitan Areas in the United States. New York: National Municipal League.

THOMAS, J. H., J. KELLY, and W. I. GARMS (1966) Determinants of Education Expenditures in Large Cities in the United States. Palo Alto, Calif.: Stanford University Press.

U.S. Bureau of the Census (1973) Census of Governments. Vol. I: Government Organization. Taxable Property Values and Assessment-Sales Price Ratios, 1972. Part I: Taxable and Other Property Values. Part II: Assessment-Sales Price Ratios and Tax Ratios. Washington, D.C.: Government Printing Office.

––– (1971) Census of Population 1970. Vol. V: Residential Finance. Washington, D.C.: Government Printing Office.

––– (1970a) Local Government Finances in Selected Metropolitan Areas, 1970. Washington, D.C. Government Printing Office.

––– (1970b) Census of Housing, 1970. Metropolitan Housing Characteristics HC-(2) Series. Washington, D.C.: Government Printing Office.

––– (1970c) Census of Housing, 1970. Residential Finance. Washington, D.C.: Government Printing Office.

––– (1962-1972) State Finances. Washington, D.C.: Government Printing Office.

––– (1962-1971/1972) Governmental Finance. Washington, D.C.: Government Printing Office.

––– (1902-1971/1972) City Finances. Washington, D.C.: Government Printing Office. (In earlier years: Financial Statistics of Cities)

U.S. Department of the Treasury, Office of General Revenue Sharing (1973) Data Elements: Entitlement I eriod 4. Washington, D.C.: Government Printing Office.

University of Amsterdam Sociographical Department (1967) Urban Core and Inner City. Leyden: E. J. Brill.

WEBER, A. F. (1967) The Growth of Cities in the Nineteenth Century. Ithaca, N.Y.: Cornell University Press.

Part II

URBAN HOUSING MARKETS
AND THEIR INTERPRETATION

Introduction

□ HOUSING MARKETS IN THE UNITED STATES are subject to a wide variety of forces. Depending upon the nature of these forces, the character of the market can be altered in either a positive or negative direction. It is clear, however, that the character of the market is not altered solely by the operation of forces recognized generally as economic; most often the changes in the quality and character of housing within a given market are the result of the interaction of social and economic factors which lead to a particular outcome. The papers in this section focus on a selected set of forces which are responsible for allocating housing to subgroups within the population and/or the effects of selected variables on housing values. Each of these papers represents a unique and/or innovative approach to the analysis of housing market functioning. The differences in the way the individual authors choose to approach their work make it somewhat difficult to compare the conclusions reached. Nevertheless, each paper is sufficiently novel in approach that it is likely to generate a future response from those researchers working in the housing field who, for one reason or another, will find the analysis and conclusions of these writers stimulating and groundbreaking or misleading and short of the point.

Harvey is likely to be the most severely criticized among the three contributors because his paper tends to focus almost totally on the character of the capitalistic system in relation to the emergence of a particular kind of urban infrastructure. But regardless of Harvey's philosophic predilections, his detailed analysis of the role of local financial institutions in delineating housing submarkets is masterfully done. The zeal with which he approaches his topic will be denounced by some, but few can criticize the insight he provides into the operation of individual housing markets within the city of Baltimore. Harvey's paper actually transcends the operation of a local housing market and addresses itself to the much larger issue of the role of economic systems in urban structure.

The second paper in this series focuses upon the operation of the dual housing market largely based on race in the Chicago Metropolitan area. Berry's analysis of housing construction swings and their impact on altering black access to improved housing is indeed interesting. Under certain conditions Berry contends that filtering works to improve the quality of the housing supply available to the black population. No doubt some will not be convinced by Berry's picture of black access to quality housing during boom periods in housing construction, but the evidence he presents in the Chicago case is persuasive. His paper—while essentially concerning itself with the conceptual notion of short-term housing swings and their effect in a dual housing market—also provides the reader with interesting data on shifts in the housing stock within the Chicago Metropolitan Area during the most recent decade. Berry further demonstrates that, although there is evidence of black progress in upgrading the housing stock occupied by the group, there is little evidence of the group's ability to establish a foothold in suburbia.

The third paper, by Nourse and Phares, examines yet another aspect of housing market dynamics. These authors have chosen to plow new ground in their attempt to explain the principal factor leading to housing value change over time. They assert that the principle variables (age and race) which have traditionally been associated with the blight-decay-abandonment cycle are less important than what they define as the income transition. All three papers employ a spatial orientation in their analyses, although in each paper the spatial scale differs. Nourse and Phares have selected fifteen neighborhoods within St. Louis County to test their hypothesis of the income transition thesis on housing values. Their results verify the strength of income in altering values over time. It is likely that critics will question the validity of the technique employed by the authors to ascertain average income levels at each time interval.

Each of the papers appearing in this section chooses to investigate an aspect of housing market behavior which has previously received a good deal of attention. Yet these authors provide new insight into the impact of forces which include both economic and social components on market operations. Harvey's analysis of the role of the public sector on monetary flow as it effects housing dynamics at the micro scale is indeed revealing. And likewise, Berry has once again provided the reader with a longitudinal approach to understanding the role of housing construction cycles in allocating housing in a racially dichotomous housing field. Nourse and Phares in their very lucid presentation have provided us with yet another explanation of changing housing values in a temporal-spatial context. Each of these papers represents new approaches in the examination of what might be termed age-old questions as they relate to aspects of housing market dynamics in American cities. These papers are welcome additions to the literature which has as its objective providing insight into the spatial dynamics of urban housing markets.

Unresolved problems include at least the following:

(1) In the examples in this section the housing markets of Baltimore, St. Louis and Chicago are each portrayed with the help of a different analytical model. Do the models emphasize or obscure the differences in housing market conditions in these three cities, each of which represents a distinctly peculiar metropolitan condition?

(2) Are in fact the social and economic conditions in each urban housing market so distinct that each will perforce generate its own distinct model of market behavior?

(3) What kinds of models of the urban housing market would facilitate genuine comparative research between different metropolitan areas? Is more of a systems model required to eliminate the biases which different investigators read into forms of correlation analysis?

3

The Political Economy of Urbanization in Advanced Capitalist Societies: The Case of the United States

DAVID HARVEY

☐ IN THIS PAPER I will seek to say something about the political economy of urbanization in advanced capitalist societies.[1] I will begin by establishing certain broad relationships between urbanization and economic growth processes and then, by focusing on the example of the United States, examine certain of these relationships in greater detail. I will seek to show that the production of the built environment involves processes of fixed capital formation which are "mediated" through a structure of governmental and financial institutions. These institutional mediations have both general and highly localized impacts upon the urbanization process. By way of conclusion I shall examine how the nature of the urbanization process as it currently exists in the United States contains certain contradictory tendencies which, potentially at least, can be the source of social, economic, and political instability.

I. CAPITAL FORMATION AND URBANIZATION

The processes of urbanization involve the creation of a built environment which subsequently functions as a vast man-made resource system—a reservoir of fixed and immobile capital assets to be used in all phases of commodity production and in final consumption. These assets have to be maintained and from time to time renewed if society is to be reproduced in its existing state. A certain proportion of social product has therefore to be laid aside as a surplus to reproduce the built environment. Expanded reproduction, whether it implies population growth or per-capita changes in material product, requires an even greater quantity of surplus for enhancing the built environment, while qualitative social changes may require not the replacement but the refashioning of the built environment to meet new economic and technological contingencies and new social wants and needs. All of this requires, of course, human effort and ingenuity. But it also requires some way of extracting and allocating a surplus to provide for the needs of the urban infrastructure.

Data assembled for various countries in what we may called the "advanced capitalist" category (Tables 1 and 2) suggest that between a fifth and a quarter of Gross Domestic Product went on average into fixed capital formation (1966-1971). Of the amounts allocated to

TABLE 1

Proportion of Gross Domestic Product Going to Fixed Capital Formation By Various Categories in Selected Advanced Capitalist Countries (average for the years 1966-1971 inclusive)

Country	Gross Fixed Capital Formation	Percent of Gross Domestic Product			
		Total Construction	Residential	Non-Residential	Transport and Machinery
United States	19.8	11.9	4.1	7.8	7.9
Canada	21.7	13.8	4.2	9.6	7.9
United Kingdom	19.2	9.8	3.4	6.4	9.4
Germany	25.5	13.8	5.4	8.4	11.9
France	25.9	14.3	6.5	7.8	11.5
Netherlands	26.8	15.1	5.2	8.9	12.6
Belgium	21.1	12.0	5.0	7.0	9.1
Sweden	22.2	14.5	5.0	9.5	7.8
Italy	19.7	11.8	6.1	5.7	8.1

SOURCE: *Yearbook of National Accounts Statistics, 1972* (New York: United Nations, 1974).

TABLE 2

Proportion of Gross Fixed Capital Formation Going to Construction and Producers' Durables in Selected Advanced Capitalist Countries

(average for the years 1966-1971 inclusive)

Country	Percent of Gross Fixed Capital Formation			
	Construction		Non-Residential	Producers' Durables (transport and machinery)
	Total	Residential		
United States	59.8	20.7	39.1	40.2
Canada	63.5	19.4	44.1	36.5
United Kingdom	51.2	17.9	33.3	48.8
Germany	54.2	21.1	33.1	46.8
France	55.3	25.2	30.1	44.7
Belgium	56.9	23.5	33.4	43.1
Netherlands	52.8	19.5	33.3	47.2
Sweden	64.9	22.2	42.7	35.1
Italy	59.2	30.7	28.5	40.7

SOURCE: *Yearbook of National Accounts Statistics, 1972* (New York: United Nations, 1974).

this end, between a half and two-thirds typically went into the built environment. The figures vary somewhat from country to country. Stone (1973) concludes, for example, that the resources invested in the built environment in Britain amount currently to "about a twelfth of the national product or about half the total annual capital formation. When costs of maintenance and improvement are added, the resources required represent about an eighth of the national product." This proportion is rather low compared to most other advanced capitalist nations, which typically put at least one-seventh of their national product into the built environment.

The inputs into the built environment also vary over time. Apart from business cycle variations (which tend to be very strong particularly in the construction industry) certain long-term trends are discernible. In the United States, for example, Kuznets (1961) reports that construction as a percentage of Gross Domestic Capital formation has fallen over the various long swings in the economy from over two-thirds in the period 1889-1918 to under one-half in the period 1946-1955 (Table 3). He concludes that construction has lessened in importance relative to producers' durables (notably machinery) in fixed capital formation. Disaggregating the construction data indicates that the most important relative decline has been in construction for the business sector (the "other" category in Table 3). This suggests that "it has become increasingly possible to

TABLE 3

GROSS DOMESTIC CAPITAL FORMATION (GDCF) IN THE AMERICAN ECONOMY FOR DIFFERENT SECTORS OVER VARIOUS TIME PERIODS

(excluding military expenditures, in constant 1929 dollars)

Time Period	Percent of Gross Domestic Capital Formation		Percent of GDCF in Construction		
	Producers' Durables	Construction	Non-Farm Residential	"Other"	Government
1889-1918	26.1	67.2	32.0	57.5	10.5
1919-1948	39.7	53.7	31.4	39.1	29.5
1946-1955	45.9	46.3	34.1	39.3	26.6

SOURCE: Kuznets, 1961: Tables 14 and 18.

produce more finished commodities with absolutely and proportionately less gross or net additions to the stock of producers' durable equipment" (Kuznets, 1961: 170). Grebler, Blank, and Winnick (1956: 14-15), in a parallel study of capital formation in residential real estate, likewise report that "the use of real resources for private residential construction has shown a marked decline since 1891 in relation to gross national product, total capital formation and aggregate consumption," adding, somewhat surprisingly, that "residential construction has suffered some decline even in relation to total new construction."

These and other studies indicate that much of the urban infrastructure in the United States was created prior to 1920 and that the fixed assets created by then are still heavily drawn upon. Since that time the rate of gross capital formation in urban infrastructure has been lower than the rate of growth in GNP and in personal income, while the rate of net capital formation in urban infrastructure has been remarkably low (Table 4). Sporadic booms and slumps in housing construction prior to World War II kept residential development somewhat below the general trend in fixed capital formation. More recently, particularly since 1945, burgeoning governmental activity and private commercial development (stores, shopping plazas, offices, and the like) and the strong subsidization of housing construction have only partially offset the steady secular decline in investment in the fixed and immobile urban infrastructure compared to investment in producers' durables, to growth in GNP, or to growth in personal income (Table 5). In general, it seems that there is a long-term shift from urban infrastructure (plant, utilities, transport facilities) to producers' durables in the production sphere

TABLE 4

NET DOMESTIC CAPITAL FORMATION (NDCF) IN THE AMERICAN ECONOMY FOR DIFFERENT SECTORS OVER VARIOUS TIME PERIODS

(excluding military expenditures, in constant 1929 dollars)

Time Period	Percent of Net Domestic Capital Formation		Percent of NDCF in Construction		
	Producers' Durables	Construction	Non-Farm Residential	"Other"	Government
First Estimates					
1889-1918	13.9	71.7	36.6	51.5	12.0
1919-1948	41.2	32.7	75.2	−57.4	82.2
1946-1955	28.8	47.8	57.8	14.8	27.5
Alternative Estimates (derived by a different treatment of depreciation)					
1889-1918	11.5	74.7	33.8	55.2	11.0
1919-1948	28.1	45.8	53.6	−12.3	58.6
1946-1955	37.6	44.6	47.2	30.3	22.5

SOURCE: Kuznets, 1961: Tables 14 and 18.

and a parallel shift from urban infrastructure (housing, transport facilities, community facilities, and the like) to consumers' durables (automobiles, household equipment, and the like) in the consumption sphere. The conclusion is inescapable that capital formation through the creation of urban infrastructure is of declining relative significance from the standpoint of the proportion of the national product which it absorbs. And it seems likely that similar conclusions would be reached from a study of other advanced capitalist nations (apart from those severely ravaged by war).

TABLE 5

NEW CONSTRUCTION ACTIVITY (VALUE PUT IN PLACE) IN PROPORTION TO U.S. GNP FOR SELECTED TIME PERIODS

	1946-1950	1958-1962	1968-1972
Total new construction	9.56	11.00	10.15
Total private construction	7.74	7.72	7.23
Total residential	4.70	5.52	3.83
New housing	3.82	3.52	3.05
Total non-residential	3.01	3.13	3.26
Commercial	0.50	0.85	1.02
Industrial	0.58	0.52	0.60
Other	1.97	1.77	1.78
Total public construction	1.84	3.30	2.90
Federal-owned	0.48	0.74	0.37
State-owned	1.36	2.55	2.54

SOURCE: *Economic Report of the President, 1974*: Table C-38.

Investment in the built environment has a significance far beyond the direct investment that it absorbs. First, these investments generate certain multiplier effects because the subsequent use (and hence the value) of an urban infrastructure depends upon the commitment of further resources. Under private property relationships and a capitalist mode of consumption, these multiplier effects can be quite substantial (most households are heavily over-equipped relative to their needs). The construction of housing, for example, requires complementary investments in transportation, household equipment, community facilities, and the like, while it also imposes a variety of recurrent operating costs (energy inputs being a prime example). Both the capital and operating costs of urban infrastructure appear to vary greatly according to design considerations (size, density, distributions of activity), and the multiplier effects of investment in urban infrastructure are therefore partially dependent upon design considerations (Stone, 1973). In addition, construction activity has certain strong multiplier effects in the employment sphere.

Second, the elements of urban infrastructure usually have a long lifespan and are typically financed by taking on long-term debts and obligations. If these debts are to be repaid, then it is essential that the elements of urban infrastructure maintain their "value," which means, under capitalism, that profitable (exchange value) uses must, directly or indirectly, be found for them. There is, therefore, a persistent problem of maintaining capital values in the tangible and objectified form these values assume in the built environment. This problem is emphasized by the tendency in advanced capitalist societies to secure long-term debts of all kinds against real property values, and is exacerbated by the predilection to look for long-term capital gains in property. Consequently, the formation of an urban infrastructure imposes both an economic and social obligation on the part of subsequent generations to find profitable uses for it.

Third, the characteristics of urban infrastructure partially determine how well the economy functions. The relations here are complex and difficult to specify, but inefficiency in the urban infrastructure for a given productive purpose can raise costs of production substantially and thereby have an impact upon wage rates, employment, inflation, and so on. If, for example, costs of circulation are relatively high (because of inefficiencies and congestion costs in the transportation system for example), then presumably the rate of exploitation of labor power must be raised if profit rates are to remain constant. Similarly, if job opportunities and housing opportunities are far distant from each other in a competitive labor market, then employers may have to raise wages to compensate for travel time which, if the costs are not passed on to the consumer, means lower profits and a slowed rate of capital accumulation and economic growth.

When we take all of these considerations into account, it becomes apparent that urban infrastructure has a significance far greater than can be calculated from the direct investment required to complete and sustain it. But exactly what the significance is, both qualitatively and quantitatively, is not easy to conceptualize, let alone to estimate.

II. INSTITUTIONS, CAPITAL MARKETS, AND THE ALLOCATION OF RESOURCES TO THE BUILT ENVIRONMENT

To a large degree (and the tendency is strongly emphasized in advanced capitalist societies) the production of urban infrastructure involves resort to capital markets. Many aspects of economic development are financed out of undistributed profits (in the private sector) and current taxation (in the public sector), but the construction of the built environment is usually financed by long-term loans (for example, residential mortgages, government bonds for capital improvement projects, and the like). This gives to investment in urban infrastructure a special character and explains why financial and monetary factors play so important a role in this sphere compared to almost all other sectors of the economy. The heavy reliance on external (as opposed to internal) financing suggests that the production of urban infrastructure will likely be very sensitive to conditions in capital markets, and that financial structure—understood as the complex of money forms (coin, paper obligations, securities, mortgages, and so on) and financial intermediaries which administer these money forms—will have some kind of impact upon what is produced. The role of financial intermediaries is, as the Hunt Commission Report (1971: 11) puts it, to "gather savings and distribute the funds to numerous borrowers, thus affecting the allocations of real resources—what is produced, how it is produced, and to whom it is distributed." Specialized intermediaries are often held to be an essential feature in any complex economy, while it is sometimes also argued that "an effective system of intermediaries is an indispensable element in promoting a high level of economic growth" (Hunt Commission, 1971: 12).

The relationships between financial structure and economic growth in general are, however, problematic, for, as Goldsmith (1969: 390-409) concludes after extensive investigation, neither economic theory nor economic history tell us with any certainty whether financial structure affects, either quantitatively or qualitatively, the path of economic growth. Adequate financial structure is, of course, a necessary condition; whether and in what respects it is a sufficient condition is a matter for debate. While the evidence on these relationships may be equivocal as a general proposition, the dominance of external financing and of capital markets in general in shaping investment in urban infrastructures makes the hypothesis of a relationship between the mediating influences of financial structure and the result in the form of the built environment appear much more firmly based. Grebler (1964: 27) suggests, for example, that

differences in institutional arrangements of mortgage financing may have influenced the physical form of real estate development and type of tenure. The early emergence of specialized home mortgage lenders in Great Britain and the United States may have fostered single-family house building and home ownership; and on the European continent, early orientation of mortgage banks to the financing of large income properties may have favored apartment building and renting.

Fisher and Siegman (1972), to give another example, examine how institutional structures affect the cyclical character of residential construction in different countries. Comparing the United States and Canada, they note that "much the same patterns of housing production have emerged in both countries even though certain important features in the mortgage markets in the two countries have been quite different." But in the case of Germany they point out that credit-associated downturns in housing production have been scarcely noticeable, and they attribute this to "the particular institutional structure of housing finance, together with the comprehensive tax and interest incentives provided by the Government which leads to a comparatively stable flow of funds for private housing transactions."

If we include in the concept of financial structure, as it appears we must, the various instrumentalities through which governments exercise influence in the fiscal and monetary spheres (including the intricate patterns of direct and indirect subsidies which governments create), then it is plain that the mediations of financial structure will almost certainly have some kind of impact upon urbanization. The difficulty here, of course, is to show which particular aspects of financial structure are crucial in shaping urban infrastructure and which aspects can readily be compensated for. Put another way, and viewed from a different perspective, we wish to know what aspects of the urbanization process in advanced capitalist societies are rendered homogeneous and invariant in all such societies by the very nature of the economic system which dominates them, and what aspects can vary within (presumably) certain bounds without conflicting with the basic necessities of profit-making and capital accumulation. These are two perspectives on the same question, for the financial structure through which the production of urban infrastructure is mediated is a superstructural form which has evolved, in its main features at least, in response to the need to maintain the rate of capital accumulation. But in the same way that there is no unique relation between superstructural forms—legal, juridical, ideological, military, governmental, administrative, and cultural—and the economic basis of advanced capitalism, so a certain autonomy to the urbanization process must be conceded within the broad constraints set by the functioning of the economic basis.

If, however, the Marxian concept of a "mode of production" is any guide (see Harvey, 1973, chapter 6; Lefebvre, 1970) then it suggests that the diversity of urbanization under advanced capitalism is a surface configuration of processes and forms which conceals the basic modus vivendi of capitalism, which is to accumulate, make profits, or, as Marx preferred to put it, to bring about the self-expansion of value. By the same token, if this line of thinking is correct, it can be anticipated that the urban problems with which we are all too familiar (inner-city decay in American cities, sky-rocketing land and housing prices in European and Canadian inner-cities, speculative developer activity, suburban sprawl, pollution and congestion, the lack of balance between housing and employment opportunities, and the like) are all surface manifestations of deep structural problems in such societies. But we cannot exclude the possibility either that the autonomous development of an urbanization process can make demands upon economic organization which cannot easily be met and which are, therefore, potentially the source of capitalist crisis.

The fundamental conception emerging here is of an urbanization process which reflects the operation of certain fundamental forces at work in advanced capitalist societies. The surplus which investment in urban infrastructure represents is mobilized through the financial superstructure, and it is this which gives to the latter its crucial mediating role. Failure to mobilize this surplus and to put it to use effectively, for whatever reason, means a failure to reproduce and enhance those man-made resources crucial for the reproduction of the expansionary capitalist mode of production itself. For no other reason the "problematic" posed by the relationships between economic basis, financial superstructure, and urbanization is crucially in need of systematic investigation.

To lay the bones of all this bare would require, however, nothing less than a truly massive comparative study of the urbanization process in all advanced capitalist societies. In what follows, a more limited problem is discussed. The Marxian posture is assumed and then applied to one aspect of urban infrastructure—housing—in only one country—the United States.

III. FINANCIAL STRUCTURE, HOUSING MARKETS, AND THE URBANIZATION PROCESS IN THE UNITED STATES

Money performs a variety of functions in society: it provides a means of exchange, a measure of value, a means of storing wealth, and a means of credit. Money also assumes a variety of forms in advanced capitalist society, comprising coin and a great diversity fo promissory notes and paper obligations. These various functions and forms are intertwined, making money a most complex commodity. From the standpoint of understanding investment in urban infra-structure—and in housing in particular—it is that part of financial structure which "manages" the savings-investment function through capital markets which is crucial. We are therefore primarily concerned with those intermediaries which draw in savings and allocate these savings to investment. There is a difference, for example, between (1) commercial banks, which in the United States primarily function to manage money as a means of exchange, (2) thrift institutions oriented to managing savings, and (3) life insurance and pension funds whose primary function is to look for means to store and accumulate wealth. But commercial banks also perform both short- and long-term savings and investment functions primarily through the operations of their savings and trust departments, while thrift institutions, if Gurley and Shaw (1956) are correct, can by their policies affect the quantity and velocity of money in circulation. Insofar as these intermediaries store and accumulate wealth, and use land and property to perform that function, they come to invest in urban infrastructure; in so doing, however, they rely upon the fact that money can function as a measure of value for fixed and immobile capital assets incorporated into the built environment.

The financial structure of the United States is fragmented into numerous types of financial intermediary, all of which are subject, however, to government (either state or federal) regulation and supervision. The overwhelming impression created by a study of financial structure is of a chaos of private activity under an incoherent and arbitrarily constructed umbrella of government regulation. From time to time there have been calls for a major restructuring of the financial system—the Commission on Money and Credit (1961) and the Hunt Commission Report on Financial Regulation (1971) made basically similar proposals. The structure of regulation, and indeed the whole financial superstructure, does not appear so arbitrary when it is viewed in historical perspective for, as the Hunt Commission Report (1971: 12) puts it, the regulatory framework has evolved as a series of "responses to financial crises—particularly those during the Civil War years, the period from 1907 to 1914, and between 1927 and 1935." The financial superstructure has undergone major transformations during each economic crisis; the current structure in the United States "still closely resembles the structure left at the end of the 1930's."

The financial structure in the United States has evolved in an ad hoc, pragmatic fashion to meet the demands put upon it by the forces of economic growth and capital accumulation. For example, new kinds of specialized intermediaries have emerged to manage special aspects of the savings-investment function. As a consequence the financial system has a patchwork-quilt structure. Savings and loan associations, mutual savings banks, credit unions, commercial banks, life insurance companies, pension funds, real estate invest-ment trusts, and mortgage bankers are all private institutions active in the provision of housing finance. Federal, state, and local governments have created an assortment of finance and credit agencies through which the flow of investment into housing is also mediated. Some of these agencies operate directly in the housing investment field, others function indirectly by insuring mortgages (the Federal Housing Administration—FHA), insuring deposits (Federal Savings and Loan Insurance Corporation—FSLIC, and Federal Deposit Insurance Corporation—FDIC), or affecting the viability of housing investment by operating in the secondary mortgage market (Federal National Mortgage Association—FNMA; Government National Mortgage Association—GNMA; Federal Home Loan Mortgage Corporation—FHLMC). Financial intermediaries are both numerous and specialized in the housing market in the United States. In what follows, therefore, I shall try to show how this complex financial superstructure has impacts at both the national and local level on the how, when, where, and how much of housing investment. These impacts will then be related (1) to an overall interpretation of the urbanization process in the United States and (2) to certain basic structural problems inherent in the form of organization of advanced capitalist societies.

1. LONG-TERM NATIONAL IMPACTS

Most housing production in the United States lies in the private sector, and public housing plays a very minor role in shaping the urbanization process in general (although it may be locally impor-tant). In this the United States is different from most other advanced capitalist societies where public housing (as in Britain, for example) is very much more significant. Housing policy in the United States therefore emerges through the complex interactions between federal, state, and local governments, on the one hand, and various private institutions, on the other. The collective result of fiscal and monetary policies on the part of federal government, investment and regulatory policies on the part of states and local governments, and portfolio management and investment policies of the various kinds of private intermediary involved in the provision of housing finance, is nevertheless a quite coherent investment process through which funds are channeled into the housing sector in general and into certain aspects of it in particular (into new investment versus the purchase of second-hand homes, into single-family as opposed to multifamily dwelling units, and so on).

We can begin by identifying the simplest but perhaps most profound consequence of financial intermediation in the housing sector. The very existence of financial institutions as intermediaries favors owner-occupancy because it would be much more difficult

to find mortgage lenders among other individuals, or possibly among builders using their accumulated savings, than it now is where these loans are made routinely in large numbers by financial institutions.... Thus the absence of financial institutions would have resulted in a quite different distribution of housing between owner-occuped and rented quarters [Goldsmith, 1971: 6].

Institutional intermediaries have become progressively more important in the United States with the advance of capitalism and, as a consequence, the noninstitutional mediation of housing finance has declined from over 50% in 1900 to less than 10% at the present day (see Figure 1). The rise of nonbanking financial intermediaries (that is, intermediaries whose prime function is other than managing money as a means of exchange) has likewise "probably been most helpful to the market for home ... mortgages and thus to the spread of home ownership in the face of rapid urbanization" (Goldsmith, 1971: 8-9). In the United States these nonbanking financial intermediaries have become much more important during the twentieth century (partly as the result of the fairly strict regulations imposed to prevent financial concentration and oligopoly among banking interests on a national scale). Nonbanking financial intermediaries have also increased their role in the provision of housing finance (see Figure 1) and if the financial intermediaries specializing in single-family mortgages (the savings and loan associations and to a lesser extent the mutual savings banks) are added, then the advantage of single-family owner occupancy compared to apartment dwelling and renting becomes quite marked by virtue of the very nature of the financial superstructure in the United States.

Federal fiscal and monetary policies have greatly emphasized this advantage. Fiscal policy produces various subsidies to homeowners. In 1864-1865 homeowners were first permitted tax deductions on mortgage interest and property tax payments, and this principle was

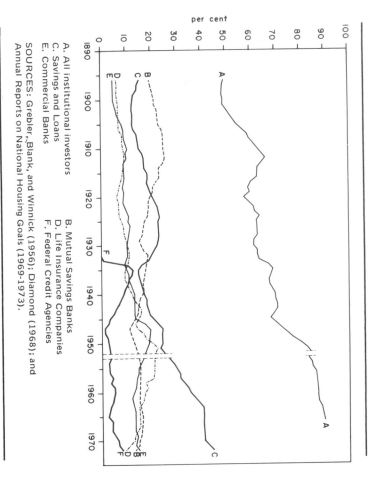

Figure 1: PERCENT OF TOTAL RESIDENTIAL MORTGAGE DEBT IN THE UNITED STATES HELD AT YEAR'S END, 1896-1972

A. All institutional investors   B. Mutual Savings Banks
C. Savings and Loans   D. Life Insurance Companies
E. Commercial Banks   F. Federal Credit Agencies

SOURCES: Grebler, Blank, and Winnick (1956); Diamond (1968); and Annual Reports on National Housing Goals (1969-1973).

confirmed in the constitutional amendment establishing the federal income tax in 1913. It is estimated that this provision cost the federal government $6 billion in 1973. In the same year an additional $1.3 billion was forgone by allowing homeowners to defer or exclude capital gains realized on the sale of their homes, while it was also estimated that another $6.04 billion was forgone by failing to tax the imputed net rental (the difference between the gross rent that an owner-occupier could receive if the home was rented and the overall costs of maintenance, repairs, and depreciation. While the last of these "subsidies" is controversial, the net benefits provided by federal fiscal policies to homeowners are very substantial indeed (see U.S. Congress, 1973b: 24-25; Aaron, 1972).

FHA (and later VA) insurance of generally middle-income home ownership effectively revolutionized and secured the mortgage market for this kind of housing and thereby did much to stimulate investment in middle-income single-family dwelling units. It is doubtful if the FHA could have succeeded, however, without the creation of a secondary market for housing mortgages. Therefore, one of the responses to the crisis conditions of the 1930s was to create FNMA as an institution to provide depth in the secondary mortgage market with respect to FHA insured mortgages. The reform of FNMA to deal in conventional mortgages and the creation of GNMA and FHLMC in the late 1960s extended these advantages. The setting up of federal savings and loan associations in the 1930s as peculiarly advantaged intermediaries in housing finance (with respect to the interest rates they can offer depositors, the insurance of FSLIC and short-term advances from the Federal Home Loan Bank Board—FHLBB), served strongly to channel savings and investments into institutions which specialized in the financing of owner-occupancy (note the rising importance of the savings and loan associations—S&Ls—as financial intermediaries illustrated in Figure 1). The special treatment meted out to S&Ls was only one of a whole gamut of measures taken during the 1930s to ensure the stability of financial institutions while simultaneously channeling funds to preferred borrowers in order to stimulate the economy.

Homeownership has in fact been viewed, ever since President Hoover's Conference on Home Building and Homeownership in 1931, as "a valid objective of public policy in and for itself" (U.S. Congress, 1973b: 12). Exactly how and why public policy came to take this stance is a difficult question to answer, although in the 1930s it was certainly seen as something that fulfilled national goals of economic and social stability in a period of severe economic and social disruption. The self-perpetuating ideology of homeownership is most clearly manifest in the history of FHA policies which, over the years, have brought homeownership within the reach of lower and lower income groups until, in 1968, the FHA was transformed so that it could become a vehicle for promoting homeownership among the poor and near-poor. Partly as a result of the post-1968 experience, the supposed benefits of homeownership (many of which result from federal fiscal policies in the first place) are being called into question. Saving in the form of real estate equity, for example, may not be a rational decision for the majority of the population —certainly "housing as an investment for low-income individuals is illiquid, risky, requires complex management, and has high maintenance costs," and it may well be that by comparison "a savings and loan account is a safer and more liquid investment and one which requires little monitoring and expertise" (U.S. Congress, 1973b: 4, 109; Marcuse, 1973).

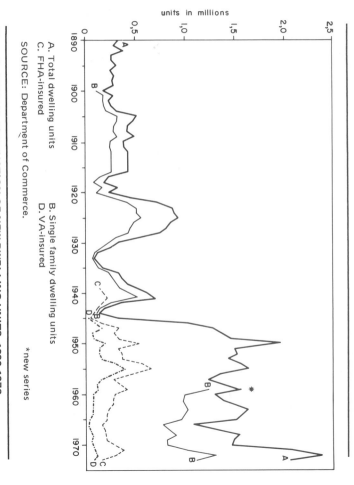

Figure 2: ANNUAL PRODUCTION OF NEW DWELLING UNITS, 1890-1973

A. Total dwelling units   B. Single family dwelling units

C. FHA-insured   D. VA-insured   *new series

SOURCE: Department of Commerce.

The consequences of all these policies taken together were startling. The proportion of dwelling units owner-occupied which stood at 47.8% in 1890 and fluctuated to a low of 43.6% in 1940 rose sharply to 61.9% in 1960, levelling off to 62.9% in 1970. Housing construction in general experienced a remarkable postwar boom, and until the early 1960s most of this construction was in the single-family dwelling unit category (see Figure 2) or in the 1-4 family dwelling unit category of Figure 3. Only during the 1960s did apartment construction rise markedly from 20% of new dwelling units to around 40% by the end of the decade. In the post-1945 period, given the financial superstructure which had emerged, low-density, single-unit homeownership became highly advantageous to the mass of the American population.

The consequences of the rise of financial intermediaries and government subsidies for urbanization in the United States have been (1) low-density suburban sprawl (emphasized by fiscal policies toward housing production and land speculation both of which encouraged sprawl and a personalized mode of transportation which made it feasible); (2) strong multiplier effects from housing

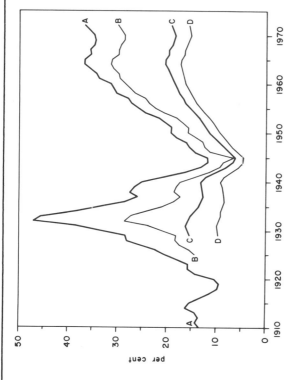

Total residential mortgage debt as a percentage of (A) Gross National Product and (C) Total Public and Private Debt. Mortgage debt on 1-4 family dwelling units as a percentage of (B) Gross National Product and (D) Total Public and Private Debt.

SOURCES: Grebler, Blank, and Winnick (1956); Annual Reports on National Housing Goals (1969-1973).

Figure 3: RESIDENTIAL MORTGAGE DEBT IN THE U.S. ECONOMY, 1910-1972

construction through complementary investments (urban infra-structure and household equipment) and high constant operating costs; (3) an assumption, which grew as the urbanization process gathered momentum, that urban expansion was inevitable, which gave rise to (4) self-fulfilling speculative activity on the part of individuals and (5) self-fulfilling growth-inducing investments on the part of many local governments which became "growth dependent" by incurring heavy debts now in order to raise the tax base in the future. Relatively low land costs, increasing geographical mobility, social and economic conflict in the central cities, have all com-pounded these trends. Only in the last few years, particularly since the mid-1960s, have rising land prices, no-growth movements in the suburbs, increasing congestion and pollution costs, the compounding problems of fiscal inequality among local governments, and the rising costs of the energy inputs necessary to sustain the urban system, begun to suggest that the current urbanization process is less than optimal. Forces are now arising in opposition to an urbanization process which since the 1930s has been shaped by the mediations of the financial superstructure into a high-consumption, high energy-consuming, individualized, suburbanization process.

This kind of urbanization has been bought at a price. The general levels of indebtedness attached to urban infrastructure have risen remarkably (state and local government indebtedness for investment in urban infrastructure being a very significant component). Residential debt as a proportion of total private and public national debt rose from 9.5% in 1947 to 23.7% in 1972, while expressed as a ratio to GNP it has now risen to levels comparable to those experienced in the crisis conditions of the 1930s when GNP tumbled (see Figure 3). The debt on single-family dwellings has become of great significance. In general, residential debt has become the largest single component underpinning the viability and security of the financial superstructure. Per-capita indebtedness (see Table 6) has also risen remarkably in proportion to personal disposable income. The social consequences of this personal debt-encumbrance are considerable—for example, the defensive attitudes of owner-occupiers toward social change has a lot to do with the fact that personal savings are locked into house values, while the commitment to a long-term mortgage places all kinds of pressures on individuals to find the wherewithal to meet that obligation.

Residential mortgages also absorb a considerable, if fluctuating, proportion of funds raised in credit markets (see Table 7). Here it seems that housing functions as a residual investment because

TABLE 6
NON-FARM RESIDENTIAL MORTGAGE DEBT PER CAPITA AND
IN RELATION TO PER CAPITA DISPOSABLE INCOME, 1890-1973

Year	Total Residential Mortgage Debt Outstanding ($ billions)	Residential Mortgage Debt Per Capita ($)	Per Capita Debt in Ratio to Per Capita Disposable Income (%)
1890	2.3	68.4	—
1900	2.9	65.1	19.7
1910	4.5	74.6	16.1
1920	9.4	126.2	12.6
1930	30.2	325.8	41.0
1940	24.9	245.7	32.9
1950	54.9	429.9	26.7
1960	161.6	894.4	46.2
1970	334.0	1630.2	48.3
1973	470.7	2237.1	53.3

SOURCE: Grebler, Blank and Winnick, 1956; *Economic Report of the President, 1974.*

corporate and government borrowings are less interest-rate sensitive —housing typically takes up whatever is left over after these "basic" needs are met. Nevertheless, the demand for residential mortgages is a very important element if only because it provides a substantial outlet for surplus funds looking for viable investment opportunities in times of capital surplus. It is, however, such a substantial sector at any time that its influence is bound to be strongly felt in credit markets.

The stability of both the social structure and the financial superstructure are to large degree tied to the health of the residential mortgage market. And it has to be remembered that the stability of the financial system, the ability of surplus capital to find outlets in real estate development, and the ability of individuals to bear the burden of long-term indebtedness are separate facets of the same thing. Any massive fall in personal disposable income would generate widespread mortgage foreclosures which would, in turn, collapse the financial system and generate chaos in capital markets as well as throughout the economy at large. Further, the structure of debt-encumbrance is such that it appears to be predicated on future economic growth. Dependence on future growth to pay off the debt—mortgaging the future to pay for present investment—is in effect a redistribution of income across generations (the next generation pays for the investment we make now), but if growth slackens then the only other mechanism for reducing past debts (without going even deeper into debt to the subsequent generation) is by lowering real incomes through a high rate of inflation (which effectively redistributes from savers to borrowers as well as unevenly across various social groups in the population). The point here, however, is that future urban growth may be essential to secure indebtedness incurred in creating past urban growth.

The dominant impression which an analysis of institutional mediation in the urbanization process creates is of a diversity of instruments and institutions, a chaos of policies and regulatory frameworks, all of which, by some miracle, impose a certain logic and coherence on the totality of the urbanization process. This logic and coherence is not the result of a premeditated urban growth strategy, for there has been no such strategy, either at the local or national level, these last forty years. But a strong de facto urban policy has in fact existed since the 1930s in the form of a direct byproduct of federal fiscal and monetary policies designed to prevent a return of severe depression conditions.

TABLE 7
PERCENT OF TOTAL FUNDS RAISED IN CREDIT MARKETS
BY NONFINANCIAL SECTORS FOR VARIOUS PURPOSES, 1965-1973

	1965	1966	1967	1968	1969	1970	1971	1972	1973
U.S. Government	2.6	5.3	15.8	14.2	−3.9	13.1	17.4	10.4	5.2
Corporate equities	0.3	1.2	2.7	−1.5	3.7	5.0	8.0	6.0	3.0
State and Local govt.	10.4	8.3	9.5	10.0	10.8	11.6	11.3	7.1	5.4
Corporate and foreign bonds	8.4	16.2	19.3	14.8	14.2	21.1	13.4	7.9	6.2
Mortgages	36.6	32.9	26.8	28.7	29.3	26.4	31.9	40.5	39.0
Home	22.0	17.2	14.0	16.0	17.2	13.1	17.7	23.9	22.8
Other residential	5.1	4.6	4.4	3.6	5.1	5.9	6.0	6.2	5.1
Commercial	6.3	8.4	5.7	6.8	5.8	5.4	6.8	8.9	8.8
Farm	3.1	2.7	2.8	2.3	2.1	1.8	1.4	1.6	2.4
Bank loans	20.2	15.8	10.3	13.8	16.7	6.6	6.3	13.1	22.3
Consumer credit	13.7	9.4	5.5	10.6	11.4	6.1	7.6	11.5	12.2
Open market paper	−0.4	1.5	2.5	1.7	3.6	3.9	−0.6	−0.9	1.3
Other	8.0	9.2	6.2	7.6	13.1	6.0	4.5	4.2	5.2

SOURCE: *Economic Report of the President, 1974:* Table C-54.

The thinking about urbanization manifest in the current debate over national urban policies has changed substantially, however, from that contained, for example, in the extensive 1937 report of the National Resources Committee on *Our Cities: Their Role in the National Economy*. Most of the present literature, from the Kerner and Douglas Commissions onward, focuses directly on social, political, and quality-of-life issues almost to the exclusion of questions of organization for production. Cities are now, in short, looked upon as consumption artifacts rather than as "workshops" for the production of goods and services. The reasons for this shift in thinking are not hard to find. The question which the 1930s posed, as Keynes (1936) showed, was not how to organize the production of value efficiently, but how to circulate the value produced (clear the market) and realize it through the consumption process (generate an effective demand). Malthus (1836: 325) had long before identified the latter as a key problem of capitalist society:

while it is quite certain that an adequate passion for consumption may fully keep up the proper proportion between supply and demand, whatever may be the powers of production, it appears to be quite as certain that an inordinate passion for accumulation must inevitably lead to a supply of commodities beyond what the structure and habits of such a society will permit to be profitably consumed.

The American city is now designed to stimulate consumption. The emphasis upon sprawl, individualized modes of consumption, owner-occupancy, and the like, is to be interpreted as one of several responses to the underconsumption problems of the 1930s (military expenditures being another). And it is in these terms, too, that we can interpret how the financial superstructure, itself created in response to the crisis conditions of the 1930s, so mediated the flow of investment into urban infrastructure, including housing, that its mediations served to transform cities once fashioned as the "work-shops of industrial society" into cities for the artificial stimulation of consumption.

In the midst of the muddled policy thinking in the United States appears a nervous belief that, left to themselves, many households would purchase and consume less housing and ancillary equipment than perhaps would be economically and socially desirable from the standpoint of economic growth and sustained capital accumulation.

The capital gains provision for homeowners is indicative of this kind of thinking—it permits capital gains to be deferred provided the owner moves to another house of equal or greater value (unless the owner is retired). It appears as if a perpetual effort has had to be mounted to counter the tendency for a declining relative rate of capital formation in the residential sector. Such a view may sound curious in a consumption-oriented society, the outward appearance of which leads us to agree with Adam Smith, that mankind has an infinite and insatiable appetite for "trinkets and baubles." But Malthus (1836: 321) was probably more correct when he argued that the history of human society sufficiently demonstrates that "an efficient taste for luxuries and conveniences, that is, such a taste as will properly stimulate industry, instead of being ready to appear the moment it is required, is a plant of slow growth." On the time-scale of human history the growth of the consumer society in the United States and of cities as artifacts for stimulating consumption may appear extraordinarily rapid. But, like any hothouse plant, it required a tremendous amount of nurturing.

2. THE FINANCIAL STRUCTURE AND LOCAL MARKETS:
THE CASE OF BALTIMORE

There is abundant evidence that the financial superstructure plays an important role in the organization of local housing markets and that many of the "urban problems" with which we are familiar —racial and class segregation, housing abandonment, neighborhood decay, speculative change, fiscal inequalities between cities and suburbs, inequality of access to services (such as education and health care)—are in some way tied to residential differentiation in cities which is, in turn, tied to the way in which investment is channeled into local housing markets. There are numerous studies which document certain aspects of this process in detail (for example, Sternlieb, 1969; Sternlieb and Burchell, 1973; Mollenkopf and Pynoos, 1972) while it is a familiar topic for Congressional hearings (see, for example, U.S. Congress, 1972a and 1969). Rather than attempt a summary of this material, Baltimore will be used as an example to show how the financial superstructure relates to a local housing market and through this to all aspects of community life and politics.

The financial superstructure is so organized that it can resolve a social aggregation problem. It links local activities to national aggregative needs by creating local "decision environments" as contexts within which individuals exercise choice. Relations between the total social structure and individual activities are in part established through the financial superstructure. I have elsewhere sought to show in some detail how and why this social aggregation process necessitates the creation of distinctive housing submarkets —residential areas with distinctive housing characteristics (Harvey, 1974, and forthcoming). The central proposition of that argument, which will be treated as an assumption here, is that the financial superstructure serves to coordinate the urbanization process in a particular locale with the overall aggregative push toward stimulating effective demand and facilitating capital accumulation. This proposition requires qualification, for the coordinations are not perfect nor are they established by way of a conscious conspiracy. The coordinating process has, in fact, been arrived at in much the same ad hoc adaptive way that has characterized the setting up of the financial superstructure in general.

An analysis of the activities of the various kinds of financial intermediary in the housing mortgage market in Baltimore City shows two things. First, different intermediaries are responsible for originating mortgages over different price ranges (see Table 8). Small-scale state-chartered S&Ls were the only institutions willing to originate mortgages on housing in the below-$7,000 category, whereas savings banks and commercial banks confined themselves to the upper price ranges. Second, different intermediaries serve

TABLE 8

DISTRIBUTION OF MORTGAGE ACTIVITY IN DIFFERENT PRICE CATEGORIES BY TYPE OF INSTITUTION, BALTIMORE CITY, 1972

	Under $7,000	$7,000-$9,999	$10,000-$11,999	$12,000-$14,999	Over $15,000
Private	39	16	13	7	7
State S&Ls	42	33	21	21	20
Federal S&Ls	10	22	30	31	35
Mortgage banks	7	24	29	23	12
Savings banks	—	3	5	15	19
Commercial banks	1	1	2	3	7
Percent of city's trans- actions in category	21	19	15	20	24

SOURCE: "Homeownership and the Baltimore Mortgage Market," Draft Report of the Home Ownership Development Program, Department of Housing and Community Development, Baltimore City, 1973.

TABLE 9
HOUSING SUBMARKETS—BALTIMORE CITY, 1970

| | Total Houses Sold | Sales Per 100 Proper-ties | % Transactions by Source of Funds: | | | | | | | | % Sales Insured | | Average Sale Price ($)[b] |
			Cash	Pvt	Fed S&L	State S&L	Mtge Bank	Comm Bank	Savings Bank	Other[a]	FHA	VA	
Inner City	1,199	1.86	65.7	15.0	3.0	12.0	2.2	0.5	0.2	1.7	2.9	1.1	3,498
1. East	646	2.33	64.7	15.0	2.2	14.3	2.2	0.5	0.1	1.2	3.4	1.4	3,437
2. West	553	1.51	67.0	15.1	4.0	9.2	2.3	0.4	0.4	2.2	2.3	0.6	3,568
Ethnic	760	3.34	39.9	5.5	6.1	43.2	2.0	0.8	0.9	2.2	2.6	0.7	6,372
1. E. Baltimore	579	3.40	39.7	4.8	5.5	43.7	2.4	1.0	1.2	2.2	3.2	0.7	6,769
2. S. Baltimore	181	3.20	40.3	7.7	7.7	41.4	0.6			2.2	0.6	0.6	5,102
Hampden	99	2.40	40.4	8.1	18.2	26.3	4.0		3.0		14.1	2.0	7,059
West Baltimore	497	2.32	30.6	12.5	12.1	11.7	22.3	1.6	3.1	6.0	25.8	4.2	8,664
South Baltimore	322	3.16	28.3	7.4	22.7	13.4	13.4	1.9	4.0	9.0	22.7	10.6	8,751
High Turnover	2,072	5.28	19.1	6.1	13.6	14.9	32.8	1.2	5.7	6.2	38.2	9.5	9,902
1. Northwest	1,071	5.42	20.0	7.2	9.7	13.8	40.9	1.1	2.9	4.5	46.8	7.4	9,312
2. Northeast	693	5.07	20.6	6.4	14.4	16.5	29.0	1.4	5.6	5.9	34.5	10.2	9,779
3. North	308	5.35	12.7	1.4	25.3	18.1	13.3	0.7	15.9	12.7	31.5	15.5	12,330
Middle Income	1,077	3.15	20.8	4.4	29.8	17.0	8.6	1.9	8.7	9.0	17.7	11.1	12,760
1. Southwest	212	3.46	17.0	6.6	29.2	8.5	15.1	1.0	10.8	11.7	30.2	17.0	12,848
2. Northeast	865	3.09	21.7	3.8	30.0	19.2	7.0	2.0	8.2	8.2	14.7	9.7	12,751
Upper Income	361	3.84	19.4	6.9	23.5	10.5	8.6	7.2	21.1	2.8	11.9	3.6	27,413

a. Assumed mortgages and subject to mortgage.

b. Ground rent is sometimes included in the sales price and this distorts the averages in certain respects. The relative differentials between the submarkets are of the right order however.

SOURCE: City Planning Department Tabulations from Lusk Reports.

different geographical areas and act to form distinctive housing submarkets as far as housing finance is concerned (see Figure 4 and Tables 9 and 10). A "snapshot" of activity in 1970 shows, for example, that small-scale, community-based S&Ls dominated housing finance in the traditional ethnic areas of South and East Baltimore, that the middle-income white areas of northeast and southeast Baltimore were largely served by the federal S&Ls, that the affluent areas drew upon the financial resources of savings banks and commercial banks, while areas of high turnover were strongly associated with mortgage company finance in association with FHA insurance.

The structure of financing across the various housing submarkets, and the residential differentiation which this implies, has a history and is constantly in the course of evolution. Changes occurring within submarkets promote boundary shifts. On occasion, too, whole new submarkets can be dramatically created. Such changes are a response to a variety of forces which stem from changing relative wage rates, changing job opportunities within a changing structure of the division of labor, migratory movements, and so on. But all of these forces are marshalled and given coherence in the urban context through the mediating power of the financial superstructure. The

TABLE 10

HOUSING SUBMARKETS—BALTIMORE CITY, 1970 (Census Data)

	Median Income[a]	% Black Occupied D.U.'s	% Units Owner Occupied	Mean $ Value of Own. Occ.	% Renter Occupied	Mean Monthly Rent
Inner City	6,259	72.2	28.5	6,259	71.5	77.5
1. East	6,201	65.1	29.3	6,380	70.7	75.2
2. West	6,297	76.9	27.9	6,963	72.1	78.9
Ethnic	8,822	1.0	66.0	8,005	34.0	76.8
1. E. Baltimore	8,836	1.2	66.3	8,368	33.7	78.7
2. S. Baltimore	8,785	0.2	64.7	6,504	35.3	69.6
Hampden	8,730	0.3	58.8	7,960	41.2	76.8
W. Baltimore	9,566	84.1	50.0	13,842	50.0	103.7
S. Baltimore	8,941	0.1	56.9	9,741	43.1	82.0
High Turnover	10,413	34.3	53.5	11,886	46.5	113.8
1. Northwest	9,483	55.4	49.3	11,867	50.7	110.6
2. Northeast	10,753	30.4	58.5	11,533	41.5	111.5
3. North	11,510	1.3	49.0	12,726	51.0	125.1
Middle Income	10,639	2.8	62.6	13,221	37.5	104.1
1. Southwest	10,655	4.4	48.8	13,470	51.2	108.1
2. Northeast	10,634	2.3	66.2	13,174	33.8	103.0
Upper Income	17,577	1.7	50.8	27,097	49.2	141.4

a. Weighted average of median incomes for census tracts in submarket.
SOURCE: 1970 Census.

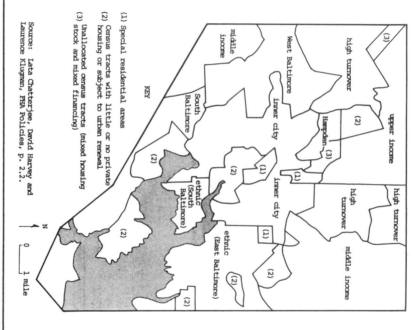

Figure 4: HOUSING SUBMARKETS IN BALTIMORE CITY, 1970

Source: Lata Chatterjee, David Harvey and
Laurence Klugman, FHA Policies, p. 2.2.

contemporary history of residential differentiation in Baltimore can be used to illustrate this point.

A. West Baltimore and the Land-Installment Contract

The snapshot of activity in 1970 (Figure 4 and Tables 9 and 10) shows West Baltimore as a quiet, almost stagnant, housing market populated largely by blacks, many of whom had achieved the status of moderate income homeowners. This submarket was not always this way. Throughout much of the 1960s it had been the scene of turmoil, rapid social and racial change, community conflict and outrage. At issue during this time was how could blacks, many of whom were experiencing modest gains in income and rising expectations, gain access to reasonable quality housing. In the early 1960s financial institutions were not prepared to provide mortgage finance to this social group. The grounds for such denial varied from scarcely concealed racial prejudice to rationalizations based on creditworthiness or the expected future value of the properties being bought. Government institutions seldom intervened and when they did so, as was the case with the FHA prior to 1964, they frequently acted to formalize discrimination.

Into this vacuum created by rising effective demand and the failure of financial intermediaries to provide credit crept the landlord-speculator. And the tool which the landlord-speculator used was known as the land-installment contract.[2] The submarket of West Baltimore was dramatically carved out during the 1960s by the use of this device. A community group, the Activists (1970, 1971), documented over 4,000 land-installment contract transactions in Baltimore City in the early 1960s and the majority of these were in West Baltimore. One organization bought 1,768 houses for $10.8 million and sold 742 of them on the land-installment contract for $9.4 million (the rest going into the rental inventory). The average purchase price for the speculator on the houses finally sold was $6,868 compared to final average sales price of $12,706. The net mark-up, called the "black tax," became the center of controversy in Baltimore during the 1960s, as did the very high rate of induced turnover in the housing stock. The latter had all the marks of a well-organized block-busting operation. As the Activists sought to track down the origin of this problem, they quickly became aware that it was to be attributed in large part to the unwillingness of financial institutions to provide the black home-seeker with mortgage credit except through the intermediary of the speculator. They learned, for example, that several small S&Ls, and not a few not-so-small ones, were using most of their financial resources (80% or more in several cases) to finance land-installment contract activity, i.e., to finance the speculator rather than the prospective homeowner directly.

The speculators' use of the land-installment contract required additional capital to that which could legally be provided by the S&Ls. The commercial banks provided short-term loans (at prime interest rates) to finance the initial purchase of the house and also lent to the speculator for longer periods to help cover the difference between the appraised value fo the property and the final purchase price to the buyer. For the commercial banks this was sound business. It yielded them prime rates of return on short-term loans to fairly secure speculative operations which were successfully pooling risks. This kind of business was obviously to be preferred to making illiquid and usually small homeownership loans at an interest rate held down by state usury laws.

By 1969 the use of the land-installment contract and the speculative activity associated with it were much diminished. The urban riots (in which discontent over housing exploitation played an important role), public pressure on the S&Ls not to finance this kind of activity (backed up by tightening Federal Home Loan Bank Board regulation of federal S&Ls), and the public pillorying of landlord-speculators who used the land-installment contract were, perhaps, the most important factors. But by 1969 the new submarket of West Baltimore had been created and the land-installment contract had served its purpose. The immediate purpose, of course, was to generate profits (and in some cases excess profits) for the speculator-landlords and to yield relatively high rates of return to the financial institutions. But we can identify a deeper purpose by relating the events in West Baltimore to the interpretation of the urbanization process in general set out in the preceeding section.

The financial institutions, by denying funds to certain groups in particular areas and channeling investment to preferred speculative borrowers, created a decision context in which speculative activity was almost bound to succeed. In the process, a new submarket was created by displacing a middle-income white population which was forced, as a consequence, to look elsewhere for new housing opportunities. Most of these new opportunities were being created on the suburban fringe where land-speculators, developers, construction interests, and the like were actively investing with the resources channeled to them through the financial superstructure. There was, consequently, a multiplier effect to investment in inner-city speculation in the form of a new effective demand for housing largely registered on the suburban periphery (Harvey, 1974). Such a multiplier effect does not have to be consciously or explicitly manipulated, although some institutions (and even individual speculators) operate at both ends of this geographically segmented process and are clearly aware of the connections. The market mechanism and the structure of "scal and monetary policies on the part of the government along with the specific objectives of the private elements in the financial superstructure and of other intermediaries in the housing market are sufficient in themselves broadly to guarantee this outcome. The creation of the submarket of West Baltimore has to be interpreted, therefore, as an example of the operation of those processes, coordinated through the financial superstructure, which give sufficient dynamism to the urbanization process to match the dynamism of capitalist accumulation in general.

B. Northeast Baltimore and the FHA

The history of West Baltimore and the land-installment contract during the 1960s was being repeated in many cities in the United States during that period. The processes described and the resentments they aroused contributed to the urban discontents which culminated in the urban riots. During the 1960s, responses were in the process of being fashioned to allow the growth of effective demand and capital accumulation to proceed more smoothly. Regulatory action—over the S&Ls, over the discriminatory red-lining practices of the FHA, over open housing—began to change the emphasis of urban development while leaving the overall goals of economic growth, price stability, and reasonably full employment intact. But something more was needed, as both the Kerner and Douglas Commissions pointed out, and this "something" was provided by the 1968 Housing Act. Among the provisions of this Act were a whole series of measures designed to bring the social and economic "benefits" of homeownership to low- and moderate-income people. One of the programs for doing this was the 221(d)(2) program which permitted the purchase of a house without a down payment.

The 221(d)(2) program basically replaced the land-installment contract in Baltimore as the means whereby relatively low-income groups could obtain access to housing. It therefore held out the prospect of continuing the processes of neighborhood change accomplished by means of the land-installment contract prior to 1968, but doing so via a very different set of financial intermediaries. It brought together mortgage companies and FHA insurance. The connection between these two institutions had been strongly developed post-1945 (Mortgage Bankers Association, 1973). Mortgage bankers do not hold mortgages. They originate them and then sell them in the secondary market mainly to life insurance companies and mutual savings banks (see Diamond, 1968). Risk-free mortgages are generally easier to sell in the secondary market and for this reason mortgage companies deal extensively in FHA-insured mortgages. With a tradition of dealing with the FHA, mortgage companies played a vital role in promoting the use of the 221(d)(2) program.

The effects in Baltimore were quite dramatic. First, there was a marked decline in the use of the land-installment contract in the city. Klugman (1974) collected data for two transects in northeast and northwest Baltimore; Table 11, taken from his work, illustrates the substitution process. The rise of the 221(d)(2) program was rapid and became highly concentrated in the areas designated High Turnover in Figure 4 (see also Table 9). Figure 5 demonstrates the rapid rise and fall of the FHA-mortgage company combination over the period 1970-1973 in these high turnover areas, while Figure 6 portrays the geographical distribution of FHA-insured mortgage finance in 1970. Most of this was of the 221(d)(2) variety (at least three-quarters) and in the areas of great FHA concentration it was almost exclusively so. Plainly, the FHA programs, with mortgage company finance, were functioning on a geographical basis consistent with long-run processes of changing residential differentiation within the city.

The FHA's powers are basically powers of denial. The Agency cannot initiate the use of programs although it does have ways of signaling to various housing intermediaries what it will or will not do. The FHA plays a vital role, therefore, in creating a "decision context" within which housing intermediaries operate. Figure 6 demonstrates, for example, that the FHA was not insuring in the

TABLE 11

**NUMBERS OF LAND-INSTALLMENT CONTRACTS AND
221(d)(2) INSURED MORTGAGES IN NORTHEAST AND
NORTHWEST BALTIMORE, 1965-1972**

Area	1965	1966	1967	1968	1969	1970	1971	1972
Northeast Baltimore								
Land-installment contracts	36	100	73	45	35	19	29	19
221(d)(2)	—	—	—	149	204	191	364	238
Northwest Baltimore								
Land-installment contracts	33	64	79	71	34	19	23	15
221(d)(2)	—	—	—	131	129	244	354	234

SOURCE: Klugman, 1974: Tables VI-2 and VI-3.

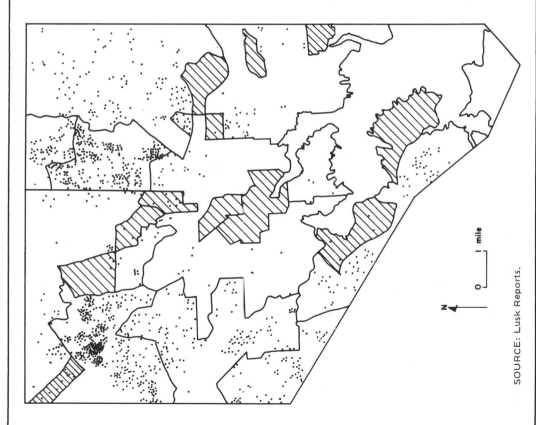

SOURCE: Lusk Reports.

Figure 5: DISTRIBUTION OF FHA-INSURED MORTGAGES ACROSS HOUSING SUBMARKETS, BALTIMORE CITY, 1970

untouchable zone of the inner city, but Klugman shows that during the period 1969-1972, at least, the FHA was receiving no requests for insurance in that area. The obvious "red-line" in mortgage finance is an excellent example of how something very firmly etched into the geographical landscape of the city can arise by the complex of interactions and expectations which exist in the housing market. The FHA received no requests for insurance because realtors anticipated that requests would be denied; and financial intermedi-

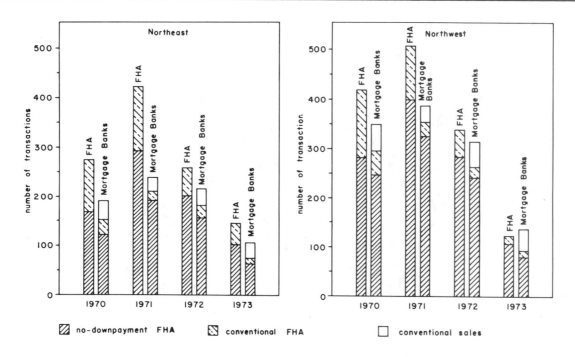

FHA-insured sales and mortgage originations by mortgage bankers in (A) Northeast and (B) Northwest high turnover submarkets, Baltimore, annually, 1970-1973, distinguishing between conventional sales, conventional FHA insurance and no-down-payment programs.

SOURCE: Tabulated from the Lusk Reports.

Figure 6.

aries would not undertake the risk in areas which were decaying—in part, because mortgage finance was not available. Nobody in particular is at fault, but the red-line nevertheless exists.

The 221(d)(2) program was viewed by all parties as mortgage finance of last resort. It was, in addition, a program which could be used to continue the processes of neighborhood change. Speculators, for example, could pick up a house for, say, $8,000, make some improvements and add all the usual expenses and the profit margin, and then sell it under a 221(d)(2) program for, say, $13,000. This was certainly happening in Baltimore.[3] Landlords (often the same persons as the speculators) could also benefit. For a variety of reasons, largely to do with leveraging practices and taxation arrangements, landlords find it advantageous to purchase a house with the assistance of a mortgage and then to sell it to recover their own part of the equity after seven years or so. The market for such housing was not particularly good, and in effect the 221(d)(2) program created a market into which landlords could sell frequently to the sitting tenant who was then faced with the option of buying or moving. There had been a large volume of landlord purchases in northwest Baltimore during the early and mid-1960s as the area was transformed from owner-occupancy to absentee-landlord ownership and by 1970 many of these houses were ripe for selling. Again, there is evidence that landlords used the 221(d)(2) program to bail out their equity from a housing stock that was rapidly depreciating in some areas—how extensive this was is difficult to judge.

But perhaps the most important agent for change, particularly in the northeast high turnover submarket, was the realtor. Under the 221(d)(2) program it was possible to move low-income black families into middle-income white neighborhoods and thereby to mount a block-busting operation. In the northeastern area, which was largely owner-occupied, the prospects for stimulating turnover by such techniques were favorable, particularly as conventional lenders were beginning to show signs of nervousness and to withdraw financing. Realtor activity, some of it quite unscrupulous, grew remarkably in the northeastern high turnover submarket after 1968, and the 221(d)(2) program and mortgage company finance provided the wherewithal. Community reaction was fast and direct. One group in particular, organized on Saul Alinsky lines, adopted confrontation tactics initially against realtors and subsequently against the FHA which, they claimed, was forcing neighborhood change and artificially depressing house prices. Community activism reached a peak in

1971-1972. But thereafter FHA and mortgage banker involvement in this submarket declined (Figure 5), and with it community activism declined also.

The reason for the gradual disengagement of the FHA-mortgage banker combination from this market had little to do with the level of community activism. By 1972, the 221(d)(2) and 235 program scandals had erupted in Detroit and in other cities, and the FHA was required to adopt stringent new consumer protection measures and to adjust its appraisal practices. Consumer protectionism meant more red tape and delays in mortgage commitments and generally made FHA insurance less attractive. At the same time, mortgage companies were finding FHA mortgages less profitable because of relative movements in interest rates. Mortgage companies began to withdraw from the less profitable small-price range market (in Baltimore, as a matter of policy, they did not finance transactions much below $7,000; but by 1973 this has moved up to $12,000) as well as from participating in the origination of FHA mortgages. Instead, they began to move into the conventional mortgage market on higher-valued homes (see Goldberg and Taschdjian, 1974). In addition, S&Ls, which had suffered a heavy savings outflow in 1969 (a process called "disintermediation") were flush with funds in 1971 and 1972 and were willing to service more loans in areas such as the northeast where there was still a substantial demand for conventional mortgage finance. The credit crunch in housing finance, which set in by the summer of 1973, quieted the turnover in the northeastern sub-markets and with it died the final remnants of confrontation tactics and activism.

It may sound odd to suggest that community activism is interest rate-sensitive, but in effect this proved to be in part the case in the northeast high turnover submarket. The general point, of course, is that events of national import were being transmitted via the financial superstructure into the decision environments of local housing submarkets. The land-installment contract successfully fashioned a new submarket in West Baltimore and in the process spawned community opposition in the form of the Activists. When the land-installment contract diminished in use, so the Activists became a rump organization without an issue to confront. The rise of the 221(d)(2)-mortgage banker combination continued the processes of neighborhood change in a different manner and in a different area; and this, too, spawned an opposition group. When the mortgage bankers began to withdraw from servicing FHA mortgages and the

FHA became much more consumer protection-minded, the problem diminished and the community group was left with nothing to fight over. But, again, by then substantial changes had already been wrought by the FHA programs, and a mixed and somewhat volatile situation had been created—particularly in the northeastern part of the city. All was quiet in 1974 because the housing market was extremely depressed. What will happen in the next boom is, however, problematic.

C. The Ethnic Submarkets and the
Evolution of National Financial Policy

The submarkets designated "ethnic" in Figure 4 are dominated by relatively small-scale state-chartered S&Ls as far as mortgage finance is concerned (see Table 9). These institutions, many of which are quite old, are rather special intermediaries; they take in savings from the community and then, making use of detailed local knowledge about both housing and people, apply these savings to foster homeownership within the community. Many of these S&Ls are ethnic in origin (as names like Golden, Prague, Kopernik, Kosciuszko, and Slavie suggest) and some continue to serve identifiable ethnic groups within the population. This helps to stabilize communities and to provide mortgage finance for not very affluent blue-collar workers who thereby gain access to reasonable quality housing at a relatively low cost compared to their incomes. The mode of operation of these S&Ls means, however, that they are often exclusionary towards other ethnic and racial groups. It also means that savings, far from entering into the mainstream of capital markets, remain locked into the community from which they originate.

Over the years these community-based S&Ls have weakened somewhat under a number of pressures. Some of them, easy prey to the penetration of unscrupulous interests, fell to financing speculative activity. Increasing geographical mobility and rising incomes forced some to expand their operations to finance migration, and in the process they lost their strong neighborhood identity. Others became expansionary-minded and profit-oriented and in businesslike fashion began to put funds where the rate of return was highest and to chase savings aggressively by offering attractive deposit terms. These last, in particular, would typically attract savings from wherever they could and lend where it was most profitable; they typically performed the savings-investment function in a fashion that fostered suburban growth while starving the inner city of financial support. These last practices were critically viewed in the Baltimore City Homeownership Program Report (1973) because they involved the export of savings from the city to enhance development (and the tax base) in the surrounding counties. The Report recommended legislative action which would put pressure on the S&Ls to do more business in the area from which savings originated, thus segmenting the market for homeownership finance even further. In 1970 there were still a substantial number of S&Ls which continued to operate either on a neighborhood basis or to serve the needs of distinctive classes of people, thereby contributing to the social stability of the city. From the standpoint of the City, these S&Ls were invaluable and action was advocated to preserve and, if possible, to promote them.

Unfortunately, this logic flies in the face of that explicitly developed in the Hunt Commission Report (1971), the recommendations of which currently lie, in modified form, before the U.S. Congress. The Hunt proposals were arrived at through a consideration of serious problems in the national flows of credit and savings in a period of inflation dominated by an apparently serious capital shortage. The Commission did not attempt to recommend monetary and fiscal policies which would channel savings into mortgage markets on a stable basis even though it admitted (1971: 78) that "past failure to meet national housing goals was largely a result of the effects of these policies on housing demand and the supply of mortgage funds." Nevertheless, the Commission was greatly concerned to eliminate barriers to the flow of credit in general and to make capital markets much more competitive. The Commission proposed a reorganization of the ways in which financial institutions would operate which, if carried out, would have far-reaching implications for mortgage finance and housing markets.

The recommendation which is of particular interest here is that which would transform S&Ls into "mini-banks" by permitting them to offer a wide range of customer services (consumer credit, for example). At the same time S&Ls would lose their privileged position with respect to interest rates and be forced to compete in the open market for savings with all the other intermediaries. Such a reorganization would, as Lane Kirkland notes in his dissenting opinion (Hunt Commission, 1971: 131), probably increase the profitability of the S&Ls without necessarily generating any additional funds for the mortgage market—"in fact, in periods of tight money, the mortgage market would be hard hit."

It is difficult to predict what the effects of the Hunt proposals would be. But we can venture a scenario of the following sort. Almost certainly, S&Ls would be forced by competition to become more businesslike and in general to increase in size and sophistication if they are successfully to compete with large commercial banks. Under these conditions it is doubtful if the small-scale, community-based S&L would survive. The S&Ls which could compete successfully would probably be those which, in the name of business rationality, are currently fostering the flow of funds into areas of greatest profit (the suburbs for the most part) at the expense of the inner city. In competition with the banks, the S&Ls would presumably become more bank-like in their behavior and chase the more lucrative business in the more affluent submarkets or lend to speculators, developers, landlords, and the like (unless compensating arrangements, such as tax credits for servicing low-value mortgages, were forthcoming).

If we project these changes into the Baltimore situation, then the effects upon the ethnic submarkets would likely be quite devastating. Financing for low-value mortgages would be hard to obtain—at best it would be "residual" finance after the more lucrative upper-income market had been served. The community-based S&Ls would disappear as would the localized structure of savings and investment in housing. Savings, which usually get channeled into housing finance, would not be protected from the depradations of corporations hungry for external sources of funds. The flow of funds to meet the needs of the not-so-affluent white ethnic and blue-collar worker would be seriously curtailed particularly in times of general credit shortage. A major mechanism for ensuring neighborhood social stability would be destroyed and the disruption of communities by external forces made that much easier.

The Hunt Commission proposals, if accepted in raw form, would in fact perpetuate and accentuate certain of the urbanization processes which we have examined in this article. The land-installment contract devastated existing communities and created new ones in their place. The 221(d)(2) program took over where the land-installment contract left off. Now before the Congress, as the FHA and all of its programs lie in disrepute, are a set of proposals for the reform of the financial superstructure which, if they pass uncompensated for, will perpetuate the processes of neighborhood change and displacement merely by encouraging the flow of investment funds to wherever the rate of return is highest (the more affluent areas and areas of new development and redevelopment) at the expense of less-well-off groups in older areas. Under the Hunt Commission proposals, the pace of urban development and redevelopment will almost certainly quicken, effective demand will thereby be artificially stimulated, and sustained capital accumulation, from this standpoint at least, will be facilitated. The proposals seem almost tailor-made to perpetuate and enhance the dynamics of an urbanization process which, since the 1930s, has been strongly directed to stimulating consumption. The irony is, of course, that these connections lie largely unnoticed, for we still live in a world where it is presumed that the urbanization process can be left to look after itself and that major reforms of the financial superstructure can be contemplated in relationship to extremely limited goals.

IV. CONTRADICTIONS

But can urbanization be left to look after itself? Or do urbanization processes themselves contain conflicts and contradictions which can spell the nemesis of sustained capital accumulation and be the source of that selfsame social and economic instability which the political, social, and economic responses devised in the 1930s were designed to counteract?

Since the mid-1960s long-submerged contradictions seem to have been surfacing, at first as fragmented and apparently unconnected localized conflicts, but later as a more generalized economic malaise. The strange combinations of high inflation and unemployment, monetary restraint and booming credit markets, fiscal and monetary instability (both national and international), resource shortages, sagging real economic growth, and so on, all suggest the existence of some deep structural problems within advanced capitalism. By way of conclusion, two major contradictions within the urbanization process will be identified. These, to some degree, contribute to or are a manifestation of the larger aggregative problems from which advanced capitalism is currently suffering.

1. INDEBTEDNESS, GROWTH DEPENDENCE, AND THE
STABILITY OF THE FINANCIAL SUPERSTRUCTURE

We have shown that the financial superstructure mediates investment in urban infrastructure because the latter makes heavy demands upon external funds. In the process, the security of the financial superstructure becomes attached to the security of the long-term mortgage debt which in turn depends upon the ability to find future profitable uses for the elements of urban infrastructure. The ability of the financial superstructure to function in any of its roles would be seriously impaired if the capital value of the various elements of urban infrastructure which secure the long-term debt were to experience any major decline. Investment in urban infrastructure consequently commits society to certain patterns of use for a long period of time, for if profitable uses cannot be found on a sustained basis, then the viability of the financial superstructure will likely to undercut.

There is also some evidence that the structure of indebtedness is such as to be predicated on future growth. This condition arises in part from the use of property as a means for realizing capital gains while it also stems from the fact that much of the investment, both public and private, takes place on a "build now and pay later" principle. The delayed costs of such investment have to be paid for out of future growth.

In order to stimulate an effective demand for product, however, the financial superstructure also coordinates activities which promote development and redevelopment. This in part involves stimulating new investment by writing off the value of the old. Rising effective demand can be generated by a number of means (including population growth) but an expanding rate of economic obsolescence in urban infrastructure is an important component. Obviously, a delicate balance must exist between writing off the value of capital assets in order to stimulate effective demand and preserving the value of debts incurred by financial institutions. A recent Congressional study on "The Central City Problem and Urban Renewal Policy" (U.S. Congress, 1973a: 7) provides an interesting perspective on this problem:

Financial institutions hold tens of billions of dollars of mortgages and bonds secured by central city residential and business properties and private and public utilities. Continued central city deterioration would lead to the devaluation of these assets with serious repercussions within the national financial structure.

The study (1973a: 299) goes on to estimate that 29% of all mortgage debt was secured against inner-city properties in 1960 and to guess that this proportion was around a quarter in 1970. When state and local debt secured against property taxes and the debts assumed by public and private utilities are added to the burden of residential debt, the total value of indebtedness secured by the elements of urban infrastructure becomes quite staggering. The report (1973a: 20-21, 7) remarks that "society would bear a sizeable economic burden if infrastructure which is unused and underutilized is duplicated elsewhere—a situation clearly occurring with the flight of middle-class families to the suburbs."

And herein lies the contradiction. If the flight to the suburbs continues at its present pace (and certainly if it quickens) then there are bound to be serious repercussions within the national financial structure. If the flight does not continue there may be equally serious repercussions for the maintenance of effective demand and a return to the underconsumption problems of the 1930s.

Of course there are various tactics to resolve such a contradiction. Some are explicit—urban renewal policy, land use regulation, and the like. Others may be devised by accident. For example, when FNMA was set up as a credit agency to deal in conventional mortgages, it was anticipated that FNMA would purchase mortgages when credit was tight and sell them when credit was easier. But as well as performing this stabilizing function, FNMA has also accumulated mortgage holdings rapidly—the mortgage debt held by Federal Credit Agencies has in fact doubled since 1968 (see Figure 1). The suspicion is that many private financial institutions are using FNMA as a dumping ground for the less secure inner-city mortgages and that the federal government is therefore managing the debt security-effective demand contradiction without knowing it. Such a tactic shifts the problem onto far broader shoulders within the financial super-structure but it does not eliminate the contradiction.

2. THE CITY AS A CONSUMPTION ARTIFACT AND AS A "WORKSHOP" FOR THE PRODUCTION OF VALUE

For a generation or more now, the American city has been promoted as a consumption artifact in the cause of promoting that consumerism which was fashioned as a response to the under-consumption problems of the 1930s. A generation raised in such an atmosphere is bound to take its consumerism seriously, to become concerned over a whole range of issues which can conveniently be subsumed in the phrase "the quality of life." Such a generation is also likely to take its own consumer sovereignty seriously and to attach great importance to the political and civil liberties which have largely substituted for economic rights in the history of capitalism in the United States.

The contemporary history of residential differentiation in any city in the United States (and Baltimore has been used as an example here) shows, however, that communities are disrupted, populations moved (often against their will), and the whole structure of the city altered as the urbanization process, coordinated in its major outlines through the financial superstructure, is utilized as a vehicle to sustain an effective demand for product. Individuals choose in the context of decision environments created by forces which they plainly do not control. Individuals have no say in what the aggregative character of the urbanization process should look like.

The need (which inevitably attaches to sustained capital accumulation) to promote an effective demand at an expanding rate exercises an increasingly disruptive effect upon whole communities (inner city and suburb alike are affected by the operations of speculators coordinated through the financial superstructure). Also, the very scale of suburbanization makes it increasingly difficult over time to sustain the myth of the suburb as an adequate "urban-rural" compromise. Further, the shaping of the city as a consumption artifact has seriously affected the efficiency of cities as organizations of space for purposes of production and circulation. The production of value and of surplus value for purposes of capital accumulation are thus inhibited because, for example, new employment opportunities are increasingly divorced from the distribution of a potential labor force. Promoting the city as an artifact for consumption purposes conflicts with the functioning of cities as the "workshops" for the production and circulation of value. Refashioning the urbanization process for the purposes of production (finding a better spatial equilibrium for employment and housing opportunities, for example) means not only containing those processes whereby an effective demand is currently stimulated but challenging the exclusionary powers of suburban jurisdictions—in other words, challenging the myths of political rights and consumer sovereignty that have for long substituted for real economic rights in the United States, particularly for that segment of society which we conventionally call "the middle class."

These processes generate political reactions in much the same fashion that community activism within submarkets in Baltimore was a response to external pressures. The response in the suburbs is complex and to some degree ambiguous, but it certainly involves a strong defense of the exclusionary powers of suburban jurisdictions and of the political rights which have been built up in support of them over a generation or more as well as a growing emphasis on issues such as the quality of life and of environment which frequently merges into a "no-growth" political stance.

And herein lies the contradiction. Having built the American suburb in response to the underconsumption problems of the 1930s and having promoted the myths of consumer sovereignty, freedom of choice, and political rights, the political power of the suburbs is now being mobilized (1) to counter the thrust toward more development to sustain effective demand, and (2) to prevent any rationalization of the structure of urban communities from the standpoint of the production and circulation of value. And should the economic necessities of capitalism reassert themselves, as they must if capitalism is to be reproduced as an on-going system, then the result can only be a destruction of the myths of consumer sovereignty and of freedom of choice, the breaking of the political power and autonomy of the suburbs, and the gradual "proletarianization" of that large segment of the American population which has long come to regard itself as "classless"—the middle class.

The substantive thesis which this paper has sought to document is that there is an intimate connection between financial superstructure and the shape and form taken by the urbanization process. Since the financial superstructure has largely been fashioned as a response to problems in the sustained accumulation of capital and in particular to crises in that process, the financial superstructure mediates the relationship between the main dynamic of sustained capital accumulation, on the one hand, and the urbanization process, on the other. Within these relationships we can see a variety of conflicts, tensions, and contradictions. The evidence is incontrovertible that urbanization manifests and perhaps contributes to many of the contradictions implicit in a dynamic capitalist mode of production. Exactly how, when, and where these contradictions erupt into conflict on a localized or national scale, and the degree to which these contradictions are exacerbated by the particular shape assumed by the urbanization process and by the long-term commitment which creating the built environment entails, is, however, something which requires a great deal more research and analysis. That it is a topic worthy of our closest attention is beyond any shadow of doubt.

NOTES

1. This paper is part of a very general investigation of the political economy of urbanization in advanced capitalist societies. This investigation was begun with Harvey (1973) and continued in a series of articles (Harvey and Chatterjee, 1974; Harvey, 1974 and forthcoming). While I recognize that readers frequently become irritated with authors who are always referring to themselves, I feel that this paper can best be understood as a contribution to a very much larger project.

2. The land-installment contract works as follows (the figures are hypothetical but typical): A speculator purchases a house for $7,000, adds a purchase commission and various financing charges, renovates and redecorates the property, allocates overhead costs, charges a sales commission, and then finally adds a gross profit margin of, say, 20%. The house is then sold for $13,000 when its appraised value is perhaps only $9,000. To finance this kind of transaction, the speculator arranges a conventional loan for the purchaser, typically from a S&L, for the $9,000 and then loans a further $4,000 which the speculator usually borrows in his own name from a bank. The speculator retains title to the property to protect his position, but allows the purchaser immediate possession. The monthly payments, which are relatively high, cover interest charges on the total $13,000 loaned plus administrative expenses, and a small part is put to redeeming the principle. When $4,000 has been redeemed, the speculator arranges refinancing so that the buyer can get full title and build equity in the property. For the duration of the land-installment contract, the financial institutions are protected against any default on the part of the purchaser because the property remains in the speculator's name. If the speculator's business is sound, then the financial institutions have nothing to fear. The land-installment contract is a perfectly legal device and, used properly, it can be an effective way for low-income families to become homeowners when other forms of financing are unavailable. But, obviously, there are ample opportunities for abuse and the gaining of quick excess profits.

3. In Detroit, the combination of poor appraisal practices on the part of the FHA, corruption of FHA officials, and sharp practices on the part of speculators led to widespread defaults on 221(d)(2) program mortgages as low-income people found themselves faced with massive repair bills shortly after moving in. The scandals of Detroit were absent in Baltimore largely because of conservative administration on the part of the local FHA office. But speculators could, and did, use the program in Baltimore to their own advantage without the scandalous consequences which ensued in Detroit.

REFERENCES

AARON, H. J. (1972) Shelter and Subsidies. Washington, D.C.: Brookings Institution.

Activists (1971) Baltimore Under Siege. Baltimore: The Activists. (mimeo)

—— (1970) "Communities under siege." Baltimore: The Activists. (mimeo)

Baltimore City Home Ownership Development Program (1973) Homeownership and the Baltimore Mortgage Market. Baltimore: Department of Housing and Community Development.

BARTELL, H. R. (1969) "An analysis of Illinois savings and loan associations which failed in the period 1963-68," in I. Friend (ed.) Study of the Savings and Loan Industry. Washington, D.C.: Federal Home Loan Bank.

CHATTERJEE, L., D. HARVEY and L. KLUGMAN (1974) FHA Policies and the Baltimore City Housing Market. Baltimore: Johns Hopkins University, Center for Metropolitan Planning and Research.

Commission on Money and Credit (1961) Money and Credit: Their Influence on Jobs, Prices and Growth. Englewood Cliffs, N.J.: Prentice-Hall.

DIAMOND, A. H. (1968) Mortgage Loan Gross Flows. Washington, D.C.: U.S. Department of Housing and Urban Development.

FISHER, R. M. and C. J. SIEGMAN (1972) "Patterns of housing experience during periods of credit restraint in industrialized countries." Journal of Finance 27: 193-205.

GOLDBERG, L. G. and M. TASCHDJIAN (1974) "Changing competition in residential mortgage markets." Bankers Magazine 157, 2: 84-87.

GOLDSMITH, R. W. [ed.] (1971) "Institutional investors and corporate stock–a background study," in Securities and Exchange Commission, Institutional Investor Study Report, Part 6. Washington, D.C.: House Document No. 92-64, Government Printing Office.

—— (1969) Financial Structure and Development. New Haven: Yale University Press.

GREBLER, L. (1964) "Financial intermediaries and the allocation of capital," in Conference on Savings and Residential Financing, 1964. Chicago: U.S. Savings and Loan League.

—— D. M. BLANK and L. WINNICK (1956) Capital Formation in Residential Real Estate. Princeton, N.J.: Princeton University Press.

GURLEY, J. G. and E. S. SHAW (1956) "Financial intermediaries and the savings investment process." Journal of Finance 11: 257-266.

HARVEY, D. (forthcoming) "Class structure and the theory of residential differentiation," in M. Chisholm (ed.) Bristol Essays in Geography. London: Heinemann.

—— (1974) "Class-monopoly rent, finance capital and the urban revolution." Regional Studies 8, 3.

—— (1973) Social Justice and the City. Baltimore: Johns Hopkins University Press.

—— and L. CHATTERJEE (1974) "Absolute rent and the structuring of space by financial and governmental institutions." Antipode 6, 2: 22-36.

HERMAN, E. S. (1969) "Conflicts of interest in the savings and loan industry," in I. Friend (ed.) Study of the Savings and Loan Industry. Washington, D.C.: Federal Home Loan Bank Board.

Hunt Commission (1971) Report of the President's Commission on Financial Structure and Regulation. Washington, D.C.: Government Printing Office.

KEYNES, J. M. (1936) The General Theory of Employment, Interest and Money. New York: Harcourt Brace.

KLUGMAN, L. S. (1974) "The FHA and home ownership in the Baltimore housing market (1963-72)." Ph.D. dissertation. Clark University, Department of Geography.

KUZNETS, S. (1961) Capital in the American Economy: Its Formation and Financing. Princeton, N.J.: Princeton University Press.

LEFEBVRE, H. (1970) La Revolution Urbaine. Paris: Gallimard.

MALTHUS, T. R. (1836) Principles of Political Economy. New York: Augustus M. Kelley.

MARCUSE, P. (1972) "Homeownership for low income families: financial implications." Land Economics 48: 134-143.

MOLLENKOPF, J. and J. PYNOOS (1972) "Property, politics and local housing policy." Politics and Society 2: 407-432.

Mortgage Bankers Association (1973) "Statement before the Subcommittee on Housing and Urban Affairs." U.S. Senate Committee on Banking and Currency, Ninety-third Congress, First Session, 1973 Housing and Urban Development Legislation, Part 1. Washington, D.C.: Government Printing Office.

National Resources Committee [of the United States] (1937) Our Cities: Their Role in the National Economy. Washington, D.C.: Government Printing Office.

STERNLIEB, G. (1969) The Tenement Landlord. New Brunswick, N.J.: Rutgers University Press.

--- and R. W. BURCHELL (1973) Residential Abandonment. New Brunswick, N.J.: Rutgers University, Center for Urban Policy Research.

STONE, P. A. (1973) The Structure Size and Costs of Urban Settlement. London: Cambridge University Press.

U.S. Congress (1973a) The Central City Problem and Urban Renewal Policy. A study prepared by the Congressional Research Service, Library of Congress, for the Senate Committee on Banking Housing and Urban Affairs, Ninety-third Congress, First Session. Washington, D.C.: Government Printing Office.

--- (1973b) Housing in the Seventies. Hearings on Housing and Community Development Legislation, Part 3, House of Representatives Subcommittee on Banking and Currency, Ninety-third Congress, First Session. Washington, D.C.: Government Printing Office.

--- (1972a) Competition in Real Estate and Mortgage Lending. Hearings before the Subcommittee on Antitrust and Monopoly, Senate Committee on the Judiciary, Ninety-second Congress, Second Session. Washington, D.C.: Government Printing Office.

--- (1972b) Report on National Growth. Report of the President's Domestic Council Committee on National Growth. Hearings before the Subcommittee on Housing on "National Growth Policy," House of Representatives, Committee on Banking and Currency, Ninety-second Congress, Second Session. Washington, D.C.: Government Printing Office.

--- (1969) Finance of Inner City Housing. Hearings before the Ad Hoc Subcommittee on Home Financing Practices and Procedures, House of Representatives, Committee on Banking and Currency, Ninety-first Congress, First Session. Washington, D.C.: Government Printing Office.

--- (1967) A Study of Mortgage Credit. Subcommittee on Housing and Urban Affairs, Senate Committee on Banking and Currency, Ninetieth Congress, First Session. Washington, D.C.: Government Printing Office.

4

Short-Term Housing Cycles in a Dualistic Metropolis

BRIAN J.L. BERRY

Great whirls have little whirls,
that feed on their velocity;
and little whirls have lesser whirls,
and so on to viscosity.

Lewis F. Richardson

□ IT IS NOW FORTY YEARS since Homer Hoyt's pioneering study of the first hundred years of Chicago's real estate cycle. Yet it is probably fair to say that we have still to explore and fully document the consequences for metropolitan form of the long-term cyclical behavior of metroland and property markets that he described (Hoyt, 1933). The gap is even greater when it comes to revealing analysis of shorter-term housing market fluctuations that are superimposed on Hoyt's longer-term real estate cycle.

I thus begin this paper by reviewing the pulsating rhythms of outward growth and inner transformation of metropolitan regions that have been produced by the longer-term fluctuations. New construction is seen as being linked to social change in the existing housing stock by filtering mechanisms that are manifested as "housing chains" during the longer-term boom periods. In such periods of growth, access by minority groups to improved housing supplies within dualistic housing markets is eased. It is found not unreasonable to argue that, for the majority, a boom period is one in which it is easier to withdraw to more distant communities if there appears to be a risk of minority entry to a neighborhood, producing continued residential segregation. On the other hand, during a downturn, when the housing supply becomes much more restricted, competition for the available housing stock is more likely to be accompanied not only by segregation but also by discriminatory behavior by the majority in the effort to restrict access to housing supplies by the minority. The period of the 1960s, in contrast to the 1950s, thus can be interpreted as one in which segregation in metropolitan housing markets remained profound, but in which many of the discriminatory manifestations of earlier exclusionary behavior were alleviated.

The analytic portion of the paper focuses on the shorter-term housing cycles that are superimposed upon the longer-term fluctuations. In particular, attention centers on rhythms of real estate transfers occurring month-by-month in a dualistic housing market during the upswing period of a longer-term boom, and as the boom slackens, in the period when mortgage money tightens and interest rates rise. Short-term cycles in the majority (white), minority (black), and transitional (white-to-black) segments of the housing market are analyzed for their periodicity and for their cross-correlations. Of particular interest is whether, at the frontier of racial change, the cycle of activity in the white housing market leads that in the transitional white-to-black (ghetto expansion) and the black housing market in a manner commensurate with filtering mechanisms and the assumed direction of housing chains, or whether the transitional cycle leads the white because a "push" factor is present as threatened whites flee what they perceive to be the black threat to their neighborhoods (Berry et al., 1975). An original data set assembled for Chicago for the years since 1967 is subjected to time-series analysis to help resolve the issue.

THE CHICAGO REAL ESTATE CYCLE

Hoyt looked at land booms and said they could be "traced to factors which led speculators to expect an extraordinary increase in the population within a locality in a relatively short time" (Hoyt, 1933: 368). "Speculative influences," he said, "may magnify the expected future increase beyond all reasonable possibilities" (p. 369). The result was what he termed the Chicago real estate cycle, defined as

the composite effect of the cyclical movements of a series of forces . . . which communicate impulses to each other in a time sequence . . . These cycles in the order in which they appear are the cycles of population growth, of the rent levels and operating costs of existing buildings, of new construction, of land values, and of subdivision activity [pp. 269, 372].

This analysis led him to conclude that the precipitating event was a spurt in population growth, which, because the housing supply cannot be increased immediately, causes the following sequence of events (Hoyt, 1933: 377 ff.): gross rents begin to rise rapidly and net rents even more rapidly, with the consequence that selling prices of existing buildings advance sharply, producing an upsurge in new construction, supported by easy credit and shoestring financing that swells the construction rate, which absorbs vacant land and produces a land boom whose consequences include a range of optimistic growth forecasts in the bullish atmosphere, and stimulates subdivision activity on the periphery, as well as lavish expenditures for public improvements. These together, Hoyt said, lead to a peak with all the real estate factors running at full tide.

But then a lull follows because speculative activity will push construction and subdivision activity far out in advance of real growth, and foreclosures will increase on the most tenuously-financed and risky ventures. There will then be an accompanying lag in the stock market and business generally that will produce further attrition, increasing vacancies so that financial institutions then reverse their liberal lending policies, and a period of stagnation and foreclosures will ensue.

SPATIAL CONSEQUENCES OF THE HOUSING CYCLE

The evidence indicated, Hoyt continued, that it takes four to five years for the wreckage of the collapse to be cleared away, and only then is the time ripe for a new upswing of activity. Five cycles of boom and collapse were identified in Chicago's first hundred years between 1830 and 1933, averaging eight to ten years on the upswing and a similar period from peak to bottom. The spatial consequences were the creation of a ring-like city consisting of successive growth bands separated in *time* from previous bands by the downturn decades of lagging activity in the real estate market, in *architectural form* by the building styles favored during the successive building booms, and in *location* by the concentration of each new flurry of growth on vacant land at the periphery.

During an upswing, the different growth bands are linked by population movements and transformed socially by filtering. Filtering takes place, according to Kristof (1972) when new construction exceeds the rate necessary to house normal growth, thereby exerting a downward pressure on rents and prices of existing housing, permitting lower-income families to obtain better housing relative to their existing quarters. This rise in housing quality should be independent of the effects of any increases in the general level of incomes or of changes in rent-income ratios. Moreover, decline in quality should not be forced by reduction in maintenance and repair expenditures because rents and prices are forced downwards. Finally, a mechanism (such as abandonment) should exist to remove the worst housing from the market without adversely affecting rents and prices of housing at the lowest level.

What is implied is a relationship between new construction, housing turnover, and the improvement of housing that has been documented in several recent studies of housing chains. Some occupants of new homes vacate older homes that then become available to others who, in turn, vacate still older houses, and so on in sequence of moves that permit *upward filtering of households* along the scale of property values and *downward filtering of housing units* along the income scale.

Each ten new units that are built permit, in all, between 25 and 30 families to make voluntary and presumably more satisfactory adjustments in their housing circumstances—10 into the new units, and 15 to 20 into units made available by the ensuing housing turnover (Kristof, 1965; Lansing et al., 1969). Those moving do so for one of four reasons: (a) residential readjustments within the local housing market due to life-cycle changes and new family formation; (b) immigration into the housing market from other regions; (c) upward mobility due to job and income changes; and (d) racially-related relocation.

About half of all the moves of American households are made possible by new construction and the housing chains they set in motion. The other sources are twofold: (a) breakup of the family due to death, divorce, or separation, and (b) migration from the local housing market to other markets in other regions.

FILTERING IN DUALISTIC HOUSING MARKETS

A dualistic housing market is one in which there is residential segregation by race, and in which a white majority preempts the outlying areas of new construction and existing zones of superior residential amenity, while the black (and/or other) minority is left the existing housing stock, usually within the central city, and frequently in the zones of greatest environmental risk. As Kristof (1972) points out, this means that two-thirds of all the moves in white suburbia originate with new construction, while most of the housing available to blacks originates from turnover in the existing stock.

We can examine the data for Chicago, which are revealing in this respect. During the building boom of the 1960s, 481,553 new housing units were built in metropolitan Chicago. In 1970, 257,590 of the new units were occupied by white homeowners and 146,029 by white renters, another 13,849 by black homeowners and 27,153 by black renters, and 27,934 of the units were vacant. It was because the number of households in the metropolitan area in the decade increased by only 285,729—a ratio of 1.7 additional housing units for each additional family—that a massive chain of housing moves was initiated down the chain of housing values and progressively inward from the suburbs to the core of the city. The consequences were that many families were enabled to improve their housing condition dramatically during the decade, downward pressure was exerted on the prices of older housing units, and discriminatory pricing—blacks paying more than whites for identical units—was eliminated. Not only was there a dramatic improvement in the housing condition of Chicago's central-city minorities, as over 128,000 units were transferred from white to black occupancy, but 63,000 of the worst units in the city could be demolished at the time that tens of thousands of additional undesirable units were being abandoned. The Chicago region thus provides a classic example of filtering mechanisms at work during a period of housing boom, contrasting markedly with the experience of the three previous decades, when whites fought aggressively to constrain minority access to the housing supply and, as a result, confined the city's minority to the worst housing and so restricted the supply that blacks in fact *did* pay more.

The evidence for the 1960s is provided in Tables 1-3. The massive growth of the suburban housing stock is shown in Table 1. In the decade, 352,057 new housing units were constructed, largely for whites. In the entire six-county suburban area, only 4,188 out of 223,845 new homes were sold to blacks and only 3,712 out of 111,290 new apartments were rented to blacks. In addition, some 3,208 blacks purchased homes previously owned by whites, and 2,153 blacks moved into apartments previously rented by whites. In contrast to the net increase of 287,000 white families in suburban Chicago, only 13,261 new black families were able to obtain residence in suburbia, and many of these residences were in or contiguous to suburban "mini-ghettos." A continuing pattern of white exclusivity thus limited minority access to suburbia.

Contrast this picture with that of the central city shown in Tables 2 and 3. There was net decline of 41,500 white homeowners and 76,900 white renters in the central city in the 1960-1970 decade as whites departed to the "safety" of segregated suburbia, or left the Chicago region altogether. Net increases in the central-city black population consisted of 37,669 new homeowners, more than doubling black home ownership in the decade, and 43,708 new renters.

TABLE 1
CHANGES IN THE SUBURBAN HOUSING STOCK, 1960-1970

	1960 Stock	New Construction, 1960-1970	Withdrawn from Stock, 1960-1970	Net Change in Stock, 1960-1970	1970 Stock
Total housing units	812,652	352,057	83,980	+268,077	1,080,729
Occupied housing units	740,508	335,135	29,841	+305,294	1,045,802
Owner occupied	561,170	223,845	26,275	+197,570	758,740
White owners	555,480	219,657	34,998	+184,669	740,179
Black owners	8,690	4,188	(3,208)[a]	+ 7,396	16,086
Renter occupied	176,338	111,290	576	+110,714	287,052
White renters	167,964	107,578	4,859	+102,719	270,683
Black renters	8,374	3,712	(2,153)[a]	+ 5,865	14,239

a. Net increase over new construction due to transfer of units from white to black occupancy.

TABLE 2

CHANGES IN THE CENTRAL CITY'S HOUSING STOCK, 1960-1970

	1960 Stock	New Construction, 1960-1970	Withdrawn from Stock, 1960-1970	Net Change in Stock, 1960-1970	1970 Stock
Total housing units	1,214,598	129,496	134,988	− 5,492	1,209,106
Occupied housing units	1,157,409	118,484	138,039[a]	−19,555	1,137,854
Owner occupied	396,727	33,745	34,115[a]	− 370	396,357
White owners	360,117	24,084	65,609	−41,525	318,592
Black owners	36,610	9,661	(28,008)[b]	+37,669	74,279
Renter occupied	760,682	84,739	103,930[a]	−19,191	741,491
White renters	564,029	61,298	138,220	−76,922	487,107
Black renters	196,653	23,441	(20,267)[b]	+43,708	240,361

a. These 138,039 units were demolished in the decade. Of the demolitions, 63,000 were in areas occupied by black residents in 1960, and 75,000 in white areas.

b. Net increase over new construction due to transfer of units from white to black occupancy.

The complex dynamics of white-to-black filtering were as follows: some 128,829 units were transferred from white to black occupancy, allowing net increases over new construction of 28,008 in black home ownership and 20,267 in black rental of good-quality flats and apartments. In addition, 63,000 black families were able to move into better-quality housing from dilapidated and other units that were demolished in the decade within the area of the 1960 ghetto. And finally, there was a net increase by 1960 of 17,554 units vacant in the black residential area of 1970, contributing to abandonment.

Thus, to the extent that people have been able to participate in Chicago's private housing market, the normal market mechanism has delivered increasing supplies of improved-quality housing and has both permitted and impelled the demolition of the worst housing in the worst neighborhoods. To the extent that unmet needs remain,

TABLE 3

DYNAMICS OF CHICAGO'S DUAL HOUSING MARKET

	White Market	Black Market
Occupied housing units in 1960	924,146	233,263
New construction, 1960-1970	85,382	33,102
Demolitions, 1960-1970	75,000	63,000
Housing stock in 1970 of 1960 market areas	934,528	203,365
Housing stock in 1970 of 1970 market areas	805,699	314,640
Transfers from white to black market, 1960-1970	(−128,829)[a]	+111,275[a]

a. Difference between these figures represents a net increase in vacancies in black residential areas of 17,554 units by 1970, a growing surplus of property associated with abandonment.

they are limited to those households excluded from normal housing market channels through poverty, age, or infirmity, or prevented from exploring a full range of options in spite of ability to pay because of the stranglehold that dualism still imposes on Chicagoland's suburban housing stock.

There may be two aspects of the experience of the 1960s that differ from those of earlier boom periods in the Chicago real estate cycle, however. In previous booms, emigration of the middle class to the new suburbs was complemented by immigration of new groups into the inner city. But in the last decade, the movement has left behind large masses of older obsolete housing no longer reoccupied by new immigrants. Immigration of rural blacks and of the Spanish-speaking population has been insufficient to take up the slack, and central city densities have been thinned.

Secondly, price decreases appear to have been so severe that disinvestment and deterioration through decreased maintenance also appear to have resulted, producing an accelerating and more contagious form of abandonment. In consequence, concentrations of welfare families usually are found in neighborhoods where the abandonment process has reached significant levels. Other characteristics are high unemployment and high levels of female-headed households with children. School problems are frequent and tend to accelerate the outflow of previous residents. Finally, drug addiction, crime, and vandalism are at high levels. The concentration of such problems among a significant proportion of the population of any given neighborhood has become the identifying mark of "collapsed neighborhoods" from which the working poor—black as well as white families, with children—have fled or will refuse voluntarily to seek living quarters. These concomitantly have been areas in which public services have deteriorated or vanished, whether as a precursor or as a consequence of the foregoing developments.

Thus, the combination of loosening housing markets, outflow of white middle-class population, and the low incomes of nonwhite successors, have resulted in deterioration and ultimately abandonment. By and large, the turnover of housing has permitted lower-income segments of the community to inherit better housing (Table 3). But the consequences of two decades of high-level output at a rate of 1.7 units of new construction for each net addition to households has become clearly visible through the filtration of large segments of the housing stock right out of the housing supply through abandonment and subsequent demolition (Kristof, 1972; Berry et al., 1975).

BRIAN J.L. BERRY [173]

THE SHORT-TERM HOUSING CYCLE AND
THE METABOLISM OF THE METROPOLIS

Superimposed upon the successive periods of boom and collapse of the Chicago real estate cycle are shorter-term fluctuations of urban activity, pulsating rhythms reflecting the metabolism of the metropolis. Some of these rhythms are seasonal, others are monthly, weekly, and daily—as for example, the ebb and flow of commuters. What is revealed about seasonal rhythms by Figures 1-4? Figure 1 provides one expression of seasonality, the regular rise and fall of natural gas deliveries by pipeline, fluctuating around the steady increase of demand taking place in a growing metropolis. Figure 2 shows one of the seasonal counter-rhythms, that of the city's airports' response to summer vacation travel. Even more remarkable is the regularity of the double-peaked rhythms of the city's marriage seasons (Figure 3). Likewise, unemployment peaks seasonally as young adults pass from school into the work force, although this rhythm is superimposed on a larger one reflecting the nation's economic health (Figure 4).

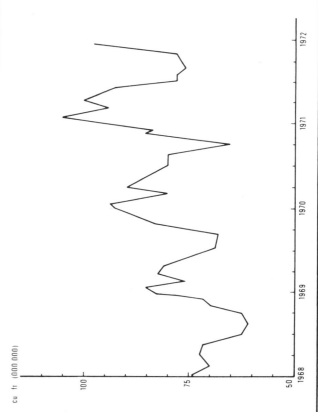

Figure 1: EXPRESSIONS OF SEASONALITY 1—MONTHLY NATURAL GAS DELIVERIES TO THE CHICAGO REGION

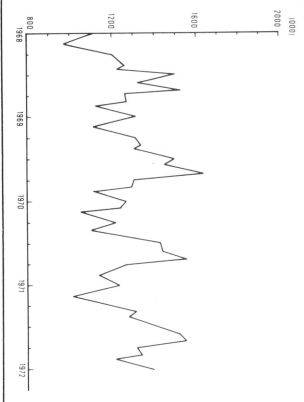

Figure 2: EXPRESSIONS OF SEASONALITY 2—NUMBERS OF SCHEDULED AIR PASSENGER ARRIVALS EACH MONTH AT O'HARE INTERNATIONAL AIRPORT

Such seasonality also characterizes the metropolitan housing market. Summer is moving season, as evidenced in Figure 5 by the fluctuations in the number of real estate transfers in Cook County; and it is the season when people decide to build new homes, as can be seen by fluctuations in issuance of building permits, also shown in Figure 5. Consistent with notions of filtering, lagging behind each moving season are the cyclical variations in the residential vacancy rate, peaking in the autumn (Figure 6).

Somewhat different rhythms from those found in the metropolitan housing market as a whole are evident in the central city, however. To obtain the time series shown in Figure 7, every real estate transfer in the city of Chicago was monitored for a period extending over several years. Included in the items recorded was such information as address, date of sale, price of the transaction, tax assessments, and name of buyer and seller. One of the summaries prepared from these data involved construction of three monthly time series, the first recording sales of single-family homes among whites within white neighborhoods (white transactions), the second sales from white to black families in changing neighborhoods (changing transactions), and the third among black families within

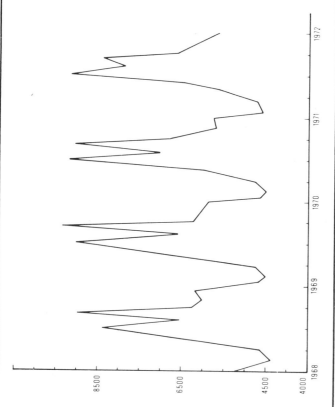

Figure 3: EXPRESSIONS OF SEASONALITY 3—MARRIAGE LICENSES IN METROPOLITAN CHICAGO EACH MONTH

the ghetto (black transactions). These three time series are graphed in Figure 7.

The graph shows cyclical upswings and downturns of housing market activity within the central city, more marked in the white than in the transitional areas and in the transitional than in the black. Yet the pulsation is more rapid than that in the metropolitan housing market as a whole (compare Figures 5 and 7). The first question to arise was thus: What is the periodicity of the central city's short-term housing market cycle? And the second question was one of sorting out a series of alternative hypotheses about the three submarkets: Do the rhythms in the different submarkets reflect the operation of filtering mechanisms, the white cycle leading transitional cycle and transitional cycle leading the housing cycle in black neighborhoods? Or do they perhaps reflect a push mechanism, black demands providing the pressure for ghetto expansion and sparking white flight—in which case the black cycle should lead the transitional and the transitional cycle the white? Or, thirdly: Do the three sub-markets simply move in concert, each reflecting a larger metabolism, neither leading nor lagging behind the others? The first question was

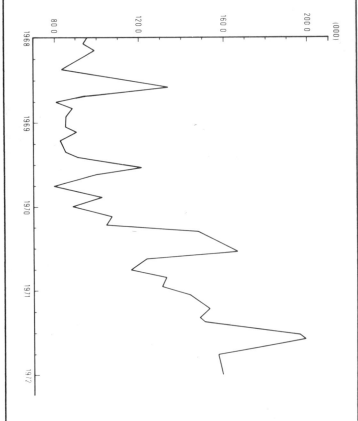

Figure 4: EXPRESSIONS OF SEASONALITY 4—NUMBERS UNEMPLOYED IN METROPOLITAN CHICAGO EACH MONTH

answered by examining the serial autocorrelation of each of the three time series. The second question was resolved by analyzing the cross correlations of the lagged series.

NATURE OF THE RHYTHMS

Figure 8 plots the autocorrelation function of each time series. A strong cyclical pattern is evident in the case of the white and transitional submarkets, but is much less characteristic of the black submarket. One question to be resolved is why the black submarket has so much weaker a seasonal metabolism.

The passage from peak to valley takes 3 to 4 months, as does the rise from a valley to the next peak, so that the cycle has a 6 to 8 month peak-to-peak periodicity. This contrasts with the cycle for the metropolitan housing market as a whole, which pulses at half that pace. Figures 5 and 7 reveal that instead of the midsummer peak that characterizes the metropolitan housing market, the city's real estate

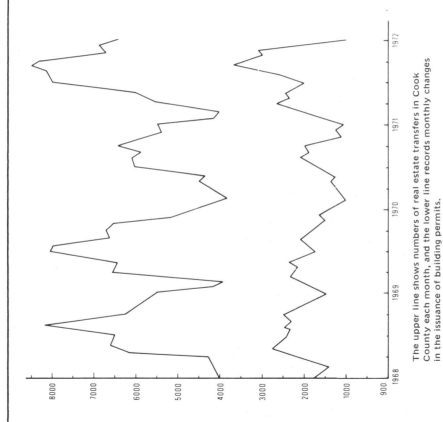

The upper line shows numbers of real estate transfers in Cook County each month, and the lower line records monthly changes in the issuance of building permits.

Figure 5: CYCLICAL ACTIVITY IN THE CHICAGO HOUSING MARKET

cycle peaks in spring and autumn and shows both midsummer and midwinter lags. No one I have questioned has any explanation why the central city's housing markets pulsate in this way, twice as rapidly as those in the suburbs, although it may relate to the stronger linkages between the rental and ownership markets in the central city, while the greater proportion of moves takes place between metropolitan areas in the suburbs and extends over the entire summer.

THE LEADS AND LAGS

Given the double-peaked central-city cycles, do the three sub-markets move in unison or do pushes and pulls create distinctive

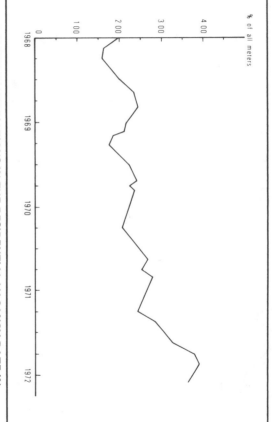

Figure 6: MONTHLY VARIATIONS IN THE RESIDENTIAL VACANCY RATE IN CHICAGO, AS INDICATED BY THE PERCENTAGE OF RESIDENTIAL ELECTRIC METERS VACANT

lags? This question can be answered by examining Figures 9 and 10, in which the lagged cross-correlations of the time series are graphed.

In Figure 9, the relationships between the white and the transitional series are shown. The unlagged correlation of the two series is 0.85, indicating a strong synchronic rhythm. The fact that the lagged cross-correlations drop off rapidly to a maximum inverse relationship of the series at three months and rise again to cross-correlations with a high of 0.65 at seven months confirms the synchronic nature of the upswings and downturns.

The synchronic relationship is weaker between the transitional and the black series, as shown in Figure 10; the unlagged correlation of the black and transitional series is only 0.70. However, a regression of the black on the transitional series and on prevailing interest rates as a second independent variable produces an R² of 0.82, with both coefficients proving to be highly significant, compared with an R² of 0.74 for the regression of the transitional series on the white and on interest rates. This reveals both that the black submarket has been far more responsive to changes in the availability of mortgage moneys than has the transitional submarket; and further, that the synchronic correlation of the two series would have been higher if there had been no shifts in the mortgage market because, controlling for interest rates, the partial correlation of the unlagged correlation of the black and the transitional series exceeds the simple correlation.

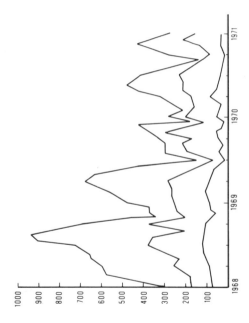

The upper line shows monthly variations in real estate transfers in the white residential area of the city; the second line charts similar trends in the ghetto expansion area, thus involving real estate transfers from white to black; and the lowest line charts real estate transfers within the ghetto.

Figure 7: CYCLICAL ACTIVITY IN THE CITY OF CHICAGO'S HOUSING SUBMARKETS

Turning now to the dimensions of the cross-correlations, to the extent that either the white or the transitional series leads the other, it is the *transitional*—for example, the correlation of the white series with the previous month's transitional series (that is, transitional lagged one month) is 0.54, compared with 0.39 for the transitional lagged behind the white, and so on for the next three lags. This

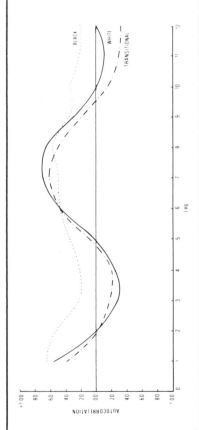

Figure 8: AUTOCORRELATIONS OF THE WHITE, BLACK AND TRANSITIONAL TIME SERIES PLOTTED IN FIGURE 7

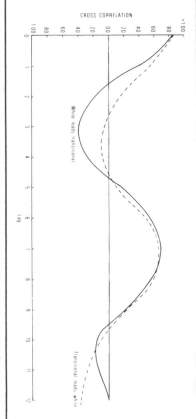

Figure 9: LAGGED CROSS-CORRELATIONS OF THE WHITE AND TRANSITIONAL TIME SERIES

would appear to indicate that in the central city, the first to move each spring and fall are most likely to be whites departing transitional areas and selling to blacks.

Comparing the lagged cross-correlations for the transitional and the black series, no such sequence is evident, however; apparently the two are quite synchronized. This would confirm the earlier conclusion: the onset of each moving season comes earliest in transitional areas where departing whites sell to incoming blacks. If the whites depart either to suburbia or the central-city apartments and the blacks come from apartments or sell homes in the ghetto, the relationships would be explained. Spring and fall are traditional moving seasons for Chicago's apartment dwellers; most leases are set to expire then. The seasonality of the central-city housing market

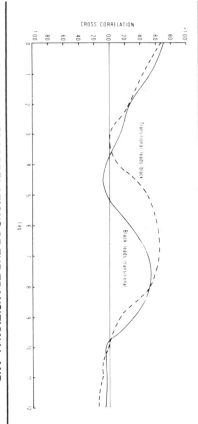

Figure 10: LAGGED CROSS-CORRELATIONS OF THE TRANSITIONAL AND THE BLACK TIME SERIES

thus reflects ingrained practices in the city's real estate industry, rather than the broader seasonal patterns found in the sales of houses in suburbia.

REFERENCES

ALBERTS, W. (1962) "Business cycles, residential construction cycles and the mortgage market." Journal of Political Economy 70, 1: 263-281.

BERRY, B.J.L. et al. (1975) "Attitudes to integration: the role of status in community response to racial change." American Journal of Sociology. In press.

BOYCE, D. E., W. B. ALLEN, A. ISSERMAN, R. R. MUDGE, and P. B. SLATER (1972) "Impact of the Philadelphia-Lindenwold high-speed line upon residential property values and zoning policies." Phase One Interim Report, Regional Science Department, University of Pennsylvania.

BRADY, E. (1963) "Regional cycles of residential construction and the interregional mortgage market: 1954-59." Land Economics 39, 1: 15-30.

DILBECK, H. (1964) "The response of local residential construction to changes in national credit conditions, 1953-59." Journal of Business 37, 3: 295-308.

FISHER, E. M. and L. WINNICK (1951) "A reformulation of the filtering concept." Journal of Social Issues 7, 1 and 2: 47-58.

GUTTENTAG, J. (1961) "The short cycle in residential construction." American Economic Review 51, 3: 275-298.

HOYT, H. (1933) One Hundred Years of Land Values in Chicago. Chicago: University of Chicago Press.

KRISTOF, F. S. (1972 and 1973) "Federal housing policy: subsidized production, filtration and objectives." Land Economics 48, 1 and 49, 2.

——— (1965) "Housing policy goals and the turnover of housing." Journal of the American Institute of Planners 31, 3: 232-245.

LANSING, J. B. et al. (1969) New Homes for Poor People. Ann Arbor: University of Michigan, Institute for Social Research.

de LEEUW, F. (1970) Time Lags in the Rental Housing Market. Washington, D.C.: Urban Institute.

MAISEL, S. (1963) "Fluctuations in residential construction starts." American Economic Review 53, 3: 359-383.

MUTH, R. (1962) "Interest rates, contract terms, and the allocation of mortgage funds." Journal of Finance 17, 1: 63-109.

SLATER, P. B. (1972) "Statistical decomposition procedures and their application to residential sales price data." Unpublished Ph.D. dissertation. Philadelphia: Department of Regional Science, University of Pennsylvania.

SMITH, W. (1958) "The impact of monetary policy on residential construction, 1948-58," in Study of Mortgage Credit. Washington, D.C.: Government Printing Office.

SPARKS, G. (1967) "An econometric analysis of the role of financial intermediaries in postwar residential building cycles," in R. Feber (ed.) Determinants of Investment Behavior. New York: National Bureau of Economic Research.

SUMICHRAST, M. and N. FARQUHAR (1967) Demolitions and Other Factors in Housing Replacement Demand. Washington: Homebuilding Press of the National Association of Home Builders.

WHITEHAND, J.W.R. (1973) "Urban-rent theory, time series, and morphogenesis." Area.

——— (1972) "Building cycles and the spatial pattern of urban growth." Transactions of the Institute of British Geographers 56 (July): 39-56.

WINGER, A. R., Discussion Papers. Program in the Role of Growth Centers in Regional Economic Development, University of Kentucky:

No. 22—"Short term activity in residential construction markets: some regional considerations."

No. 24—"Residential construction, acceleration and urban growth."

No. 25—"Residential construction and the urban housing adjustments of the migrating household."

No. 26—"Mover origin, residential construction and the urban form."

5

Socioeconomic Transition and Housing Values: A Comparative Analysis of Urban Neighborhoods

HUGH O. NOURSE
DONALD PHARES

MILIEU FOR THE STUDY OF

URBAN CHANGE

☐ A POPULAR AREA OF ACADEMIC INQUIRY of late is the phenomenon of neighborhood transition. Every major urban area is experiencing significant upheaval in existing social and economic patterns. Such changes elicit a wide array of responses within both the public and private sector. Taken en masse, they offer a scenario of urban dynamics that is confusing at best.

One facet of transition that is focal to issues of urban development is the blight-decay-abandonment cycle. This cycle of neighborhood decline is most closely manifested in the prevailing value of property in a community. Shifts in the desirability of a community as a place of residence can be linked to relative and absolute shifts in its value status. A viable, new, expanding area benefits from an upward value

AUTHORS' NOTE: *We would like to acknowledge financial support from the Department of Housing and Urban Development. Full responsibility must, of course, rest with the authors.*

[183]

trend while an older area first encounters stabilization and then falling values. This value cycle has been incorporated into guidelines for appraisal procedures and neighborhood analysis (AIREA, 1964: 87; Andrews, 1971: chapter 5).

A moment's reflection on what has been offered by professionals and academics alike shows that severe weaknesses exist in the methods employed in deciphering neighborhood decline-abandonment and its impact on property values. Numerous explanations have been put forth, and some tested empirically, as determinants of the change in value of residential property. These have run the full gamut from race, air pollution, and highways to open space, various public services, and the age of the housing stock.[1] This last factor—age—is often blamed single-handedly for the blight-decay-abandonment of an area (Muth, 1969: 116).

Lowry (1960) has challenged whether age, per se, must inexorably elicit property deterioration and depreciation in value. With proper maintenance and necessary improvements, he noted, the quality of any structure could be maintained and need not suffer economic obsolescence. Dwellings deteriorate in quality because the income associated with the property begins to fall. If the occupant is the owner, lower income leaves fewer resources for maintenance and improvements. If he is a renter, less is available for rental payments and the owner receives a decreased income flow. In order to maintain a certain level of profit, he must cut back on expenditures for maintenance and improvements. Lessened upkeep becomes an economically rational adjustment to a decline in the income associated with a particular property. Since property value represents the discounted present value of earnings from property, value will decline in response to lower income.

What we propose to demonstrate is that income is the prime factor, rather than age or race, in determining property value and that income transition spreads from spatially contiguous areas in which income has already begun to decline. We are hypothesizing that a contiguous neighborhood with lower income than a given neighborhood influences the trend of income and therefore property values in that neighborhood. The influence is one of proximity and the expansion of a path of change.

Our purpose is to show that shifts in the income status of neighborhoods do indeed become manifested in property values and that such change is much more likely to occur in proximity to income transition areas, regardless of the age of the housing stock or race of the residents. We use the St. Louis area as a study base.

AN ARBITRAGE[2] MODEL OF
URBAN TRANSITION

In the context of urban renewal and zoning, Martin Bailey (Bailey, 1959; Muth, 1969: 130-135) has developed a demand model depicting the conversion of housing from high- to low-income occupancy. While his model is an attempt to explain block-busting in segregated housing markets, it can be employed in a much broader context by noting that housing for the poor is fixed in supply and spatially segregated and that the private market does not build new housing for the poor. Stringent building codes push the price of new housing beyond the reach of the poor and many middle-income families. Their only source of shelter from the private sector is the extant stock of housing.

Market separation is further intensified by a prevalent tendency to live with others of similar income, race, or ethnic background. For illustration, consider a city with housing, identical except for location, occupied by two distinct groups of people—rich and poor. People with high incomes prefer to live in exclusive neighborhoods with a majority of other high-income families. Poor people prefer to live in neighborhoods adjacent to high-income families. This results in four prices for property—the price of housing for the rich in exclusive neighborhoods, one for the rich in neighborhoods adjacent to poor people, one for the poor in neighborhoods exclusively occupied by other poor, and one for the poor in neighborhoods adjacent to rich people.

The boundary separating these markets moves one way or the other, depending upon whether it is profitable to convert boundary property from high to low income or vice versa. Prices in turn depend upon the available stock of housing relative to growth in the number of households in each group.

We need to account for shifting demand (due to factors such as migration, birth, marriage, and death) relative to shifting supply (due to factors such as demolition, destruction, and new construction) for each group. During the last couple of decades, immigration of poor into large metropolitan areas has been high. In addition, the stock of low-income housing has been reduced through condemnation for building code violations, new highways, urban renewal, public housing, arson, and undermaintenance leading to abandonment.[3] Additions to the housing stock for the poor have been substantially less than demolitions. This growth of demand relative to supply exerts an upward pressure on the price of housing, making the conversion of housing for the rich to housing for the poor more profitable.

As the boundary moves, housing formerly worth a premium due to its location adjacent to high-income families no longer elicits a premium; the income associated with the property declines. This exerts downward pressure on property values. Although it is true that there is no necessity for property to deteriorate with age, if income falls there will be fewer resources available for normal repairs. Maintenance is reduced because other property expenses remain relatively fixed (e.g., taxes and mortgage payments). In a period of inflation, the decrease may not be absolute, although this is possible: nevertheless, value declines relative to new housing costs. Declining maintenance inexorably leads to a deterioration in the quality of the housing stock.

The end of this cycle is abandoned housing. The boundary is pushed too far, driving the price of housing in the exclusively poor neighborhood below that for rich. This drives the income on some housing units so low that it is no longer feasible to meet building code standards, tax and mortgage payments, and other fixed costs. The least-cost solution is abandonment.

If we incorporate racial factors into this market process, it is altered only slightly. Prices in exclusively black neighborhoods would eventually fall to a level below that for the same quality white housing. This occurs when the boundary between segregated areas shifts to the point where there is no longer a sufficient demand for additional black housing. From some of the evidence in the National Urban League's (1971) survey on abandoned housing, the sequence was, in fact, from white to upper-income black and then to low-income black.

A FRAMEWORK FOR THE ANALYSIS OF NEIGHBORHOOD CHANGE

SELECTION OF STUDY AREAS[4]

Testing the arbitrage model requires selection of a set of communities located throughout both St. Louis City and County. We can then contrast the time trend in income status, property values, racial composition, and age of the housing stock across areas. The neighborhoods are purposely located at varying distances from sections of the City and County that have already undergone income transition. Each area is identified in the map constituting Figure 1.

Neighborhoods were chosen to accomplish several purposes. First, five areas were defined to portray a range of experiences within the City of St. Louis. Second, an attempt was made to examine a ring of inner-suburban communities at varying degrees of proximity to income transition in the city. The areas chosen in the county ring the city nearly completely. Third, the age of the areas varies considerably, as do both the level and change in income, the level and change in racial composition, and proximity to transition. Median-year-built ranges from 1889 to 1952; income decile rank, from 2 to 10 with changes up and down by as much as 4 ranks; racial composition varies between 0 and 100% nonwhite with changes from 0 to 94%; and proximity to income transition, between "immediate" and "far removed." Variation in the critical dimensions of age of housing, income, racial composition, and proximity to shifting income status are well represented by these fifteen communities. Finally, the study areas enable us to examine specifically the impact of contiguity to low income on a given area. We can, therefore, investigate the influence of proximity to changing income on community housing values.

If age or race were, in fact, more important factors in determining housing values than income, we would expect, generally, property values in older or racially changing areas to drop more and more rapidly than values in other areas. Our purpose is to address this question directly.

DATA AND METHODOLOGY

To undertake the task set out above, we must contrast time trends for income, property values, race, and age across communities. Within each community a sample of residential structures was randomly selected from the stock of housing existing in a year when the area was nearly completely developed. Using these as observations, the occupation-income history of each unit was traced over the time period 1926-1930/1970-1971.[5]

While federal census data include estimates of median family

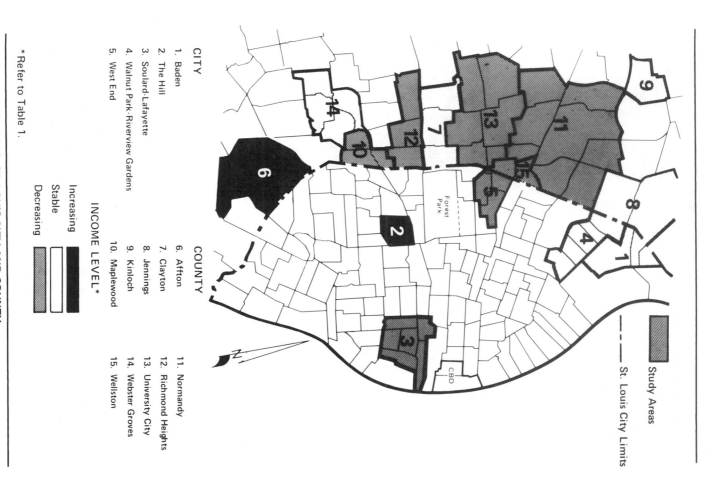

Figure 1: STUDY AREAS—ST. LOUIS CITY AND COUNTY

* Refer to Table 1.

CITY

1. Baden
2. The Hill
3. Soulard-Lafayette
4. Walnut Park-Riverview Gardens
5. West End

COUNTY

6. Affton
7. Clayton
8. Jennings
9. Kinloch
10. Maplewood

11. Normandy
12. Richmond Heights
13. University City
14. Webster Groves
15. Wellston

INCOME LEVEL*

Increasing
Stable
Decreasing

Study Areas

St. Louis City Limits

income, they are available only at ten-year intervals. To develop a more continuous data series on community income level, we used occupation data contained in the R. L. Polk Directory for St. Louis City and County (Polk, various years). While this source is not available every year, it was published every second or third year prior to 1960 and has been published yearly ever since.

Having selected a random sample of dwelling units in each of fifteen communities, we proceeded by looking up these units in a Polk Directory for every year it was available. Information on the occupation of the head-of-household and changes in occupancy were recorded and converted into an income decile rank for the median earnings of the occupation. An occupation's income decile rank was derived from federal census data for the years 1939, 1949, and 1959 (Miller, 1966: 96-97). Each occupation was assigned a decile rank from 1 to 10—the lowest 10% of the wage earners a 1, the highest 10% a 10. The rank for the nearest available census year was given to an occupation recorded between census years.

By converting an occupation to an income decile, we avoid the problem of small changes in median income indicating a change in our estimate of neighborhood income status. Furthermore, this measure is only an approximation of income status, and the decile rank seems more reasonable and stable than median dollar income. Housing value is more a function of permanent income than the vagaries of current income. Our decile rank is more representative of status or permanent income.[6] Since computation of the decile rank is based on median wages and salaries, we make the assumption that nonwage income is proportional to wage income.

The end product is a set of time series data on the occupation-income decile rank of the head-of-household for approximately 300 dwelling units in each of our fifteen communities.

Employing these data we were able to construct a measure to relate the process of neighborhood income transition (Guy and Nourse, 1970; Nourse, Phares, and Stevens, 1973). The median income decile rank was determined for *each* neighborhood for *each* year data were available, based on our random sample of dwelling units. This measure of income transition provides a sound basis for relating the changing income make-up of a neighborhood to the value of property.

To complete the analysis, we needed data on housing prices; two sources were considered. First was that available from Federal Revenue Stamps affixed to property deed transaction.[7] The reliability of these data has been found to be quite high (Tonty et al., 1954). These tax stamps were no longer required by federal law after 1967 and the State of Missouri, unlike many other states, did not pick up their use. Since post-1967 value information is important, it was decided not to use this source.

It was decided instead to rely on median housing value as reported in the decennial federal censuses for 1940, 1950, 1960, and 1970. The obvious difficulty is that these data are much more discrete than desirable. They do, however, provide an index based on *all* housing in a neighborhood, not just that changing hands.[8] While the discreteness of the data prevents us from making precise or definitive statements about the timing of change in value relative to changes in income, we can trace the time path of income transition across study areas and relate this to observed variations in value.

SOCIOECONOMIC CHANGE AND NEIGHBORHOOD HOUSE VALUES

INTRODUCTION

In examining the role of income in determining values, it is convenient to look at neighborhoods in toto when they cluster in direct line with the path of low-income movement. The remaining areas can then be discussed individually in the context of their particular attributes. Several questions will be addressed in the discussion to follow. First, does age of housing or race of resident relate to value? Second, is income transition associated with value? Finally, does the arbitrage model adequately portray the process of neighborhood transition?

What we are considering is the proposition that changes in housing values in an area are not predominantly a function of age or race. Rather, values respond to income in a particular way. As low income spreads in a wave-like manner—out from the City and into the County—this elicits a response in the housing market. In essence, there becomes less income available to effectuate the repairs, improvements, and upkeep necessary to maintain property. The inescapable end result is a drop in value and a commencement of the cycle of blight-decay-abandonment.[9] Transition is most intimately

TABLE 1

INCOME TRANSITION BY NEIGHBORHOOD (1926-1971)

Income Status (decile rank)	Neighborhood[a]
Increasing (1-5)	The Hill
Increasing (5-7)	Affton
Stable (9-10)	Webster Groves
Stable (9-10)	Clayton
Stable (2)	Kinloch
Stable (6-7)[b]	Baden
(5-6)	Walnut Park-Riverview Gardens
(6-7)	Jennings
Decreasing (5 to 3)	Soulard-Lafayette
Decreasing (7 to 4)[b]	Maplewood
(9 to 7)	Richmond Heights
Decreasing (8 to 4)[b]	West End
(6 to 4)	Wellston
(8 to 6)	Normandy
(8 to 7)	University City

a. Refer to Map 1 for their location and spatial proximity to each other.
b. Treated as a "cluster" in the textual discussion.

linked with proximity to other areas that have already experienced falling incomes.

Given the sizable number of areas examined, it is convenient to begin by identifying those areas with similar income transition experiences. They can then be grouped spatially to examine the influence of proximity. Such a procedure produces a clustering of neighborhood areas as shown in Table 1.

What is found are three distinct spatial "clusters" of neighborhoods, one stable and two declining in income, comprised of nine individual communities, and six others that are spatially isolated and vary across the full range of income change from decreasing to stable to increasing. Table 2 summarizes the historical experience of each community between 1926 and 1971.

TRANSITION IN URBAN NEIGHBORHOODS[10]

Soulard-Lafayette

The most spatially isolated neighborhood is Soulard-Lafayette, located in one of the oldest parts of the City, with the median year for housing construction being 1889. Currently, it has a great many severely blighted and abandoned structures, but there are recent signs of attempts at rehabilitation.

TABLE 2

AGE OF HOUSING STOCK AND INCOME CHANGE BY NEIGHBORHOOD

Neighborhood	Median Year Built	Period of Change in Income[a]	Income Change (decile rank)	
			From	To
St. Louis City				
Baden	1932	1936-37	5	6
		1939-40[b]	6	7
		1952-55[b]	7	6
The Hill	1910	1941-42[b]	1	2
		1946-48[b]	2	3
		1963-64[b]	3	4
		1970-71	4	5
Soulard-Lafayette	1889	1936-37	5	4
		1942-44	4	5
		1952-55	5	4
		1966-68	4	3
Walnut Park-Riverview Gardens	1923	1926-30[b]	5	6
		1952-56[b]	6	5
		1965-66	5	6
		1969-70	6	5
West End	1900-10	1952-55	8	7
		1955-56	7	6
		1958-59	6	5
		1960-61	5	4
St. Louis County				
Affton	1952	1934-36	5	6
		1941-43[b]	6	7
Clayton	1929	1939-41	9	10
		1953-59[b]	10	9
		1962-63	9	10
Jennings	1941	1941-43	5	7
		1953-55	7	6
Kinloch	1940	1953-55	3	2
Maplewood	1920	1959-61	6	5
		1970-71	5	4
Normandy	1929	1930-32	8	7
		1966-67	7	6
Richmond Heights	1926	1928-30	9	8
		1941-43	8	9
		1953-57[b]	9	8
University City	1938	1943-46	8	9
		1953-55	9	8
		1961-62	8	7
Webster Groves	1920	1934-36	9	10
		1961-67[b]	8	7
		1968-69	9	10
Wellston	1918	1938-53	4	5
		1959-61	6	5
		1961-69	5	4

a. This is when a change in income took place during the period 1926-71.

b. Period defines a trend; there may be an inversion in the decile rank for a year or two.

The income level in Soulard has fallen steadily from a peak decile of 5 to a current level of 3. Housing values increased both absolutely and relative to the metropolitan area median between 1940 and 1960 (see Tables 3 and 4). It was during this period, however, that income remained at a median level of 5. Commencing in the mid-1950s, income began to recede—reaching a decile rank of 3 by 1968. Following the drop in income, housing values responded. Soulard is the *only* area of the 15 examined in which there was an *absolute* (and significant) drop in value between 1960 and 1970. It appears as though income status prior to the mid-late 1950s was adequate to permit property in the area to be kept up. Housing values increased more than 50% between 1950 and 1960, keeping pace with the general appreciation for the entire SMSA. Once income declined too far, however, it became economically unfeasible to devote sufficient resources to property maintenance and improvements. Thus began a precipitous drop in absolute and relative values (see Tables 3 and 4) between 1960 and 1970.

TABLE 3

MEDIAN HOUSING VALUES BY NEIGHBORHOOD: 1940-1970

(in dollars)

Neighborhood	1940	1950	1960	1970
St. Louis City				
Baden	5,319	11,589	14,999	15,552
The Hill	2,456	6,751	11,429	13,047
Soulard-Lafayette	1,914	6,374	9,802	7,961
Walnut Park-				
Riverview Gardens	3,006	8,795	11,696	12,180
West End	4,371	12,659	12,800	13,019
St. Louis County				
Affton	4,223	12,112	16,037	17,803
Clayton	14,905	20,000+[a]	25,000+[a]	38,965
Jennings	3,221	9,493	12,668	13,838
Kinloch	1,196	2,310	5,300	7,825
Maplewood	4,056	9,656	11,934	12,863
Normandy	3,790	9,826	12,718	14,083
Richmond Heights	5,922	13,809	17,256	17,902
University City	8,510	15,000+[a]	18,400	18,600
Webster Groves	5,690	12,823	15,044	17,529
Wellston	2,487	5,642	7,600	7,669
St. Louis Metropolitan Area				
Total SMSA	3,856	8,919	12,900	16,300

a. Falls in an open-ended class, no median computed.

SOURCE: Compiled from Census tract data contained in *Census of Population and Housing* for 1940, 1950, 1960 and 1970. Tracts were aggregated to most closely approximate the study areas.

TABLE 4

HOUSING VALUE RELATIVES[a] AND RELATIVE RANKS[b] FOR NEIGHBORHOODS: 1940-1970 (relative ranks in parentheses)

Neighborhood	1940	1950	1960	1970
St. Louis City				
Baden	138 (5)	130 (7)	116 (6)	95 (6)
The Hill	64 (13)	76 (12)	89 (9)	80 (9c)
Soulard-Lafayette	50 (14)	72 (13)	76 (12)	49 (12)
Walnut Park-				
Riverview Gardens	94 (10)	99 (11)	91 (11)	75 (11)
West End	113 (6)	142 (5)	99 (7c)	80 (9c)
St. Louis County				
Affton	110 (7)	136 (6)	124 (4)	109 (4)
Clayton	387 (1)	—d (1)	—d (1)	239 (1)
Jennings	84 (11)	106 (10)	98 (8)	85 (8)
Kinloch	31 (15)	26 (15)	41 (14)	48 (13)
Maplewood	105 (8)	108 (9)	93 (10)	79 (10)
Normandy	98 (9)	110 (8)	99 (7c)	86 (7)
Richmond Heights	154 (3)	155 (3)	134 (3)	110 (3)
University City	221 (2)	—d (2)	143 (2)	114 (2)
Webster Groves	148 (4)	144 (4)	117 (5)	108 (5)
Wellston	65 (12)	63 (14)	59 (13)	47 (14)

a. Median for area relative to median for the entire metropolitan area.
b. The relatives for a year ranked from 1 (high) to n (low), shown in parentheses.
c. Tied ranking.
d. No median computed.
SOURCE: Computed from data in Table 3.

It should also be noted that Soulard did not experience a drastic, sudden change in nonwhite population concentration during the period of major income and value decline (see Table 5). Other areas experienced relatively far greater and more rapid racial transition during the period 1960-1970 without an absolute drop in housing values (e.g., University City, Walnut Park, and Normandy).

The Hill

The Hill, located just south of Forest Park and quite far west of the Soulard area, has been an all-white, Italian, working-class neighborhood for decades.

For a long period of time, the Hill had a decile rank in the range 1-2, defining it as a low economic status residential area. After World War II its income level appreciated steadily, reaching 5 by 1971. Property values moved in the same direction, nearly doubling

TABLE 5

PERCENTAGE OF POPULATION NON-WHITE BY NEIGHBORHOOD:
1940-1970

Neighborhood	1940	1950	1960	1970
St. Louis City				
Baden	a	1.18	a	5.86
The Hill	a	a	a	a
Soulard-Lafayette	8.99	10.65	11.47	28.76
Walnut Park-				
Riverview Gardens	a	a	a	34.53
West End	1.42	1.30	59.78	95.54
St. Louis County				
Affton	a	a	a	a
Clayton	3.63	2.84	1.20	1.46
Jennings	a	a	a	a
Kinloch	100.00	100.00	100.00	100.00
Maplewood	a	a	a	a
Normandy	a	a	a	15.19
Richmond Heights	7.11	6.97	6.81	12.82
University City	a	a	a	20.04
Webster Groves	13.12	14.33	8.15	10.17
Wellston	5.40	6.10	8.56	71.12

a. Less than 1%.
SOURCE: Same as Table 3.

between 1950 and 1960. Relative to the metropolitan area, the Hill has improved overall, but slipped somewhat after 1960. It has not dropped relatively by nearly as much as several other areas. In fact, its relative rank[11] has gone from 13 in 1940 to 9 in 1970. Also to be noted is the fact that it is one of the oldest residential areas in the sample: the median-year-built is 1910. Here we have an instance of older housing going hand in hand with rising income and an increase in absolute value and relative status.

While the Hill has been gaining in income status, absolute values have followed suit. Relative to the entire metropolitan area, however, it has stabilized. This implies that the association between income and values on the up side may not be symmetrical with changes downward. Values do inexorably decline in response to falling income, especially as you approach the low end of the income spectrum: they do not seem to increase in a similar manner as income rises.[12]

Affton

Affton is the newest community studied with a median-year-built of 1952. It was developed, in large part, in response to the upsurge in demand for new suburban housing after World War II. Its location, south in St. Louis County bordering on the far southern part of the city, effectively isolates it from the out-movement of low income that has become the paradigm of change in the "central" part of the city and county. Affton has remained an all-white residential area with above-average income status.

The income rank for Affton has risen from a 5-6 in the 1930s to a 7 for every year since 1941 except one. Housing values have increased over the period 1940-1970 from $4,223 to $17,803. While recent appreciation has not kept pace with the general rise in values for the entire SMSA, it has maintained its above-average position.

Compared to other communities that began at an equal or higher status relative to the SMSA, Affton has tended to stabilize. In terms of relative rank, Affton improved from 7 to 4 and remained there while others have fallen. Affton may decline even more relative to housing values for the SMSA, due to the general increase in values in the west-county, but its income status and distance from transition will prevent it from following the route of an extreme like Soulard, at least for a while.

Kinloch

Before discussing the "clusters" of communities (shown in Table 1), it will be useful to look at the three remaining individual areas that delineate extremes in income level. Kinloch is an all-black municipality located to the west and north in the county. Its existence as a black enclave for decades has exposed it to considerable scrutiny for its unique social and political characteristics (Kramer and Walter, 1968a; 1968b; 1969). While data here are less complete than for other areas, we were able to put together enough information to examine its history.[13]

A most interesting facet of Kinloch is that it has remained stable at an income rank of 2 since 1955. It offers an example of a stable community at the very bottom of the income distribution. Of even more interest is the fact that median values in the area have not dropped either absolutely or relatively, as one might anticipate. In fact, they appreciated more between 1960 and 1970 than any other area examined. Values rose nearly 48%, outstripping a 26% increase for the entire metropolitan area–this in the context of an all-black racial composition. Apparently, the stability of its income (albeit low) and its unique racial-political status have enabled Kinloch to avoid the path to abandonment. While a majority of its housing stock is of low quality, it does not have the serious abandonment problem found in the central part of St. Louis City (Institute for Urban and Regional Studies, 1972; National Urban League, 1971).

At the other extreme of the income distribution are two stable, high-income/status areas, Clayton and Webster Groves. While these communities differ drastically in some respects, they share the common denominator of being older areas that have stabilized at the upper end of the income distribution.

Clayton

Clayton is a municipality with some unique attributes (Kersten and Ross, 1968). It has evolved into a displaced central business district (CBD), in direct competition with the CBD in St. Louis City, with a major concentration of office buildings and high-class shopping facilities. This affords it a very competitive position as a place to work *and* live. Next, and crucial to its stability, is the fact that Forest Park lies directly in the path of low-income migration out from the City. Clayton is fortunate enough to be spatially positioned with an extremely large natural barrier protecting it from the main thrust of low-income migration that has affected the status of many of its neighbors.

Clayton's income rank has remained at 10 for most of the period since 1962 (it has been a 10 during 17 of the 27 years for which data were available). While its housing stock is quite old–median-year-built is 1929–housing values have consistently been far above the SMSA level; median value for 1970 was nearly $39,000. Relative to the entire SMSA, Clayton has fallen from 387 to 239, yet remains considerably above the other 14 communities: it continues unequivocally as a high-status, virtually all-white residential area.

One can but speculate, but it seems apparent that Clayton's location directly west of Forest Park has protected it from the throes of income transition in contiguous areas that has so heavily influenced the cluster of communities just to its south (Richmond

Heights and Maplewood) and north (Normandy, Wellston, West End, and University City). The park has broken the contiguity and thereby the spread of low income and associated drop in housing values.

Webster Groves

Webster Groves is an old residential neighborhood, serving as a bedroom suburb for the city since before the turn of the century, located far south in the county. Median-year-built for its housing stock is 1920, ranking it one of the oldest communities examined. Its income status, like that of Clayton, has remained high; it is currently a 10 (it has been a 9 during 19 of the 27 years recorded, but a 10 for 7 of the past 9).

Housing values have appreciated over the period 1940-1970, although it should be noted that Webster has never attained the position of Clayton. Relative to the SMSA median, Webster lost ground, falling from 148 to 108. But compared to those areas in line with the main thrust of low income, it has not fallen nearly so far. University City and Richmond Heights were previously higher, but both have fallen relatively more (221 to 114 and 154 to 110) due to their proximity to income transition. Also, absolute values in Webster have continued to rise while those in University City and Richmond Heights have virtually stabilized. As found with both Clayton and Affton, distance from the main thrust of income decline has enabled Webster to hold onto its income status and maintain its housing stock, despite the age of its housing.

The results thus far indicate that neither race nor age is the sine qua non in determining value that has so often been assumed. Proximity to income transition emerges as a basic motive force behind absolute and relative shifts in housing values and community status. This will become more clearly defined as we shift attention to the three clusters of neighborhoods (refer to Table 1).

Baden, Walnut Park-Riverview Gardens, Jennings

The first cluster—Baden, Walnut Park-Riverview Gardens, Jennings—lies in the extreme northern part of the city and bordering the county. The two areas located in the city are relatively older, with a median-year-built of 1923 in Walnut Park and 1932 in Baden. Jennings, on the other hand, is a newer suburban area developed mainly after World War II (median-year-built is 1941).

None of these communities has been absolutely stable over the entire study period, but they have settled around the mid-range of the income distribution (5-7). Stability is related to their location out of the main thrust of downward income transition to their south. Values in each area have risen over the period 1940-1970 with the largest absolute increase taking place in Jennings, the area farthest removed from low-income concentration in the city.

Relative to the SMSA, Baden has fallen the farthest from 138 to 95, Walnut Park comes next dropping from 99 to 75, while Jennings has settled back to near its 1940 level of 84, peaking at 106. Even more descriptive of what has occurred is the shift in relative status of these communities over the period 1940-1970. Baden dropped from 5 to 7 and then rose back to 6; Jennings went from 11 to 8; while Walnut Park dropped from 10 in 1940 to 11 in 1950-1970. Stability in income has had its impact on housing values. Jennings, the farthest removed from income transition, remained stable relative to the SMSA median and rose in relative status. Baden and Walnut Park declined relative to the SMSA median and fell one relative rank. It appears that income stability produced stability in the relative status of these communities.

It should be noted that considerable racial transition occurred in Walnut Park; it went from less than 1% nonwhite in 1960 to almost 35% in 1970. Absolute values and relative status did not drop, however, in response. Walnut Park's relative rank is the same in 1970 after racial change as it was in 1950 and 1960 before the change. Relative to the SMSA, it fell by about as much as Baden and Jennings—both of which remain virtually all-white.

Maplewood, Richmond Heights

The second cluster of communities is situated farther south in St. Louis County. Interestingly, it is positioned between the two highest income areas, Clayton and Webster Groves. Richmond Heights and Maplewood have, in effect, served as a buffer zone, isolating Clayton and Webster from the spread of income transition. Both areas are relatively older, with a median-year-built of 1920 in Maplewood and 1926 in Richmond Heights.

Median income rank in both areas has dropped. Maplewood fell from an above mid-income rank of 7 to a 4. Not only has it moved downward but it has reached the threshold of low income that begins to seriously threaten the capacity of an area to maintain its housing stock. Next to the West End, Maplewood lost the most in income status of any community examined. Richmond Heights also fell but from a much higher level. In 1926, it was ranked in the ninth decile, placing it on a near-equal status with its neighbor Clayton. Managing to hold on to a 8-9 decile rank until the early 1960s, it then dropped and stayed at a 7. Richmond Heights has experienced transition in income status, but not to the point at which the pressure of low income on values seems to be the most intense (4 or below). Both areas have begun to manifest the impact of the encroachment of low income, especially since 1961-1962.

Housing values responded to the drop in income. Absolute value in Richmond Heights leveled off after 1960, although remaining above the SMSA median. Relative to the SMSA, Richmond Heights has fallen from 155 to 110. It continues as a higher-status residential area, but fell from a relative rank of 3 in 1940 to 4 in 1970. Maplewood, on the other hand, dropped well below the SMSA median. While absolute values have increased slightly in Maplewood between 1960 and 1970, its relative status declined. Relative to the SMSA, it went from 108 to 79, falling at the same time from 9 to 11 in relative rank.

Sharp income decline in Maplewood has exerted its influence on the relative position of Maplewood. It has fallen into the income range where values begin to suffer the most. Richmond Heights lost some of its status as a high-income area, but remains above the SMSA in median value. It has not yet lost enough in income rank to manifest the impact on values found elsewhere.

Racial composition changed slightly in Richmond Heights while Maplewood remains all-white. The percentage nonwhite in Richmond Heights has almost doubled since 1960, but only went from 6.8 to 12.8. Once again, however, changing racial composition seems to bear much less relationship to values than income. Maplewood, an all-white area, has experienced a far greater drop in relative value and status than many areas examined.

Normandy, Wellston, West End, University City

The final cluster of communities—Normandy, Wellston, West End, and University City—lies directly in line with the main thrust of low-income movement out from the city. While these are not the newest communities examined, neither are they the oldest. Median-year-built ranges between 1900 and the 1930s.

Income status began to depreciate earliest in the West End. Change began in the early 1950s with a drop from 8 to 4 taking place over the ten-year period 1952-1961 (see Table 2). This was in response to income transition that had already commenced farther east near the CBD. Next to experience the impact of change was Wellston. During the period 1934-1953 Wellston rose in median rank from 4 to 6, placing it firmly in the mid-income range. By 1959-1960, however, transition in the contiguous West End began to spread in influence. Wellston dropped from 6 to 5 between 1959-1961 and then to a 4 between 1961-1969. Low-income migration had pushed far enough west to affect Wellston.

Next to feel the effect were University City and Normandy. University City had hovered around the 8-9 decile rank until 1969 when it fell to and remained at 7. By 1969, income transition to its east and north in the West End and Wellston had spilled over into the northern part of University City. During this period, the process had already manifest itself in Normandy. With a consistent decile rank of 7 prior to 1966, Normandy dropped to a 6 in 1967 and has remained there.

The timing and movement of income change follows a clear pattern. It began earliest in the West End in response to low income farther east in the city. Then it spread to Wellston and outward into Normandy and University City. The drop in income has been the greatest in the West End (8 to 4) and Wellston (6 to 4) where the process has been in operation the longest. Normandy and University City have started to decline, but only recently and just one income rank thus far.

As anticipated, values in each area responded to the changing income status. The West End showed an absolute rise between 1940 and 1950, but then began to level off; the median has remained virtually unchanged since 1950, even though the period 1950-1960 is one during which housing values were appreciating significantly in *every* other area. In relative terms it fell from a peak well above the

SMSA median at 142 to far below it at 80. This represented a drop in relative status from 5 to 10.

Income decline took its toll first in the West End. Wellston, next in line, appreciated in absolute value between 1940 and 1960. Between 1960 and 1970, however, the impact of contiguous lower income became manifest; values virtually stabilized. In 1940 Wellston was far below the SMSA median at 65. By 1970 it had fallen to 47, a drop in rank from 12 to 14. This coincides with the onset of income change. Once again, declining income becomes translated into values and status.

The next areas to be influenced were Normandy and University City. The process of change reached them during the middle to late 1960s. Values in both areas rose over the period 1940-1960, but began to level off between 1960-1970 in University City and increase much more slowly in Normandy. Both communities dropped relative to the SMSA: Normandy fell from 110 to 86; University City, from 221 to 114. Since the fall in income has been recent, relative status has not yet been affected.

The impact of change has influenced Normandy's values by far less. The reason lies in their relative spatial setting. University City sits directly in the westward path of low-income migration out through the West End and Wellston. Normandy, on the other hand, is contiguous to the area of transition but much more of it is well removed from the immediate thrust, lying to the west and north.

Income change in this cluster has, to a large extent, gone hand in hand with a shifting racial composition. The out-movement of low income from the city contained a high incidence of nonwhites. The West End felt the major impact between 1950-1960, while the other three communities were affected later on during the period 1960-1970. Wellston had already lost what relative status it had by 1950, well before racial change began. University City ranked second over the entire period 1940-1970, while Normandy increased from 9 to 7.

The West End is the one area of the four that has been hit the hardest and the fastest. Falling four decile ranks in a ten-year period, from a high-status 8 to a near-poverty 4, has played havoc with the housing market. While the period of change is unequivocally coincidental with racial transition to nearly all-black, a fall in income of this magnitude and rapidity and to such a low level would have elicited an equivalent response in any housing market, irrespective of race or age of housing.

Two independent studies on the association between race and property values in Normandy and University City have concluded that racial transition had no impact on housing values (Phares, 1971; McKenna and Werner, 1970). Thus, while income and race are intertwined in this group of communities, evidence from this and other studies supports income as being the prime motive force behind shifts in value, not the color of resident or age of the housing stock.

SUMMARY AND CONCLUSIONS

Urban areas have experienced rapid and fundamental shifts in their form and structure since their original establishment, particularly as reflected in suburbanization and exurbanization. At the very core of these changes lies the individual neighborhood or local community from which much of the transition process emanates. Getting a firm grasp on the factors affecting neighborhoods permits development of a more credible scenario of intraurban dynamics. We have set out here to scrutinize three dimensions basic to any neighborhood in an attempt to discern the impact each exerts on an area's evolution.

The factors considered are, first, age of the housing stock. Very often one finds age singled out as a prime cause of urban blight (Muth, 1969: 116-121), manifested as declining property values and finally abandonment. Abstracting from the much too obvious instances such as Georgetown, we were interested in the relation between age of a community and the prevailing level of value, trend in value, and relative status.

A second factor often assigned a role in neighborhood deterioration is racial mixture (AIREA, 1964). "The myth of declining property values" seems to retain a tenacious hold on conventional wisdom despite a plethora of empirical analyses that has consistently shown statistical refutation of any association between race per se and property values (Phares, 1971; McKenna and Werner, 1970).

Finally, we turn to the socioeconomic evolution of a community vis-à-vis its income status. Using a demand-oriented model of the local housing market developed by Bailey (1959) and further refined by Muth (1969) and Nourse (1973: chapter 11) we were able to link the process of income transition to the value status of a neighborhood. Specifically, it was postulated that the prime mover behind urban deterioration (as manifested in housing values) is income, with its link to the relative profitability of putting the housing stock to alternative uses. Community evolution becomes a function of profit and a boundary of low-income moving, in a wave-like fashion, across an urban area, affecting one segment and then spreading into contiguous areas. Housing values respond to fluctuations in income. As low income spreads, this elicits a response in the local housing market. Values decline in response to falling income and the onset of blight and decay forewarns a suitable environment for abandonment. Low income spreads, wave-like, leaving in its wake a path of blight, decay, and abandonment.

Nowhere in this process does age or race per se play a primary role in the socioeconomic evolution of a neighborhood, either reflected by values or income. Rather, transition is most intimately linked with proximity to another area that has already experienced the throes of declining income. Needless to say, the arbitrage model is offered as a paradigm of urban transition requiring empirical substantiation. To accomplish this we selected fifteen communities located in St. Louis City and County, an urban area having a well-documented problem with neighborhood decline and abandonment (Williams, 1973; National Urban League, 1971; Institute for Urban and Regional Studies, 1972; Nourse and Phares, 1974; Little, Nourse, and Phares, 1974).

Our empirical findings support the arbitrage model as an accurate paradigm of urban transition. Low income does indeed spread in a wave-like manner: neighborhoods in the path (contiguous) are influenced. Housing values respond to variations in the income level of a community. Those suffering the greatest drop in value status were those that had undergone income transition for the longest period of time (e.g., West End, Wellston). The Soulard area epitomizes the process in its extreme.

Income decline was found to prevail in *all* communities in the wake of low-income movement. Communities that were spatially isolated from the thrust of low-income expansion or otherwise protected (for example, by natural barriers, such as Clayton with Forest Park) did not experience a drop in income or housing values. Witness, for example, the cases of Affton, Jennings, Clayton, or Webster Groves.

Not only was there a clearly defined association between income changes and value and proximity to low-income concentration, but the appropriate timing of the boundary movement can also be documented. This was found with the temporal chain of movement from the West End to Wellston, then Normandy and University City (refer to the timing of change in Table 2). In short, the empirical findings are completely consistent with the arbitrage, moving-boundary paradigm of neighborhood transition.

Several other facets of the arbitrage model also emerge from our empirical findings. It appears that the moving boundary can be interrupted, and even halted, by a natural or man-made barrier. As in the case of Clayton, Forest Park served to block the spread of low-income westward. The same condition was found to obtain *within* the Normandy area. When we examined transition within Normandy, it was found that income change occurred along the eastern border contiguous with Wellston. Further low-income movement westward from there has been retarded by a large land mass (comprised of cemeteries, country clubs, and public and private institutions) located in central Normandy (Nourse and Phares, 1974: 94-102; Nourse, Phares, and Stevens, 1973: 113-116).

Additional confirmation for the influence of barriers was found in University City. In this instance man-made barriers have retarded the spread of low income to the southern part of University City. A main thoroughfare combined with rather extensive use of private and gated streets, cul-de-sacs, and limited entry subdivisions prevent the unhindered flow of people. These barriers have protected the southern part of the community from the low income that has influenced the northern area (Nourse and Phares, 1974: 94-102).

The flow of low income also seems to be deterred when it does not exist in a mass sufficient enough to permit its expansion. Low-income migration westward out of the city into the county is the prevailing scenario of transition in the St. Louis area. The mass behind this migration is considerable and closely follows the arbitrage model. Low-income pockets in two areas within the county, however, have not spread from their original location. Kinloch has been a very low-income, black residential enclave for decades and yet it borders on much higher-income areas such as Normandy. There has been no spread of low income, with its associated impact on housing values, into contiguous areas simply because of an insufficient population mass. The same was found true for an area within Webster Groves (an area to the northwest which contained residences for servants and domestics working in Webster). Again, the mass was inadequate to produce an expansion of its impact to contiguous parts of the county.

The arbitrage model works as a description of the process of urban transition. It is a process, however, which depends upon a substantial population mass for its continuance. Further, it can be interrupted and even halted by barriers, either natural or man-made. There is also some evidence, albeit far from conclusive, that ethnic solidarity might halt its spread. This is suggested by empirical findings for the Hill, an Italian working-class enclave, that has risen in income status despite proximity to low income.

The association between income and housing values has been found to be of prime importance. The same cannot be said of age of housing or race per se. Comparison of age, race, and housing values across fifteen communities reveals one fact clearly: neither age nor race can be assigned a primal role in neighborhood deterioration. While both can be found to exist hand in hand with declining values, this is far from consistently the case. Rather, it is income that provides the motive forces behind transition and changes in the value of property. The process is predicated on proximity to low income and follows the arbitrage model outlined earlier.

NOTES

1. The literature in this area is quite extensive; see Phares (1971), Muth (1969), and Nourse (1973) for some of the relevant citations.

2. The word arbitrage is used to denote the shift in usage of the housing stock from one income stratum to another, the motivating force being relative profitability.

3. For background in St. Louis, see Institute for Urban and Regional Studies (1972).

4. Study areas in St. Louis County were defined by municipal boundaries. The one exception being Normandy, which is actually a school district encompassing 25 "postage stamp-sized" municipalities. In the city, areas were defined through a combination of socioeconomic characteristics and personal knowledge using census tract boundaries.

5. Sampling was done on an address basis. In most instances a sample of 300 was chosen. The actual sample size was above 300 in cases where an address had a multiple-dwelling structure and below when observations were lost for various reasons. Beginning and ending years vary due to the frequency with which Polk Directories were available.

6. Statistical substantiation for this can be found in the appendices to Little, Nourse and Phares (1974).

7. Use of this data source for analysis of the St. Louis area can be found in Phares (1971); McKenna and Werner (1970); and Bailey, Muth and Nourse (1963).

8. The reliability of this data source is documented in Kish and Lansing (1954) and Kain and Quigley (1972).

9. Refer to Institute for Urban and Regional Studies (1972) for detail on this cycle in St. Louis.

10. The discussion to follow revolves around the data contained in Tables 1-5; reference should be made to them as necessary.

11. Relative rank refers to the ranking from high to low of the value relatives in Table 4. It gives a *ranking* of the relative value status of each community.

12. One would, of course, need a much larger number of "increasing" areas to make any definitive statements.

13. We were unable to construct a median income decile rank prior to 1953.

REFERENCES

American Institute of Real Estate Appraisers [AIREA] (1964) The Appraisal of Real Estate. Chicago: American Institute of Real Estate Appraisers.

ANDREWS, R. B. (1971) Urban Land Economics and Public Policy. New York: Free Press.

BAILEY, M. J. (1959) "Note on the economics of residential zoning and urban renewal." Land Economics 35 (August): 288-292.

--- R. F. MUTH, and H. O. NOURSE (1963) "A regression method for real estate price index construction." Journal of the American Statistical Association 58 (December): 993-1010.

GUY, D. and H. O. NOURSE (1970) "The filtering process: the Webster Groves and Kankakee cases." Papers and Proceedings of the American Real Estate and Urban Economics Association 5 (December): 33-49.

Institute for Urban and Regional Studies (1972) Urban Decay in St. Louis. Springfield, Va.: National Technical Information Service, Document PD-209 947.

KAIN, J. F. and J. M. QUIGLEY (1972) "Note on owner's estimate of housing value." Journal of the American Statistical Association 67 (December): 803-806.

KERSTEN, E. W. and D. R. ROSS (1968) "Clayton: a new metropolitan focus in the St. Louis area." Annals of the Association of American Geographers 58 (December): 637-649.

KISH, L. and J. B. LANSING (1954) "Response errors in estimating the value of homes." Journal of the American Statistical Association 49 (September): 520-538.

KRAMER, J. and I. WALTER (1969) "Political autonomy and economic dependence in an all-Negro municipality." American Journal of Economics and Sociology 28 (July): 225-248.

--- (1968a) "Economic growth and human resources in a Negro community." Business and Government Review 9 (July-August): 5-17.

--- (1968b) "Politics in an all-Negro city." Urban Affairs Quarterly 4 (September): 65-87.

LITTLE, J., H. O. NOURSE, and D. PHARES (1974) "The neighborhood succession process." Report submitted to the Department of Housing and Urban Affairs (September).

LOWRY, I. S. (1960) "Filtering and housing standards: a conceptual analysis." Land Economics 36 (November): 362-370.

McKENNA, J. P. and H. WERNER (1970) "The housing market in integrating areas." Annals of Regional Science 4 (December): 127-133.

MILLER, H. P. (1966) Income Distribution in the United States. Washington, D.C.: Government Printing Office.

MUTH, R. F. (1969) Cities and Housing. Chicago: University of Chicago Press.

National Urban League (1971) National Survey of Housing Abandonment. New York: Center for Community Change.

NOURSE, H. O. [ed.] (1973) Effect of Public Policy on Housing Markets. Lexington, Mass.: D. C. Heath.

——— and D. PHARES (1974) "The filtering process in the inner suburbs," pp. 80-104 in Sutker and Sutker (eds.) Racial Transition in the Inner Suburb: Studies of the St. Louis Area. New York: Praeger.

——— and J. STEVENS (1973) "The effect of aging and income transition on neighborhood house values," pp. 107-119 in H. O. Nourse (ed.) The Effect of Public Policy on Housing Markets. Lexington, Mass.: D. C. Heath.

PHARES, D. (1971) "Racial change and housing values: transition in an inner suburb." Social Science Quarterly 51 (December): 560-573.

POLK, R. L. (various years) St. Louis City and County Directory. Detroit: R. L. Polk Co.

TONTY, R. L. et al. (1954) "Reliability of deed samples as indicators of land market activity." Land Economics 30 (February): 44-51.

WILLIAMS, B. A. (1973) St. Louis: A City and Its Suburbs. Santa Monica, Calif.: RAND Corporation #R-1353-NSF.

Part III

ANALYZING THE DYNAMICS OF URBAN SPATIAL CHANGES

Introduction

□ THE TITLES OF THE THREE PAPERS which constitute this sector seem to bear little in common, but common threads do permeate each of these contributions. These common threads are the forces operating in urban systems that lead, on the one hand, to residential location assignments and consequently, on the other, to the intervention of forces which make the original assignments socially undesirable or threatening and/or economically disadvantageous. In each of these papers the role of selected decision-makers is portrayed as influencing the extent to which one's locational niche remains functional—in the sense of promoting satisfaction—or becomes dysfunctional, as it weakens the individual's opportunity to find a secure position in the overall social economy of which he/she happens to be a part. While this theme is common to each of the presentations, the style, the environments, and the emphasis on the operation of quite different intervening forces results in a general submergence of their single commonality and instead focuses intensively on a particular phenomenology.

The attempt to isolate phenomena influencing spatial behavior in an urban setting is a focus of increasing interest both in North America and elsewhere, although works of the type conducted by Christian and Ley have received greatest emphasis in the United

States. In fact, cultural differences establish the research agenda and tend to dictate the direction of the research focus in a given nation. Gonen's paper, for example, at least in terms of his approach to the problem, reflects a social system that has emerged in a quite different way from that discussed by Christian and Ley. It is likewise clear that the topics chosen by these individual researchers are influenced by a set of concerns which evolve out of the environment in which they themselves are actors. Thus, the prevailing value system dictates the research agenda, and the problem focus of each of these writers is at the interface of economic and social systems. The individual writers have chosen to emphasize the role of one or more decision units in resolving the conflict associated with the allocation of resources in individual environments.

The topic which Christian has chosen to pursue is no doubt the one which has received the greatest recent attention by a variety of American social scientists, for it is of major interest to public officials, particularly big city mayors, as well as citizens whose economic security is threatened by corporate decisions. While there seem to be few in-depth studies of the hardship imposed upon blacks and other low-income groups by the decision of firms to locate in environments which are financially and socially inaccessible, it is generally assumed that this is part of an emerging national pattern. Most analyses of this problem area have tended to focus on the impact of industrial relocation on the general state of the economy of central cities rather than on its impact on specific subpopulations. The works of Kain, Mooney, Harrison and Gold represent exceptions; but even among this battery of authors, there is much disagreement.

Christian supports the position of those who contend that the black unskilled and semi-skilled worker is victimized by the relocation decision of the firm and, consequently, by the economic and social barriers which impede the worker's ability to secure improved access to new workplace locations. Critics of Christian are likely to emphasize his failure to present information which actually details the fate of those minority workers in response to a firm's decision to relocate. Nevertheless, while data on worker response are generally missing, Christian does describe with precision the actions of the firm and likewise illustrates that blacks are seldom residents in the communities in which the firm relocates. Thus, Christian's work can be added to the emerging body of literature which is attempting to specify both the economic and social dimensions of this phenomenon.

The paper by Ley, a geographer, is a nontraditional representative of his discipline. Ley, like the other writers in this section, is primarily concerned with impact of the interaction of a variety of actors in the environment on the perceived quality of life of the population residing therein. Ley's concern with the postulated relationship between the environment and the emergence of youth gangs is in the tradition of the human ecologist. But by emphasizing the role of national culture, he demonstrates that similar physical environments may or may not spawn the evolution of gangs. Thus, a Canadian city (Vancouver) where there are environmental settings possessing some of the criteria thought to promote gang genesis appears to be without gangs of the traditional type. But in the area of specifying how individuals regard the safety of their environment, the decision to utilize the environment in a given way is often based on a perceived threat. The gang phenomenon which is assumed to be a sociocultural phenomenon casts a shadow throughout its behavioral setting, but its dominance is contined to specific channels. These channels, as well as the larger gang-delineated domain, are affected in terms of both social and economic decisions rendered by all whose lives are played out within this behavioral setting. If Ley is correct, the role of the national culture should be looked at more closely to ascertain its impact on the evolution of this phenomenon.

The final paper in this section provides the reader with insight into the spatial evolution of housing patterns in Israel. Gonen demonstrates the role of various agents in determining the location and residential composition of housing in a non-free enterprise economy. His longitudinal analysis of the emerging pattern of urban form tends to highlight the contribution of different forces during specific time intervals on spatial residential structure. The resulting pattern tends to reflect the inverse of U.S. urban residential spatial structure: the process of suburbanization as it exists in the United States does not appear to represent a common urban developmental strategy in terms of the forces which appear to be at work in Israel. Likewise the incongruence between the objectives of the state and that of selected individuals produces housing market behavior that is basically dissimilar from the modal pattern in U.S. cities. Gonen's lucid analysis of the spatial dynamics of the Israeli housing market clearly demonstrates the role of social and economic forces in an alternate cultural setting; but it is evident that the evolving spatial-residential mix is deeply rooted in history.

Several important questions still to be examined in this area include the following:

(1) Is a "rugged empiricist" approach the most effective way to emphasize the role of decision units in determining the outcome of resource allocation processes representing social and economic forces? Does a sense of dissatisfaction with these outcomes lead to the determination of appropriate urban research agendas?

(2) Does the emerging tradition of the human ecologist provide a guide to the study of the relationship of socioeconomic behavior in distinct urban environments? Do we possess adequate measuring techniques to effectively "map" socioeconomic behavior?

(3) Is it inevitable that the differences in the historical and cultural parameters of an urban place will require that this particular phenomenology has to submerge the more general study of common elements operating in most, if not all, urban environments?

6

Emerging Patterns of Industrial Activity Within Large Metropolitan Areas and Their Impact on the Central City Work Force

CHARLES M. CHRISTIAN

☐ THE COMPETITION BETWEEN CENTRAL CITIES and their respective metropolitan hinterlands is becoming more intense as the urbanization of geographic space continues. One such competitive aspect is that of attracting industrial activities for building an effective tax base in order to provide the services demanded by the residential population. Another clearly associated competitive aspect is that of attracting job opportunities in order to secure the economic and social well-being of the residential population within their respective areas. Although this competitive phenomenon is not new, its growing intensity is of increasing importance as it is consistently the central cities of our nation which are the recipients of the spoils. In other words, major central cities throughout the nation have not been successful in attracting their share of the increasing industrial activity of metropolitan areas, nor have they been successful in retarding the rapid and increasing outmigration of firms and employment opportunities from their respective environs to the surrounding hinterlands.

AUTHOR'S NOTE: *The author would like to acknowledge the help and criticism of Connie Williams, Sari Bennett, and Tina Fried in the development of this paper.*

Historically, central cities have been considered balanced communities, with an equitable share of residential, commercial, and industrial land uses for self-sustaining economic and social viability and growth. In recent years, however, most major central cities have undergone and are still undergoing varied stages of ecological land use succession in which residential and industrial land uses have generally been the most seriously affected. Residential land uses, primarily single-family dwellings, have suffered as a result of outmigrating middle- and upper-income groups, creating numerous vacancies in the zones of abandonment. Also, there are increasingly fewer housing starts in central cities; and urban renewal continually takes its toll in the clearance of residential units in slum and blighted areas. As single-family residential land uses decline, there is an increasing dominance of higher-density residential and commercial land uses developing. On a different level, industrial land uses have suffered from industrial firm outmigrations from the central cities, primarily resulting from the inabilities of central cities to offer adequate space for industrial firm expansions, reasonable land costs, and efficient transportation networks for the necessary flows of goods, services, and employees. The decline of industrial activity within major central cities has resulted from industrial relocations from the central cities to suburban and outlying rural areas; limited and/or declining industrial firm expansions; firm closures and failures; and declining new starts of industrial establishments within central cities. From this ecological framework, the central question emerges: What is the impact of declining industrial activity in the central city on the central city resident-employee and his immediate environment?

This paper will address itself primarily to industrial outmigration and its consequences for the central city. In an effort to delimit some of these consequences, primary focus will be given to a discussion of the historical importance of central cities as centers of population, industrial activity, and employment activity. A cursory examination will be made of six selected central cities, and their metropolitan areas; a more in-depth analysis will be made of industrial and employment activity in only the Chicago metropolitan area, with special emphasis devoted to the central city.

HISTORICAL SIGNIFICANCE OF
INDUSTRIAL AND EMPLOYMENT
ACTIVITIES IN THE CENTRAL CITY

Several theoretical constructs have depicted the nation's cities as suppliers of goods and services to the city's population as well as to a larger area—the city's hinterland (Ullman, 1970: 58-67). Accordingly, cities as central places served as trade centers for a tributary area which was primarily agricultural. As a marketplace, an administrative center, or transportation node, the city was indeed the hub of vitality for its adjacent region. However, among these centers or central places were many specialized cities performing many of the above functions but influenced by manufacturing or some other specialized activity (Harris, 1943: 88-89; Nelson, 1955: 189-210). As these centers grew, so did their functions—some becoming more specialized while others were becoming more diversified. In each case, urban centers continued to provide the goods and services for an increasing and demanding population, and thus many of these centers were able to maintain their significance as interdependent nodes in a larger and more complex federation of general and special centers.

It was indeed the industrial centers which provided opportunities for an underemployed rural population, for numerous foreign migrants seeking employment, and for the immediate residential populations of the centers. Cities as centers thus became concentration nodes for specialized services and functions for a larger area, as well as for immigrants and native migrants.

Even as early as 1899 the industrial decentralization and suburbanization processes were evident in major central cities. Creamer (1963) and McLaughlin (1938: 186) document industrial activity as growing more rapidly in suburban sections surrounding many manufacturing cities than within the cities themselves. Moses and Williamson (1967: 211-222) further document this suburbanization trend by an analysis of Chicago for the period 1908-1920. In their findings concerning 473 identifiable firms, they concluded that over half of these firms had shifted location during the period. The importance of these historical trends highlights not only the historical decentralization and suburbanization processes, but the historical breadth of the ecological succession of land use within total metropolitan districts.

The rapidity of ecological land use succession could not have sustained its momentum without the transportation innovations of the nineteenth century. Immediately prior to and during the Civil War, the horsecar dominated the major urban areas, probably having the first real significant impact upon the dispersion of industrial activities and land use change throughout the metropolitan districts (Handlin, 1959: 18). Elevated steam railcars entered the urbanization process in the 1870s, and in the 1890s most metropolitan districts had adopted the electric trolley system (Warner, 1962; Nelson, 1969: 199). These innovations definitely affected the urban landscape by allowing decentralization to take place, as well as extending the geographic limits of the city. It was not until the 1920s that major cities such as Chicago, Philadelphia, and New York experienced a period of rapid growth. Precipitated by cheap and rapid highway transportation coupled with such other significant innovations as mechanical refrigeration, transmission cables, and horizontal processing (assembly line), the urban scene in major metropolitan districts began to show a noticeable impact of residential, industrial, and commercial dispersal (Walker, 1938: 132-134). More importantly, however, it was the automobile and subsidized highway construction which precipitated the most massive decentralization and suburbanization of activities from central cities. With about 8,000 automobiles registered in 1900, rising to well over 8 million in 1920, and 25 million in 1945 (Nelson, 1969: 200), it is no wonder that populations and factories would rapidly envelop the suburban landscape.

It is also during the 1920s that massive migration was taking place from rural areas to major northern urban nodes. Banfield (1973: 16) notes that "heavy as it was, migration to the city seldom fully offset the decentralizing effects of the commuter railroad and the trolley and the expansion of commercial and industrial land uses near the city's center." In an effort to alleviate serious problems resulting from heavy migration to central cities, the Federal Housing Administration and Veterans Administration further facilitated outward expansion by an extensive subsidization program directed toward the acquisition of housing beyond the city's boundaries. Beneficiaries of these programs were predominantly the middle-income white populations; and almost by design, these programs excluded blacks and other low-income disadvantaged populations destined or already living in central city enclaves. As the city was rapidly being abandoned by a white population, industry, and commerce, residential land use was increasingly being left to immigrant and incoming migrant populations. Portions of the cities began to stagnate and underwent a decline in services offered. Slums, high densities, crime, tenements, and shanties were rapidly dominating the city's landscape. The central cities were losing their monopoly of amenity accessibility as numerous amenities were being offered in suburban locations beyond the city's boundaries.

With the increasing urbanization of the suburbs and their offering of more attractive sites and amenities for residential, industrial, and to some degree commercial activities, the central city has evidently long been losing its competitive advantages. Today the city is no longer the only transportation nexus as efficient transport networks are increasingly becoming ubiquitous in the total metropolitan area; it is no longer the only administrative center as political power has decentralized in favor of fragmented political nodes throughout the metropolis; it is no longer the only marketplace as the city now must share with numerous other dispersed market nodes; and it is no longer the only hub of vitality and growth as industrial, commercial, and residential growth is increasingly being found in the more spacious sites beyond central city boundaries.

INDUSTRIAL AND EMPLOYMENT TRENDS
FOR SELECTED CENTRAL CITIES

Any description of the United States today must include the overwhelming aspect of urbanization and its concomitant domestic problems. Increasing urban growth, indicated by statistics throughout the literature, is sometimes perceived as "urban crisis" because of the tremendous influence such a process is having in conditioning American social and economic life. Also inherent in this "urban crisis" connotation is our inability to cope with and solve urban problems. Urbanization has presented numerous disparities, primarily of metropolitan character, which have produced disagreement and confusion toward the development and implementation of multidimensional policies for the solution of such disparities. The spatial distribution of people and their economic and social activities characterizes the overwhelming disparity resulting from urbanization.

In an effort to highlight the above disparities, six metropolitan areas will be analyzed and discussed with specific concern given to an historical synthesis of population, industry, and employment transformation within these areas. Primary consideration for the selection of these metropolitan areas was given to the historical significance of manufacturing as the primary source of employment and the significance of a sizable and increasing black population in the central cities of these metropolitan areas. These factors are considered important because of the potential impact of industrial decentralization upon a largely immobile central city black population. The cities selected for this investigation were Chicago, Philadelphia, Baltimore, Detroit, Pittsburgh, and Buffalo.

REDISTRIBUTION AND GROWTH PATTERNS OF THE CENTRAL CITY POPULATION

In analyzing the historical aspect of metropolitan growth, the 1900-1920 period revealed that central cities grew more rapidly in population than their surrounding hinterland (Advisory Commission on Intergovernmental Relations, 1971: 8). This growth reflected the large immigration from both rural areas and foreign countries (Advisory Commission on Intergovernmental Relations, 1971: 8). The period of the 1930s to World War II is characterized by significant growth throughout metropolitan areas. However, it is further noted that major central cities began their rapid relative and often absolute decline in population growth during the period following World War II.

Populations were indeed bypassing the major central cities as well as relocating their residences from central cities in favor of the more spacious and less congested environment offered by suburban areas. Central cities, however, continued to attract populations, but their growth was less than equal to their surrounding suburban municipalities. This whole process of intrametropolitan growth differential was well established after World War II. Since 1950 the central city's population growth has been less than one-third of the growth of its surrounding metropolitan area (Advisory Commission on Intergovernmental Relations, 1971: 5). It is also during this period that the racial dimension has become a more critical element in this urban growth process.

During the 1950s major metropolitan areas throughout the nation underwent major population distributional transformations in which the major central cities were becoming increasingly black, while areas outside the city were increasingly becoming the home of middle- and higher-income whites. In 1940 nearly three-quarters (72%) of the nation's black population lived in the South (Schnore, 1965: 256-257). Between 1940 and 1966 a net total of 3.7 million blacks left the South in favor of other regions of the United States (U.S. Department of Commerce, 1967). Reception areas for these migrants were the central cities of major metropolitan areas, primarily in the northeastern, the north central, and western regions. The 1950-1960 decade revealed that blacks were indeed an urban population, but more importantly, blacks were a central city population (Schnore, 1965: 256-257). In fact, 58% of the total black population now live in central cities of major metropolitan areas, while only about 28% of the white population of the nation now live in these cities.

Blacks now comprise about 11% of the national population which amounts to more than 23.4 million persons (U.S. Department of Commerce, 1972: 1). Although more than 50% of the black population still lives in the South, the continuing outmigration from southern rural areas and the subsequent concentration within major central cities presently depict a critical problem demanding resolution. The "problem of adequate accommodation" primarily as it relates to housing and employing black populations is perceived as critical. The problem arises because job opportunities are increasingly being developed in the surrounding suburban hinterland and affordable housing is seldom found in close proximity to these jobs. In essence, blacks are generally underrepresented in suburban areas.

According to Figure 1, nonwhite populations are frequently the only growth populations found in central cities. While nonwhites presently account for slightly more than 20% of all persons living in the central cities of Pittsburgh and Buffalo, they account for as much as one-third to one-half of the central city population in Chicago, Philadelphia, Detroit and Baltimore. In contrast, the nonwhite population accounts for less than 6% of all persons living in metropolitan areas outside these central cities. Generally, these central cities have always been the place of residence of the largest percentage of nonwhite population within their respective metropolitan areas. However, during the period from 1930 to the present, variations in population growth within these metropolitan areas have occurred. From 1930 to 1950 percentage growth of nonwhites in the

Per Cent Non-White Population

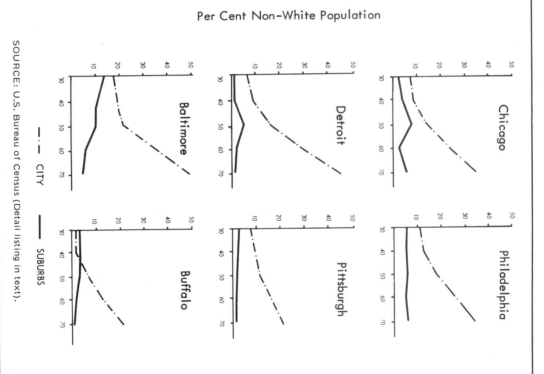

CITY ——   —— SUBURBS

SOURCE: U.S. Bureau of Census (Detail listing in text).

Figure 1: NONWHITE POPULATION IN THE CENTRAL CITY AND SUBURBS

selected central cities was moderate; and since 1950 the percentage growth of nonwhites had doubled. Suburban areas during the 1930-1950 period experienced unstable percentage growth of non-white populations. Chicago, Detroit, and Buffalo experienced some increases in the percentage nonwhite population in suburban areas; while the remaining suburban areas showed some decline, with Baltimore experiencing the most significant percentage decline. After 1950 percentage nonwhite population within these selected suburban areas has been stable or declining. Baltimore recorded the highest consistent decline, and Pittsburgh, Detroit, and Buffalo showed moderate declines in percentage nonwhite population growth. Chicago and Philadelphia during this twenty-year period experienced both increases and decreases in percentage nonwhite population in their suburban areas. Chicago was characterized by the greatest instability during this period. This is indicated by a sharp decrease in percentage nonwhite population during the 1950-1960 decade which was offset by a similar increase during the 1960-1970 decade.

Blacks are increasingly becoming majority populations of major central cities throughout the nation as a result of increasing black migration to these areas and the rapid outmigration of whites to suburban zones. Although blacks are showing an increased propensity to settle in suburban areas, a larger proportion of blacks are excluded from suburban areas because of barriers restricting the supply of housing at the low end of the cost spectrum and practices of racial discrimination. This disparity which grows out of differential population growth, primarily along racial lines, coupled with the outmigration of job opportunities suggests a most complex and difficult task for major central city governments to resolve. However, this responsibility should not be permitted to fall on the shoulders of the central city alone.

OVERVIEW OF

INDUSTRIAL ACTIVITY TRENDS

Industrial activity has been characterized as showing a general shift out of the central cities of major metropolitan areas. The increasing rapidity of declining industrial activity for central cities highlights a critical disparity between central cities and the suburbs. Several studies have detailed particular aspects of the problem resulting from decentralization of economic activities. Vernon (1959) depicts the increasing economic and social disadvantages of the central city as it continues to function in the larger metropolis. In a more recent study of metropolitan growth dynamics, Birch (1970) seeks an understanding of particular disparities in metropolitan growth by focusing on the general impact of increasing specialization in the central city and the increasing residential and economic growth of

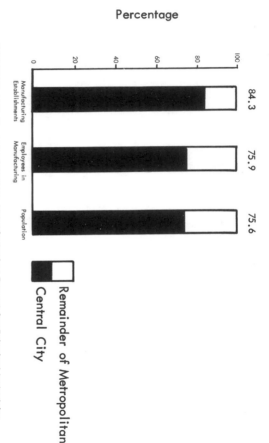

Percentage

Manufacturing Establishments 84.3

Employees in Manufacturing 75.9

Population 75.6

Remainder of Metropolitan
Central City

SOURCE: U.S. Bureau of Census, Manufacturing Report for Principal Industries, Volume X, U.S. Government Printing Office, Washington, D.C., 1913: 904-954.

Figure 2: PROPORTIONS OF SELECTED CHARACTERISTICS OF METROPOLITAN AREAS, 1909

the suburbs. In general, each of these studies portrays the cumulative effects of metropolitan growth as being a transfer of urban problems and industrial activity from the central cities to their suburban hinterlands.

This significance of any disparities within metropolitan areas today must consider the historical aspects of industrial activity. Figure 2 shows that of the thirteen metropolitan districts defined in 1910 by the Census, the major share of manufacturing establishments, production workers in manufacturing, and population were concentrated in the central cities of metropolitan districts. In fact, almost 85% of the manufacturing establishments, 76% of the employees in manufacturing, and 76% of the metropolitan population were found within the central cities of these metropolitan areas.

In the early 1930s industrial growth was increasing throughout metropolitan areas. Increases in plant size and improvements in material-handling techniques hastened the outmigration of manufacturing establishments from central cities (Banfield, 1973: 21). Figure 3 shows that during the 1939-1947 period, five of the six selected metropolitan areas experienced rapid industrial growth

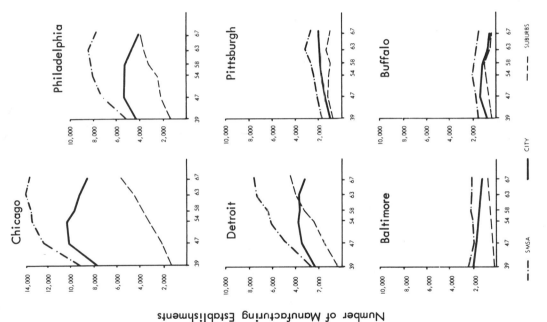

Figure 3: MANUFACTURING ESTABLISHMENTS IN METROPOLITAN AREAS BY CENTRAL CITY, SUBURBS AND SMSA, 1939-1967

SOURCE: U.S. Bureau of Census (Detail listing in text).

activity. Decentralization of manufacturing establishments had not especially affected the economic well-being of central cities as they continued to attract a sizable share of the metropolitan industrial growth to offset losses to suburban areas. Baltimore was, however, the exception as both the city and the total metropolitan area declined in industrial activity, while the suburbs showed a slight increase. During the 1939-1947 period, the selected metropolitan areas were dominated by their central cities in the number of manufacturing establishments. After 1947, only the Chicago, Detroit, and Philadelphia metropolitan areas attracted any large number of manufacturing establishments. Also, central cities, with the exception of Pittsburgh, began their dismal absolute decline in manufacturing establishments, while their suburban areas, with the exception of Pittsburgh and Buffalo, were increasing phenomenally in manufacturing establishments. In fact, increases in manufacturing activity for suburban areas as the central cities were experiencing static or declining manufacturing growth. This suggests that a greater and increasing number of the manufacturing establishments were simply changing locations from central cities to numerous suburban areas.

The changing locational patterns of manufacturing activity within the six selected metropolitan areas depict some general consistency. Chicago, which definitely had an overwhelming advantage as a manufacturing city because of its superior relative location in a developing industrial economy (Mayer, 1970: 11-26), continued its growth in manufacturing establishments up to 1954 and thereafter has experienced a consistent decline. In contrast, its suburban counterparts have shown consistent increases almost equal to the number of establishments lost from the city. In fact, suburban areas have increased their number of manufacturing establishments from about 1,000 in 1939 to almost 5,500 in 1967. It must, however, be kept in mind that the city still maintains the highest number of industrial establishments in the metropolitan area. According to the observed trend, it is projected that the city of Chicago will continue to lose ground vis-à-vis the suburban ring, and by 1975 the central city will no longer be able to maintain this manufacturing dominance.

The metropolitan areas of Detroit and Philadelphia tend to spell out the long-range trends for central city manufacturing activity. While the metropolitan areas have shown growth in manufacturing establishments after 1947, only the Detroit metropolitan area continued to grow after 1963, while Philadelphia declined sharply. In each of these areas, the central cities have shown abrupt declines in manufacturing establishments while the suburban counterparts are experiencing large increases in establishments. While Detroit relinquished its dominance of manufacturing activity in the early 1960s, the city of Philadelphia has consistently lost manufacturing establishments since 1947, and in 1967 the city and the suburbs had an equal proportion of manufacturing establishments. It is suspected, however, that given the existing industrial growth trend already established for the Philadelphia metropolitan area, the early 1970s will find that the suburbs have already forged ahead of the central city in the superiority of manufacturing activity.

The smaller industrial metropolitan areas—Pittsburgh, Baltimore, and Buffalo—have experienced less consistent manufacturing activity behavior than the larger ones. In essence, these metropolitan areas have remained almost stable after 1947; however, the largest relative manufacturing establishment increase was experienced by Pittsburgh. Even so, after almost thirty years of industrial growth, each of the metropolitan areas tends to have about the same number of manufacturing establishments that they had in 1939. There are, however, some internal changes in the manufacturing activity of these metropolitan areas.

In Pittsburgh metropolitan area, the central city has gradually attracted manufacturing establishments while the suburbs have shown only a very slight increase since 1947. Although Baltimore remained stable since 1939, the central city has declined since 1939 while the suburbs experienced a similar counteractivity in manufacturing establishments. Within the Buffalo metropolitan area, both the central city and the suburbs have declined sharply in manufacturing establishments since 1958.

In general, these selected metropolitan areas are presently experiencing stable or declining manufacturing growth activity according to the number of manufacturing establishments. The central cities tend to be experiencing the largest decreases in manufacturing activity and, hence, tend to depict potential social and economic problems for central city governments. These statistics also strongly suggest that a large proportion of the metropolitan areas' decline in manufacturing activity is the result of the central cities' industrial losses and that many of these losses are primarily suburban gains.

OVERVIEW OF DISPARITY IN

EMPLOYMENT OPPORTUNITIES

Central cities have always been the home of high concentrations of populations. These concentrations were the result of numerous migrants seeking the many employment opportunities offered by central cities. The population composition of central cities was highly diverse, although by and large these areas were the place of residence of a disproportionate number of unskilled persons. Since residential concentrations were, to a large extent, dictated by the location of employment opportunities, unskilled populations often resided in undesirable areas near their place of employment.

The central city long enjoyed a monopoly of employment opportunity. However, employment opportunities have recently become far less numerous. During the early decentralization process, the spatial shifts in employment opportunities resulting from industrial decentralization have generally waited for a subsequent shift in the residential movements of population to accommodate relocated jobs. The location of manufacturing activity plays a significant role in the formation of urban structure as the location of jobs influences the location of urban households (Kain, 1970: 20). There yet appears to be some disparity in the evolving pattern of suburban residence and the location of selected employment opportunities in this environment. This mismatch between jobs and workers has been discussed elsewhere (Gold, 1972). Many manufacturing firms in Chicago, after relocating to a suburban location, have erected large signs describing the type of employees needed (primarily unskilled and semi-skilled workers); and numerous advertisements have been recently directed to black residential areas in order to overcome this deficit (Christian, 1973).

An historical perspective of employment is presented in Figure 4 for the six selected metropolitan districts. Production workers in manufacturing for these metropolitan areas previous to 1939 were concentrated in central cities. Figure 4 illustrates that during three recent periods changes in manufacturing location have been underway which have had a negative impact upon these selected central cities. Since 1958, however, increases in manufacturing production workers have been greatest in suburban areas. There is no reason to believe that increases in employment opportunities in suburban areas will discontinue, in spite of the many efforts of central cities to halt the flight of industry out of the central cities.

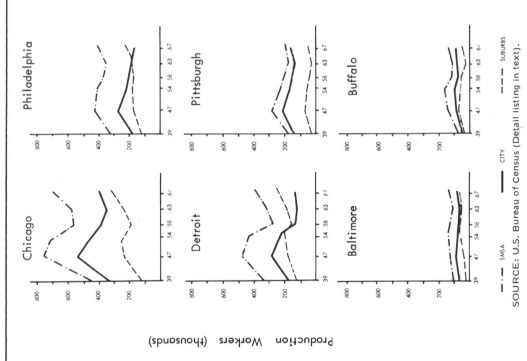

SMSA — · —   CITY ——   SUBURBS – – –

SOURCE: U.S. Bureau of Census (Detail listing in text).

Figure 4: NUMBER OF PRODUCTION WORKERS FOR METROPOLITAN AREAS BY CENTRAL CITY, SUBURBS AND SMSA, 1939-1967

AN HISTORICAL OVERVIEW OF THE
REDISTRIBUTION OF MANUFACTURING
ACTIVITY IN CHICAGO

Because of manufacturing's significance to the total economy of the city and the fact that the city of Chicago has depended heavily on manufacturing since it was incorporated as a city (Mayer, 1970:

11-26), it is suggested that any spatial changes in the locational patterns of manufacturing within the city, as well as in the total metropolitan area, will have a serious impact on numerous aspects of the city's social and economic life.

Many of Chicago's urban problems, especially those of the inner-city areas, have been linked directly or indirectly with the decline of manufacturing activity (Mayor's Committee for Economic and Cultural Development of Chicago, 1970: 2-5). Other intracity dynamics, such as immigration of low-income and unskilled groups and the high outmigration of high-income groups, also feed the inner city's deterioration and high unemployment. The force which directly affects unemployment is the outmigration of manufacturing firms. This loss of employment potential contributes significantly to the intensity and magnitude of many inner-city problems. These forces also effectively reduce the amount of revenue necessary to combat inner-city problems, hence suggesting a continuing cycle of deterioration, decay, and high unemployment for the central city.

Several empirical studies depict the historical trend of industrial relocation of manufacturing activity in the Chicago Metropolitan Area. One of the first detailed studies of industrial decentralization from the city of Chicago was that by Mitchell (1933). By dividing the Chicago Metropolitan Region into the city of Chicago, and Zones B and C (suburban areas), Mitchell found approximately 127 industrial establishments lost from the city of Chicago to the two suburban zones during the 1923-1931 period. A directional bias was noted as firms relocating out of the city moved to western and southern suburban communities adjacent to the city's boundaries.

The industrial activity trend continued almost unabated during the 1936-1940 period. Relocations from the city, as well as new industrial firms locating in the total metropolitan area, were still directed to the adjacent western and southern suburban communities (Chicago Plan Commission, 1942: 47-49).

During the 1936-1940 period, the city lost a total of 47 industrial firms, while two industrial establishments were relocated from the rest of the SMA to the city. The city also attracted approximately 30% (42) of the industrial establishments new to the SMA during this period; however, because of the decline in number of industrial establishments, the city's net change was −3.

Industrial relocations from the city of Chicago continued its decentralization trend during the 1940s. However, the city still continued to attract some industrial firms. Reeder (1952), studying the Chicago Metropolitan Area's industrial climate for the period 1941-1950, notes that during 1941-1945 the city showed a net gain of 71 industrial establishments. However, during the 1946-1950 period the city experienced a net loss of 86 firms. Although the city remained especially attractive to some industries during the 1941-1946 period, incoming industries did not offset the losses from the city during the 1946-1950 period. The increase during the 1941-1946 period in the number of establishments for the city reflects the influence of the war on the locational patterns of industrial firms.

Suburban communities realized their greatest gains in industrial establishments during the 1946-1950 period by attracting more than 332 firms new to their area, plus at least 233 industrial relocations from the city of Chicago. The new gains for the suburban communities were 565 industrial firms. Not only is there a greater portion of industrial establishments relocating from the city of Chicago, but these firms are selecting sites farther out from the city's boundaries (Reeder, 1952: 226).

The 1947-1957 period is one characterized by an even larger number of industrial establishments relocating from the city of Chicago. Chicago's Department of City Planning (1961) delimited a total of 2,618 industrial movements occurring during this period. A brief summary of their findings shows that 1,342 industrial establishments were new to the SMA, of which 533 located within the boundaries of Chicago, 528 were to suburban Cook County, and 281 were to the rest of the SMA. Although the city attracted industries during this ten-year period, the city lost approximately 591 industries to the suburban areas through relocations, while only receiving 19 relocations from the suburban areas.

While the spatial magnitude of industrial activity changes from core areas of the city toward the numerous dispersed suburban communities, internal movements were also undergoing a similar dispersion within the city. During the ten years, there were 651 internal industrial relocations within the city of Chicago. Although these movements represent neither losses nor gains to the city, they potentially pose problems of increasing inaccessibility to many central city employees who are not able to move closer to these new relocated employment sites or who cannot afford the increase in commuting cost.

More importantly, the city is increasingly declining in the number of manufacturing employment opportunities to be offered to its residential population. Furthermore, it is especially noteworthy to elaborate upon those areas of highest decline in manufacturing activity within the city of Chicago. Employment opportunity losses are especially critical to a city's population, but a more critical impact may depend on the spatial variation of these losses. According to a previous study of the black community (Christian and Bennett, 1973: 14-20), it was found that areas of highest employment opportunity losses during the 1955-1963 period were also the areas of highest population density, highest percentage unemployment, and highest percentage of black population. In essence, the impact is viewed as being more critical when these factors are taken into consideration as such relocations of employment opportunities can only increase the poverty already present within these areas.

RECENT TRENDS AND POTENTIAL CONSEQUENCES OF INDUSTRIAL AND EMPLOYMENT RELOCATIONS

In view of the large number of industrial firms and employment opportunities relocating from particular areas of the city of Chicago, and the fact that these areas generally coincide with areas of high black residential concentrations within the city, it appears logical, given the numerous constraints affecting their mobility, that this population would be potentially the most affected by such out-migrations. This hypothesis is further based on the fact that the greatest increases in blue-collar employment opportunities are not found in the city, but are increasingly being found in the numerous suburban areas far distant from black residential areas. Much recent literature highlights the economic disadvantages confronting black central city residents. Constraints such as discrimination, low wages, inadequate and lack of low-cost housing around new suburban industrial developments, directly influence the residential mobility of segments of the central city black population. In essence, the outmigration of jobs from the central city to suburban locations highlights the increasing erosion of economic well-being for black central city residents. In an effort to provide further insight into the

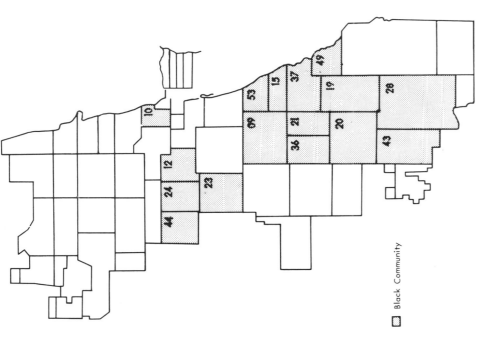

Black Community

SOURCE: U.S. Bureau of Census: Census of Population and Housing, Chicago and Northwestern Indiana Standard Consolidated Area, Reproduced by Chicago Association of Commerce and Industry, 1970.

Figure 5: THE BLACK COMMUNITY OF CHICAGO, 1970 (Defined by Postal Zones)

consequences of industrial outmigration, this part of the paper will deal with an extension of industrial employment relocation trends for the period 1965-1971 from the most economically disadvantaged areas of the city—the black community of Chicago.

The black community of Chicago is operationally defined to include those postal zones consisting of 30% or more black residents in 1970 (Figure 5). It is within this area of the city of Chicago that the most rapid decline in industrial activity has occurred; and likewise this area includes the largest number of low-income and low-skilled residents within the city. The choice in delineating the black community is based on the assumption that the white population perceives less than 30% black population as only mildly threatening within the context of a predominantly white neighborhood; and thereby many will remain in the area as long as blacks do not exceed this level. When the percentage black population reaches more than 30%, a significant threshold value is reached which elicits a response from white persons in the form of acceleration of outmovement (Rose, 1972: 51-52).

In order to ascertain the recent trends of industrial relocations out of the black community and the resultant employment losses, the following assumption is made concerning these data: an address change listed for an industry in either directory[1] is considered a relocation for the year in which the address change was first noted. Several classifications of industrial relocations are defined in order to discuss relocation patterns within and from the black community of Chicago. The classification schema is as follows:

(1) In-Zone Movements—Origin and destination of firm relocation takes place within the same postal zone of the black community.

(2) Between-Zone Movements of the Black Community (Type 1)—Destination of firm relocation was outside of the origin postal zone within the black community.

(3) Between-Zone Movements of the City (Type 2)—Destination of firm relocation was to a postal zone outside the black community but within the city of Chicago.

(4) Suburban Movements—Destination of firm relocation was outside the corporate limits of the city of Chicago but within the Chicago Standard Metropolitan Statistical Area (excluding Lake and Porter Counties, Indiana).

(5) In-State Movements—Destination of firm relocation was outside the Chicago SMSA but within the state of Illinois.

(6) Out-of-State Movements—Destination of firm relocation was to any location outside the State of Illinois.

In each classification, the number of firms, and more importantly, the number of employment opportunities moving within and from particular zones of the black community are of primary interest.

MOVEMENT PATTERNS OF INDUSTRIAL

FIRMS AND EMPLOYMENT OPPORTUNITIES

WITHIN AND FROM THE BLACK COMMUNITY

OF CHICAGO–1965 THROUGH 1971

IN-ZONE MOVEMENTS

Industrial firm and employment opportunity relocations within the same zone are not considered within this analysis to be especially disadvantageous to the resident employees of the city because these moves are short. They generally result in only small increases in commuting costs and inconveniences, and oftentimes do not pre-cipitate basic residential or journey-to-work decision-making. As shown in Table 1, of the 1,011 firms relocating during the study period, 135 industrial firms moved within the same postal zone. Furthermore, of the more than 50,000 employment opportunities relocating, an estimated 5,397 jobs were relocated within their zone of origin. These movements made up approximately 13.4% of the total firms and approximately 10.8% of all the spatial changes in job addresses of the black community.

TABLE 1

INDUSTRIAL FIRM AND EMPLOYMENT OPPORTUNITY RELOCATIONS WITHIN AND FROM THE BLACK COMMUNITY OF CHICAGO, 1965-1971

(by classification of movement)

Movement Classification	Firms Relocated (n)	% of Total Firms Relocated	Employment Opportunities Relocated (n)	% of Total Employment Opportunities Relocated
In-zone	135	13.35	5,397	10.76
Between-zone[a] (type 1)	150	14.84	4,379	8.73
Between-zone[b] (type 2)	346	34.22	16,239	32.40
Suburban movements	345	34.12	22,285	44.46
In-state	5	.45	47	.09
Other states	30	3.00	1,772	3.54
Totals	1,011	99.98[c]	50,119	99.98[c]

a. Between-zone (type 1) refers to the movement of firms and employment opportunities within the black community of Chicago.
b. Between-zone (type 2) refers to the movement of firms and employment opportunities to zones of the city outside the black community.
c. Column percentage totals do not equal 100% due to rounding.

SOURCE: Computed from data taken from the *Illinois Manufacturers Directory*, 1965-1971; and the *Chicago, Cook County and Illinois Industrial Directory*, 1965-1971.

Three zones (9, 10, and 12) accounted for the highest number of firms relocating within zones. These same zones, plus one other zone peripheral to the central area (zone 23), accounted for the highest number of relocating employment opportunities. The relocation of firms and employment opportunities within these zones indicates that particular areas still appear to be attractive to some industrial firms, as shown by a large number of firms which maintained a location close to the central portion of the city.

BETWEEN-ZONE MOVEMENTS WITHIN THE BLACK COMMUNITY (TYPE 1)

Between-zone movements of the black community comprised 150 (14.8%) of the total industrial firm movements and 4,379 (8.7%) of the total employment opportunities spatially relocated between zones of the black community (Table 1). The movement of these firms and their employment opportunities potentially presents a more serious problem to the resident-employees of the black community than do in-zone movements. Between-zone movements (Type 1) may possibly be out of reach for many employees who previously depended on a short journey-to-work trip as a necessary condition of employment. Further, the individual accustomed to a short distance between residence and workplace may, for the first time, be confronted with a substantial increase in the economic costs incurred in his new journey-to-work trip.

As a result of the employment opportunities and firms lost particular origin areas are more seriously affected than others, Zones 21, 23, 12, and 24 each lost more than 500 jobs or a total of 2,913 jobs as a result of industrial relocations between zones of the black community. These employment opportunities accounted for approximately 66.5% of all jobs relocating within the black community.

Destinations for these between-zone relocations were generally to postal zones 12, 20, 9, and 36. These zones received 2,671 (61%) of the employment opportunities and 55 (37%) of the firms. Generally, the zones which lost the highest number of firms and jobs were not necessarily the reception zones of between-zone relocations. Only zone 12, which lost 14 industrial firms and 552 employment opportunities, was able to attract enough new firms and jobs to offset its losses. Zone 12 received 16 firms and 1,041 employment opportunities from other zones of the black community for a net gain of two firms and 489 jobs.

BETWEEN-ZONE MOVEMENTS IN THE CITY (TYPE 2)

Movements of firms and employment opportunities to zones outside the black community potentially provide a more adverse impact on the black community. A movement from a border zone of the southern portion of the black community to a zone in the extreme northern portion of the city can be as great as 25 miles—a considerable distance and inconvenience for many central city residents to travel. Such a commuting distance may further affect an employee's journey to work considerably in terms of time and overall economic costs.

Movements to destinations outside the black community but within the city accounted for 346 (34.2%) of the total firms and 16,239 (32.4%) of the total employment opportunities relocated (Table 1). In fact, firm movements to nonblack postal zones were more than double that of black community between-zone and in-zone movements combined. Job opportunities which relocated outside the black community to other zones of the city were more than four times that of relocations in and between black postal zones. To compound the problem of lost firms and jobs for black community residents, there appears to be a direct positive relationship between the number of firms moving and the distance firms moved; and the number of employees affected by firm movements also increases with distance of movement. Again, the adverse consequences on black area jobseekers from increased travel time and costs to relocated employment sites are obvious.

All zones in the black community lost firms to zones outside the black community. Three of these zones (9, 10, and 12), each located adjacent to the central business district, lost a total of 188 firms, accounting for 54.3% of the total firm movement to zones outside the black community. These zones also contributed the largest employment opportunity losses to zones outside the black community. In fact, these zones accounted for 10,356 (63.8%) of the jobs lost. Other zones also contributed heavily to losses of employment opportunities from the black community, but the magnitude of the loss was much smaller. Seven zones accounted for 14,258 (87.8%) of the total employment opportunities relocated to zones outside the black community.

Summarizing the movement of firms and employment opportunities from the black community to other areas of the city, it is shown that 631 of 1,011 black community industrial firms relocated within the city of Chicago, accounting for 62.4% of the total pertinent firm relocations of the 1965 through 1971 period. Of these city movements, approximately 21.4% of the industrial firms relocated within the same black area postal zone; 23.8% moved to other zones (Type 1) within the black community; and 54.8% of the firms relocated from the black community to other zones (Type 2) of the city (Table 1).

Of the employment opportunities relocating within the city of Chicago, a total of 26,015 jobs were spatially redistributed during the 1965-1971 period. Of these employment opportunities, approximately 21.0% relocated within the same zone of the black community; 16.7% relocated to other zones of the black community; and 62.3% relocated to zones of the city outside the black community. Overall employment opportunities relocating within the city of Chicago accounted for 51.9% of the total employment opportunities relocating within and from the black community during the 1965-1971 period (Table 1).

SUBURBAN MOVEMENTS

Relocations of firms and employment opportunities are producing the most adverse consequences for the central city and its residents. Several critical aspects of these relocations are (1) the fact that the great majority of these job relocations are skilled and unskilled blue-collar jobs (in essence, these are the job opportunities which initially attracted the large number of black migrants to the city); (2) industrial firms which are relocating are primarily the firms which traditionally have employed the greatest number of black employees; (3) the fact that employment in most suburban areas demands ownership or accessibility to a private automobile readily excludes many blacks from maintaining jobs or, for that matter, even seeking employment in these areas; (4) the low wages often paid to central city residents related to the cost of commuting presents an insurmountable barrier for many; and (5) the perceived level of discrimination practiced in the numerous suburban municipalities has the effect of diminishing the desire to work in alien locations.

Firm movements from the black community of Chicago to suburban locations are many. During the 1965-1971 period, 345 industrial firms relocated from the black community to suburban locations within the Chicago Metropolitan Area (Table 1). These movements accounted for 34.1% of all the movements of industrial firms. A total of 22,285 jobs were lost from the black community during this period, accounting for 44.5% of all employment opportunities which moved within and from the black community (Table 1). Relative to other movements, suburban locations are the reception areas of the larger firms, as indicated by the fact that 34.1% of the relocated industrial firms accounted for 44.5% of the relocated employment opportunities. Each of the firm movements averaged approximately 65 employment opportunities, whereas movements within the city averaged less than 40 employees per firm.

Six zones (9, 10, 12, 23, 24, and 44) were the hardest hit by industrial firm losses to suburban locations. These zones lost 258, approximately 75%, of the 345 suburban movements from the black community. Zones 24, 44, and 12 situated in the West Central District, lost 13,208 jobs, accounting for over one-half (59.3%) of all jobs resituated to suburban locations. Other zones with large losses of employment opportunities were zones 9, 10, 23, and 28 which lost a total of 5,432 jobs or 24.4% of total jobs removed from the black community to suburban locations.

The destinations of these relocated industrial firms and their respective employment opportunities display a definite distance bias as the largest number of industrial firms and the largest employment opportunities lost to the black community were resituated in suburban Cook County. In fact, 17,058 (76.5%) of the employment opportunities relocated to places within suburban Cook County. DuPage County was the second most attractive county, receiving 4,202 employment opportunities. Other counties—Lake, Will, McHenry, and Kane—each received less than 240 employment opportunities from the black community.

Given the 345 industrial firms and the more than 22,285 employment opportunities relocating from the central city to suburban locations, and the potentially higher commuting cost and inaccessibility confronting central city residents to these job opportunities, numerous central city blacks are finding it increasingly difficult to transfer to jobs at new job locations.

For the central city black resident who is searching for a job, a minimum level of access to and information about transportation and employment opportunities is particularly important in his search. Since workers find jobs primarily through informal channels of communication—through knowing others who work at the same plant or through living nearby and hearing of job vacancies—the spatial segregation of residences can be extremely critical to a worker's employment potential. The spatial segregation between central city black residents and suburban job opportunities, coupled with the workers' lack of contact with other suburban workers, may prevent them from ever learning about job openings for which they are qualified. In essence, the relocation of industrial firms and their related blue-collar employment opportunities are likened to a wage tax on the poor, which could take the form of (1) higher cost of the journey to work; (2) acceptance of job—and pay—beneath one's capabilities; (3) longer periods of unemployment while seeking appropriate jobs; and (4) higher rates of unemployment (Wingo, 1972: 4).

The exodus of thousands of unskilled and skilled blue-collar job opportunities is definitely producing increasing difficulty for some central city black residents. DeVise (1967: 158) gives further emphasis to this disparity:

The availability of public transportation and the paucity of black car ownership (between 10 percent and 20 percent own cars) are listed as additional impediments to black employment in new industrial communities . . . the dual housing market comprises a dual labor market . . . as a result an almost fully employed white labor force coexists alongside a black labor force where unemployment rates vary from five percent to 37 percent depending on community or residence.

From these points, it appears evident that in order for blacks to be employed in the numerous suburban employment nodes, they must live near these jobs or public transportation, or other forms of transit must be convenient and available at reasonable costs. Presently, public transportation is not an excellent nor adequate commuter system for the central city reverse commuter who works in a suburban location. Confirming this from numerous case studies, the Urban League of Chicago points out in a recent publication that public transportation accessibility, or the unavailability of private automobiles, prohibits many jobseekers from accepting available jobs at suburban job sites (Chicago Urban League, 1971).

A closer view of industrial firm and job relocations from the central city to suburban locations tends to further highlight the acute disadvantages imposed upon the central city black residents. Table 2 shows that of the 345 industrial firms relocating to the suburban locations only 15.7% (54 firms) moved less than five miles from origin sites in the city; 30.4% (105 firms) relocated within six and ten miles, and the remaining 53.9% (186 firms) relocated more than ten miles from their origin locations within the central city. The most critical aspect is found when viewing the job movements. More than 62% (14,007) of the job opportunities relocated more than ten miles beyond their previous locations in the central city. It must be noted that, although distance is only part of the inaccessibility problem, it is clearly shown that the firms with fewer employment opportunities are locating closer to their origin central city locations while the firms with larger employment potentials are locating farther from their origin central city locations. Hence, the greater number of jobs relocating from the central city are increasingly being found at greater distances from the place of residence of central city black populations.

Compounding this distance constraint affecting the central city jobholder and jobseeker is the inadequate and inconvenient transport system. Even greater barriers to employment opportunity for central city black residents are posed by the fact that these employment opportunities are dispersed over no fewer than 84 suburban municipalities throughout the Chicago Metropolitan Area. Figure 6, however, shows the distribution of suburban zones which received 200 or more job opportunities during the study period. Thirty-one of the 84 suburban reception areas received at least this number. Of

TABLE 2

DISTANCE OF SUBURBAN MOVEMENTS FROM THE BLACK COMMUNITY OF CHICAGO, 1965-1971

Distance Firm Moved[a]	Firms (n)	% of Total Firms	Employment Opportunities (n)	% of Total Employment Opportunities
0- 5 miles	54	15.7	1,322	5.9
6-10 miles	105	30.4	6,956	31.2
11-15 miles	90	26.1	6,688	30.0
16-20 miles	21	6.1	4,418	20.0
21-25 miles	52	15.1	1,268	5.6
over 25 miles	23	6.6	1,633	7.3
Totals	345	100.0	22,285	100.0

a. Distance of firm movements is calculated according to straight line air miles from centroid of origin postal zones to centroid of destination postal zones.
SOURCE: Computed by author.

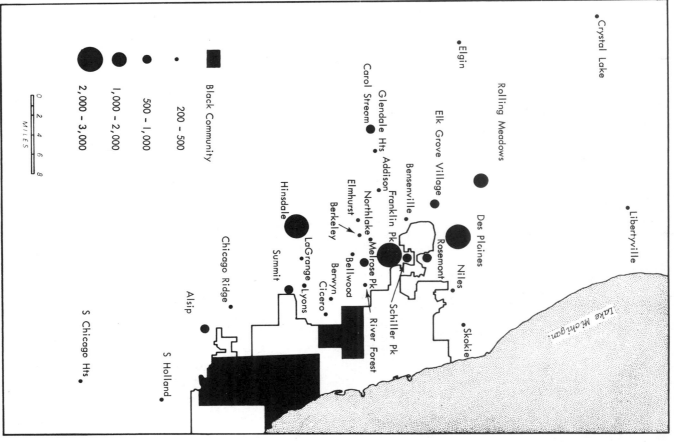

Figure 6: JOB OPPORTUNITIES RELOCATING TO SUBURBAN DESTINATIONS
FROM CHICAGO, 1965-1971

these reception areas, three suburbs, Franklin Park, Des Plaines, and Hinsdale, each received over 2,000 job opportunities. In fact, these three areas received a total of 7,083 employment opportunities, almost one-third of the total employment opportunities relocated from the black community of Chicago to the suburbs. Rolling Meadows, a northwestern suburb, was the only member of the second largest suburban reception group, receiving 1,508 job opportunities from the black community. Other suburban receptions groups receiving fewer than 1,000 job opportunities tend to be found dispersed throughout the Chicago Metropolitan Area; however, the larger number of reception areas was generally directed toward the northwestern and western metropolitan area. Other areas of the metropolitan district were recipients of employment opportunities; however, they received smaller firms of fewer than 200 employees. In essence, although there appears to be some tendency toward relocated employment opportunities to locate in the northwestern and western suburban areas, a dispersed pattern of relocated job opportunities predominates.

Given the spatial dispersion and directional tendency of relocations from the black community of Chicago, there appears little doubt that such relocations are producing insurmountable difficulties for many blacks and low-income workers of the city. Although no evidence has been presented to suggest that these industries are relocating to avoid employing these workers, Table 3 does, however, present evidence that relocating employment opportunities are not being resituated in suburbs with large black populations. In fact, only three of the largest reception areas Summit, LaGrange, and Elgin, have more than 1% black populations. In the two largest reception areas, Franklin Park and Des Plaines, there appears to be no black population. In essence, relocated job opportunities are being resituated in suburban areas void of black populations and it is strongly suggested that blacks are not relocating their residences to these areas to maintain their jobs. Hence, it is further suggested that relocated employment opportunities are becoming increasingly inaccessible to a population dependent on accessibility.

The dispersed pattern of these relocated job opportunities and the fact that blacks are not residing in these suburban areas suggest that the worker is inevitably compelled to purchase an automobile for longer journey-to-work trips or participate in a ride-sharing behavior, or use several costly modes of transportation if he is to remain with his relocated job; or the employee may simply quit and take his chances of being employed closer to his residence or at a location more accessible to public transportation. The last and more disruptive of alternatives, which the employee may be forced to accept, is that of unemployment.

TABLE 3
BLACK POPULATION IN SUBURBAN AREAS
RECEIVING MORE THAN 200 JOB OPPORTUNITIES, 1970

Suburban Area	Job Opportunities Received (n)	% of Black Population[a]
1. Glendale Heights	200	0.3
2. Elgin	202	4.8
3. Berwyn	210	0.0
4. Chicago Ridge	216	0.0
5. Skokie	220	0.1
6. S. Chicago Heights	221	0.1
7. Bellwood	252	0.6
8. LaGrange	258	2.5
9. Niles	262	0.0
10. Lyons	266	0.0
11. Berkeley	286	0.0
12. Elmhurst	296	0.3
13. Crystal Lake	300	0.0
14. Libertyville	300	0.1
15. Addison	332	0.1
16. River Forest	334	0.3
17. Bensenville	408	0.0
18. Cicero	418	0.0
19. South Holland	446	0.0
20. Northlake	468	0.0
21. Melrose Park	563	0.6
22. Schiller Park	632	0.1
23. Elk Grove Village	727	0.1
24. Rosemont	779	0.1
25. Carol Stream	815	0.9
26. Alsip	884	0.2
27. Summit	992	19.1
28. Rolling Meadows	1,508	0.1
29. Hinsdale	2,188	0.5
30. Des Plaines	2,256	0.0
31. Franklin Park	2,639	0.0

a. Only Chicago Ridge has absolutely no blacks among their population, while other areas have so few that they did not appear in the percentage calculations. Also figures were not rounded.

SOURCE: Data for number of job opportunities received by suburban communities was collected and computed by author; data for percentage black population in selected suburban areas collected from *Suburban Factbook*, Northeastern Illinois Planning Commission, July 1971.

IN-STATE AND OTHER-STATE MOVEMENTS

Relocations of industrial firms and employment opportunities from the black community to other cities within the state of Illinois and movements to other states of the U.S. are comparatively few. Five firms with approximately 47 employment opportunities moved from the black community to other cities within the state. A total of 30 industrial firms and 1,772 employment opportunities relocated completely out of the state, both accounting for 35 industrial firms and 1,819 employment opportunities lost from the black community and from the Chicago Metropolitan Area. Because these were long-distance movements generally involving a distance too great for commutation, the impact is almost definitely one in which most employees are severed from the firm. This is based on the fact that long-distance firm movements generally do not involve a firm's commitment to relocating its blue-collar employees. Hence, it is surmised that these employees became either temporarily or permanently unemployed.

CONCLUSIONS AND IMPLICATIONS

The central city labor force is becoming a rapidly diminishing resource for metropolitan economic growth, based not only on the individual qualities of central city residents—such as education or skill level demanded by industrial firms throughout metropolitan areas—but, more importantly, on the increasing inaccessibility of blue-collar jobs, which produces higher levels of unemployment and lack of labor force participation in central cities. In a general economic sense, the labor force resource must be demanded and maximum accessibility qualities must exist in order for it to be utilized efficiently. In recent years, the demand for unskilled and semi-skilled labor has declined rapidly throughout the nation and most particularly within major industrial metropolitan areas. This lack of demand has definitely affected the employment potential of black populations housed in numerous large central cities. It appears that this trend will continue as numerous industries adopt technological innovations throughout their individual plants which replace a significant number of lower-skilled employees.

The accessibility factor, although critical, appears more soluble for central city residents than the absolute decline in demand by industries for unskilled and semi-skilled employees. Although inaccessibility to job opportunities potentially has the same disadvantageous effects—loss of job, less job stability, and lack of efficient job-search capacity, it appears that the flight of industrial firms and their blue-collar employment base to suburban communities will continue almost unabated until these reception communities are saturated. This trend found within older major central cities establishes "points of justification" for the creation of public policy with regard to hostage central city black populations who find themselves segregated residentially and increasingly becoming segregated from employment. In essence, it is the blue-collar job opportunities which have offered blacks their most direct means for escaping the ghetto environment. However, it is being noted that the rapid and increasing outmigration of these employment opportunities is producing insurmountable barriers for blacks housed in central city locations. The economic, social, psychological, and geographical costs incurred by central city populations to avail themselves of the suburban job opportunities are generally too great for many to absorb. This separation between work and residence runs counter to the economic well-being of the black population and is becoming more intense as central city black residences and workplaces are being separated farther and farther apart. This not only suggests a critical problem now, but also suggests a critical and troublesome future for major central cities throughout the nation.

Findings of this paper reveal the following growing disparities which adversely affect blacks and nonwhites in central cities:

(1) Central cities are rapidly losing their preeminence as the place of location of industrial firms and production workers.

(2) Firms and employment opportunities lost from the central city are primarily those which are relocating to dispersed suburban locations.

(3) Although central cities have historically lost manufacturing firms and employment opportunities, the critical nature of recent losses is related to the continuing immigration of black and nonwhite populations seeking those job opportunities which are most rapidly relocating to suburban locations.

(4) Job relocations to dispersed suburban municipalities appear to produce insurmountable barriers as black and nonwhite central city residents are not relocating in large numbers to suburban locations to avail themselves of these relocated job opportunities.

(5) The majority of job opportunities relocating from the central city to suburban locations is increasingly being found farther and farther from the residences of black and nonwhite central city populations.

Hence, a greater distance disparity is produced between residence and workplace, a consequence severely hindering many black residents who do not have access to personal automobiles or public transportation to these employment nodes.

A more comprehensive, long-run approach to the employment dilemma is the development of "new towns" within central city areas. Although an extremely expensive approach, these new towns will attract firms into a planned industrial surrounding and offer other amenities which affect locational decisions. Such developments will allow central cities to better compete with suburban municipalities in attracting industrial firms. Since firms have now translated beautiful and clean surroundings and community prestige into economic factors of location, new towns appear to be a viable approach to the total economic development of blighted black communities.

NOTE

1. Data for analyses of industrial relocation were taken from the *Illinois Manufacturers Directory* and the *Chicago, Cook County and Illinois Industrial Directory, 1965-1971*.

REFERENCES

Advisory Commission on Intergovernmental Relations (1971) "The pattern of urbanization," pp. 3-40 in L. K. Loewenstein (ed.) Urban Studies. New York: Free Press.

BANFIELD, E. C. (1973) "The logic of metropolitan growth," pp. 11-33 in A. Shank (ed.) Political Power and the Urban Crisis. Boston: Holbrook.

BIRCH, D. L. (1970) The Economic Future of City and Suburb. New York: Committee for Economic Development Supplementary Paper No. 30.

Chicago Plan Commission (1942) Industrial and Commercial Background for Planning Chicago. Chicago: City of Chicago Plan Commission.

Chicago Urban League (1971) Linking Black Residence to Suburban Employment through Mass Transportation: Research Report. Chicago: Chicago Urban League.

CHRISTIAN, C. M. (1973) Information presented is that based on the author's in-depth, personal canvassing, and interviewing of owners, officers, and managers of industrial firms in the Chicago metropolitan area.

——— and S. J. BENNETT (1973) "Industrial relocations from the black community of Chicago." Growth and Change 4 (April): 14-20.

CREAMER, D. (1963) Changing Location of Manufacturing Employment. New York: National Industrial Conference Board.

Department of City Planning of Chicago (1961) Industrial Movement and Expansion, 1947-1957. Chicago: Chicago Department of City Planning.

DeVISE, P. (1967) Chicago's Widening Color Gap. Chicago: DePaul University Press.

GOLD, N. N. (1972) "The mismatch of jobs and low-income people in metropolitan areas and its implications for the central city poor," pp. 443-486 in S. M. Mazie (ed.) Population Distribution and Policy, Volume V, Research Reports. Washington, D.C.: Government Printing Office.

HANDLIN, O. (1959) Boston Immigrant. Cambridge: Harvard University Press.

HARRIS, C. D. (1943) "A functional classification of cities in the United States." Geographical Review 33 (January): 86-99.

KAIN, J. F. (1970) "The distribution and movement of jobs and industry," pp. 1-43 in J. Q. Wilson (ed.) The Metropolitan Enigma. New York: Doubleday.

MAYER, H. M. (1970) "Chicago: city of decisions," pp. 11-26 in J. Cutler (ed.) The Chicago Metropolitan Area: Selected Geographic Readings. New York: Simon & Schuster.

Mayor's Committee for Economic and Cultural Development (1970) A Partnership of Action: Mid Chicago Economic Development Project. Chicago: Mayor's Committee for Economic and Cultural Development.

McLAUGHLIN, G. E. (1938) Growth of American Manufacturing Areas. Pittsburgh: University of Pittsburgh, Bureau of Business Research.

MITCHELL, W. M. (1933) "Trends in industrial location in the Chicago region since 1920." Ph.D. dissertation. University of Chicago.

MOSES, L. and H. F. WILLIAMSON, Jr. (1967) "The location of economic activity in cities." American Economic Review, Papers and Proceedings (May): 211-222.

NELSON, H. J. (1969) "The form and structure of cities: urban growth patterns." Journal of Geography 68 (April): 198-207.

——— (1955) "A service classification of American cities." Economic Geography 31 (July): 189-210.

REEDER, L. F. (1952) "Industrial location in the Chicago Metropolitan Area with special reference to population." Ph.D. dissertation. University of Chicago.

ROSE, H. M. (1972) "The spatial development of black residential sub-systems." Economic Geography 48 (January): 43-65.

SCHNORE, L. F. (1965) The Urban Scene. New York: Free Press.

ULLMAN, E. L. (1970) "A theory for the location of cities," pp. 58-67 in R. G. Putnam, F. Taylor, and P. Kettle (eds.) A Geography of Urban Places. Toronto: Methuen.

U.S. Department of Commerce, Bureau of Census (1971) Census of Population and Housing, 1970, Chicago and Northwestern Indiana Standard Consolidated Area: Statistics of Census Tracts. Chicago: Reproduced by Chicago Association of Commerce and Industry.

U.S. Department of Commerce, Bureau of Labor Statistics (1967) Social and Economic Conditions of Negroes in the United States, Current Population Report No. 332, Series P-23, No. 24. Washington, D.C.: Government Printing Office.

VERNON, R. (1959) The Changing Economic Function of the Central City. New York: Committee for Economic Development.

WALKER, M. L. (1938) Urban Blight and Slums. Cambridge: Harvard University Press.

WARNER, S. B., Jr. (1962) Streetcar Suburb. Cambridge: Harvard University Press and MIT Press.

WINGO, L. (1972) "Introduction: some public economics of social exclusion," pp. 1-8 in L. Wingo (ed.) The Governance of Metropolitan Regions: Minority Perspectives. Washington, D.C.: Resources for the Future.

7

The Street Gang in Its Milieu

DAVID LEY

Whatever we may think of the strength of virtue, experience proves that the higher orders are indebted for their exemption from atrocious crime or disorderly habits chiefly due to their fortunate removal from the scene of temptation. [Archibald Allison, 1840][1]

INTRODUCTION

☐ JUVENILE DELINQUENCY occurs in settings within neighborhoods which are parts of cities comprising collectively a nation-state. None of these spatial scales are incidental in an explanation of street crime by youngsters; the variable nature of social processes through the several tiers of space affects the incidence of delinquent behavior. At the block scale, differential degrees of social control and surveillance provide settings either more or less permissive of deviant activities; at the national scale, questions of ethnic pluralism, class rigidity, unequal opportunities, and national ideologies define states such as cultural lag between host and immigrant populations and blocked opportunity which have always favored the appearance of indigenous institutions like the street gang. Place is not a constant, and studies of the street gang culture should show sensitivity to its variabilities. Moreover, once local conditions exist propitious to the emergence of the gang, there is a feedback mechanism whereby the gang adds its own distinctive contribution to the meaning of its environment. Thus, there is a spatial ecology which leads to the formation of the gang, and then an ongoing social ecology which lends a meaning to space. These two interlocking states will provide the major themes of this essay.

The ambience between space and society was scarcely absent in the early writing on the street gang. The meticulous investigations of the Chicago human ecologists continually emphasized the reciprocal relationship between the environment and deviant group behavior. Robert Park, in his preface to Frederic Thrasher's *The Gang*, wrote that, "It is the slum, the city wilderness, as it has been called, which provides the city gang its natural habitat" (Thrasher, 1963: ix). The detailed statistical research of Shaw and McKay (1969) in a wide range of American cities had made the same discovery: that the gang was a characteristic of a particular environment, inner-city neighborhoods of aging housing, high densities, and nonconforming land use (particularly industry and transportation rights of way), with large immigrant populations and attendant high levels of social pathology.

THE SPATIAL ECOLOGY OF
THE STREET GANG

Perhaps the most intriguing finding of Shaw and McKay was that gang behavior is fixed in particular neighborhoods, not particular social groups. In the revised (1969) edition of their book, they show that the areas of acute gang delinquency in Chicago have remained relatively stable over sixty years, despite the movement of a succession of ethnic waves through these neighborhoods. In the 1920s in Chicago, the most recurrent ethnic origin of gangs was Polish; by the 1960s this dubious honor had passed to the black population. In New York in the 1970s, this stage in turn is being challenged by the formation of a tenacious Puerto Rican gang structure. If the sequence were to be traced back retrospectively into the nineteenth century, the same generalizations would hold. Philadelphia delinquency patterns for 1840-1870 (Johnson, 1973) contained the seeds of the distribution in 1930, which showed continuity with the spatial pattern of the 1970s. But the occupants of the high crime neighborhoods in the mid-nineteenth century city preceded the arrival of immigrant groups from southern and eastern Europe. Their ethnic background was a constant source of despair to contemporary observers. Thus John Watson in his *Annals of Philadelphia* for 1856 lamented:

These combinations of lawless lads in the cities of Philadelphia and New York, under indicative names—signifying outlawry and mischief—is wholly a new manifestation of progress.... We are sorry to say, that such boys are peculiarly belonging to the Saxon race ...
[Smart, 1973]

The spatial regularity of teenage crime in the city through time can lead to awkward questions of causality. A cartographic analysis emphasizes local variables at the expense of more global constructs. Shaw and McKay's correlation analysis of map distributions continues to be a popular methodology in the study of the geography of crime (Harries, 1974), but this methodology treats variables which operate at only one spatial scale. At worst, such analysis leads from conclusions such as "there is a high correlation between group X and delinquency" to an extrapolation that "there is crime wherever you find group X" to the stereotyped assertion "group X cause crime." Harvey's (1972) criticism of such blunt forms of ecological analysis is well-taken. There is no fatalistic inevitability that inner-city neighborhoods in American cities will become the core of high levels of criminal activity. A delinquent act is an adaptive response to an existential definition of a situation. It is an act which incorporates both a personality and an environment, and the environment contains components which are national as well as local. The cross-cultural examples of teenage gangs which follow show that the violent gang is primarily a phenomenon of American but not Canadian cities; and consequently a causal explanation must include variables and ideologies which are national, and not simply local, in their range.

Thomas and Znaniecki (1927) provide a useful starting point for discussing the ecology of gang behavior. They pointed to four general classes of individual desires: "the desire for recognition or status; the desire for safety or security; the desire for power; the desire for new experiences" (p. 73). In other hands these categories shift somewhat; Park restates Thomas' four needs as security, new experience, affection, and recognition (Park and Burgess, 1968: 119). Implicitly, if a society in a given location fails to satisfy these needs, then the society is redefined in some manner by those who are deprived. Gang formation is one such form of societal redefinition by inner-city youth to make good perceived inadequacies.

This broad statement has formed the main theme of social science theorizing since the early work of Thrasher and Shaw and McKay. Though there are several nuances between individual theories, the consensus has been to deal specifically with the problem of status deprivation among inner-city adolescents. Inner-city society is characterized by an absence of status, and the street gang is an indigenous institution created to supply status in a subculture of blocked opportunity where recognition cannot easily be won legitimately. The street gang is an organization to distribute status rewards. Recognition in an organization is achieved by excelling; in the gang it is achieved by excelling at acts perceived as deviant by larger society.

Cohen (1955) was among the first to develop this theme in his analysis of the problems confronting the inner-city youth.

> These problems are mainly status problems: certain children are denied status in the respectable society because they cannot meet the criteria of the respectable status system. [1955: 121]

Cohen conceptualized gang delinquency as a conscious reaction against the middle-class values from which the youth felt excluded. This line of thinking was supported by Miller (1958) who saw "a dominant concern with status" as a central if more abstract dimension of a lower-class milieu which regarded such attributes as toughness and trouble as integral parts of their lifestyle.

Status compensation theories continue to be broadly supported in more recent research (Short, 1964; Cloward, 1968; Kobrin et al., 1968), though there is now an awareness that the audience before which gang members perform is internal rather than neighborhood-wide. Group rather than class motivations appeared critical in an extensive Chicago study:

> The gang provides the audience for much of the acting out which occurs in situations involving elements external to the group, and it is the most immediate system of rewards and punishments to which members are responsive much of the time. [Short and Strodtbeck, 1968: 215]

This recent shift of emphasis is a return to a position held by Thrasher in his pioneer work: "Internally the gang may be viewed as a struggle for recognition" (Thrasher, 1963: 230). Even Yablonsky (1970), who places more emphasis than most participant observers upon the role of abnormal personality factors in the conflict pattern of the violent gang, returns repeatedly to the catalytic effect of a threat to a gang member's 'rep,' his perceived status within the group, in precipitating violent behavior.

There has been less definite discussion of the role of the gang in meeting the other three needs suggested by Park—affection, new experience, and security. An analysis of affection is more usually subsumed under the rubric of self-identity in a peer group environment; whilst gang contacts might follow the inner-city code of "tough love," dispassionate respect and adherence, they are not usually characterized by strong and intimate trust relationships. Although gang life has its share of dreary repetition, it holds promise of new adventures, of an enticing world of make-believe, a uniform, a new identity, a new name, a new home, a world where much becomes possible which is denied in the austerity of inner-city living. In this world on the edge of reality the gang member "lives among soldiers and knights, pirates and banditti" (Thrasher, 1963: 85).

The fourth basic need suggested by Park was security, and this variable is of particular interest to geographers, for in the gang world security means safe space. Territoriality seems to be a primitive mechanism of social regulation in environments of abnormal flux and uncertainty (Esser, 1971). For the gang member, his turf and particularly his hangout act as a haven of safe space. The role of the turf and the hangout as refuge is a key to the existential quality of inner-city space. A gang's clubhouse, typically in an abandoned structure, is a place smothered in identity. Decoration is often elaborate, with photographs, psychedelic posters, and Day-glo paintings suggesting the collective fantasies of the group; for an outsider, his own sense of intrusion is marked because the sense of possession by gang members is so strong. For some gang members, as for this Puerto Rican youth in New York, the hangout is a surrogate home.

Tony, another Cypress Bachelor, says his gang provided what the world had failed to give him. "This is our home," he says, gesturing at the five-room basement clubhouse in an unheated apartment house. [Tolchin, 1973]

Beyond the hangout is the space over which the gang claims control, its territory:

> The area immediately surrounding this cherished spot [the hangout] is home territory, beyond whose borders lie the lands of the enemy and the great unknown world. . . . Most of the activities of the group have a definite relationship to this geographical division of its world.
>
> [Thrasher, 1963: 90]

As Thrasher indicates elsewhere in his book, the division of space was not this simple even in Chicago in the 1920s, and it has certainly been complicated by the availability of the motor car to inner-city youth in the contemporary American city. Nevertheless, the bold profile of his schema is still valid. The street gang remains a place-based subculture; the majority of Philadelphia gangs take their name from a street intersection near the core of their turf (e.g., 15-Clymer, 58-Osage, 12-Poplar), and the most common precedent for a gang fight is one small group of boys challenging another: "Where you from?" Territory is usually the rationale, if not always the cause, for gang fighting (Yablonsky, 1970: 108).

The advantage of this sociological approach to understanding the street gang is that it does not restrict the investigator to a consideration of local variables alone. Shaw and McKay's cartographic analysis of juvenile delinquency admitted only those factors which operated at the urban scale. Similarly, when Park wrote that it is the slum which is the natural habitat of the gang, he was rejecting any factors which originated at a scale greater than the neighborhood. In contrast, the theory of the social marginality of ethnic and working-class groups which demands such compensatory organizations as the street gang implicates a national culture, and not local pathologies. As the section on Vancouver will demonstrate, there is nothing inevitable or "natural" about inner-city environments which cause them to be the habitat of the violent gang.

At the same time it would be short-sighted to disregard entirely the role of local variables upon gang formation in the American city. Just as inner-city groups suffer a *social marginality* at the hands of the mainstream culture, so they also suffer a *spatial marginality*. Typically, they are groups with very weak discretion over their neighborhoods (Wolpert et al., 1972); they are marginal bidders for land in the face of more powerful competitors. Inner-city neighborhoods are areas of nonconforming land use, where industry, transportation lines, and institutions are intermingled with working-class homes. This interpenetration and the nuisances which accompany it—noise, pollution, traffic congestion—give the inner-city resident a far weaker sense of spatial mastery than his suburban counterpart. In addition he has more limited discretion over even his own home. Frequently a tenant, he has no control over the vagaries of an absentee landlord or a land-hungry urban renewal agency (Keyes, 1969). It is he who is regarded as a transient by City Hall, and not the institutions and other nonconforming land uses which challenge the quality of his neighborhood. When the institution wishes to expand, it is always the resident who pays the cost; in West Philadelphia, for example, the black inner-city neighborhood of Mantua has been razed by the expansion of the University of Pennsylvania and Drexel Institute (Wolpert, 1970). Shaw and McKay noted that the zone of the gang was also the zone of industrial invasion, a high proportion of rental units, and a high level of demolition. In the same manner the gang habitat of the present generation is characterized by institutional expansion, rental public housing, and the federal bulldozer, active in urban renewal and highway construction. In each instance, then, the spatial marginality of the inner-city resident is reinforced; it is others who control the spatial options in his neighborhood.

Such a lack of spatial mastery provides the mandate for a territorial imperative (Suttles, 1972; Ley and Cybriwsky, 1974), for the establishment of a small secure area where group control can be maximized against the flux and uncertainty of the confusing and sometimes manipulative city. This is certainly the rationale for the defended neighborhood—the community defended from external predators by informal vigilante groups or more formal community associations. The gang territory might be seen as a similar attempt to impose some form of predictability over a small part of an uncertain environment.

There is yet a third level of marginality which contributes to the formation of the gang subculture. Rather than a social or spatial marginality, this is a *life-cycle marginality*, the confusion of adolescence itself. This contributory cause of the gang was mentioned by Thrasher, but has scarcely been developed by more recent writers. Though probably of lesser salience than the other forms of marginality, the adolescent stage of the life cycle does raise its own uncertainties which can best be accommodated in the context of a peer group social unit.

These three levels of marginality represent a synthesis of a broad literature theorizing upon the causes for the emergence of the street gang. Still other levels might be added. A few writers, notably Yablonsky, have stressed personality marginality, particularly socio-pathic factors, as individual variables encouraging membership in fighting gangs, though this is certainly a minority position. The two major variables are here regarded as being social and spatial marginality, with life-cycle marginality a necessary but not a sufficient cause. The major variables comprise a perceived or *behavioral environment* (Kirk, 1963) conducive to the emergence of a gang structure. Moreover, this behavioral environment confronting teenagers is to some degree place-specific. As the following discussion will show, it is necessary to admit variables which act at a range of spatial scales to understand the incidence of gang behavior. The variables, both individually and collectively, vary in their salience in different areas. The whole spatial ecology of the street gang must be encountered.

THE STREET GANG IN VANCOUVER [2]

Vancouver, British Columbia, is a young metropolis of over one million people. It is about the same age as the Chicago of which Thrasher and Shaw and McKay were writing during the interwar years, and is growing at the rate of around 35,000 persons a year, with 40% of this number immigrants, non-Canadian nationals. The city has undergone drastic physical change in the past twenty years to accommodate this influx, which has resulted in land-use instability with considerable demolition of inner-city neighborhoods. It would seem as if, at the urban scale, the ecology which Shaw and McKay claimed necessary for the formation of the street gang is present.

But unlike other newly emergent West Coast cities, most notably Los Angeles, Vancouver scarcely has a gang problem. Though there are sporadic reports from social service groups in inner-city com-munities of a "gang problem" (Sarti, 1973; Gibbs, 1973), these reports are so irregular that the question has been raised in a forum of social workers as to whether there are bona fide gangs in the city at all (Barling, 1973). A researcher in the neighboring city of Victoria, investigating the behavior of a loose group of mildly delinquent adolescents who gathered around a drive-in restaurant, had to invent his own name for the "gang"; they recognized no such formalization of a gang identity for themselves (Porteous, 1973). If we accept Thrasher's definition of the gang as including the components of territory, conflict, and planning (Thrasher, 1963: 46), then, as we shall see, there are no groups in Vancouver conforming to such a definition.

The most likely candidate for gang status in the city is a group we shall call the Derby Park gang. This group of adolescents might be known to a vigilant urban dweller from their magic-marker graffiti in bus shelters and public buildings over a broad expanse of the city; their rather inconspicuous graffiti have been noted as far as ten miles from Derby Park. But this group is by no means an occupant of the decaying inner city; although a few of the members live on an adjacent Indian reservation, about 40% live close to Derby Park, in a neighborhood which includes the homes of several city aldermen and is within the top quartile in terms of median family income. The neighborhood is residential with a bountiful acreage of parkland, and is totally free of noxious industrial and transportation land uses; there has been almost no new building over the past decade, and no identifiable immigrant minority groups occupy the area.

The emergence of Derby Park to a notoriety above that of their inner-city contemporaries is incongruent in the light of results by researchers of the American street gang. It would appear that the main variables identified in the American literature which make the inner city the "natural habitat" of the gang, while perhaps necessary, are certainly not sufficient in at least one Canadian city.

In fact, the origins of the Derby Park gang seem to stem from a form of social marginality which is not place-specific. Core members of the group are all high-school dropouts—indeed dropping out is a requirement for entry to the inner core. Derby Park is a classic case of Cohen's (1955) theory of the gang as a reaction-formation against middle-class values, a theory which is probably most appropriate among middle-class boys raised on such values, rather than for inner-city youngsters. Members of Derby Park delight in giving an impression of being bad; they work at it. Their prime targets are the symbols of middle-class respectability: their dress is self-consciously nonconformist, their graffiti are always decorated with swastikas, and a popular target are buildings on a nearby university campus. While being interviewed, members of the group enjoyed being perverse; when asked how they perceived themselves and their friends, their most common response was "with my eyes"; when asked if they had to do anything to join the group, two members replied "kill an old lady"; when asked for a listing of group activities, the youngsters gave only answers which were designed to shock, introducing obscenity wherever possible.

At one level, Derby Park is a gang. It has identified itself as such, and sought to reinforce this identity through its graffiti and impression management. But this public image disregards the private world of the youth. There is no leader and no gang hierarchy; there is no recognition of territory, no staking out of it, and no territorial conflict. There is limited consensus in the group; during the interview, half the respondents said there was a code of honor, half said there was none; some said there were group nicknames, some said there were not. The gang claim is a posture, and a theatrical posture at that, for a dropout group who spend much of their time drinking and taking soft drugs. The Derby Park group is truly a phenomenon of a middle-class neighborhood. The boys cannot simply drop out to the withdrawal lifestyle centered around alcohol and drugs. They must drop out with distinction, "really" drop out; they have perverted, but not rejected, the middle-class values of achievement which they malign so freely.

A second Vancouver group, centered around the Pacific National Exhibition grounds (PNE), resides in a neighborhood which resembles more closely the milieu supposedly conducive to gang formation. The area is a blue-collar neighborhood, with many immigrant families, on the outer edge of Vancouver's inner city. The PNE is a large facility that has some blighting influence on the adjacent residential area. High noise levels associated with its varied entertainments constitute one polluting factor, but more serious objections are raised by the illegal parking of patrons on neighborhood streets. There are other nonconforming land uses, mainly industrial, particularly along and near the waterfront.

This group is more localized than the free-ranging youth from Derby Park. The majority of them live within a mile of the PNE, and their four main loitering sites—a hamburger drive-in, the area behind a high school, a small park, and a community center—are all adjacent to the PNE grounds; in contrast there is a distance of up to four miles between the most popular hangouts of the Derby Park group.

It is the group's presence at the community center which is regarded as most troublesome by caretaker groups, and led to their designation as a "gang" by the center's new director. Another center worker lamented on their intimidation of younger boys in the games room and their loud and aggressive behavior. But closer examination showed that much of this ribaldry was intragroup; in the weight-lifting room, for example, the boys' behavior was competitive and status-seeking before their own peer group. Nevertheless, there are symptoms of a more durable social unit among the boys. They met at high school and a number have been together for over two years; they claim to have been involved in shoplifting and sporadic breaking and entering in the past, and their socialization into illegality is maintained by their ongoing alcohol and drug consumption.

Once again, however, it is important to distinguish between a public and a private image. For example, there was a good deal of talk among girls on the edge of the group that one of the boys had been in a knife fight some three years before; but subsequent inquiry revealed that reality was more pallid, as one of the boys had nicked his opponent on the thumb with a small penknife. The image of assertive loudmouthing and aggressive individuality is equally super-ficial, though none the less disturbing for other community center visitors. Several observations showed the boys to be anxious conformists looking for leadership. On one occasion a community graffiti board was put on display, with an open invitation for messages. One of the boys, after some thought, wrote "Polack" on the board, and this was mimicked by several others; then a second drew a submarine, and two of his friends followed suit, drawing a submarine of exactly the same size and shape. A fourth modified slightly the submarine design into a hashish pipe, and with a laugh the eight or nine boys left the board. As one of the community center workers commented: "These kids are just a group of lambs looking for a shepherd."

There is little about this group to suggest the solidarity of a genuine gang. There is no formal social organization, no recognized turf, no conflict behavior. The majority of the group is still at high school; and though first- or second-generation Canadians, the group shows a bewildering ethnic diversity—a miniature United Nations with 12 different ethnic backgrounds represented among 17 core group members (Table 1).

With such a pallid gang identity, it is scarcely surprising that the group makes an extremely languid claim on space. It makes no claim to territory, and talk of such elicits blank stares from group members. Even in the community center the group is scarcely an invasion force and commonly occupies corridors and other inter-stitial areas rather than activity centers.

This group of youngsters is responding to a behavioral environ-

TABLE 1

SELECTED CHARACTERISTICS OF THE PNE GROUP

Age	Sex	Years of Schooling	Work Status	Ethnic Origin
15	F	9	At school	Finnish
17	M	11	At school	Irish
16	M	9	Looking for work	Polish
17	F	12	At school	Scottish-German
16	M	9	Works	Dutch
17	M	11	At school	Danish
16	F	11	At school	English
17	M	9	Dislikes work	Portuguese
17	M	9	Relaxing	Chinese
17	M	10	Works	Scottish
17	M	10	Works	German
15	M	10	At school	German
17	M	11	At school	German
17	M	10	Works	French
17	M	11	At school	English
17	M	11	At school	Italian
17	M	11	Out of work	Scottish
15	M	9	Dislikes work	Irish

SOURCE: John Purdy, Department of Geography, University of British Columbia, unpublished data.

ment which does not impose upon them a position of intense marginality. To use rather old-fashioned language, they are eminently "recoverable" to mainstream norms. They respond readily to adults who acknowledge them. Their main grievance with the new director at the community center was his neglect of them: "He just sits up there in his office talking to old fogies and doing his bookwork. He never comes down here and talks to us."

This is a position of marginality which is easily corrected. In Vancouver a mild behavioral environment has led to a languid, stillborn gang structure with a more diffuse spatial distribution which is only weakly territorial. However, in Philadelphia pronounced marginality has created a tenacious and abrasive gang structure, concentrated in the inner city and strongly territorial. There is a direct correspondence between the level of marginality and the clarity of the social and spatial identity of the gang structure. This difference is more the difference between two nations than between two cities.

THE STREET GANG IN PHILADELPHIA

The gang in Philadelphia is also striving for recognition and peer group status. But here inner-city youth encounter a behavioral environment which imposes an intense marginality upon them and in turn elicits from them a perverse response. In Philadelphia the needs of the boys are greater, as their marginality is more acute; at the same time options for legitimate fulfilment are more limited and the cost of illegitimate outlets is much higher.

It is not at all difficult to document the role of status deprivation in motivating adolescents into gang membership. In the words of gang member 'Mousie':

Some people try to stay out of gangs. . . . They stay in their house all the time, but they get tired of it. They don't want to stay in and be nothin'. When you're in a gang you're somethin', man. You've got respect.

This reality is understood by many gang workers. One of the greatest problems in turning gang members "conservative" is to open new channels for winning prestige and status.

[Existing gang conversion processes] are not enough because they do not deal with the prestige problem. Old forms of behavior are not given up easily, particularly if there is no corresponding gain in another direction. [Austin, n.d.]

The more brazen the behavior, the greater the return in status. In June 1971, two gangs from the inner northern suburbs carried their feuding almost ten miles to the south. Six members of the Pulaski Town gang fired into a crowd leaving a rock festival in South Philadelphia which included six members of a rival gang; luckily, there were only three injuries. Members understand the illegality of their acts and expect arrest; in gang homicides arrest rates are close to 100%. Yet gang honor and prestige demand individual sacrifice and loyalty to an unrelenting code.

This point can be emphasized by describing additional incidents drawn from an inner-city neighborhood of Philadelphia we shall call Monroe (Ley, 1974b). Early in the summer of 1970, a member of the 39th and Sutton gang was walking down the street holding three fingers high and chanting 3-9-S when a shotgun blast rang out and tore off a finger. The next day in school a member of the rival 45th and Richmond gang proudly displayed the missing finger.

Later that summer, two outsiders, O. and M., journeyed to 39th and Sutton for a gang party.[3]

While there O. and M. reportedly bragged they were a "hit and wheel man team" meaning that O. could drive a vehicle while M., a passenger, could aim a shotgun at the window.

A few of the gang members (39-S) asked the two visitors, as a favor, to drive around the streets, locate and kill some members of an enemy gang in the area (45-R).

L.H., 22, a member of 39-S street gang, O., and M. drove around the west side of 46th Street for some time but could find no potential victims.

It was suggested that the trip cross to the east side of 46th Street. As they approached Bates and Malcolm Streets, they spotted A. and R. who were walking towards R's home . . .

A. and R. were members of the 45th and Richmond gang. Both were shot, and one later died from his wounds.

It is instructive to analyze the case more closely. It begins with teenagers denied access to legitimate outlets for recognition in the wider society. They are forced to invent alternative sources within their own life space to meet this need, and their solution is the enacted status of gang membership. But membership implies role-playing. Amongst unknown gang members it was natural for O. and M. to assert themselves by boasting of their prowess in activities esteemed by the peer group present. It was equally natural for the peer group to challenge the legitimacy of that boast and demand to see it demonstrated. Winning status in the inner city carries a heavy price. The gang code precluded backing down once the challenge had been issued.

Violence and aggression are inevitable products of the gang code. Between October 1, 1962, and December 31, 1970, 158 gang-related homicides, and over 850 incidents involving a stabbing or shooting, were reported in Philadelphia. This is certainly an underestimate as, for various reasons, many assaults are not reported. The finger-shooting incident reported above was learned from Monroe teen-agers; no record exists in the city files of this incident having taken place. Though the number of such incidents may have peaked, the trend until recently was upward. In 1970, 35 gang-related homicides had occurred; but by the end of 1971 there had been another 45 fatalities.

The size of the problem is uncertain. A frequently quoted figure places the number of gangs at 93, with 5,300 members. However, a

1971 listing of the Youth Conservation Service contains 190 groups, and officials speak of the existence of "at least" 220. The groups engage in differing levels of activity. Many are dormant or sporadic in their acts of violence, but hostility is easily kindled. A number are classed as active, protecting a turf, involved in neighborhood disturbances, prepared to use lethal weapons, having a hangout and organizational power structure, and hostile to all authority.

There is no constant gang size. In North Philadelphia sizes of the active groups range from the Valley gang, a sometimes loose-knit coalition of 250 members, to smaller groups like the 20-member band which claims the turf around 15th and Oxford Streets. Fissioning is a common process among the larger gangs. Cool World Valley was formerly a part of the Village gang, but then it moved north of Lehigh Avenue and assumed a separate identity. The Village has since become one of Cool World Valley's main antagonists.

Neither is there any constant turf size (Figure 1). Turfs range in area from the 60 blocks claimed by the 200-member Zulu Nation to only one or two blocks controlled by smaller groups like 24th and Redner.

The map shows only the northern quadrant of Philadelphia's inner-city zone. A tenacious gang structure exists in the equivalent sections of South and West Philadelphia (King, 1973). In all three areas the majority of the gangs are currently black, but the presence of fighting gangs long precedes the onset of heavy black immigration (Laurie, 1973); one West Philadelphia gang, the Barbary Coast, has survived through several generations of Philadelphia history. These zones of heavy gang activity coincide with the general pattern of Shaw and McKay's data for the city in the 1920s. But there has been one important departure from this pattern not shown on the map. The formation of fighting gangs, while still concentrated in the inner city, has spread beyond this zone. In the past decade new gangs have formed 60 blocks from downtown in middle-class residential neighborhoods like West Oak Lane. Gang killings have occurred on the city line, and suburban gangs are appearing (Lynch, 1973; Nordland, 1973). Clearly the behavioral environment of perceived marginality is extending through the city's socioeconomic spectrum —another indication that gang behavior can be instigated by variables less parochial than those of the inner city.

Gang incidents are often precipitated by the intrusion of one group upon the territory of another. As movement across turfs becomes so difficult, turf infringements are most likely to be made

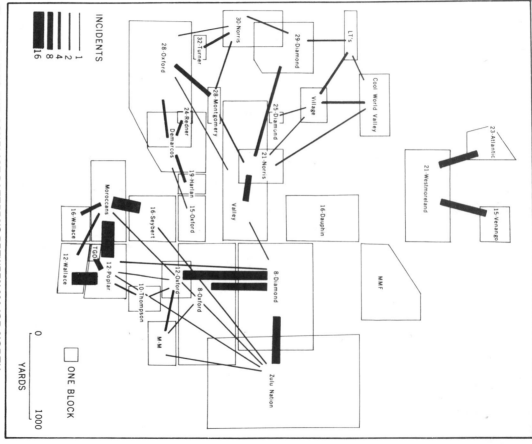

Figure 1: TURFS AND CONFLICT PATTERNS BETWEEN MAJOR NORTH PHILADELPHIA GANGS, 1966-1970

INCIDENTS

	1
	2
	4
	8
	16

0   YARDS   1000

☐ ONE BLOCK

by neighboring gangs (see Figure 1). There is an overwhelming tendency for groups to fight other groups located close to them. Among 32 of the most active street gangs in North Philadelphia it is possible to identify 5 conflict cells. Intergroup fighting is largely confined to these cells of close neighbors; hostilities against a more distant group are irregular.

Propinquity emerges as of critical significance in the choice of enemies. It is possible to distinguish 188 incidents between the 32 groups between 1966 and 1970. Incidents are defined as a homicide, or reported stabbing, shooting, or gang fight. This figure is a gross underestimate of actual conflict, for, apart from unrecorded incidents, not all of the recorded data include full gang identities of both assailant and victim. As each interaction represents a conflict for two groups, the number of individual incidents is doubled. Of these 376 incidents, 226 or 60% occurred between gangs who shared a common boundary, and another 23% between gangs whose territories were two blocks or less apart. Only two incidents occurred between groups whose turfs were separated by more than ten blocks (Figure 2). If we were to graph the number of incidents against the nearest neighbor ranking of adversaries, we would find again a very marked distance decay effect. Of the 376 offenses, 166 are perpetrated between nearest neighbors, and another 137 between second or third

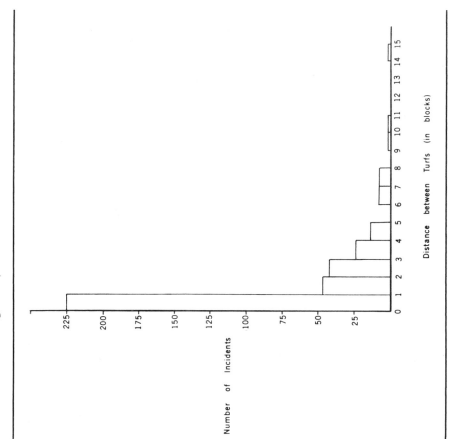

Figure 2: PROXIMITY AND AGGRESSION

nearest neighbors, so that 80% of intergang conflict is conducted between close neighbors. In the perverted world of Philadelphia's inner-city gang culture, one's neighbor is one's enemy.

THE SOCIAL ECOLOGY OF A
GANG NEIGHBORHOOD

Just as extreme marginality leads to alternative organizational structures like the street gang, so the gang adds its own stamp to the nature of neighborhood space. Despite rhetoric to the contrary, neighborhood gangs are more usually predators than protectors of the local community. To have a local gang is to have a local problem. The salience of the problem, of course, varies from place to place, but in many inner-city areas the local gangs are regarded as a critical dimension of the behavioral environment. In the Monroe area of North Philadelphia, a neighborhood outside the most intense area of gang feuding, juvenile problems were perceived as easily the most serious in a sample of community responses (Table 2).

Even more significant were a series of questions which were designed to discover how inner-city residents differentiated various other neighborhoods in the ghetto from Monroe. Three categories of discriminators were noted: general evaluative statements (such as "It's better here"); statements comparing the physical environment (housing, traffic, and so on); and statements comparing the behavioral environment. Surprisingly, the largest group of discriminators fell into the category of the behavioral environment, and the first-ranking subgroup comprised phrases describing the level of gang activity (Table 3). The role of the street gang should clearly not be underplayed in understanding the experiential reality of inner-city life.

TABLE 2
LEADING PROBLEMS PERCEIVED BY MONROE RESIDENTS

Community Problem	Number Times Mentioned
Gangs/teenagers	55
Drugs	33
Graffiti	18
Crime	15
Bars/drunkenness	11

SOURCE: Ley (1974b)

TABLE 3

DIMENSIONS OF NEIGHBORHOOD DIFFERENTIATION

Evaluative			42
Housing quality	18		
Physical condition	9		
Descriptive land use	8	Physical Environment	44
Income	5		
Traffic density	4		
Gang activity	21		
Quietness	16		
Violence	15	Behavioral Environment	64
Bad habits	6		
Privacy	6		
Total			150

SOURCE: Ley (1974b)

Fear of gangs is a major deterrent upon adolescent movement, including the journey to school. The threat of gang intimidation is believed by educators to be the major cause of the high absentee rate of inner-city schools. The significance of this variable is affirmed by the Philadelphia School Board's practice of assigning caseworkers to areas designed to coincide with gang turfs. Inner-city school districts experience a high number of student transfers each year, and caseworkers believe that the majority of the transfers are gang-related. Existing evidence would support this hypothesis. A map was prepared showing the home address of male twelfth-grade absentees from one high school for a single day in May 1971 (Figure 3). On that day there were almost 250 absentees in this category. The six blocks with the highest number of absentees coincide with the home corner of six of the most active street gangs.[4]

Gang members are not the only group from the adolescent community who are impeded in their movement patterns. Virtually all teenagers are potential gang victims. Consider the statement of Tyrone Warlow, a junior at one of the city universities who had always lived in Monroe. He worked when he could for a community organization trying to dissolve the three local gangs, and regarded the danger as less severe than it had once been.

It used to be terrible, man. It used not to be safe to walk beyond 41st Street without some guys getting at you. I live in 37th's territory but I was never with them. I knew cats on Sutton and some on Richmond. Like I was between all three, and that's cool.

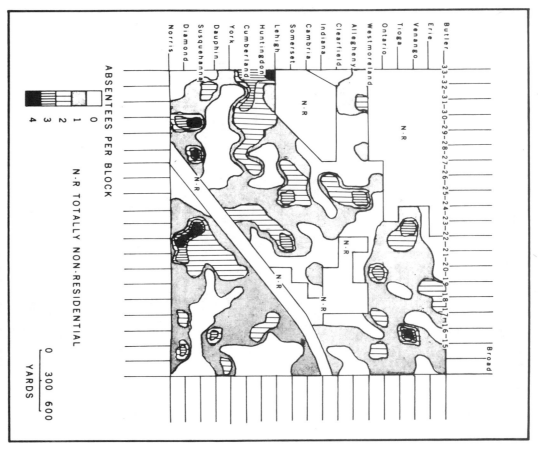

Figure 3: ABSENTEES FROM A NORTH PHILADELPHIA HIGH SCHOOL

ABSENTEES PER BLOCK

0
1
2
3
4

N·R TOTALLY NON-RESIDENTIAL

0   300   600

YARDS

There were three active street gangs in Monroe with a maximum total of 200 members: a small compact group in the western end of the neighborhood (37-Henfield), the largest gang with an elongated territory in the center (39-Sutton), and an eastern group (45-Richmond) generally in alliance with 37-H against 39-S. Gang membership tends to be concentrated near turf cores with a notable transitional zone of low membership along the boundary zones. The marchland between territories, though unoccupied by gang members, is certainly contested. Around their home corners, gang members are secure, but on the periphery they are vulnerable if alone, and hostile if in a group. In the border zone, gang graffiti also takes on an aggressive posture against rivals (Ley and Cybriwsky, 1974). Predictably, the majority of gang fighting occurs along or near this marchland. Again we see the importance of the varying qualities of an existential space in directing behavior, a conclusion which will be repeated as we turn from the teenage gang to the larger community.

STRESS AND SPATIAL BEHAVIOR IN MONROE

A sample of 116 neighborhood adults were questioned to discover if there was any differentiation or bias in their perception of neighborhood space, and to see if their actions were constrained by that differentiation. They were asked if there were any blocks with a bad reputation in the neighborhood and, if so, to identify them. These responses, aggregated, were used to build up a stress surface for Monroe (Figure 4).

Monroe residents have an acute differentiation of space. There is considerable agreement as to the location of good and bad blocks. The blocks equally perceived as most noxious are 39th and Sutton, and 45th and Richmond, the territorial core of the two major gangs. Over 30% of the residents named these blocks, a remarkably high figure when it is considered that none of the respondents lived on these corners, and a number lived up to 13 blocks away. Setting the most noxious blocks at 100%, the contours of a stress surface have been interpolated, so that the rest of Monroe is rated relative to its two worst blocks.[5] Central Monroe receives the highest rating, with ridges running along Sutton and Richmond Street. Not only do these streets contain gang corners, but they are also the scene of regular drug transactions and are often terrorized by junkies and winos. Richmond Street contains a number of abandoned stores and is physically dilapidated. A spine of higher scores runs north up 40th from Sutton, but the streets between Richmond and Sutton generally have lower stress levels. Below Southampton Avenue and east of 46th Street scores of 2-35% continue over all blocks. But the steepest gradient is northward. From the high stress scores on Richmond Street, levels drop abruptly, and north of Rochester Avenue almost every block has a score of 0%. In the space of two blocks, from 45th and Richmond north to 45th and Milton, stress levels fall from 100% to 0%.

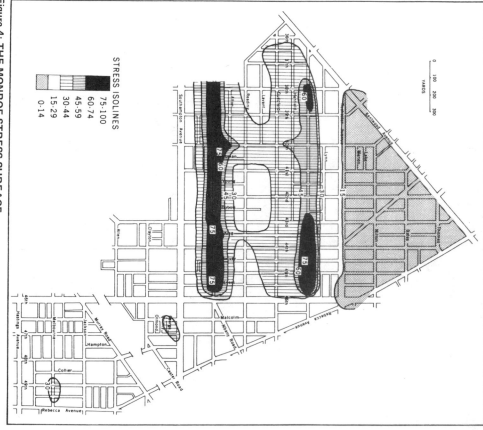

Figure 4: THE MONROE STRESS SURFACE

STRESS ISOLINES

75-100
60-74
45-59
30-44
15-29
0-14

The stress topography generated from this question suggests a consensus of a moderate general surface, with two notable peaks around gang corners in central Monroe and an area of low relative stress north of Rochester Avenue. If, as hypothesized, it is this behavioral environment which is salient, then spatial behavior should be governed by the contours of this surface. Zones of high stress should be zones of avoidance, zones of low stress should encourage approach.

This hypothesis was tested first with a question asking residents which block they would prefer to move to if they were given the chance. Of the 21 answers naming a Monroe address, 16 fell north of Rochester Avenue, in the low stress area, confirming the hypothesis.

A more thorough testing was achieved by a second migration question, which presented each respondent with a card containing the names of the ten main east-west streets. Each was asked which, if any, of the street he or she would like to live on. Again, as expected, there was a preference for the area north of Rochester; predictably, Sutton Street, a gang and drug street, received the least favor. In terms of migration potential, spatial patterns do show themselves sensitive to the configuration of the local behavioral environment.

As a further test of the effect of the stress surface upon spatial behavior, residents were asked to specify pedestrian routes at night from their home to given points in Monroe. It was expected that routes would be selected in accordance with the topography of the stress surface.

The hypothesis was confirmed on the whole. Although residents were informed that they had to make the trip, 56 responses were refusals to make the trip by foot under any circumstances, while another 14 were unable to supply a route, although the destinations were in no instance more than six blocks from the block of residence. Thus 31% of answers represented failures to make the trip at all. Some residents were emphatic in their resistance: "I wouldn't go for the world," "I wouldn't go even in the day time," "I'd take an airplane. Wouldn't go there at any time." Only 27% of responses selected a direct route between home and destination, and some of these added qualifiers: "I'll tell you the route, but it's not a safe one", "I'd go the direct way, but I'd take chains with me!"

The largest group of responses (42%) did name a route, but the path selected was an indirect one, adding a quarter, a half, or even more to the length of the journey. These diversions were made to avoid perceived dangerous blocks, and mark a preference for the safer, better-lit, skeletal pattern of thoroughfares.

The high stress area of Richmond Street, particularly around the gang corner at its eastern end, prompted persistent avoidance by residents. The pattern of avoidance was practiced, for example, when the journey began at King Street (Figure 5). To arrive at 45th and Guildford, seven routes are direct, crossing Richmond. But nine paths skirt the danger of Richmond Street by following 46th, even though almost one half the distance is added to the journey. Similarly, when the destination from King shifts to Reading and King, only five people would travel directly down King, crossing Richmond. Two would cross Richmond on 43rd, a wider and better-lit street on a bus route, while six would make the crossing on

Figure 5: PEDESTRIAN ROUTES FROM KING STREET

46th Street, the main north-south artery, although this path would add four additional blocks to the original five-block journey. Six residents would not make the journey at all.

We conclude that the spatial behavior of residents reflects a response to their discrimination of blocks on a stress index. Blocks which they perceive as having a high stress score are avoided, and blocks with lower stress scores become the preferred avenues of movement. Notice also that the stress surface of residents is the reverse of the surface for gang members. For the latter, security is maximal at the core of the turf; but for the rest of the community, stress reaches its peak on the gang corners, and movement patterns are designed to avoid them.

CONCLUSION

In this essay we have stressed that the juvenile gang cannot be discussed independently of its milieu. It is a milieu, a behavioral environment with salient variables at a range of spatial scales, which forces the emergence of the gang as a solution to culturally imposed marginality. Once formed, the gang provides its own imprint to space, adding a salient dimension to neighborhood perception and a powerful constraint to neighborhood behavior by nongang members.

In seeking to understand the gang, the search for causality must not end with the inner city. In the wise words of Jacob Riis: "The gang is a distemper of the slums; a friend come to tell us that something has gone amiss in our social life" (Thrasher, 1963: 342). The gang, its hangout, and its territory are attempts by inner-city adolescents to impose order upon an uncertain world and even to squeeze some glory from it. But as the cross-cultural examples showed, there is nothing inevitable or "natural" about an inner-city experience of confusion and deprivation. It exists because it has been imposed and sanctioned by wider as well as local society in a myriad of large and small decisions and nondecisions within a national ideology where the maximization of social fulfilment has never been uppermost; the distemper of the street gang is a pointer to a socially irresponsible ideology (Harvey, 1973; Ley, 1974a).

NOTES

1. The quotation is taken from Shaw and McKay (1969: 8).
2. Part of this data is taken from unpublished papers by John Purdy and John Harbick.

In the absence of comparative studies, the extension of these comments to other Canadian cities should not be pressed too firmly.

3. This extract is quoted verbatim from the files of the Youth Conservation Service, Welfare Department, City of Philadelphia. I am grateful to Roman Cybriwsky for several sources in this section of the paper.

4. The map distribution is of gross figures; no allowances for density variations were made.

5. This methodology followed in part the procedure of Gould and White (1974) for the construction of place perception surfaces.

REFERENCES

AUSTIN, D. (n.d.) "Goals for gang workers." City of Philadelphia, Department of Public Welfare, Division of Youth Conservation Service.

BARLING, A. (1973) "When a gang is not a gang." Vancouver Sun (March 13): 41.

CLOWARD, R. (1968) "Illegitimate means, anomie, and deviant behavior," pp. 156-178 in J. Short (ed.) Gang Delinquency and Delinquent Subcultures. New York: Harper & Row.

COHEN, A. (1955) Delinquent Boys. New York: Free Press.

ESSER, A. [ed.] (1971) International Symposium on the Use of Space by Animals and Men. New York: Plenum Press.

GIBBS, J. (1973) "Police blame parents for delinquency." Vancouver Sun (June 1): 11.

GOULD, P. and R. WHITE (1974) Mental Maps. Harmondsworth, Eng.: Penguin.

HARRIES, K. (1974) The Geography of Crime and Justice. New York: McGraw-Hill.

HARVEY, D. (1973) Social Justice and the City. London: Edward Arnold.

——— (1972) "Revolutionary and counter-revolutionary theory in geography and the problem of ghetto formation," pp. 1-26 in H. Rose and H. McConnell (eds.) Geography of the Ghetto. DeKalb: Northern Illinois University Press.

JOHNSON, D. (1973) "Crime patterns in Philadelphia, 1840-70," pp. 89-110 in A. Davis and M. Haller (eds.) The Peoples of Philadelphia. Philadelphia: Temple University Press.

KEYES, L. (1969) The Rehabilitation Planning Game. Cambridge: MIT Press.

KING, W. (1973) "In West Philadelphia, the gang wars are a way of death." New York Times (June 11): 30.

KIRK, W. (1963) "Problems of geography." Geography 47: 357-371.

KOBRIN, S. et al. (1968) "Criteria of status among gang groups," pp. 178-208 in J. Short (ed.) Gang Delinquency and Delinquent Subcultures. New York: Harper & Row.

LAURIE, B. (1973) "Fire companies and gangs in Southwark: the 1840s," pp. 71-88 in A. Davis and M. Haller (eds.) The Peoples of Philadelphia. Philadelphia: Temple University Press.

LEY, D. (1974a) "The city and good and evil." Antipode: A Radical Journal of Geography 6 (April): 66-74.

——— (1974b) The Black Inner City as Frontier Outpost: Images and Behavior of a Philadelphia Neighborhood. Washington, D.C.: Association of American Geographers Monograph Series, No. 7.

——— and R. CYBRIWSKY (1974) "Urban graffiti as territorial markers." Annals, Association of American Geographers 64: 491-505.

LYNCH, D. (1973) "Gang life spreads to suburbs . . ." Philadelphia Inquirer (July 29): 1-B.

MILLER, W. (1958) "Lower class culture as a generating milieu of gang delinquency." Journal of Social Issues 14: 5-18.

NORDLAND, R. (1973) "... Boasts, beer, battles." Philadelphia Inquirer (July 29): 1-B.

PARK, R. and E. BURGESS (1968) The City. Chicago: University of Chicago Press, rev. ed.

PORTEOUS, J. (1973) "The Burnside teenage gang," pp. 130-148 in C. Forward (ed.) Residential and Neighbourhood Studies in Victoria. Victoria, B.C.: Western Geographical Series, No. 5.

SARTI, R. (1973) "Chinatown gangs 'approaching crisis'." Vancouver Sun (October 28).

SHAW, C. and H. McKAY (1969) Juvenile Delinquency and Urban Areas. Chicago: University of Chicago Press, rev. ed.

SHORT, J. (1964) "Gang delinquency and anomie," pp. 98-127 in M. Clinard (ed.) Anomie and Deviant Behavior. New York: Free Press.

——— and F. STRODTBECK (1968) "Why gangs fight," pp. 246-285 in J. Short (ed.) Gang Delinquency and Delinquent Subcultures. New York: Harper & Row.

SMART, J. (1973) "Gangs wrote their names on Phila. walls in 1850s." Philadelphia Inquirer (January 12): G-3.

SUTTLES, G. (1972) The Social Construction of Communities. Chicago: University of Chicago Press.

THOMAS, W. and F. ZNANIECKI (1927) The Polish Peasant in Europe and America. New York: Alfred A. Knopf.

THRASHER, F. (1963) The Gang. Chicago: University of Chicago Press, rev. ed.

TOLCHIN, M. (1973) "Gangs spread terror in the South Bronx." New York Times (January 16): C-1.

WOLPERT, J. (1970) "Departures from the usual environment in locational analysis." Annals, Association of American Geographers 60: 220-229.

——— A. MUMPHREY, and J. SELEY (1972) Metropolitan Neighborhoods: Participation and Conflict over Change. Washington, D.C.: Association of American Geographers, Commission on College Geography Series, No. 16.

YABLONSKY, L. (1970) The Violent Gang. Baltimore: Penguin.

8

Locational and Ecological Aspects of Urban Public Sector Housing: The Israeli Case

AMIRAM GONEN

INTRODUCTION

□ PATTERNS OF TWENTIETH-CENTURY URBANIZATION take divergent paths in various sociopolitical forms of society. The major distinguishing attribute among sociopolitical forms "is increasing public involvement in deliberately managing urbanisation" (Berry, 1973: 165). At one extreme is the free enterprise, decentralized market-oriented system; on the other, a bureaucratized and standardized urban development in centrally directed socialist economies; while in between are systems with a varying impact on the free choice of individuals, the collective power of organizations (such as business corporations and labor unions) and of government. Public involvement in the different sociopolitical forms is carried out in a variety of ways, from overall urban planning through supply of the basic infrastructure such as transportation, utilities, education, and health networks, to active public participation in housing and industrial development. With increasing public involvement there is an increased restriction of the range of individual choice of urban life patterns, which inevitably affects the spatial distribution of activities within an urban network or an urban area.

One of the most direct forms of involvement of public organizations in urban development is in the sphere of housing. In many countries, state and municipal governments have become powerful factors in the housing market, by acquiring and developing land for the purpose of providing housing for quite a large segment of the population. In Britain, public involvement in housing development was mainly represented by the municipal level: municipal housing often accounting for one-third or more of the total housing in a city (Herbert, 1972: 181). The role of British municipal councils in housing and its impact on the social space of British towns had led to the suggestion that the classical models of urban residential structure are ineffective in describing the new ecological patterns that emerge (Robson, 1969: 132). An alternative view expressed doubts whether a municipal activity in housing should completely invalidate the classical models and that it could be taken merely as a distortion factor (Johnston, 1971: 161). Nevertheless, in recent British studies, municipal housing emerges as a powerful variable in describing urban residential structure (Herbert, 1972: 166), thus adding another dimension to those generally present in studies of towns largely developed by the private sector.

In Israeli cities, too, urban residential structure is closely related to the spatial configuration of public-sector housing (Gradus, 1971, and Gonen, 1972). This observation was the basis for a suggestion to introduce to the comparative study of urban ecology an additional scale, that of the level of direct public involvement in urban development (Gradus, 1972: 808).

Since public involvement in urban development is in general on the increase in most countries, and in some, as in Israel during the 1950s and 1960s, is even a decisive factor, there is a need to develop a theoretical framework that deals with the locational behavior of public organizations responsible for housing development. Indeed, urban planning theories provide us with some insight, but often they recommend to public organizations what should be their advisable *locational strategies*. They usually do not offer explanatory or even descriptive generalizations of the *locational behavior* of public organizations with regard to urban development. The purpose of this paper is to offer such an explanation.

Urban residential patterns and their dynamics are often described and explained on the basis of the locational behavior of the individual household. Many of the spatial models dealing with socioeconomic and demographic attributes of the urban population implicitly assume that there is a close relationship between demand and supply, thus visualizing a housing market in which suppliers are acting to match demand as near as possible. This assumption has some validity in cities of the "free enterprise, decentralized market-oriented system." Where the involvement of public organizations is strongly evident, such an assumption cannot be very useful. In their locational decision-making, public organizations are not necessarily trying to match the spatial distribution of the individual households demand, but very often are trying to thwart the potential spatial impact of such demand and direct it toward spatial patterns which comply with some notions of what is beneficial for the public as a whole and not necessarily for the individual household. Moreover, public organizations have their own framework of considerations, whether economic or political, and formulate accordingly their own locational policies, which often are far from matching the locational behavior as assumed by theoretical models of urban residential patterns.

Herbert (1972) suggests that the myriad of considerations that affected the decision-making processes leading to the location of public housing in British cities includes "the availability of public utilities and services and the diplomatic balance amongst the wards of the city" (p. 181). This generalization calls for a rather dispersed pattern of the public sector housing, especially the need for a diplomatic balance. However, in the British as well as the Israeli studies cited above, the recurring spatial pattern that alters or distorts the classical ecological models is that public sector housing estates are usually found in peripheral locations, thus forming "a girdle of low status population around the outskirts of the town" (Robson, 1969: 133). The relationship between status and public housing estates is rather evident, since, as a rule, the public sector is mainly involved in supplying housing for households of the lower brackets of the income scale, even though one author claims this is not the case in some British cities (Mabry, 1968). Less evident are the factors behind the tendency of the public sector to develop housing in outer areas of cities.

This paper suggests, on the basis of the experience with public sector housing in Israeli urban areas, that the basic difference between the individual household and the public organization with regard to residential location is in the weight assigned to accessibility considerations. The housing organization tends to base its locational decision largely on the minimization of site costs, namely, the costs of land and the costs of construction, while the household is engaged in some form of substitution between accessibility and site costs. Site costs are often the only costs the organization has to pay, being solely concerned with the production of housing, and not with residing in it. It is the long-range residence in a particular location which brings about an accumulation of accessibility costs, which the household cannot ignore, as so often does the institutional decision-maker, operating within the constraints of short-term (sometimes even annual) budgets. In urban areas, low site costs, accompanied by high accessibility costs, are to be found in outer locations. There, land values are low and large tracts of land are available. Such tracts make possible lower construction costs due to economies of scale. Thus the household depending on the public organization for housing is forced to move to locations less accessible than it might have otherwise preferred. However, when this dependence is reduced, as a result of upward economic mobility, the household will tend to improve its accessibility even if this change will mean an increase in site costs. The vacated housing unit, though, stays in the same location, available for another household, dependent on the public organization for its housing. Thus the spatial impact of public-sector housing construction on the social space of urban areas endures.

The emphasis on outer areas in the locational decision-making of public housing organizations should leave a spatial impact not only on the ecological structure of cities but also on their density patterns. In cities with a substantial proportion of housing developed by public organizations, the spread of the residential areas should be more pronounced than in cities where private housing is the dominant factor, and consequently, the density gradient from the center is less steep. This pattern is enhanced by the relatively high density in which public housing estates are developed, even in peripheral locations, where private residential areas are often of lower densities and their population of a higher socioeconomic status. The relatively low density gradient in cities with increased public involvement in housing has substantial implications on the overall distance costs in the urban system, particularly with regard to the infrastructure of transportation and utilities.

The rest of this paper provides some empirical evidence on the involvement of public organizations in the supply of urban housing in Israel and its impact on the residential structure of Israeli urban areas. After a discussion of the conditions that led to substantial involvement of public organizations in the supply of housing for the

Jewish population in the country, the paper proceeds to present the major types of public housing suppliers: the ethnocultural group, the political-ideological organization, and the governmental agency, each prevalent in a different period, each leaving its impact on the social space of Israeli cities to this date.

IMMIGRATION AND
RESIDENTIAL DEVELOPMENT

Much of modern urban growth in Israel was a result of continuous Jewish immigration. The immigrants, in spite of strong ideological and organizational efforts by the Zionist movement to emphasize agricultural and rural development, flocked into the old and newly emergent urban centers. The immigration of Jews was in the form of waves, each wave bringing a large number of immigrants in a short span of years (Sicron, 1957: chapter 3). Often these waves were not so much a result of economic development in the country itself, but were rather associated with emergencies abroad. Each wave of immigration brought with it a large demand for urban housing, causing a critical shortage and soaring prices of housing services. The huge demand for housing led private developers to build mainly for upper and upper-middle income groups, leaving the lower-income population without an adequate supply. Each wave usually established hastily built neighborhoods, sometimes of a temporary nature, situated wherever a reasonable supply of land was available. In many instances doubling-up in apartments occurred, with two households sharing one small apartment. In a few instances there was even evidence of squatting taking place.

The lower-income groups who were suffering most from the shortage of housing and its high prices made up the largest share of the population. The country was still in its incipient stage of economic development, but continued to face high rates of immigration. The veteran Jewish population was characterized by a rather low income, but so was the majority of the immigrants. The immigrants constituted a heterogeneous collection of young pioneering workers, refugees from persecutions, poor orthodox persons from the backward towns of Eastern Europe and the Arab countries. Added to this group was a small number of immigrants of moderate to substantial means. The latter were able to take care of their housing needs without great difficulty, while the rest had to wait for a long time for a satisfactory residential solution.

This substantial gap between supply and demand as a result of immigration called for the emergence of a variety of public suppliers of housing, in addition to the private construction industry. Since the 1920s ethnic groups and political-ideological organizations have existed whose primary goal was to assist their members or followers in the procurement of housing. These public organizations were involved in acquiring land, recruiting capital or credit, planning neighborhoods, and taking care of their construction. The multiplicity of the various housing agencies, formed in order to provide housing to a large part of the Jewish population, reflects the ethnocultural, socioeconomic, and political-ideological diversity that characterized that population between the 1920s and the 1950s. The diversity was mainly a product of the nature of immigration. On the one hand, there were the Jews coming from traditional societies in the Middle East, North Africa, and Eastern Europe, and carrying with them strong ethnocultural ties that forcefully affected residential segregation. On the other hand, there were the Jewish immigrants who had already gone a long way in the transition from traditionalism to modernity that took place mainly in Europe and also in some of the larger cities of the Middle East and North Africa. These modernizing immigrants searched for a place of residence not so much on the basis of ethnocultural belonging but rather with regard to socioeconomic achievement. It was the modernity-oriented immigrants who were responsible for the emergence of status neighborhoods in the growing urban areas, while the traditional streams of immigrants congregated in old as well as new quarters of a specific ethnocultural homogeneity.

Each of the two large blocks of the Jewish population, the traditional and the modern, were faced with the problem of housing shortage, and each has developed its own institutional framework to take care of that problem, in addition to the activity of the private housing market. Within the traditional block of the population, ethnocultural grouping was the foundation of neighborhood development. The modern block was characterized by the emergence of a political-ideological basis for most of the public housing organization.

ETHNOCULTURAL HOUSING

Prior to the first two decades of the twentieth century the cities of Israel possessed a traditional residential spatial structure; that is, each town consisted of a variety of ethnocultural neighborhoods. The variety of such neighborhoods was a function of the heterogeneity of the population. Perhaps the most striking example of ethnocultural segregation was in the old city of Jerusalem in the middle of the nineteenth century where each major group had its own quarter: Armenian, Christian (largely Arab and Greek), Jewish, and Moslem. Whatever socioeconomic differentiation existed in these quarters, they were seldom expressed spatially.

In almost all the older towns of the country, Arabs and Jews had separate residential sections which in turn were internally differentiated along religious and ethnic dimensions. The Arabs were segregated mainly according to their religious affiliation; the Jews, by their ethnocultural division into community groups based largely on their country of origin. This segregation had continued in the larger towns in the early decades of urban expansion that had begun during the third quarter of the nineteenth century.

The Jewish neighborhoods were divided primarily between Eastern (Oriental) and Western (Ashkenazi) ethnocultural groups. Some of these neighborhoods still carry names which indicate the ethnocultural affinity of the founders. In Jerusalem one can still find the Bokharian, Hungarian, Vilna, or Warsaw quarters (Kimhi, 1973: 112). Other neighborhoods, though their names are associated with biblical quotations, such as Residences of Israel, Tent of Moses, or Gates of Charity, were also inhabited by a particular Jewish group and for quite a while have preserved their ethnocultural homogeneity.

Most of these ethnocultural neighborhoods were not just spontaneous agglomerations of individual households belonging to the same ethnocultural group, but rather a collective and institutionalized effort on the part of the leadership of such a group, thus assigning the neighborhood to the specific group right from the beginning. The leadership, often with the help of a philanthropist or on the basis of fund-raising abroad, acquired land and initiated the development of a residential neighborhood, the housing units of which were sold, leased, or rented to members of the ethnocultural group. This leadership was involved also in setting up a mortgage system, wherever it was necessary.

The collective involvement of the ethnocultural group in the development of a neighborhood was not foreign to its traditional tasks. Mutual aid within the group in matters of social welfare and community services was extended also to housing. The rising number of Jewish immigrants and the mounting shortage of housing only emphasized the need for an increased involvement of the group in the supply of housing for its members. The members became more dependent on the group for the provision of residence.

It was the enterprising leadership of the ethnocultural group which made the decision with regard to the location of the neighborhood. Availability of a sizable tract of land suitable for the development of a whole neighborhood was often the overriding consideration. Also, the facts that the financial resources were rather limited, the prospective tenants were of rather a low income, and the pressure on housing was high and often led to the purchase of land in peripheral locations. This geographic pattern later repeated itself in other forms of residential development initiated by public organizations.

Religious orthodoxy was and still remains a strong and evident segregatory force among Jews. Although a large proportion of the Jewish orthodox population is rather dispersed over the urban areas, there are many neighborhoods which are overwhelmingly orthodox. While modernity tends to break down segregation based on ethnicity, orthodoxy is less affected by it. This is partly due to some spatial constraints that are religiously embedded. The frequent (twice-a-day) attendance at the synagogue and the curb on the use of a vehicle on the Sabbath limit the geographic dispersion of residence. So do the desire to offer an orthodox education to the young and the need to shop in stores that abide by the orthodox code. Such considerations, which relate to the range of goods and services and to the threshold of establishments, call for a concentration of orthodox population.

Spatial segregation allows the orthodox to maintain the desired character of a neighborhood. In a homogeneous orthodox neighborhood such a character, which may include a curb on traffic and business on the Sabbath and holidays, can be imposed formally or informally. Territoriality becomes then an important element in the locational considerations of the orthodox. To ensure such territorial distinction through residential segregation, orthodox groups, and later even orthodox political organizations, were active in developing neighborhoods of their own, thus again creating a situation in which the orthodox individual was encouraged to follow the locational policy of the organization. Again, it was rather the peripheral locations which were preferred by the organizations. The result is that orthodox neighborhoods are found presently in a variety of locations, each location representing a former periphery of the expanding urban area.

Not all the orthodox neighborhoods were established in this way. Some orthodox neighborhoods came into being through the process of infiltration and expansion. This is particularly evident in the inner areas of Jerusalem where orthodox neighborhoods form a contiguous sector, north and northwest of the commercial center (Buchspann, 1972).

SECTORAL HOUSING

Since the 1920s more new Jewish neighborhoods were formed, not so much on the basis of ethnocultural affiliation, but rather in terms of socioeconomic status. This is more apparent in Tel Aviv and the new coastal towns than in Jerusalem and the older inland towns. It was also largely limited to the upper part of the socioeconomic scale. Nevertheless, it represented a substantial transition toward a new organization of the urban social space, one which is based on socioeconomic differentiation.

However, while the ubiquitous transition from ethnocultural to socioeconomic segregation was going on, a new type of residential development—the sectoral neighborhood—emerged in the late 1920s to become an important component of Israeli cities until the 1950s. The sectoral neighborhood was based largely on political-ideological affiliation of its residents, though often ethnocultural or socioeconomic dimensions were not absent. The developer of the sectoral neighborhood was a public organization of a particular political-ideological sector of the Jewish population. The main sectors were three: (1) the Federation of Labor (Histadruth); (2) the various organizations of the middle-class, free-enterprise, liberal-conservative sector; and (3) the politicized-orthodox sector. Sometimes the public organizations represented a combination between a *landsmanschaft* and a particular political-ideological sector; in other cases it was an orthodox political organization, inclined toward one of the two other main sectors. Some of the organizations were based on an occupational group; others shared the same employer, but were almost invariably tied with one of the main three sectors. The combinations were quite numerous, each producing a specialized organization engaged in providing housing to their members and followers and, as a rule, in the form of a sectoral neighborhood.

The years between the 1920s and the 1950s were marked by a high degree of ideological and political fragmentation of the Jewish population. This period was characterized by an intensive ideological atmosphere. Much of the economic development of the country was organized along sectoral lines, especially in the rural areas. Rural settlements were established by a particular sector which was also responsible for continuous economic and political support. The sector was the major and often only channel of the rural community to the national resources of the Jewish population. These resources were concentrated in the national institutions set up by the Zionist Movement for the purpose of mobilizing financial support and of creating tools of economic development. The Zionist Movement was basically a coalition of the various political-ideological sectors of the Jewish population, and the allocation of resources for economic development was largely done on the sectoral basis. An individual had no direct access to these resources.

The high rates of immigration and the rather severe economic conditions in a country which was just setting out on the path of economic development made it advantageous and sometimes necessary for the Jewish immigrant, especially the one with almost no economic resources of his own, to join or affiliate with some kind of a sectoral organization. This provided him with indirect access to the Jewish national resources. The need to rely on the sectoral channels to the Jewish national resources was emphasized by the rather conservative policy of the British mandatory government in Palestine with regard to public expenditures on economic development and social services (Halevi and Klinov-Malul, 1968: 36-37). Consequently, the political-ideological organizations became important instruments in the provision of jobs, housing, and services for many of the Jewish immigrants. Gradually these organizations developed their own powerful economic functions, thus not only acting as a power group channeling resources toward their constituency but also as organizations controlling their own investment resources and directly engaged in economic activities, the provision of housing being a major one.

In most cases sectoral neighborhoods were developed outside the built-up area of towns. A number of factors collaborated in establishing this locational pattern, the main one being the low cost of land. The soaring land values within the built-up area, mainly the result of continued high rates of immigration, pushed the rather poorly financed organizations to look for outlying locations. Moreover, because of their limited resources, many of the sectoral organizations relied on land purchased by the Jewish National Fund (JNF), the land-purchasing institution of the Zionist Movement. The policy of the JNF was to purchase mainly agricultural land, when and where possible, for the purpose of establishing Jewish agricultural communities. This was in line with the general Zionist ideology that emphasized the "return to the soil" of the Jewish population and the development of an occupational structure more inclined toward primary and secondary industries than was the case of the Jewish population in the Diaspora. However, the JNF also purchased land near urban areas, and it was often this land that the various sector organizations demanded from the JNF for establishing neighborhoods of their followers with no adequate housing. The JNF, run by a directorate which in a way represented a coalition of the various sectors of the Jewish population taking part in the Zionist Movement, alloted this land, though sometimes reluctantly, to sectoral organizations for housing purposes. The dependence on the nonurban land of the JNF was a major factor in the dispersal of the sectoral neighborhoods, thus contributing to a substantial decentralization or urban development.

The sectoral organizations were interested not only in solving the housing needs of their followers, but also in establishing what could be termed a "sectoral territory" within the urban area. Such a territory in the form of a sectoral neighborhood enabled the development of life patterns in line with the general ideology of the respective sector. In self-contained neighborhoods, community institutions could be set up which not only would be shaped according to the sector's ideological principles, but would be controlled in some way by the central bodies of the nationwide sector. Moreover, the establishment of a neighborhood, much more than financial help to the individual follower, represented an observable political gain for the sector and symbolized its ability to draw resources from the national institutions. This interest in developing a distinct neighborhood was an additional factor in the tendency to locate such neighborhoods in peripheral locations. It was on the urban periphery that land tracts large enough to permit the development of a whole neighborhood were available.

Since the dependence on the sectoral organization for housing was concentrated mainly in low-income households, the locational behavior of the housing agencies of the ideological-political sectors had an important impact not only on the dispersal of urban development, but also on the spatial patterns of the urban social space. Low and lower-middle income neighborhoods in many of the urban areas during the 1930s and 1940s were located in outer locations, while the middle and upper-income population tended to concentrate in the inner neighborhoods. The low and lower-middle income households had to comply with the locational decisions of the sectoral organizations if they wanted to solve adequately their housing problems. The other households were in a position to make their locational decision more in line with their individual preferences.

The labor sector represented by the General Federation of Labor, known as the Histadruth, was the most active in providing housing for its members in the form of Workers' Neighborhoods. The idea of Workers' Neighborhoods involved not only the solution of the housing needs of the members of the group, but also enabled the evolution of a new urban way of life in line with forms of the kibbutz and the moshav. Cooperative forms in consumption and land holding were basic elements of Workers' Neighborhoods in their early years. However, it was the attempt to fuse agricultural ingredients into a basically urban community (Cohen, 1970: 23-29) that later had an important locational impact. The settlers of Workers' Neighborhoods were expected to be involved in some form of agriculture, and for that purpose each household had a plot of land around it to be cultivated. This rather sparse pattern of development necessitated the availability of large tracts of land. Consequently, most Workers' Neighborhoods, in areas where agriculture was possible (Jerusalem being a notable exception), were often established well beyond the periphery of the built-up area.

Many of the suburban cities in the metropolitan region of Tel Aviv and Haifa have their origin in Workers' Neighborhoods. Giv'atayim, to the east of Tel Aviv, and Holon, to the south, started as a series of Workers' Neighborhoods which later amalgamated into one municipal entity. Giv'atayim tried to preserve its original ideological basis long after its formation in the twenties. Up to the early seventies, the city has maintained a high level of allegiance in elections to labor, though the population long ago lost its working-class status and now enjoys one of the highest levels of median income per person among the municipalities of Israel.

The Haifa urban region furnishes another example of intensive sectorialization of residential development. In the 1920s a large part of the coastal plain adjoining Haifa Bay and the city of Haifa, and known as the Bay Area, was gradually bought by the JNF and other Jewish organizations. In the 1930s, when the settlement of the Bay Area was underway, the sectoral structure of the Jewish society was already quite developed, and sectors were exerting their influence on the JNF for a portion of the purchased land in order to develop suburban communities for their members in Haifa. Consequently, one after another, several sectoral communities were established, each populated according to organizational affiliation. Kiryat Haim was a labor community developed by the Histadruth and based on cooperative principles as well as on a mixture of industrial and agricultural economy, at least in its early years. Kiryat Mozkin was developed by the "Organization of the Middle Class," backed by political parties, representing at the time the liberal right of the Zionist spectrum. The neighboring Kiryat Bialik was established by middle-class Jewish immigrants from Germany, thus offering the town a particular ethnocultural distinction. However, the sponsor of Kiryat Bialik was the Association of Immigrants from Germany, which, though having an ethnic basis, acted as a political organization and, like other landsmanschaften, served as mediator between the individual and the sectoral organizations in matters of housing. The JNF has also alloted a small area of Hapoel HaMizrahi Movement—a labor-oriented organization of orthodox Jews—for the purpose of setting up Kiryat Shmuel, an orthodox suburb alongside the other suburbs. The JNF in doing so responded to a request by the orthodox political organization in the same way it did with regard to the other sectoral organizations.

These are the most outstanding examples of the suburban communities in the Haifa Bay Area, representing four major types of public organizations that were involved in securing land and developing housing for their members and followers. In a matter of half a decade (1931-1936) the Bay Area was destined to become a suburban mosaic as a result of the parallel efforts of the various sectoral organizations (Hayut, 1974).

The constitutions of many of the sectoral neighborhoods often included strong measures to protect the sectoral identity of the neighborhoods, at least in the first decade or two. Thus Kiryat Haim in the Haifa Bay Area admitted only members of the Labor Federation, and the residents had to continue their membership in the Labor Federation as long as they lived in the neighborhood

(Hayut, 1974). Alternatively, adjacent Kiryat Mozkin insisted on membership in the "Middle Class Organization" and took great efforts to exclude members of the Labor Federation. Each community strove to preserve its ideological homogeneity, as well as its socioeconomic character, and to keep the reins of local power in the hands of the founders of their ideological followers.

Even the small towns in Israel witnessed a proliferation of sectoral neighborhoods, most of them being Workers' Neighborhoods. With intensification of agriculture and nascent urbanization of these towns, workers flocked into them and an increasing shortage of housing developed. Again, as in the case of the large cities, the Workers' Neighborhoods that were established were located on low-priced land on the periphery (Grajcar, 1973: 54-56).

Each of the political-ideological sectors of the Jewish population developed also some of the necessary organizational tools for the purpose of providing housing for its followers. A variety of financing arrangements has emerged in the form of loan and saving associations and mortgage banks. However, the main tool was the sectoral housing corporation that handled the acquisition of land, development of infrastructure, construction of housing, and setting up some of the public services. The Federation of Labor established in 1928 the "Shikun" Corporation for the purpose of developing Workers' Neighborhoods, mainly in the larger cities. "RASSCO" was established in 1934 by the Jewish Agency to provide housing mainly for middle-class households and was associated with liberal middle-class political parties. "Mishkenot" was established in 1945 as a housing corporation of Hapoel HaMizrahi—a religious party with some labor orientation. "Sela" was a corporation affiliated with the Revisionist movement, on the right of the Jewish political spectrum. "Hamifde Ha'ezrahi" was involved in providing housing to followers of the General Zionists, a conservative free-enterprise party, later renamed the Liberal Party. The Shikun Company, now known as "Shikun Ovdim," was the largest of the sectoral housing companies and was responsible for a substantial number of small and large Workers' Neighborhoods in many urban areas of the country.

As in case of the ethnocultural neighborhoods, often the names of the sectoral neighborhoods identify them as such. In some instances it was the name of the housing corporation that was given to the neighborhood. Thus, many towns have a RASSCO or a Mishkenot neighborhood. In other instances the neighborhood carries the name of a leading figure in the organization. Kiryat Haim is named after

Haim Arlozoroff, a prominent labor leader in the 1920s and early 1930s. Often it is the name of the organization itself which identifies the affiliation of the neighborhood.

The sectoral fragmentation of housing development continued even during the first years of statehood, when the government became an important housing developer. However, with a decreasing dependence on the sectoral organization due to the general rise in income, the decline of ideological differentiation, the role of government in housing new immigrants, and the increasing importance of private developers in the housing market, the importance of the sectoral housing has dwindled. The sectoral housing corporations, though still providing for the needs of their sectoral organizations, have stepped into the private housing market, where they compete with the other private developers in satisfying demand of the lower-middle to upper-middle income groups. Presently, it is mainly the orthodox housing agencies that are the most actively engaged in establishing sectoral neighborhoods. The old-style Workers' Neighborhoods are not reappearing on the urban scene. Those that were established in earlier years are themselves undergoing a substantial social transformation, many of them turning into rather middle-class neighborhoods.

The sectoral neighborhoods were, then, a temporary though substantial phase in the evolution of Israeli urban areas. For at least three decades, they introduced a new principle in the spatial allocation of households, that of the political-ideological-institutional affiliation. This phase offered a substantial impetus to suburbanization of a variety of socioeconomic groups, giving rise to the first major stage in the evolution of the Israeli suburban areas into a social mosaic. The second stage was initiated by the massive construction of immigrant housing estates by the government in the first decade after the state of Israel was established.

STATE HOUSING

The establishment of the state of Israel in 1948 introduced a new kind of residential developer to the urban scene, the government itself. The emergence of the government as a major factor in the supply of housing was precipitated by the high wave of Jewish immigration that followed the establishment of the state.

In the first two years of statehood, 1948-1949, the various

sectoral housing agencies continued to take care of much of the new housing needs, both of recent immigrants and of veteran residents. However, it was soon obvious that the task was beyond their capabilities, in view of the huge demand and their commercial nature. Strong pressure developed to get the new government to assume direct responsibility for the construction of housing, a process which was formalized by the creation of a Housing Wing in the Ministry of Public Works. Later, the Housing Wing was turned into a full-fledged ministry. Apart from coordinating and contracting housing projects to the various developers, the government became directly involved in construction, especially in the country's periphery or with regard to immigrants with no financial resources.

By the early 1950s there was a widely accepted attitude that the government should undertake economic development and the provision of housing and social services as its major tasks. This acceptance was associated partly with the experience of the Jewish population in developing its own network of public institutions that were designed to develop the country more than the conservative policy of the British mandatory government had permitted. Another factor which moved the government to assume this responsibility was the growing input of socialist ideology into the Jewish society and the eventual control of the Jewish national institutions, and later the Israeli government, by socialist or socialist-inclined parties.

The increased role of the government and other public institutions, such as the Jewish Agency and the Histadruth, in the provision of housing in the early years of the state, as in economic development in general, was also associated with the change in the nature of capital import into the country. During the years of the British Mandate, a large part of Jewish capital accumulation was the result of capital transfers made by immigrants. Much of the economic development in the Jewish urban sector was based on these capital transfers, especially the development of lower-middle to upper-middle income housing in the form of socioeconomic neighborhoods. After 1948 there was a substantial change in the nature of immigration. The predominant majority of the immigrants came without financial means, and in many cases, as with many of the immigrants coming from Middle Eastern and North African countries, lacked modern skills and education. In the first five years of statehood (1948–1953) the number of immigrants surpassed the number of Jewish inhabitants in the country at the beginning of the period. Many of these recent immigrants were only able to earn a rather low income. Even high rates of domestic saving could not have provided the necessary investment resources to supply the hundreds of thousands of immigrants with jobs, housing, and basic services. Consequently, the country became involved in an intensive importation of capital. It was almost invariably government and other public institutions that were engaged in such importation of capital through loan and grant-raising abroad. By controlling capital-raising abroad, these public institutions eventually grew to dominate most of the investment resources of the country and consequently much of its economic activity, especially in agriculture and housing.

The role of the government in residential development was enhanced also by its control of much of the urban land. This position of the government as a major owner or administrator of urban land is associated with several historical developments in the country. As was pointed out earlier, the JNF was involved in purchasing land outside the urban areas, while purchasing of land within urban areas was to a large extent the domain of private capital. In later years with the expansion of the urban areas, the JNF lands, now under the control of the state's Land Administration, became major parts of the urban fringe. These publicly owned tracts of land were a decisive factor in the locational policy of the government with regard to its housing projects. Another source of government-controlled land were the Arab villages and towns adjacent to Jewish towns. After the exodus of the Arab population during the 1948 war, the Israeli government assumed control of these areas and in many instances used parts of the land for setting up immigrant housing projects. This land was often far apart from the built-up area of the Jewish towns.

The locational policy of the government with regard to immigrant housing was similar to that of the sectoral developers; only the magnitude of the problem differed. The constraints were time and budget. Hundreds of thousands of immigrants were without adequate shelter, while the economic resources of the country were quite meager in view of the huge demand brought about by mass immigration. The locational solution was found in the periphery, and sometimes well beyond the edge of the built-up area. In the periphery the government was in control of its own land, a control which affected not only costs but also the speed of planning or rezoning. Moreover, in the periphery large tracts of land were available on which large-scale housing projects could be constructed, thus cutting down on costs.

Short-range considerations were the rule. The governmental

housing agencies were mainly trying to minimize construction costs so as to maximize the number of housing units in a given annual budget. Such a policy led, to a certain extent, to a tendency to disregard accessibility costs (or distance costs), a considerable part of which is accrued over a long period of time. For while residences decentralized, industry was far slower in adopting a decentralized location policy.

The spatial result of such a locational policy was a considerable spread of the residential areas of towns (Gonen, 1972: 407). This spread was not the result of the locational preferences of thousands of households, not even the result of locational decisions of a number of private developers, but rather the result of a centralized institution that was substantially remote from the kind of locational considerations that guide the individual household. In such a situation theoretical frameworks of urban spatial structure based to some degree on interrelationship between supply and demand do not apply very well.

The governmental housing estates were designed largely for immigrants. An immigrant household, being often of very low income with little or no savings, had very little choice but to reside in one of these estates. As was the case in many instances, the immigrant household was placed by the housing agency in a particular estate where housing was available. Consequently, the involvement of the government in the provision of housing introduced a new differentiating variable in the social space of Israeli urban areas: the distinction between recent immigrants and veteran residents. The veteran population concentrated in the older, more central sections of towns, while the housing estates at the periphery consisted almost exclusively of new immigrants. Since the recent immigrants were as a rule persons of low income, the peripheral estates were also marked by a low economic status. High-income suburbs in Israel's larger cities are in many instances isolated "islands." This low status was enforced by the rather low-quality housing that characterized government immigrant estates.

LOCATIONAL READJUSTMENT

In the late 1950s and 1960s the low socioeconomic status of the peripheral housing estates was accentuated as differential outmigration deprived the estates of the more upwardly mobile population (Cohen, 1966: 123). In medium-sized cities this kind of migration took on a centripetal character, namely from the outlying estates to more centrally located neighborhoods (Gonen and Hason, 1974). Perhaps the centripetal migration within the medium-sized towns in Israel represents a dramatic illustration of the difference between the locational behavior of individual households and that of the public organizations or institutions dealing with housing. This centripetal pattern can be viewed as an effort on the part of the individual households to readjust their residential location, previously dictated to them by the institutional decision-makers, so that the new locations will conform more with the households' preferences.

The households that left the government housing estates in the medium-sized towns of Israel found in the more central sections of these towns a supply of housing provided by private developers. Unlike the institutional developers, the private developers are building mainly for those households whose financial resources enable them to search for satisfactory housing in the private sector. These private developers tend to build in more accessible locations which correspond better to the kind of locational considerations of the individual households.

In recent years the private developers have taken a larger share of the housing market in Israel, mainly as a result of the rise in the level of income and the accumulation of capital by an increasing number of households. In 1950 private developers were producing less than half of the number of housing units completed by public organizations, the government being the major supplier. In 1970 the ratio was 2 to 1 in favor of the private construction industry (Statistical Abstract of Israel, 1973: 468). This has often resulted in recompaction of urban areas (Gonen, 1970: 409) because of the tendency of private developers to build in more central locations (Gonen, 1972: 409). The emerging pattern of recompaction is expressed through the steepening of the density gradient (Grajcar, 1973: 71-88).

Associated with recompaction is the increased socioeconomic differentiation between the low-status housing estates on the periphery and the inner residential areas, where middle to upper-middle income groups continued to give much weight to accessibility and thus did not choose to decentralize. In the fast-growing Tel Aviv metropolitan region, this preference for central locations pushed up land values in the core to a very high level, causing an exodus to suburban cities, mainly of lower-middle income groups. Thus, middle and upper-middle income population forms only a small fraction of the suburban ring of Tel Aviv. A considerable part of present-day middle to upper-middle suburban neighborhoods around the metropolitan core of Tel Aviv has its origin in sectoral housing neighborhoods. These neighborhoods are frequently composed of clerks, foremen, army officers, and workers, who, since the late 1940s and early 1950s, have moved up the socioeconomic scale but have elected to stay in the same place. Nearby, low-status immigrant housing estates continued to be developed by the government. Consequently, an intricate spatial mosaic of socioeconomic status, compounded by ethnic differentiation, has emerged on the suburban ring of the metropolis.

CONCLUSION

The substantial involvement of public organizations in urban housing necessitates a reconsideration of models of urban residential structure based on the assumed interrelationship between demand and supply within a private housing market. Such models incorporate considerations of space and accessibility which are embedded in the preferences of the individual household. However, such an assumption is inadequate with regard to the locational behavior of housing organizations operated by the public sector. On the basis of evidence from Israeli towns, it is suggested that public organizations give much less weight to accessibility costs than do individual households, and, considering other factors as well, tend to seek outer locations for housing estates so as to minimize space and construction costs. Such locational behavior results in the emergence of outer neighborhoods of relatively low socioeconomic status which are not accounted for by classical models of urban residential structure; nor is such development explained by preindustrial-industrial or nonwestern-western scales. Whether this generalization on the locational behavior of public-sector housing organizations is valid in a variety of urban areas in other parts of the world and in other national contexts can only be judged by the results from additional comparative studies.

REFERENCES

BERRY, B.J.L. (1973) The Human Consequences of Urbanization. London: Macmillan.

BUCHSPANN, S. (1972) "Spatial organization of the orthodox population in Jerusalem, 1959-1971." Hebrew University of Jerusalem, Department of Geography, M.A. thesis. (in Hebrew)

COHEN, E. (1970) The City in the Zionist Ideology. Jerusalem: Hebrew University, Institute of Urban and Regional Studies.

—— (1966) "Problems of development towns and urban housing estates." Economic Quarterly (Riv'on Lekalkala) 13: 117-131. (in Hebrew)

GONEN, A. (1972) "The role of high growth rates and of public housing agencies in shaping the spatial structure of Israeli towns." Tijdschrift voor Econ. en Soc. Geographie 64 (November/December): 402-410.

—— and S. HASON (1974) "A centripetal pattern of intra-urban mobility in Israeli medium-sized towns." Geografiska Annaler 56B, 2: 144-151.

GRADUS, Y. (1972) "The interaction of two scales within the ecological structure of metropolitan Haifa, Israel," in International Geography 1972 (papers submitted to the 22nd International Geographical Congress, Canada) 2: 808-810.

—— (1971) "The spatial ecology of metropolitan Haifa, Israel: a factorial approach." University of Pittsburgh, Ph.D. dissertation.

GRAJCAR, I. (1973) "Development of built-up area and population distribution in Rishon LeZion." Hebrew University of Jerusalem, Department of Geography, M.A. thesis. (in Hebrew)

HALEVI, N. and R. KLINOV-MALUL (1968) The Economic Development of Israel. New York: Praeger.

HAYUT, Y. (1974) "Urban development in the northeastern fringe of the Haifa metropolitan region." Hebrew University of Jerusalem, Department of Geography, M.A. thesis. (in Hebrew)

HERBERT, G. (1972) Urban Geography, A Social Perspective. Newton Abbot, Eng.: David and Charles.

JOHNSTON, R. J. (1971) Urban Residential Patterns. London: Bell.

KIMHI, I. (1973) "The human ecology of Jerusalem," in D.H.K. Amiran, A. Shachar and I. Kimhi (eds.) Urban Geography of Jerusalem. Jerusalem: Massada.

MABRY, J. H. (1968) "Public housing as an ecological influence in three English cities." Land Economics 44: 393-398.

ROBSON, B. T. (1969) Urban Analysis. London: Cambridge University Press.

SICRON, M. (1957) Immigration to Israel 1948-1953. Jerusalem: Falk Institute.

Part IV

PERSPECTIVES ON URBAN STRESS AND POPULATION CHANGES

Introduction

☐ MOVEMENT TO THE CITY by populations seeking to escape the poverty of their environments of origin is a widespread phenomenon. By most indices of economic progress, these decisions are generally regarded as economically advantageous for the mover population. The papers in this section each attempt to evaluate the success of the migrant and/or poverty population in the urban environment in a variety of cultural contexts. The authors do not choose to focus upon a single overriding problem that such groups encounter in urban environments, but single out those which reflect national, philosophic, and disciplinary concerns. The concerns expressed here are complex, and thus it is unlikely that representatives from any single discipline or a single philosophical perspective will be able to provide a satisfactory interpretation. But it is apparent to some that these problems need to be cast in alternative contexts. Thus these papers range over a rather broad domain, although there are obvious similarities in both problem orientation and topic orientation. In terms of problem orientation, there exists a great deal of congruence between the work of Rose and Mier, Giblen, and Vietorisz. Yet there are sharp differences in their approach to the problem, differences which grow out of differences in disciplinary background and, consequently, differences in emphasis relating to their common concern. Three of the papers have as their central focus the role of migrants (or immigrants) in the urban economy; but beyond their interest in the problem of migrants, there is little remaining similarity in the manner in which each of the authors chooses to approach this topic.

The first paper in this section is explicitly concerned with the role of migration in altering the economic status of the migrant and indirectly concerned with reducing the employment opportunity of the resident population. Rose postulates that limited-skill black migrants, who chose to move to those metropolitan areas which were the targets of the largest number of black movers during the sixties, are likely in many instances simply to compete with resident blacks for jobs requiring limited skills. More importantly, Rose is concerned with the stress which builds up in these black communities resulting from the intensification of pressure on limited resources. As this work is exploratory and burdened by problems of determining appropriate measures of stress, it is likely to be viewed as trivial by some. Nevertheless, the author has raised some issues, however conjectural, and posited some relationships which are worthy of further investigation.

The Mier-Giblen-Vietorisz paper touches upon issues similar to those raised in the Rose paper. These authors attempt to relate the functioning of the economic system to the functioning of the urban social system: they contend that poverty leads to social distress and then set out to establish the relationship between indicators of poverty and several measures of "social pathology." This paper, like Rose's, attempts to specify relationships by employing indicators that are subject to criticism—in particular, their effort to derive an effective measure of subemployment. It is the conclusion of these writers that an effective measure of poverty is the best indicator of social distress.

The Santos paper touches upon problems of social economy in a third world context. His erudite treatment of the notion of the periphery in the pole is enlightening and should be welcomed by those whose interest is the space economy. Santos challenges the validity of the position of a number of previous writers on this topic. Not only does the author develop an effective conceptual treatment of the dual economy within the city of Lima, Peru, but he also describes in detail the role of the participants in the periphery of the economy. Santos focuses on what he defines as the lower circuit of the economy. Like other papers in this volume, this one is likely to stimulate rejoinders on the part of those scholars whose philosophical perspectives are at odds with that of the author.

The final paper in this section devotes its attention to the settlement of colored immigrants in selected British cities during the sixties. Peach, Winchester, and Wood provide us with a detailed view of the social and political problems associated with a sudden increase in minority populations. The diversity of the origin of these populations no doubt complicates things when they are lumped together as simply "coloured immigrants." These writers tend to focus on the strength of the pattern of residential segregation among the various immigrant groups as an index of host community receptivity. The approach employed in this study is modeled after that of the Taeubers, whose work on patterns of black residential segregation in American cities is universally known. Some problems of dissimilar units of analysis, in terms of scale, make comparisons between segregation levels in U.S. and British cities difficult. Thus, these authors show that inequality in ascriptive social status is a factor in the urban economy of British cities, as it is in American cities. Not only do the authors attempt to measure differences in the level of segregation of individual immigrant groups, they also attempt to explain both the role of the host society and cultural attributes of individual immigrants on the residential patterns which have emerged.

The papers in this section are thought-provoking because they raise questions and issues for which the answers are as yet unclear. The issues raised focus on the impact of social status on economic functioning, and these papers treat only a few of the myriad problems which grow out of the urban social and economic interface which led the editors of this volume to entitle it *The Social Economy of Cities*.

Some questions of importance remain to be asked with regard to research on urban populations. Three of these are:

(1) Can the problems of urban population shifts generate a more adequate interdisciplinary treatment of the migrating phenomenon? Does the notion of "social stress" offer an appropriate outcome variable that different social scientists can incorporate into their disciplinary models?

(2) What are the problems in developing and using more effective indicators of the manifestations of social and economic forces? How can we best treat the multidimensional aspects of poverty in urban environments?

(3) Do the transnational perspectives of the periphery of the urban social economy in Peru and Britain provide a more distinct understanding of the urban condition in which migrant populations represent the actors caught between the forces of poverty and progress?

9

Urban Black Migration and Social Stress: The Influence of Regional Differences in Patterns of Socialization

HAROLD M. ROSE

□ BLACK MOVERS TO CENTRAL CITIES of the larger Standard Metropolitan Areas in the nation continued apace during the decade of the sixties. This continued movement has done much to alter the racial composition and the subsequent population size of these larger core cities, but it has likewise exerted a pervasive impact upon the life chances of the mover. The principal focus of this essay is on the competitive position of black migrants with limited skills who were most recently residents of the southern United States. Interest is not simply confined to the economic progress of the mover population, but will focus upon the strain which builds up in the total black community if resources are inadequate to serve the larger black population, whether as a result of their being shifted elsewhere or their absolute lack of availability. This essay represents a crude first attempt to identify the impact of black migration on the generation of stress in the core city's black community.

MIGRATION AND URBAN STRESS

Black movement from the South to the non-South during the sixties was of slightly smaller magnitude than it had been during the previous decade. But, needless to say, this slowdown did not effect the principal targets in the same way. For instance, there was a doubling of black Detroit residents reporting that they had lived elsewhere in 1965, in comparison to those entering the metropolitan area during the period 1955-1960. New York, on the other hand, represented the inverse of this situation, with more than twice as many black residents identifying themselves as newcomers during the last five years of the fifties as did so during this same period in the sixties. It is generally held that the black mover from the southern United States improves his economic status by abandoning the South, and particularly the rural South, for the large city North. A recent study showed a net gain in annual earnings accruing to black males migrating from the rural South to the urban North of $1,760 (Iden, 1974). This was the maximum economic gain accruing to black male movers based on destination of move. Recent revelations showing that southern black migrants to northern urban areas earned higher incomes than black nonmigrants in those areas, holding age and education constant, prompt one to favor the notion of intragroup competition for scarce economic resources. If this notion is correct, it would be expected to lead to increased stress in those areas where the pressure of competition is intensified as a function of changing migration levels. The difficult task is to identify an appropriate set of stress-related outcomes that can be shown to be an outgrowth of intragroup competition.

It is generally agreed that stress is an internal response to an external stimulus which leads to both physiological and psychological changes in the human system. Aakster (1974) defined stress as any environmental force which leads (or is expected to lead) to disequilibrium upon one or more essential variables of the system. Aakster analyzed stress in terms of a set of physiological responses or health disturbances incurred in a sample population. But in attempting to explain the etiology of these illnesses, he derived a set of sociocultural patterns through the use of factor analysis. The five derived factors were: (1) disintegration of parental family; (2) dissatisfaction with living place; (3) affective dissatisfaction; (4) status dissatisfaction, and (5) worries. The derived factor structure provides clues to the identity of stress-inducing forces. It is the independent variables which lead to stress that are of principal importance here, rather than the specific health disturbances associated with these variables which were the focus of Aakster's research.

Additional insight into attempts to measure stress has been provided by Dohrenwend (1973). Dohrenwend utilized information describing the number of stressful life events incurred by a sample population during a given year and combined that with a derived score describing the difficulty of readjusting to these events as an indicator of the severity of stress. The amount of stress generated in a given community is a function of the number of stressful events encountered by its population and the adaptive mechanisms selected to restore the equilibrium condition. It should be stated, however, that stressful events are culturally determined; and it appears logical to assume that the means of readjusting to the onset of stress is likewise culturally learned.

Extensive work on the level of stress within individual communities is generally unavailable, since studies of stress are most often directed at stressful symptoms in the individual. But, obviously, individuals interact within a social and spatial milieu, and thus the number of severely stressed individuals has a direct bearing on the nature of the residential environment occupied by the group. One recent attempt to identify high- and low-stress neighborhoods was undertaken in Detroit (Kasl and Harburg, 1972). It was noted in this study that high-stress neighborhoods were characterized by indices of low socioeconomic status, high rates of crime, marital instability, and residential mobility. Another study in which the interaction between the individual and his environment was of central concern focused on New York City Health Areas (Struening, Lehman, and Rankin, 1969). In the latter work, nine predictor variables were chosen to measure their influence on a set of behavioral outcomes. The outcomes can be thought of as representing indicators of stress levels, whereas the predictor variables describe a set of statuses that lead to increased levels of stress. These and other studies are indicators of the growing interest in the relationship between individual behavior and the behavioral environment.

There is, however, a body of literature which has as its central concern the role of the environment on the problems of migrant adjustment. Over the years a number of works have been produced which were directed to the problems of adjustment of black migrants to northern cities. Conflicting positions have been taken on the issue of the black migrants' adjustment problems relative to those of indigenous urban black populations. There now seems to appear some uncertainty regarding the validity of the culture shock thesis which assigned a higher incidence of mental health problems to black migrants from the South. Pettigrew (1964) has indicated that the relationship between migration and mental health is unclear, but he indicated that there are data which indicate a downturn in the incidence of crime committed by black migrants of southern origin since World War II. A more recent assessment of this situation has led to the conclusion "that life in the North is so disruptive that migrants from the South are actually better off than those who were born in the North" (Crain and Weisman, 1972).

Concern with migrant adjustment continues to represent an integral interest to numerous behavioral scientists, and the experience of blacks in American cities likewise continues to receive much attention. Brody, who recently carried out an investigation of the adaptive problems of migrants to Rio de Janeiro, Brazil, spoke of the prospective outcomes in terms of the urban black American experience. He stated:

Rio's problems and those of the Cariocas transcend national boundaries. In varying measure and qualities they may be recognized as similar to those of New York, Baltimore, Washington or San Francisco. Baltimore and Washington, in particular, without the leavening impact of mountains or beaches, struggle under the weight of overwhelming migrant populations, mainly black, underemployed and partly literate. In their humid summer heat, the tension has climbed to points of violent eruption not yet reached by Rio's swollen population [Brody, 1973: xiv-xv].

Brody suggested that the pressures associated with massive black migration to American cities might be partially associated with the eruptions of violence which took place in those cities during the sixties. He also noted that, were it not for the stabilizing effect of culturally reinforced passivity, similar responses might be expected in Rio. It is not the objective of this essay to investigate the adaptive proclivities of black migrants to northern cities, but instead to attempt to ascertain the impact of migration on the build-up of stress in selected communities growing out of competition for limited economic resources. With this as an overriding concern, some attention will be given to operation of both social and economic forces which impinge on the social economy of cities.

The movement of southern blacks to northern urban centers is well documented and is thought to represent one of the major population movements of the twentieth century. While the total volume of black movement out of the South slackened during the sixties, more than four million blacks abandoned the region during the generation that spanned the period 1940-1970. The largest number of movers originated in the states where blacks were an integral part of an agricultural system which was in the throes of a major transformation (see Figure 1). During this period Mississippi, Alabama, South Carolina, Georgia, and North Carolina contributed more than 60% of the black migrants to northern urban centers. Louisiana and Arkansas, the other two principal source regions for black migrants, fell somewhat short of the level of the major contributing states. The major contributing states to black migration from the South, with the exception of Georgia and South Carolina, have served as principal source areas for black migrants only since 1940.

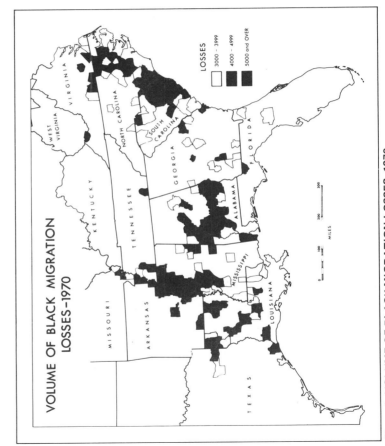

Figure 1: VOLUME OF BLACK MIGRATION LOSSES, 1970

THE STATUS OF MOVERS OF
NONMETROPOLITAN ORIGIN

While blacks are leaving both rural and urban areas in the South, attention has generally been focused on the mover of rural origin. It has been demonstrated previously that blacks from nonmetropolitan areas represent a decreasing share of all southern black movers. During the period 1955-1960 movers of nonmetropolitan origin represented only slightly more than half of the total (Taeuber and Taeuber, 1965). But the impact of rural outmovement is thought to be more pervasive than their shares imply. Beale estimated that blacks 25-29 years old living in rural counties in 1960 represented only one-third of the original cohort present in these counties ten years earlier. This would tend to indicate that two out of every three blacks in the principal mover age group tend to move during a ten-year interval. The apparent decline in economic opportunity in rural counties in the South, coupled with high incidences of poverty, can be expected to continue to decimate the South's black rural population.

During the last decade, the movement of blacks to the South from other regions was also detected. Himes (1971) indicates that 197,000 blacks left the South during 1969-1970, but during the same period 97,000 entered the region. This would tend to indicate a level of black southern immigration that is far in excess of normal return flow. Lee (1974) recently reported that only 15% of the blacks who had lived outside of the South five years preceding the 1970 census had returned to their state origin by 1970. She estimated that three-fourths of the black inmovers to the South were return movers. If this estimate is reasonable this would tend to imply that many black movers who originally entered migration streams for the North and West have either found these areas unattractive or have had second thoughts about the region of origin. On the other hand, this might reflect the growing attractiveness of a few select southern locations. At this point, however, it is not possible to ascertain if this trend will lead to a lessening of stress-inducing factors in northern urban environments. Until more detailed information is available on the residence histories and work-related characteristics of the movers from North to South, its impact on resident northern black populations is a matter of conjecture.

Rural blacks are declining as a percentage of all blacks migrating from the South to other regions of the country. During the period 1965-1970 approximately 25% of the total black migrants from the South were of rural origin. Most black migrants from the South have had some experience with urban life prior to migration. There is evidence which suggests that size of the place of origin plays a significant role on the migrants' success in his new environment. It was recently reported that blacks who had been educated in the rural and small-town South received higher incomes in northern labor markets than blacks who had been educated in large southern SMSAs or in the North (Weiss and Williamson, 1973). This finding led its authors to remark:

thus southern rural blacks suffer no competitive disadvantage in urban labor markets, North or South: on the contrary, if anything it appears to be the ghetto-educated young black who suffers the competitive disadvantage [Weiss and Williamson, 1973: 376].

Masters (1972) likewise found that lifetime migrants from the South fared better in northern labor markets, but he indicated that this finding applied mainly to those migrants with less than twelve years of education. Findings such as these imply the existence of a set of traits acquired in the environment of origin which assist the black migrant with limited education to adapt effectively to northern labor markets.

REGIONAL DIFFERENCES IN SOCIALIZATION PATTERNS

It has been said the blacks of southern origin are more likely to have been involved in the process of passive socialization than their northern counterparts (Crain and Weisman, 1972). This form of socialization, although dysfunctional in some aspects of life, increases the level of black deference to whites, which might possibly favor the black worker possessing this trait in specific job situations. Williams, in evaluating the causes of success and failure among a group of hard-core job trainees, had this to say about the job adjustment made by one worker: "In the South, Banks developed attitudes that Carter didn't develop in the West. He was taught a passive attitude—to be polite to his superiors, to obey" (Padfield and Williams, 1973). Given the inability of assumed regional differences in educational quality to explain the success of the migrant in the labor market, one then begins to focus on the role of black culture and its impact on personality development as an explanatory variable. Although the traits described above might favor the migrant in the world of work, such traits are thought to be problem-producing in the individual's personal world.

Emphasis here on size of place of residence of southern migrants is prompted by the notion that the nature of the socialization process in the metropolitan South differed from that in the nonmetropolitan South. In the latter zone passive socialization was probably more widespread, growing out of the necessity for more direct contact with whites in a superordinate-subordinate context. Passive socialization was simply a strategy of adaptation employed by segments of the dependent group. In southern metropolitan centers the need to employ this adaptive strategy was lessened as a result of the reduction in the number of direct contacts between the races. Thus, it is postulated that the size of place of origin of southern black migrants leads to the acquisition of culture and/or personality traits which provide advantages or disadvantages in securing employment in northern urban centers, where post-high school training is not critical for job success.

PRINCIPAL NONSOUTHERN DESTINATIONS
OF SOUTHERN MIGRANTS

Now one can turn to the pattern of movement that led to black settlement in the leading targets of black migration during the sixties. The principal northern destinations of black movers during the period were Los Angeles, Detroit, New York, Chicago, Philadelphia, San Francisco-Oakland, Newark, Cleveland, and St. Louis.[1] The method chosen to define migration influences the rank of places of destination among the target set. Likewise rank is also influenced by one's choice of the metropolitan area or the central city as the target unit. Three available sources of information permit one to develop three different measures of migration level. Two of these are not migration measures, with one being derived by subtracting the natural increase over a ten-year period from the total population. The residual is a measure of net migration. The alternate net measure is derived by totaling the difference between number of persons living elsewhere five years earlier from the population that resided at the target destination at the same date but currently resides elsewhere. The measure which appears to be utilized most frequently is the number of persons living elsewhere five years preceding the date of the census. It is useful to employ each of these measures of migration if one is concerned with the impact of migration on labor force participation during a ten-year period. Here the metropolitan area has been employed as the basic unit for the purpose of evaluating changes in the migration status of the individual migrant destinations (see Table 1).

Utilizing method one, the ten-year net additions accruing as a result of migration appear to be positively related to both SMSA size and the size of the black population at the beginning of the period. But it is apparent from Table 1 that this relationship fails to hold in the Philadelphia case. A weakness in the selection of the SMSA as the unit of evaluation emerges when black movement to the suburban ring accounts for a significant share of all migrants. In those instances the central city's role in attracting migrants is not accurately portrayed. This pattern is illustrated in the case of Cleveland and St. Louis, both of which suffered a net loss in the central city, while registering significant gains in the SMSA.

The ranking of principal destinations is altered when method two

TABLE 1
ESTIMATES OF BLACK MIGRATION LEVEL IN LEADING MIGRANT DESTINATIONS BY METHOD OF ANALYSIS

Destination	Method 1		Method 2	Method 3
	Central City	SMSA	SMSA	SMSA
Los Angeles	119,552	266,878	72,945	+40,437
Detroit	97,533	108,359	55,869	+44,547
New York	435,840	482,822	55,651	−13,125
Chicago	113,194	148,005	49,837	− 2,101
Philadelphia	39,648	44,556	35,546	+11,299
San Francisco-Oakland	37,485	65,190	32,842	+14,977
Newark	31,506	71,669	21,323	+ 6,145
Cleveland	−2,769	33,350	19,101	+ 5,519
St. Louis	− 948	28,817	16,701	+ 1,054

SOURCE: General Demographic Trends for Metropolitan Areas, 1960 to 1970, 1970 Census of Population and Housing, PHC(2)—For Individual States, Mobility for Metropolitan Areas.

is employed to evaluate migration levels. Utilizing this procedure Los Angeles becomes the principal destination for black migrants, followed by Detroit, New York, and Chicago. The latter method is based on data describing change during the period 1965-1970, while the former covers the entire ten-year period. The greatest discrepancy between these two sets of data is the migration levels, describing migrant movement in the New York case and secondarily in the Los Angeles case. While these two sources of data are not really measuring the same thing, it does appear that they serve as a rough index of the variation in the magnitude of migration during the two five-year intervals of the decade. The results appear to indicate that movement to New York, Los Angeles, and Chicago largely took place during the early years of the decade, while a more uniform flow took place in the other centers.

When method three is employed in evaluating migration levels, yet a third pattern of ranks emerge. Utilizing method three it is possible to determine the effect of outmigration from the individual SMSAs on the net contribution of migration to population change during the five-year interval. Outmovement from both New York and Chicago during the latter half of the decade exceeded inmovement, leading to a net loss in population. Detroit emerges as the center that witnessed the greatest net addition, followed closely by Los Angeles. San Francisco-Oakland and Philadelphia are a weak third and fourth, with even more modest additions occurring in Cleveland and St. Louis. From the perspective of job competition, method three appears to provide the most useful information, while method two lends insight into the extent to which an individual place is the recipient of a population which recently resided elsewhere and might thus encounter some problems of adjustment. Likewise, it is method two that provides the basic information on place and size of place of origin of the new residents to the SMSA.

The South's contribution to the volume of migration to this selected set of metropolitan areas is quite variable, ranging from a high of 68% in New York to slightly more than 40% in the case of San Francisco-Oakland (see Table 2). It was once thought that the black population on settling in individual northern centers tended to become nonmobile. Thus, northern blacks were described as stable residents who seldom became involved in the migration process. During the past twenty years this notion has been overturned as black movement between northern centers has accelerated. This new movement pattern has led to the increased importance of inter-

TABLE 2

REGION AND CHARACTER OF PLACE OF ORIGIN OF BLACK MIGRANTS TO METROPOLITAN AREAS AND THEIR CENTRAL CITIES, 1965-1970

Destination	N of migrants to:		N from South	% from South	% from Non-Metro Area
Los Angeles	CC[a]	48,524	23,370	47.1	28.1
	SMSA	72,945	34,370		
Detroit	CC	48,870	29,015	59.2	37.2
	SMSA	55,864	33,081		
New York	CC	45,036	31,158	68.5	44.8
	SMSA	55,651	38,206		
Chicago	CC	39,554	27,474	66.8	56.0
	SMSA	49,837	33,375		
Philadelphia	CC	21,303	18,773	50.3	39.8
	SMSA	35,546	17,885		
San Francisco-Oakland	CC	9,190	4,251	42.3	38.5
	SMSA	32,842	13,907		
Newark	CC	11,939	7,465	56.5	45.1
	SMSA	21,323	12,055		
Cleveland	CC	15,028	9,033	56.6	39.9
	SMSA	19,101	10,820		
St. Louis	CC	9,567	5,820	56.4	54.6
	SMSA	16,701	9,088		

a. CC = Central City

SOURCE: U.S. Bureau of the Census, Census of Population: 1970, Subject Reports, Final Report PC(2)-2C, Mobility for Metropolitan Areas and Census of Population and Housing, Final Report, PHC(2)—Selected Numbers.

metropolitan migration and its subsequent displacement of rural-urban migration as the modal migration pattern of black Americans. Intra- and interregional migration involving the non-South is essentially intermetropolitan migration. The intermetropolitan mover has generally been socialized in an environment that is strikingly different than that which served as the socializing milieu of the nonmetropolitan mover. From Table 2 it becomes apparent that the percentage of movers of nonmetropolitan origin is positively related to the South's contribution to the total volume of migration.

In order to foster additional insight into the nature of migrant adjustment in the target destination, a set of passive socialization scores were developed for each of the major migrant centers. These scores were computed on the basis of a given southern state's share of total migration to an individual center. Here one has assumed that the higher the percentage rural in a given origin-state's population in 1960 and the fewer metropolitan centers in that state, the higher the level of passive socialization. Given the migrant origin mix of the individual destinations, Chicago registered a score of 304, while San

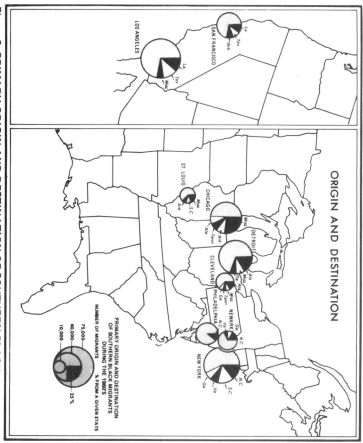

Figure 2: PRIMARY ORIGIN AND DESTINATION OF SOUTHERN BLACK MIGRANTS DURING THE 1960s

Francisco-Oakland registered a score of only 46 (see Figure 2). Chicago's high score reflects the highly rural character of the place of origin of a large percentage of its migrants, whereas intraregional migrants are more important contributors to population growth in the San Francisco-Oakland metropolitan area and thus contribute to its low score. High passive socialization scores are associated with places where a large percentage of the migrants originate in Mississippi, Arkansas, North Carolina, and South Carolina. Thus in those states where the black population until recently served as the backbone of a commercial agricultural system that was dependent upon it as a source of inexpensive labor, black deference is thought to be highly internalized. South Carolina is the only state among this group that has been a major contributor of northern migrants for more than a single generation.

MIGRATION AND JOB COMPETITION

It is assumed that high passive socialization scores favor that segment of the population seeking jobs which require only limited skills. For if, as some writers conclude, life styles acquired in northern urban environments constitute a barrier to black employability, then some aspects of this handicap should be minimized through socialization in an alternative environment. This situation is expected to lead southern black migrants of nonmetropolitan origin to settle for less in the job market. A presumption of lower esteem works to influence one's perception of what constitutes satisfactory employment. It has been stated that "Black youngsters growing up in Northern cities in the fifties and sixties, then were likely to think better of themselves and to expect more of their society than did their parents" (Boesel, 1970: 267). Thus, when participating in a job market where jobs perceived as meaningful vis-à-vis one's training and experience are disappearing, the new entrant to the job market whose base of socialization is the nonmetropolian South may find his opportunity for employment enhanced. This situation is further augmented by the migrant's potential job status in the area of origin.

In a recent description of the pattern of black migration from Holmes County, Mississippi, the migrants were identified as refugees and purposeful migrants (Shimkin, 1971). The refugees were those who had been employed in a system of agricultural tenancy and had become uprooted as a result of changes occurring in that system. The term "refugee" appears to be an appropriate one in describing a segment of the migrating population, but more than that, it has implications for migrant behavior in the job market at the place of destination. It is generally agreed that refugees possess a greater willingness to accept jobs that are normally thought to be least attractive. This possibly suggests a generational lag in the aspirations of youth possessing similar socioeconomic status, but growing up in different regions during the post-World War II period. The perceptions of the existence of a more open society on the part of black yough socialized in large northern centers is thought to have influenced the attitude of those youths regarding what constitutes acceptable employment. Friedlander (1972), in attempting to assess the backgrounds and attitudes of a small sample of inner-city youth in terms of their impact on participation in the job market, concluded the following: "There is no incentive to hold a marginal job with subsistence wages because of the lack of opportunity to rise in the occupational and income classes." The import of the passive socialization scores is to provide a crude index of the extent to which black migrants are likely to be willing to accept employment in the secondary labor market. More significant, though, is this question: To what extent does employment in the secondary market on the part of a segment of this population lead to reduced labor force participation rates on the part of the nonmigrant population and subsequently to participation in nonlegitimate economic activity?

There is also concern for the kind of environment that the migrant who is engaged in marginal employment is forced to reside in because of his/her economic marginality. Shimkin (1971) has suggested that those identified as refugees tend to settle in the poorest sections of the city. These are the sections that are generally characterized by the highest levels of disorganization. On the other hand, as Friedlander (1971: 183) suggests, "the option to hustle is always available." The principal issue here then is—does the job market operate in such a fashion as to be able to absorb a large segment of low-skilled black workers of southern origin, while being unable to provide opportunity to those black youth socialized within inner-city environments, thereby leading to an increase in illegitimate economic activity which promotes negative feedback through the total black community? This negative feedback is thought to lead to increases in victimization in the form of property loss, physical injury, incarceration, and increased welfare dependency.

Changes in the volume of new jobs created per unit of time via-à-vis changes in new entrants to the labor force (minus those leaving the labor force) can lead to a number of possible outcomes. The outcome which is the focal concern here is the ability of individual labor markets to absorb the increase in the potential black labor market entrants during the previous decade. But more specifically, the concern is one of assessing the ability of the market to absorb the net increment fostered by migration. Obviously, the ability of the market to absorb the incremental increase is partially conditioned by changes in the supply of specific kinds of jobs and the nature of the skills that the prospective workers brings to the job market. Another factor which is thought to be at work, which influences the scope and size of the potential job market for black workers, is the extent to which that market is segmented. Some economists contend that there exists a dual job market, one which caters to black and lower-class white workers. These two facets of the market have been labeled the "core" and the "periphery" of the economy.

Harrison (1972), in describing the basic components of the core and periphery of the economy, suggests that some of the major distinguishing criteria between the primary labor market and the secondary labor market are the extent of investment in capital equipment, wage levels, and employment security. The secondary market which appears to be more open to blacks tends to invest little in capital equipment, pays low wages, and fails to provide job security. Yet Flanagan (1973) is of the opinion that the dual labor market concept is applicable only to a limited segment of the black labor force. It appears to be unclear at this point the extent to which blacks tend to operate in a quasi-closed labor market. But there is evidence which suggests that blacks are more likely than whites to be confined to a selected set of job categories or industrial categories, and the emerging pattern cannot be explained simply by level of educational attainment. It has been suggested that blacks are crowded into certain job categories leading to depressed wage rates in those categories. If this is true, then the size of the potential black labor force competing for jobs in the dual market would be directly related to the incremental contribution of migration to the total supply of labor. Lack of access to a wider range of job opportunities in those labor markets where the contributions of migration intensify pressures under conditions of restricted opportunity can be expected to lead to higher incidences of stress in the black community.

It was noted that during the fifties the destination chosen by black migrants seemed not to be related to unemployment levels in the target cities. Kaun (1970) contended that the goal of black migrants is to maximize income over time rather than minimize unemployment. Some might argue that, since many of these movers were refugees, they were insensitive to conditions in the labor market of destination. Another argument which might be put forth is that migrants were in contact with relatives who were employed by firms whom they knew would provide employment for the migrant. While this latter contention cannot be verified, it has been held to represent a major factor in the hiring of low-skilled white workers. Thus, an informal communication network composed of persons with knowledge of the availability of jobs in individual firms may have the effect of facilitating labor market entry even when the general level of unemployment is high.

This leads one to investigate unemployment levels at different intervals during the previous decade at the selected target destinations, as a means of ascertaining if the patterns prevailing at the beginning of the period were similar to those at the end of the period. By 1966 there was evidence of a general downturn in unemployment levels, a condition which prevailed until 1969. This general improvement in employment opportunity was associated with the Vietnam buildup during the latter years of the decade. For blacks the highest levels of unemployment prevailing at the beginning of the period (1960) were found in Detroit (18.2%), Cleveland (12.7%), and Philadelphia (11.1%). The lowest levels occurred in New York (6.9%) and St. Louis (8.9%). While improvements had begun to show themselves by 1966, there were also some reversals in the conditions prevailing at the beginning of the period. In 1966 the highest levels of unemployment were to be found in Newark (14.2%), St. Louis (11.3%), and Cleveland (10.1%). Obviously blacks, like the general population, were able to acquire jobs during the latter three years of the sixties as unemployment rates were down in each of the migrant destinations.

By the end of the decade the lowest male unemployment rates prevailed in New York (3.9%), Chicago, and Philadelphia (4.8%). By 1970 the highest levels of unemployment occurred in Los Angeles-Long Beach (7.5%), San Francisco-Oakland (7.9%), and Detroit (7.4%). As was true during the previous decade, it does not appear that unemployment levels had a serious dampening effect on migration to the target destinations—at least in most instances this appears to be true. Although the most serious levels of unemployment prevailing in 1970 were to be found in those places where net migration during the latter half of the decade was highest (Detroit, Los Angeles, and San Francisco), those places characterized by net outmigration (New York and Chicago) or minimum net gains either had relatively low unemployment levels or made improvements over their 1965 unemployment levels. While changes occurred in the pattern of unemployment in the general population, there was not a single instance in which white unemployment rates were found to be higher in either 1965 or 1966 than they were in 1960.

It does appear on the surface, then, that blacks might have been competing among themselves for jobs, if the unemployment patterns described above are meaningful. On the other hand, some improvements might have occurred as a result of whites abandoning central city jobs during this period. The net loss of jobs in the central cities between 1959-1969 by white males totaled 900,000 (Nelson, 1974).

Nelson's position on this phenomenon is as follows: "The cross-migratory pattern of whites to suburban areas and Blacks to central cities is the greatest single factor contributing to the general improvement in employment opportunities for Negro workers in the decade of the sixties" (Nelson, 1974: 36). If Nelson is correct, this is further support for the existence of a dual labor market segmented by race and, thus, indicates that blacks were essentially competing among themselves for jobs abandoned by white workers who had become a part of the suburban job market.

A more precise way of ascertaining the impact of migration on the labor market is to determine the contribution of migration to new labor force entry. This can be done by measuring changes in the black population 10-19 years of age in 1960 to that 20-29 years of age in 1970. If there were no migration, the initial population would be expected to be smaller in 1970 than in 1960. Since the survivorship rate is high for this population, one can simply use the 1960 population as a base. The number of additional jobs held by blacks at the lower end of the job spectrum in 1970, over and above those held in 1960, divided by the number of new entrants to the labor force, should provide some indication of the pressure on the job market resulting from migration. Among the target cities new entry-level jobs (operatives, service workers, and laborers) exceeded the net increase of first-time entrants to the labor force in Los Angeles, Detroit, and St. Louis; while in most of the other cities there was a disappearance of entry-level jobs in the face of an increase in new first-time entrants to the labor force. The most extreme situations occurred in New York and Chicago, suffering a loss of 12,000 and 4,000 semi-skilled to unskilled jobs, while adding 3,900 and 1,200 first-time entrants to the labor force during the decade. In the latter two destinations migrants were largely from nonmetropolitan settings and thereby increased the intensity of competition for jobs at the lower end of the job ladder.

It should be made clear, however, that not all migrants entering the labor force for the first time during the previous decade were seeking employment in the blue-collar service sector. Approximately 70% of the urban place migrants in the 17-29 year age range in 1967 had completed approximately 9-12 years of schooling (Bowles et al., 1973), with more than half having completed 12 years of education. The nonmigrant segment of the urban population in the same age category showed a similar distribution in terms of educational attainment, with the exception that the nonmigrant was less often a high school graduate. While nonmigrants were less often high school graduates, those who did graduate from high school more often tended to engage in post-high school training. Thus the competition for jobs in a segmented market is most likely to favor the more often certificated migrant than the less often certificated nonmigrant.

In this instance, depending upon the level of segmentation, the most severe problem of intragroup job competition would take place in New York and the least severe problem would occur in Detroit. Not only was the ratio of entry jobs to initial labor entrants highest in New York, but the proportion of initial job entrants who were migrants was higher (30%) than that for any of the other target cities. Given the job situation in New York, Newark, and Chicago, it appears that such migrant movement might have been ill-advised. Philadelphia, unlike its neighbors (Newark and New York), did not attract nonmetropolitan migrants on a similar scale. This might reflect rational judgments on the part of migrants, as Philadelphia with a net loss of blacks as a result of migration still had fewer new jobs than it had persons vying for jobs. What appears to reflect blind movement of nonmetropolitan movers to northern urban centers is not easy to evaluate. As a rule nonmetropolitan movers do less well in the job market than do nonmovers when measured in terms of unemployment rates, but appear to be better off when evaluated in terms of labor force participation rates.

There is no way of knowing the extent to which migrants of nonmetropolitan origin compete directly with nonmigrants for jobs. This no doubt depends partially on the job market, its requirements, and the extent to which there exists a single queue or double queue for jobs with similar skill demands. But whatever the nature of the market the nonmetropolitan migrant tends to find the largest number of job openings in the job category described as "operative." Since this is a rather diverse category, it is not easy to ascertain if the worker is engaged in activity that is aimed at providing goods and services for an export or import market. In those instances where the worker is providing goods and services to be consumed by the larger community and/or nonlocal community, he is said to be employed in the export economy. Likewise, blacks who are hired to provide goods and services for the local black community might be thought of as being involved in the import economy. It appears that blacks possessing a different set of subjective characteristics might be chosen for one job setting or the other. Often black youths with limited skills have been provided jobs wherein they deliver services to the black community as a cooptation strategy designed to deflate preexisting levels of militancy. On the other side of the ledger, black youths of migrant origin with limited skills, but showing a willingness to engage in deferential conduct, find employment opportunity in the export economy.

Much of the occupational progress witnessed by blacks during the sixties, at least in terms of movement out of the blue-collar sector of the labor force, was related to the acquisition of professional and managerial jobs designed to provide services for a black constituency. Since blacks have less aversion to providing services to other blacks than do whites, the increase in size of the black population in any major urban center lends itself to the development of new jobs, as well as replacement jobs for black professional and quasi-professional workers. The more rapid the growth, the greater the relative number of such jobs evolve. This situation leads low-income migrants to find employment in the export economy, low-income or limited skill nonmigrants to increasingly find opportunity in the import economy, and skilled migrants to be attracted to the opportunity made available by the rapid growth of the black population. The extent to which all of the participating actors are able to adjust to shifts occurring in the social economy of individual metropolitan systems, as influenced by decisions at the national level, will determine who is advantaged and disadvantaged as a result of migration to northern urban centers.

The most severely disadvantaged are those who are unable to successfully adapt to these complex machinations so that they choose to basically participate in the irregular segment of the secondary sector of the economy. While a few escape the bonds of pauperism through success in the irregular economy, most simply choose hustling as an alternative survival mechanism. Krisberg (1974: 34), in describing a group of young hustlers whom he interviewed, had this to say:

Their ages ranged from 18-22 years. Each gang-leader was a public school dropout. Their combined arrest records totaled more than 175 contacts with the law. Most of theses arrests were for offenses involving violence, weapons use and assaultive behavior. Many in the group had served up to 36 months in state and local prisons. Their employment histories had been transitory and primarily unsuccessful.

It would appear that the more intense the competition in an individual labor market, the higher the propensity to engage in the hustle. Thus, when the labor market is unable to adjust to the pressures exerted on it by those who are potential entrants—a situation that in some instances is aggravated by net immigration—a larger percentage of the prospective jobholders engage in acts which victimize either individuals or institutions found in close proximity to the actor.

The previous discussion focused essentially on conditions in entry-level jobs vis-à-vis first-time entrants to the labor force. While the largest number of migrants are found to represent persons in the 20-29 age category, they seldom constitute more than 50% of all migrants. So entry-level jobs in the blue-collar service sector simply represent one segment of the total job picture. Migrants of older age and with previous job experience are also vying for jobs in the individual labor markets. In some markets the net outmigration of younger workers results in older workers constituting a greater percentage of all migrant workers than in cities where this is not the case. It would appear that the older limited-skill migrant would be even more amenable to accepting jobs in the secondary economy than would new entrants to the labor force, regardless of place or origin. On the other hand, older migrants who have had previous experience in large urban environments move more often, with a large percentage of them constituting the cadre of professional and quasi-professional workers who seek employment in the import economy which provides services to the black community. They, too, often constitute a growing segment of migrants who move directly from one metropolitan area to the suburban ring of another. Most of the progress enjoyed by blacks in labor force improvement during the last decade was associated with movement into the white-collar and in the upper blue-collar class (craftsman). Among the nine cities employed in this analysis, six suffered losses in entry-level jobs during the decade, whereas all recorded increases in the percentage of blacks found in upper-level blue-collar and white-collar jobs. The most serious decline in blue-collar jobs occurred in Philadelphia and New York (see Table 3). The percentage change in the other two job categories tend to be related to the percentage of growth of the total black population, as well as the extent to which migration contributed to that growth. Cleveland, Philadelphia, and St. Louis were among the cities with slow-growing populations during this period, and likewise they were the places to

TABLE 3

PERCENTAGE CHANGE IN THE NUMBER OF BLACK MALES HOLDING JOBS IN THREE MAJOR JOB CATEGORIES, 1960-1970

Place	Lower Blue-Collar & Service Occupations	Upper Blue-Collar Occupations (Craftsmen)	White-Collar Occupations
Los Angeles	+40.0	+ 70.6	+127.1
Detroit	+24.8	+ 84.2	+ 76.4
New York	− 6.3	+ 95.9	+ 71.2
Chicago	− 2.8	+208.3	+ 67.3
Philadelphia	− 4.0	+ 65.6	+ 55.8
San Francisco-Oakland	+29.6	+ 82.9	+ 64.3
Newark	+ 6.6	+103.4	+108.4
Cleveland	+ 2.0	+ 67.0	+ 68.7
St. Louis	− 3.2	+101.5	+ 67.9

SOURCE: U.S. Bureau of the Census, Census of Population and Housing: 1970, Census Tracts, Final Report PHC(1)—Selected Numbers; Census of Population and Housing: 1960, Census Tracts, Final Report and PHC(1)—Selected Numbers.

witness the smaller percentage shifts in white-collar jobs. St. Louis did show one of the higher-level shifts in the craftsman category, but this might simply reflect greater previous barriers to entry into this job category, as was generally typical of cities with a southern tradition. The higher rates of change in the upper blue-collar category probably reflects those places where blacks were able to make major penetrations in the primary economy, although in some instances this indicates a lack of a significant number of black workers in this job category in the base year. Though it will not be attempted here, it would be interesting to ascertain the percentage, as well as the percentage change, in the number of black males employed in the export (serving larger community) and import (serving the black community) economy of individual metropolitan areas. For at this point it is quite evident that black professional workers have found increased opportunity in those cities where the black population is growing the fastest, although additional white-collar jobs are rapidly becoming available to the female sector of the labor force in those cities where the greatest work opportunity is emerging in the suburban ring.

Blacks captured almost 12% of the new jobs created during the sixties (Brimmer, 1972), but most of these jobs were in the semi-skilled operative category. The manufacturing sector of the economy has traditionally provided opportunity for unskilled workers and semi-skilled workers in this country, but manufacturing jobs are decreasing both as a percentage of the total and absolutely.

They are abandoning central city locations at a rapid rate, thereby reducing the possibility for blacks with limited skills to enter the primary economy, although in 1971 more than one-fifth of all black workers were still found in manufacturing jobs (Brimmer, 1972). But even so, Brimmer indicates that blacks in manufacturing tend to be concentrated in five manufacturing categories—transportation equipment, primary metals, electrical equipment, wood and related products, and textile mill products. These hardly represent the growth sector of the economy. The loss of manufacturing jobs in the cities of traditional migrant attraction and a transformed southern agricultural economy result in the intensification of competition among blacks for limited jobs in the primary sector of the economy. The situation might have become even more acute were it not for the 730,000 jobs in the central cities abandoned by white males during the decade in the operative and craftsman category (Nelson, 1974). It was precisely in these two categories where blacks found the greatest number of openings.

The growing dependency of blacks on low-level service jobs possibly fosters negative attitudes on the part of the young black worker who had come to expect more, and thus the hustle takes on added attraction. It has been said that "jobs in service activities do not provide the job training ladder for unskilled workers in manufacturing. Service jobs can be dead end jobs for those without skills" (Ganz and O'Brien, 1973: 111). A difficult situation is made even more difficult by employers showing a preference for migrants over the ghetto-socialized youth for limited-skill jobs. This simply leads to a piling up of youths whose adaptive repertoire, while congruent with the demands of their social world, is incongruent with the demands of the larger world of work. Thus to the extent that southern black youths of nonmetropolitan origin exhibit behavioral traits which are less at odds with the perception of what constitutes desirable traits, holding level of educational attainment constant, then ghetto-socialized youths will be at a disadvantage even in a racially segmented job market. It seems that the shock waves associated with ghetto uprisings produce jobs in the import economy for youths who demonstrated a proclivity to participate in such activity. But once the shock had faded, and these jobs which were often subsidized from Washington disappeared, the only course left open was the irregular economy, and the black community is subsequently the principal victim.

Migration continues to serve an important role in providing

improvements in economic opportunity for blacks when measured against the opportunity structure prevailing in the environment of origin and the environment of destination. It is evident, however, that the prospective migrant often times misreads the level of difference between these environments. This misreading of the opportunity structure leads to significant return flow to the area of origin. During the period 1965-1970, black movers from New York to nonmetropolitan areas was 63% of the combined movement to and from these two environments. On the other hand, Detroit represented essentially the opposite situation, where movers from nonmetropolitan areas greatly exceeded those destined for such areas. This pattern is thought to reflect the migrants' relative perception of opportunity and the development of a rational adaptive strategy. The huge movement of blacks to New York from the South during the first half of the sixties probably stimulated a larger-than-usual counterflow during the latter half as an outgrowth of the discovery of declining opportunity in the unskilled job market. One is less certain about the adaptive strategies developed by that segment of the population moving from one central city to another seeking employment in this segment of the market. If the strength of attachment to home is as strong as it is among nonmetropolitan movers, then a similar return movement pattern should evolve. While patterns of countermovement represent an adaptive strategy, the basic issue here is: Does incremental growth of the black population resulting from migration contribute to differential stress levels in those cities where unskilled and semi-skilled migrants moved in large numbers? The principal target cities were those with already large black populations at the beginning of the period.

ENVIRONMENTAL STRESS AND SCARCE ECONOMIC OPPORTUNITIES

The problem of specifying stress levels within individual urban places has not yet been confronted. More difficult still is to attempt to parcel out the contributions of migration to existing levels of stress, however defined. As was pointed out earlier, the work of Dohrenwend (1973) as well as others provides an approach which might prove useful in this context. She has identified a number of life events which are generally recognized as stress-inducing. These are events which generally refer to a change in status, altered relationships, or a threat to one's security. Dohrenwend identified 26 specific life-events which requires the individual to make some adjustments. Not all of these events are thought to be undesirable in character, although 15 of the events used in the Dohrenwend questionnaire were thought to represent culturally defined losses. A major problem arises when one attempts to follow this lead by considering the objective as one of evaluating environmental stress rather than individual stress. The principal problem is the selection of ecological surrogates of stressful life-events. In this instance the surrogates chosen were violent crime (homicide, assault, robbery, and rape); family disintegration; and a measure of economic insecurity.

A stress score was derived for each of the target cities on the basis of the incidence of the individual variables involved. The greatest weight was assigned to that cluster of variables which constitute an index of violent crime. This weighting reflects an overriding emphasis on the part of blacks for environmental safety in terms of residential preference. Hinshaw and Allott (1972) have shown that blacks, like most other ethnic groups, indicate that a safe place to live is critically important in neighborhood choice. The weight given the items in this cluster reflect the relative weight given such items by Holmes and Rahe in their Social Readjustment Rating Scale. Holmes indicated that death of spouse, divorce, marital separation, death of close family member, and personal injury require the most extensive readjustment, or conversely it can be assumed that these variables contribute to higher levels of stress (Holmes and Masuda, 1974).

Violent crime characteristics are available in the Uniform Crime Reports for individual metropolitan areas. But metropolitan area rates tend to be lower than those for central cities. Because of these differences an adjusted crime rate for cities was employed as the basic data source (Municipal Performance Report, 1973). These data were further adjusted on the basis of the black percentage in the population. Blacks, because of demographic, socioeconomic, and cultural differences, tend to account for more than two-thirds of all the crimes of violence in cities. Since most crimes of violence committed by blacks are against other blacks, the derived measures are an approximation of violent crimes in the black community. There is no attempt in this instance to identify criminogenic zones within the black community, although it is apparent that internal differences prevail. The incidence of violent crimes is lower than that for the other two measures of stress; but when multiplied by an appropriate weight, it becomes the principal contributor to stress in the black community. This paper simply focuses on the adaptive response resulting from migration as a contributor to stress levels, rather than attempting to unravel the effects of a complex system on behavioral outcomes.

The surrogates employed for family disintegration and economic insecurity are the percentage of women separated from their husbands (divorced and separated) and the percentage of male unemployment.

The variables that constitute these three dimensions indirectly reflect the occurrence of events which are thought to be the most stressful. If the index developed to measure stress is reasonably valid, then the highest stress levels are to be found in Newark and Detroit, and the lowest stress levels occur in Philadelphia, New York, and San Francisco (Figure 3).

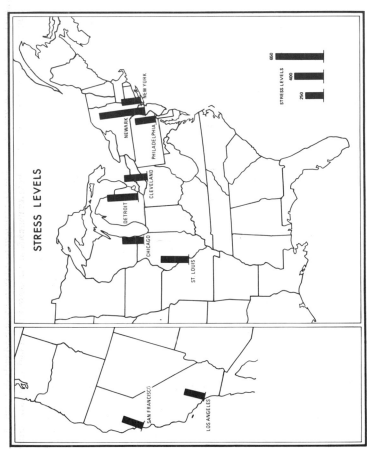

Figure 3: PREVAILING STRESS LEVELS IN PRINCIPAL MIGRANT DESTINATIONS, 1970

The implication throughout this paper has been that levels of stress are likely to be intensified as a result of the additional pressures imposed upon the black community resulting from migration, if the opportunity screens within a given labor market are not sufficiently large to accommodate the incremental population. This is a follow-up on the notion that migration should be eliminated as a causal factor in the civil upheavals during the latter half of the sixties. Since migrants were seldom found among those who were riot participants, it was generally thought that migration was unrelated to the riots–at least as a causal factor. If migration did not increase pressure on a system that was unwilling to accommodate a segment of the black population, namely, ghetto-socialized youth, then one could generally agree that these analyses were correct. But if indirectly pressures are built up in the black community which lead to increased stress, then the validity of the judgment of these analyses must be questioned.

It seems that a good deal more attention has been aimed at the tangle of pathology that is said to pervade the black family than has been given to the openness of the economy and the adaptive practices that are instituted as coping mechanisms. The orientation of much of this research leads to self-serving results–that is, the maladaptive practices employed by a segment of the black population accounts for their plight; and if these practices were altered, then one's status would likewise be altered in a positive direction. A recent example of this type of research orientation is found in a paper by Mogull (1973), who contended that, because of a number of behavioral and educational attributes, white employers are exercising rational and educational judgment when they show a preference for white workers with the same level of educational attainment as black workers. Mogull (1973: 45) further notes: "Because the education of blacks as compared with whites is over valued by the Census statistics, it follows that job discrimination against Negroes is also overestimated in the research." Here the writer justifies an economic decision which is undergirded by social status. Research orientations of this type leave us no further ahead in remedying the problems of the cities than would be the case with no research at all.

In order to determine the contribution of migration to stress in the target cities, a regression model with five independent variables was operationalized. The independent variables were:

X_1–percent increase in black males 20-29 years old

X_2—net migration

X_3—new residents

X_4—percent migrants from the South

X_5—Passive Socialization Scores.

These independent variables were regressed on the dependent variable, Stress Scores (Y). A multiple correlation coefficient of .905 was extracted with a corrected coefficient of determination of .520. More than half of the variance was explained by these five variables. Three of the variables were found to be significant at the 10% level. "New residents" with the highest partial correlation coefficient had a negative effect on stress (.858). "Percent increase in black males 20-29" and "net migration" both had a positive effect on stress (.811 and .822). Both "migrants from the South" and "Passive Socialization Scores" showed only a weak positive association with stress (.205 and .386). These five measures of migration's effects on the black communities of destination were essentially attempts to determine how migration influenced participation in the job market and consequently its feedback effect in terms of indirect impact upon the community.

The model in its original form does not directly substantiate the position attributed in this paper to the role of migrants of nonmetropolitan origin in raising stress levels in the black community. The two independent variables selected to illustrate this influence were the weakest of the five variables employed.

Because of the weak contribution of variables X_4 and X_5, a modified regression procedure was employed which permitted the entry of variables sequentially, and thus provided an opportunity to observe the contribution of each variable as it enters into the model. This procedure revealed a problem of multicollinearity between X_4 and X_5. When variables X_1 through X_4 are entered into the model, X_4 (migrants from the South) shows a strong positive association with stress. In the linear model (nontransformed variables), X_4 shows a partial correlation coefficient of .781. While this is a slightly weaker partial correlation than shown by the other variables, it is significantly higher than the level shown when X_5 enters the model. Variable X_5 leads to a washing out of X_4 and a strengthening of variables X_1 through X_3.

When the model is modified to show the impact of individual variables in a sequential format, it is shown that X_4 makes a significant contribution to stress. It is apparent that the Passive Socialization Scores developed here do not differ significantly from migrants from the South in its impact on stress; and when the two variables are included in the model, the effect of both are reduced, but the effect is to seriously depress the significance of X_4, such that its partial correlation coefficient is reduced to .034. To demonstrate effectively the role of nonmetropolitan socialization requires a rethinking of the procedure employed to derive the initial scores. It appears that passive socialization must be represented by some direct index of oppression as opposed to the indirect measures employed here. [Stinner and De Jong (1969) earlier attempted to measure the impact of discrimination on black outmigration from the South. They employed as measures of discrimination "Percentage Dixiecrat Vote, 1948," and "Presence of White Supremacy Organization, 1960."]

These variables indirectly measure the propensity for oppression and thus promote passive socialization. One might employ such measures as percent black male population employed as sharecroppers, percent black males incarcerated for periods of time in excess of three yeras for committing minor offenses, and percent black population registered to vote in 1960. The above measures tend to reflect a greater intensity of oppression than the simple measure of passive socialization employed here. The role of passive socialization remains unclear as a contributor to levels of stress partially as a result of the weakness of the measure employed and its collinearity with migrants from the South. As it stands, one is unable to distinguish the role of socialization in the nonmetropolitan South from socialization in the metropolitan South.

The variable whose contribution is more problematic is that of "new residents." This variable exhibited the highest association with stress, but in a negative direction. This tends to imply that in those places where net migration is of limited magnitude, the entry of persons from the outside as a replacement for those who leave leads to positive results. If this interpretation is correct, does it imply that limited-skill residents whom the new migrants replace are more likely to engage in acts which increase community stress than members of the replacement population? This question cannot be easily answered, but it is one which requires additional attention.

Net migration and percent increase in black males of labor force entry age during the decade showed the strongest association with stress. This is not unexpected. McNown and Singell (1973) previously demonstrated that a factor which they labeled "Demographic Change—Aggregate Demand" explained the second largest percentage of the variance in a regression model designed to parcel out the influence of variables on riot and nonriot behavior in a set of 129 large metropolitan areas. Their comment on the effect of this factor is as follows:

The positive coefficient on this factor indicates that cities experiencing the greatest population growth and influx of people, possibly as the result of greater job opportunities, are most likely to have experienced collective violence. This is consistent with the view that substantial demographic change destroyed social cohesion, put pressure on housing stocks; resulted in overcrowding public facilities like education and so on, and thus increased the likelihood of anti-social behavior [McNown and Singell, 1973: 9].

The above writers had assumed that riot behavior might be a substitute for crimes against property. Of the independent variables employed to test this assumption only the demographic factor was strongly associated with both sets of behavior. The results of the regression model developed for this study tend to question the notion that antisocial behavior is associated with newcomers to the community. Among the cities which form the observations for this study only St. Louis was not listed among the riot cities of the sixties, yet it had the third highest stress score among the nine cities. The other two cities with high stress scores were Newark and Detroit, both sites of major riots during this period. What emerges is that stress levels cannot be adequately explained by models that simply incorporate variables related to migration. But it is clear that the various migration variables do contribute to existing levels of stress.

A number of different regression models were employed to determine which format explains the greatest contribution to stress. The regression format in which none of the variables were transformed produced the best results. This format produced a multiple coefficient of correlation of .952 and a corrected coefficient of determination of .725. This format provides a 25% increase in the explanation of the individual variables, compared to the case when three of the five variables were log transformed. The predicted stress levels among the nine cities did not vary greatly from the observed levels. The greatest discrepancy occurred in Cleveland, where the observed level was more than 71 points lower than the predicted level. Other cities for which the model overpredicted levels of stress were Chicago and Los Angeles. Stress levels for all other cities were underpredicted.

SUMMARY AND CONCLUSIONS

This paper has focused attention on the impact of migration on the condition of life in black communities in those urban places that were the basic destinations of black movers during the sixties. Because of recent revelations showing that black migrants from the South were frequently better off than nonmigrants in the communities of destination, a number of questions emerged. The principal question was: Are limited-skill black migrants from the South essentially forced to compete with resident blacks, with 12 or fewer years of education, for jobs which are on the decline? A second question was: Are blacks from the South with similar levels of educational attainment favored over northern blacks for jobs requiring little skill, as a result of having developed a set of traits in the environment of socialization which white employers find more attractive? And, finally, the question was raised: What are the implications of these external decisions on the level of stress which emerges in the black community? These questions have not been answered definitively. But from the evidence assembled, it does appear that migration leads to increased pressure on scarce resources. The evidence, though, is often indirect. The differential labor force participation rates of migrants and nonmigrants, holding age constant, does offer some support for the latter position.

A number of methodological issues arose in trying to operationalize several concepts central to the study itself. The most troublesome problem was the development of an index of stress. Following the lead of several psychologists, a number of events were identified which are known to lead to stress. For most of these events surrogates had to be selected which described the level or incidence of the event within a given city. Another problem in attempting to deal with stress in an environmental context is the transfer problem. Most of the work which has been undertaken in this area focuses on the problems of individuals. In this paper one has attempted to arrive at levels of stress found within the environment resulting from interaction with the world beyond the confines of one's residential community, but leading to behavior that was basically confined to that community. Since this represents a preliminary attempt to specify the nature of a complex relationship, the results are more tentative than they might otherwise be. Nevertheless, it is evident that rising stress levels in many of the nation's black communities present a problem of serious magnitude.

A major shortcoming of this research is the spatial scale on which it has been conducted. By not identifying differential levels of stress within the black community there is no way of specifying whether migrants or nonmigrants are most often confined to the high-stress environments. A more detailed analysis of the association of migrants with a given set of characteristics with environments of differential stress is required if more than gross generalizations describing the migrant destination are to emerge. Thus, it is not readily possible to specify the impact of the move of the black migrant from the nonmetropolitan or metropolitan South to the metropolitan North or from one metropolitan location to another within the North, if that migrant has completed 12 or fewer years of education. The ultimate question is: Who benefits as a result of the decision to migrate, or better still, who suffers the most as a result of these decisions, and how can the impact of the decision be handled so that everyone will gain?

One additional oversight of this research is the lack of emphasis on the black female migrant. Because of the complex nature of the problem, it was decided to emphasize the role of the black male migrant. The plight of the black female is in large measure tied to the fate of the male. In 1970 the percentage of married black females ranged from five to ten percentage points lower in all major cities than had been the case in 1960. Likewise, there was a steady increase in the number of marital failures during the decade, with almost one-quarter of all marriages having failed by 1970. Rainwater (1971: 371) speaks to this situation as follows: "Limited access to resources places a couple in a constant risk of tense and conflicting marital relations. Since neither partner can do his job properly with the resources available the inducements for self- and other-blame are continually present."

Thus the inability of the black male to secure a stable role in the urban economy tends to result in the decision to defer the act of family formation. This, nevertheless, does not reduce contact between the sexes and often the consequences of this contact leads to additional stress. The few studies which have focused upon migration of black females tend either implicitly or explicitly to concern themselves with the role of migration on welfare case loads. Since much stress in the black community grows out of male-female relationships, a more direct examination of the role of migration on sex roles might reveal an added dimension to the emergence of stress which has been overlooked in this paper.

NOTE

1. The Standard Metropolitan Area is employed as the basic data-collecting unit of migrant destination. As a result of using the SMSA as the basic data-collecting unit the role of the central city is in some instances distorted. This is especially true in those situations where there is more than one principal city in the SMSA. Similarly in those SMSAs characterized by large-scale black suburbanization, movement to noncentral locations tends to disguise limited movement to the central city. The San Francisco SMSA, which is listed as a principal destination of migrants, is accorded its rank as a result of its being the location of two major central cities. San Francisco City during the 1965-1970 period was less attractive as a migrant destination than both Boston and Milwaukee.

REFERENCES

AAKSTER, C. W. (1974). "Psycho-social stress and health disturbances." Social Science & Medicine 8: 77-90.

BEALE, C. L. (1971) "Rural-urban migration of blacks: past and future." American Journal of Agricultural Economics 53 (May): 302-307.

BOESEL, D. (1970) "The liberal society, black youth and ghetto riots." Psychiatry (May): 265-281.

BOWLES, G. K., A. L. BACON, and R. P. NEAL (1973) Poverty Dimensions of Rural-to-Urban Migration: A Statistical Report. Washington, D.C.: Government Printing Office.

BRIMMER, A. (1972) "Economic situation of blacks in the United States." Review of Black Political Economy (Summer): 34-54.

BRODY, E. B. (1973) The Lost Ones. New York: International Universities Press.

CRAIN, R. L. and C. S. WEISMAN (1972) Discrimination, Personality, and Achievement. New York: Seminar Press.

DOHRENWEND, B. S. (1973) "Life events as stressors: a methodological inquiry." Journal of Health & Social Behavior (June): 167-175.

FLANAGAN, R. J. (1973) "Segmented market theories and racial discrimination." Industrial Relations (October): 253-273.

FRIEDLANDER, S. L. (1972) Unemployment in the Urban Core. New York: Praeger.

GANZ, A. and T. O'BRIEN (1973) "The city: sandbox, reservation, or dynamo?" Public Policy (Winter): 107-123.

HARRISON, B. H. (1972) Education, Training, and the Urban Ghetto. Baltimore: Johns Hopkins Press.

HIMES, J. S. (1971) "Some characteristics of the migration of blacks in the United States." Social Biology (December): 359-368.

HINSHAW, M. and K. ALLOTT (1972) "Environmental preferences of future housing consumers." Journal of the American Institute of Planners (March): 102-107.

HOLMES, T. H. and M. MASUDA (1974) "Life change and illness susceptability," pp. 45-72 in B. S. Dohrenwend and B. P. Dohrenwend (eds.) Stressful Life Events. New York: John Wiley.

IDEN, G. (1974) "Factors affecting earnings of southern migrants." Industrial Relations (May): 177-189.

KASL, S. V. and E. HARBURG (1972) "Perceptions of the neighborhood and the desire to move out." Journal of the American Institute of Planners (September): 318-324.

KAUN, D. D. (1970) "Negro migration and unemployment." Journal of Human Resources (Spring): 191-206.

KRISBERG, B. (1974) "Gang youth and hustling: the psychology of suvival." Issues in Criminology 9, 1 (Spring): 115-131.

LEE, A. S. (1974) "Return migration in the United States." International Migration Review (Summer): 283-300.

MASTERS, S. H. (1972) "Are black migrants from the south to northern cities worse off than blacks already there?" Journal of Human Resources: 411-423.

McNOWN, R. F. and L. D. SINGELL (1973) "A factor analysis of the socio-economic structure of riot and crime prone cities." Annals of Regional Science: 1-13.

MOGULL, R. C. (1973) "Is job discrimination among blacks overestimated?" International Journal of Group Tensions 3: 40-44.

Municipal Performance Report (1973) Aggregate data from May-June issue. New York: Council on Municipal Performance.

NELSON, J. E. (1974) "The changing economic position of black urban workers." Review of Black Political Economy (Winter): 35-48.

PADFIELD, H. and R. WILLIAMS (1973) Stay Where You Were. New York: Lippincott.

PETTIGREW, T. F. (1964) A Profile of the Negro American. Princeton: Van Nostrand.

RAINWATER, L. (1971) "Work, well-being and family life," pp. 361-378 in P. B. Doeringer and M. S. Piore (eds.) Internal Labor Markets and Manpower Analysis. Lexington, Mass.: Heath Lexington.

SHIMKIN, D. B. (1971) "Black migration and the struggle for equity: a hundred year survey," pp. 77-116 in J. W. Eaton (ed.) Migration and Social Welfare. New York: National Association of Social Workers.

STINNER, W. F. and G. F. DeJONG (1969) "Southern Negro migration: social and economic components of an ecological model." Demography (November): 455-471.

STRUENING, E. L., S. LEHMAN, and J. G. RABKIN (1969) "Context and behavior: a social area study of New York City," in E. Brody (ed.) Behavior in New Environments. Beverly Hills: Sage Publications.

TAEUBER, K. E. and A. F. TAEUBER (1965) "The changing character of Negro migration." American Journal of Sociology (January): 429-441.

WEISS, L. and J. G. WILLIAMSON (1973) "Black education, earnings, and inter-regional migration: some new evidence." American Economic Review (March): 372-383.

10

The Periphery at the Pole:
Lima, Peru

MILTON SANTOS

THE NOTION OF POLE-PERIPHERY

☐ EVER SINCE MARK JEFFERSON (1939) coined the word "primacy,"[1] the term has had manifold uses in the social sciences. Primarily a statistical notion, it gave birth to the rank-size rule[2] and even more ambitious concepts often presented as theories. These developments distorted the analysis and blocked efforts to raise new interpretations of spatial processes.

If the "core-periphery" (Friedmann, 1963) as well as the "heartland" and "hinterland" (Perloff and Wingo, 1961) approaches are geographic representations of Shil's model of center-periphery (Slater, 1968: 27), there is no doubt that the idea of primacy has been strongly influential. It has often been presented as case of territorial disparities or geographic dualism where the big city dominates the smaller ones.

As Friedmann (1963) has acknowledged, other authors had already considered the issue and prepared the foundations for the theory, albeit at different levels: (1) local (Schultz, 1953), (2) regional (Perloff and Wingo, 1961), and (3) international (Meier and Baldwin, 1957: part 2). But Friedmann should be credited with systematically developing the notion with contributions beginning in 1955. Without accepting the whole of his predecessors' ideas, he incorporated an important and original personal contribution.

Friedmann is of the opinion that the center-periphery structure appears at the national scale during the first stages of industrialization. It implies a true "colonial" relationship,[3] with the periphery contributing more to the growth of the center "than it receives in return" (Friedmann, 1963: 43). The periphery continues to be primarily a producer of agricultural materials and leads to an unfavorable situation in terms of interregional trade. The only important manufacturing industries created in the periphery are related to fixed external economies, such as raw materials or power. In terms of national industrial product and employment, the share of local industries "tends to be reduced to a very small fraction" of economic growth (Friedmann, 1963: 50). When a polarized structure emerges in the center it provokes a series of shifts from the periphery toward the center, resulting in the cascading of manpower, capital, entrepreneurial talents, foreign currencies, and raw materials.

Concentration of the most important production factors again increases marginal productivity in the center (Friedmann, 1963: 45). It is when one moves from description to explanation that the primacy problem becomes rather ambiguous. Descriptions of primacy always take into account the international elements. But in theory building these variables are often omitted. Surely this is not a unique case in the history of scientific elaboration, since between the descriptive stage of the phenomena and the theory-building stage much that is vital is often eliminated.

When describing the center-periphery opposition, Friedmann considers a wide range of variables which includes the international system. Yet the explanation proceeds as if a nation were a closed system, a black-box. According to Dobb (1965: 55) this weakness can only be understood if we remember that until only a few years ago economists still believed in the national character of the capitalist system of economic development.

To combine "internal colonization" (Gonzalez Casanova, 1969)[4] and "unequal exchange" between nations (Emmanuel, 1969) is improper as a means of explaining the pole-periphery notion within the context of a single nation. This last concept does not fit the relations between sub-units within the same country (Jalee, 1969: 161). It omits external dependence as a locational factor and assumes that the primate city is an "autonomous" national core controlling "dependent" peripheral sub-systems[5] (Friedmann, 1970: 8-9). But in fact, asymmetric regional relations imply flows which drain off resources to a receiver located outside the country. One can hardly ignore that regional economies in Third World countries are progressively tied to exports. Thus national boundaries are unable to prevent resource transfers of the type described by Friedmann.

The importance of the national core upon peripheral sub-units must be acknowledged, but it must be stressed that the imperatives of the world economy interfere and outweigh those of the internal economy. In underdeveloped countries, evolutionary regional processes have been and still are specifically linked to direct or indirect demands from the pole of the international system.

Links with the international system provide the economic conditions for geographical resource accumulation. Although it is no longer fashionable to oppose this position, there is some thinking which tends to contradict those of economic theorists who view primacy as inevitable. Myrdal (1957) suggested a process of circular causation, where wealth and poverty are concentrated and/or associated with specific social classes, economic activities, and places. For Hirschman (1964: 213) the "polarization" effect is followed by a "contagion" effect: growth poles would then lead to greater decentralization, providing growth was sustained.

There is evidence which counters Hirschman's thesis of spontaneous diffusion. In Third World metropolises, the spatial concentration of economic activities is spontaneous and cumulative. No country has been able to change this trend, although governments have sometimes attempted to do so (Norro, 1972). Even sustained growth does not modify the tendency[6] (Norro, 1972: xxix). Friedmann's (1963) position, however, is that diseconomies of scale do not prevent increased growth in metropolitan areas. The phenomenon is not transitory but structural and rooted in the technological constraints linked to economic growth (Merhav, 1969: 48-49). Modern industrial structure leads to a high degree of economic and spatial concentration.

Diffusionists employ a very different reasoning. Williamson (1965, 1968), for example, is convinced that in the countries that reach a high level of development regional disparities are smoothed out, but his examples, taken from an international set of data, are relevant only for developed countries. Pedersen and Stohr (1969: 31) extended Williamson's argument to the global scale. For Stohr (1971: 10) there is a dualistic spatial behavior until the "core" undergoes new structural change and spreads out its growth to other areas. If one is to accept the bold statement made by Gauthier (1971: 9-10), this process is already achieved in Brazil: "There is very little inequality in the growth of state income and there appears to be a tendency for the little inequality that does exist to disappear."

Berry's (1971) position on economic growth is neither intermediary nor eclectic. He does not think that macrocephaly is irreversible, but neither does he believe that it is only a step in the growth process: "Growth cannot decentralize naturally" (1971: 139). But first a strong decision structure, and secondly a planning effort, are required. Thus his position differs from that of J. R. Lasuen and Friedmann, who are both identified by Gauthier (1971) as diffusionists. This group's point of view relates the idea of diffusion of innovation (Hagerstand, 1967) to growth pole theory (Perroux, 1955).

The growth pole is formed by an "ensemble" of key industries. These industries have multiplier effects upon other economic or territorial "ensembles." Growth poles are thus defined as "points of growth"—"it cannot be confused with a place, for it is part of the formal economic space" (Beguin, 1963: 581).

Friedmann (1963, 1966, 1970) has cited the operational developments of both theories (diffusion of innovation and growth poles). He sees the investment process as representing a practical solution to be adopted (1966: 61) but he states that the avenues for regional economic growth are limited (1970: 14). He has suggested a list of nine points leading to modernization of production, marketing and information systems, improvement of banking and service structures, and optimization of human and natural resources.

However, modernization alters society and the economy at every level. Changes in locational advantages take place, which leads to shifts in activities, employment, and population (elites and poor) in favor of the core. Rural-urban migration increases exponentially and the modern activities which created the disequilibrium are not able to provide adequate jobs. Pushed back by the modernized sector, most of the population seeks refuge in the lower circuit of urban economy.

The peripheral impoverishment creates a real *periphery in the pole.* Thus, the notion of a "geographic" periphery stands in contrast to the notion of a socioeconomic periphery. One must, at the same time, consider places which become marginal to the development process and the people marginalized by economic growth. These persons constitute the social periphery in the economic pole. Thus, in the same locus in space, increased aggregate wealth fails to prevent an increase in poverty.

Nevertheless, spatial theory does not take into account the existence of a "lower urban circuit" paralleling an "upper circuit," the former dependent on the latter (McGee, 1971, 1973, 1974; Santos, 1971, 1972b, 1974, and forthcoming). Unfortunately, classical economics is only concerned with the mechanisms of the modern economy. This bias prevents the understanding of the whole economic process and its spatial implications.

LIMA: GROWTH AND POVERTY

THE GROWTH OF LIMA

Lima, the primate city of Peru, has maintained its central position since the country entered world history. Its expanding external trade between the end of the nineteenth century and the beginning of the twentieth century assured it that it would retain its position, a factor presently strengthened by industrialization. But the point is that the economic growth rate as well as income per capita are increasing while the population experiences impoverishment, absolute and relative. The economic modernization of Peru began during the postwar period. Thus the phenomena associated with modernization succeeded abruptly in promoting economic concentration and urbanization.

The intensification of primacy has also led to increased growth of tertiary employment, but associated with this change has come increased unemployment, underemployment, and an increase in poverty. Between 1961 and 1967 industry grew at a rate of 8% per annum, and since then growth has continued (Gianella, 1970: 206). This has led to an increase in jobs in commerce. Between 1940 and 1967 there was a shift in the economically active population in this sector (13.1% to 18.1%). Likewise the share of the modern sector in the gross national product has shown a steady improvement (see Table 1).

In the two most recent decades, the modern sector has represented the single growth sector, while the traditional sector and the

TABLE 1
MODERN SECTOR SHARE OF GROSS NATIONAL PRODUCT

	Modern Sector (million soles at 1963 prices)a	% of GNP
1950	10,410	26.7
1961	24,041	34.7
1968	35,423	37.1
1970	39,313	38.1

a. $1 = 43 soles
SOURCE: Modelo de Largo Plazo, 1973, p. 21.

traditional urban sector have clearly recorded a relative decline. The average increase per annum of the modern urban sector was 7.9% from 1950 to 1961, and 5.7% from 1962 to 1970 (Instituto Nacional, 1973: 16). The evolution of the national product per capita is strongly in favor of the modern sector (III) and against the rural sector (I) and the traditional urban sector (II) (see Table 3). Economic mutations involve corresponding mutations in national income and the employment structure. From 1950 to 1960 national income increased by 22.6%, but in the private sector the increase was 30.4% for capital and 29.2% for employees, well above that for workers (17.2%). Thus, growth implies a new division of GNP which is unfavorable to the state and the workers. During the same period, the remuneration of private capital increased from 20.2% to 25.1% of national income, and that of land property from 3.5% to 5.2%. On the contrary, worker salaries merely reached 14.7% of national income in 1960, while they were 16.5% in 1950. Similarly the share of small farmers and independent artisans decreased from 37% to 30.9%. A true proletarianization process has occurred resulting from the deepening trends toward economic concentration, with independent labor decreasing from 37.0% to 30.9%.

New trends in production as well as consumption overload both the state and taxpayers, that is, the entire population. Modern industries tend to locate in metropolitan areas, and this leads to an increase in state expenditures for economic and social infrastructures

TABLE 2
SECTOR SHARES OF GNP (in percentages)

Sector	1950	1970
Traditional agricultural sector	18.8	12.0
Traditional urban sector	54.5	50.0
Modern urban sector	26.7	38.0

TABLE 3

GNP PER CAPITA BY SECTOR IN RELATION WITH THE GNP

Sector	1950	1961	1968	1970
(I)	32	31	26	25
(II)	195	158	146	140
(III)	213	216	219	226
GNP	100	100	100	100

SOURCE: Modelo Largo Plazo, 1973, p. 23.

linked to modernization and urban expansion. These expenditures reduce capital for other social purposes both in the metropolis and in the rest of the country.

CHANGES IN POPULATION DISTRIBUTION
(1940-1971)

Between 1940 and 1971 the total population doubled, while the urban population increased sixfold (see Table 4). In 1940 Lima was the nation's only urban agglomeration with more than 100,000 inhabitants. But by 1972 the country included eight urban centers with populations in excess of this threshold. In 1971, besides the capital, one city had more than 250,000 inhabitants and the other seven included 100,000 to 250,000 inhabitants. The number of cities having 20,000 to 50,000 inhabitants passed from 8 to 26, and those having over 80,000 inhabitants from 10 to 39, during this thirty-year interval (Fox, 1972: 12-13, Tables 3, 4, 5). The main flow of migrants from the countryside and from other cities was directed principally toward Lima, rising from 26,000 to 80,000 people per year between 1946 and 1965.

Beside the demographic imbalance, the expansion of the modern economy looms to aggravate other forms of historical imbalances between Lima and other cities as well as the rest of the country.

TABLE 4

RURAL-URBAN POPULATION CHANGES, 1940-1971

Year	Urban	Rural
1940	15%	85%
1961	30	70
1971	43	52

Most of the nation's economic wealth is concentrated in Lima. Its GNP per capita is constantly increasing vis-à-vis the national average rate of increase. Its income per capita is more than 7 times higher than that of the poorest provinces. The wholesale trade turnover in the capital city is 22 times higher than that of Arequipa, the second city of the country. Its dominance of the retail sector is less exaggerated, being only 10 times higher. More than two-thirds of the nation's manufacturing industries are concentrated in Lima. This has led to a concentration of 65% of the total output, 67% of the workers, and 73% of the salaries (Arnao, 1972: 12-18). The trend toward employment concentration in Lima is obvious. During the period 1941-1960 employment increased by 185% in Lima, but by only 10% nationally. During the same period, out of the 650,000 new jobs created in Peru, 68% were in the capital. As a result, Lima had 10% of the jobs in 1940 and 22% in 1961.

However, economic expansion in Lima tends to hide the weakening dimensions of the total economy. As was previously noted, modernization in Lima induces a squeeze on family and independent labor while promoting improvements in wage-labor relations. Unemployment and underemployment are much in evidence. Although widespread throughout the country, especially in areas of seasonal agriculture, underemployment is also in evidence in Lima. A recent survey showed that Lima had an underemployment rate of 30%. In the Bariadas the rate was even higher (Gianella, 1970: 171). The 841,131 negative job balance for Lima province in 1970 is expected to be doubled in ten years (Hargous, 1972: 141).

Modernization, which leads to a tightening of the employment market, also demands that the work force have more formal education. In 1950 an average 4.08 years of education was all that was necessary to permit one to function satisfactorily in the work force, but by 1965 one needed 4.85 years of training (Maddison, 1971: 48). In other words, the integration of migrants and the urban poor into the formal labor market is becoming more and more difficult.

Modernization and unemployment induce strong socioeconomic disparities, especially in urban income distribution. The top 5% of the families control 30.5% of total income, while the lowest 50% of the families control only 12.4% (Gianella, 1970: 80).

THE BIG CITY APPEAL

The role of migration in Lima's demographic growth is impressive. In 1972, 1,709,724 people were born in Lima and 1,592,799 were born elsewhere in Peru. Between 1961 and 1972 the natural increase of Lima's population was 2.3% per annum while migration has added an average of 3.5% per annum to the city's growth. Among the population under 15 years of age, 81.7% were born in the capital.

One must acknowledge the power of attraction of the metropolis—a power that receives support from both the modern and poverty sector of the urban economy (Santos, 1971, 1972b, 1974, and forthcoming). In fact, if rich or educated people are attracted by the modern economy only, the poor are often absorbed by elements of the poor urban economy, as well as the modern sector. Facility of entry and higher incomes act as a magnet. Low level of education is a minor handicap in a big city. Even illiterates have wide possibilities of finding employment in some economic activity. It is significant that among Lima's population 8% consider education as completely useless in preparing them for entry into the world of work (Algunas Características, 1971). Among the illiterates, 35% are sellers, 6% are artisans or workers, 50% are in the general service economy; among those who did not complete primary school, 17% are sellers, 28% artisans and workers, and 34% in services (Algunas Características, 1971: 69; see Table 5). Moreover, earnings are higher in the capital than in any other place. Workers whose salary is under 2,000 soles/month reach as high as 63% of the work force in other cities, but only 24.0% in Lima (see Table 6).

Almost 28% of the household workers, who are among the more poorly paid in the work force, earn more than 1,000 soles per month in Lima. Among this worker group only 5% reach this earnings level in Trujillo and 1% in Arequipa.

TABLE 5

**ILLITERATES AS A PERCENTAGE OF
LIMA'S ECONOMICALLY ACTIVE POPULATION**

Total in active population	3.4
Trade	14.4
Services	1.7
Hawkers	15.0
Servants	12.3
Construction	6.5

SOURCE: Santos, 1974.

TABLE 6
EARNINGS OF INDEPENDENT WORKERS

City	Under 1000 Soles[a]	Over 5000 Soles
Lima	28%	15%
Iquitos	—	8
Puno	44	7
Chiclayo	41	7
Piura	56	7
Other cities	—	7

a. $1 = 43 soles

If we consider the educational factor, Lima still offers better salaries for equal levels of educational attainment. For people with university training, whose earnings are greater than 7,500 soles per month, we find a similar pattern prevailing. Lima has a much larger percentage of this educational group earning in excess of this threshold income than is true of other cities in the country. This confirms our assumption that in underdeveloped countries the "value" of the individual as a producer or as a consumer depends on his situation in space (Santos, 1974 and forthcoming). Thus, the big city appears to serve as a magnet attracting the poor from other cities and the countryside. Let us stress that it is not a matter of step migrations, for 83.5% of the migrants from other cities, as well as 81.7% of those coming from the countryside, came directly to Lima (Desco, 1967: 16).

During the first stage, industrial expansion is the magnet which attracts migrants into the city. But on arriving in the city many are forced into the poor sector of urban economy, which at best only allows them to survive. The developing modern sector of the economy, with its emphasis on technology and capital intensity, is experiencing a reduced need for workers. The higher educational levels required of workers in the modern sector excludes the urban poor and migrants from the modern sector of the labor market. The lower circuit, which is self-inflationary (Armstrong and McGee, 1968), absorbs this "excessive" manpower. The migrant who comes to the city is often aware of the workings of the modern sector. But he is likewise aware that it is possible to find some kind of employment in the city. The general attitude of the migrant is that it is better to be poor in Lima than anywhere else.

THE POOR AND THEIR ECONOMY

This mass of poor gathered in the capital city is allied with a low-income class of civil servants, employees, and small entrepreneurs, and it is they who support the dynamism of the lower circuit parallel to the expansion of the modern circuit. This lower circuit, through its own specific forms of commercialization, allows the under-favored classes to have access to modern goods. It also produces some goods which resemble those produced by the modern sector, as well as traditional goods.

THE SMALL RETAIL TRADE

In Lima, with a population estimated at 3,100,000 inhabitants in 1972, about 60% were shopping exclusively in traditional markets. Most of those shopping in the traditional markets were people from the "barriadas" and other poor zones of the city. But equally important, approximately 1,000,000 people from other districts were also using the lower circuit for most of their buying. Only 300,000 people, or less than 10% of the population, were customers of the supermarkets (Estudios No. 23, 1972: 43). More than 1,800,000 people essentially shop in establishments with poor accommodations and poor equipment. These facilities are found in little shops, market booths some of which are covered and others are open air. Almost two-thirds of the retail distributors are found in facilities like those described above. Another 35% are identified as street vendors.

Markets are of two types: the proper market (mercado), usually built by the state or by private companies, and the "mercadillo" or little market, built by the traders themselves. In Lima 118 mercados, 96 mercadillos, and 76 "paraditas" have been identified. The paradita is an improvised commercial structure that is operated by small traders; it represents a commercial unit that will eventually evolve into a mercadillo. It has multiplied since the government prohibited meat and fish sales by street vendors.

In the mercados the rent or the cost of a booth is quite high, which explains that many of them are unoccupied. The idle capacity in these markets averages 24% (Espada, 1971: 12), but it often rises as high as 31% in the poorest districts (Estudios No. 23, 1972: 41). Most of the small trades have a very modest monthly output.

TABLE 7

OUTPUT OF SMALL TRADES

Monthly Output (in soles)	Percentage of Traders
Under 8,000	18
8,000 to 50,000	44.6
50,000 to 200,000	28.1
200,000 to 1,000,000	9.3

a. $1 = 43 soles

SOURCE: SERH–CEMO, October 1971.

Considering as most the output ranging from 8,000 to 50,000 soles, their repartition is as shown in Table 7.

As for other activities of the lower circuit, trade depends on labor more than on capital. Among the traders employing less than twenty people, merely 37% had access to bank or monetary credit and only 20% of them used it effectively. The remaining 63% dealt with wholesalers or bigger retailers to renew their stocks. Unfortunately, there are no data which describe the smallest trader, whose access to money is still more difficult and whose dependence upon suppliers is heavier.

The number of customers for each trade unit depends on the socioeconomic level of the population. For an average of 65.9 persons, we find 39.9 persons for the poor districts, 62.7 for the middle class and upper class district. Likewise, the area of market influence increases with the socioeconomic level of the population (Espada, 1971: 53, 55).

Foodstuffs are the main products sold by the small traders. This is an outgrowth of the income structure of the population. The poor represent the largest single income group in the population, and their principal expenditure is food. It has been shown that in the traditional markets 77% of the goods for sale were foodstuffs, and among the street vendors this percentage was 62%. One finds that the poor trade sector almost monopolized the distribution of fresh vegetables and fruit (95.8 and 97.9% respectively).

Finally, we could include in this category of small retail traders the small and cheap restaurants in the center of the city, as well as the street vendors who prepare quick meals in the street. Both of these retail types are dependent upon the demands of the moderate-income strata who beleaguer the city center during the day and are unable to return home for meals.

THE STREET VENDOR

It is not easy to specify the exact number of street vendors in Lima. The figure proposed by the municipality is approximately 200,000. A local newspaper reported a figure of 250,000, of which 109,000 engage in business in the very middle of the streets (El Comercio, 14 Marzo 1973). One must acknowledge this is a striking phenomenon: they are everywhere—a number of streets surrounding a market in the center have been totally filled, thus becoming spontaneously and definitively closed to automobile traffic. Many shopkeepers of those invaded streets find it profitable to sell their goods through the hawkers.

An attempt has been made by the municipality to eliminate the chaos associated with street vendors cluttering thoroughfares, but it has been unsuccessful. The street vendor must be mobile in order to maximize sales. For most of the hawkers the number of individual sales are small, as is their profit. Fewer than 77% average more than 800 soles per month, and 44% average under 200 soles (see Table 8).

One might consider these highly profitable operations if only invested capital were used in measuring profit. But the point is that profit must not be identified with surplus value because debts are heavy and lending rates are unusually high.

In contrast to the situation observed in other cities, the hawker's job is relatively stable. It was found that 16% remained in this occupation for only one year, 37% between one and five years, and 45% for more than six years. The majority of the hawkers questioned indicated that they viewed their occupations as permanent (SERH-CEMO, 1971: Dec.). This is not a line of work to which the newcomer to the city finds easy entry. The overwhelming majority of the street vendors had lived in Lima for at least five years. Only 15% of the total were without any formal education while 65% and 20%, respectively, had completed primary and secondary school (Patch, 1973). Most street vendors work seven days a week (68%);

TABLE 8

SALES OUTPUT OF STREET VENDORS

Total Sales Output (in soles)	% of Vendors
0 - 499	30
500 - 999	24
1,000 - 1,999	16
2,000 - 4,999	15
5,000 and over	5

almost 30% work more than eight hours a day. In the center of Lima, their principal sales period corresponds to the peak hours when people leave their offices or their place of work. On Saturday evenings the main commercial streets belong to the hawkers. This is the maximum period for "impulse buying" on behalf of the walkers, but also many poor or lower class people deliberately journey to the center of the city to make purchases.

There are three general types of street vendors found in Lima's commercial core. The classical type is an independent, who pays cash for his goods or at least a part of them and secures the rest through personal credit from his suppliers. The nature of this arrangement creates strong links between the supplier and his debtor.

Others receive goods in deposit; they usually work in the center of the city, and each night they return to the store or the factory warehouse the goods they were unable to sell. Most of the time they work directly for a wholesaler or a factory; although they do not receive wages, they do receive a share of the amount of sales. This system has two advantages for the employer: he avoids the fixed costs of wages and escapes paying taxes on the sales. The loss in uncollected taxes has been valued at about 1,000 million soles annually (approximately $24 millions U.S.). This is a profit that accrues to wholesalers, small factories, and shopkeepers who utilize the hawkers to distribute their product.

There is an additional kind of street vendor who represents an extension of the modern circuit form into the lower circuit. Large food companies (for sweets, ice cream, hot dogs) are owners of a fleet of small carriages which street vendors employ to distribute their wares. These vehicles are returned to their owners at the end of the day. This is actually a form of adaptation on the part of the modern circuit to permit it to reach the widest potential clientele of the low and middle classes and thus to compete with the lower circuit on its own ground, the street.

Foodstuffs are the main items for sale (62%). But hawkers also sell new clothes (11%), used clothing (1%), as well as simple household appliances (8%), and many other items varying with the season or the demand. To explain the importance of street vendors in Lima, let us quote a local researcher: "They represent the more plastic form of supplying a great part of the population and restrain the total pauperization of a multitude who takes refuge in such occupations to get an income. This is the expression of a socioeconomic situation which cannot be changed in the short term" (Espada, 1971: 18).

The vendors work as a double-way channel. They bring down modern-sector goods to the low-income population, while they bring up to the upper circuit the people's savings through wholesalers and banks. Effectively, capital circulation is an easy metric, with only a thin capital flow moving down to the poor sector of economy parallel to the flow of goods in the opposite direction; capital flow is thick, for prices are strongly increased by each intermediary.

HANDICRAFTS AND SMALL INDUSTRIES

Peru is one of the Latin American countries where artisans have resisted modernization, thanks to the existence of an enormous poor urban population as well as a low-income middle class. Modernization pushes artisans off the countryside and the small and middle-size towns into the metropolis. There their activities are in great demand. The consumption pattern of low-income populations are characterized by purchases of modern goods of low quality or goods which are described as secondhand. The multiplicity of repairs necessary to keep old cars, radios, and televisions working provides an outlet for the services of these artisans. Items which would be thrown on the rubbish pile in a rich country provide work to many artisans without capital. In Lima there is said to be 1,000 tailors, 656 shoemakers, 306 shoe repair workers. These numbers probably underestimate the actual numbers engaged in these occupations.

Moreover, industrial expansion is dependent upon the handicrafts and small-scale industries to provide spare parts and other accessories required of such development. Finally, the presence in the city of a majority of Indians allows the survival of an "ethnic" artisan. These artisans are devoted to producing items of personal consumption, but they also produce items for the tourist market.

Small repair or production activities have a large locational plasticity because of their modest accommodation requirements. They can also easily respond to demand and change according to population needs. These businesses often go undetected by the taxation agents because they are frequently housed in residential accommodations.

A better picture of small-scale industries might emerge if we focus exclusively on the district called "La Victoria" in the center of Lima. La Victoria represents the main concentration of small and middle-size industries in the city. In this area stands the wholesalers' market

(La Parada) and numerous transportation lines connecting the city to the hinterland. The market provides these industries both with raw material and clientele who are traders and hawkers. Manpower is found among the poor and lower middle classes of the district, the most voluminous in the city. The poorly maintained buildings offer the cheap housing these activities require.

In 1963, out of 506 enterprises identified less than 10% had more than 40 workers, with the total number of workers established at almost 39% of all workers in the district worked in establishments located in this district. Those which employed fewer than 40 persons constituted 12% of the total (Deler, 1973). These activities are in a general way small and middle-size industries, although activities belonging typically to the lower circuit stand alongside the bigger industries. This can be explained partly because the heterogeneous population of the area allows the coexistence of factories of different size and nature, but on the other hand, we have to consider the destination of the output. A share of the output is consumed by local customers; another share is destined for the hinterland, while still another serves as intermediate goods for modern industry.

These industries represent a variety of types of goods production, although more than 92% of the firms are associated with five groups of enterprises. These enterprises include: (1) textile and shoes; (2) metallic goods; (3) wood and paper goods; (4) foodstuffs; and (5) chemical products. These enterprises are ranked by the number of units included in each. When ranked by number of workers, then wood and paper products take on added importance.

Most of these small industries provide goods for low-income consumers. It feeds the lower-circuit trade which is its principal outlet. But some other activities are also found to exist in this area in order to take advantage of the socioeconomic environment and to try to compete with modern industries. This is the situation of both paper and chemical products. To this group can be added the canned food industries. The localization within the city is an essential factor for all these activities.

LIMA'S UNIQUENESS

Lima is a full participant among Latin America's family of cities. Nevertheless, it distinguishes itself by reflecting the successive modes of Peru's participation in the world economy. Compared with other

Latin American countries, Peru has been precocious in many ways, some of which are related to the building of railways, economic and demographic centralization, and modern mining activity. On the other side of the ledger, the "tempo" of other aspects of socio-economic development and its associated spatial features has been tardy. Among them are a belated demographic revolution (around 2,000,000 inhabitants in 1880; 4,000,000 in 1920; 6,200,000 in 1940; 10,000,000 in 1961; 13,600,000 in 1972); belated urban-ization (85% of rural population in 1940); belated territorial and national integration; belated industrialization (47,688 workers in 1910); and the recent decision to redistribute income.

The demographic explosion is parallel to a migratory process that modern means of transportation both facilitate and encourage. The urban revolution is generalized, but Lima is the recipient of the largest share of urban migrants.

Modern industrialization, although essentially formed by light industries, works with modern technologies. Idle capacity is generally high, averaging 38.2% in 1970, but 43.6% in the consumption goods and capital goods branches. The creation of capital-intensive indus-tries provokes underemployment and unemployment, while the middle class is small and poorly paid and the poor eke out a bare existence. Centuries of social and economic segregation of the Indian population is strengthened by particularisms and traditional con-sumption patterns which are resistant to change, even in the metropolis. Artisans thus were able to maintain a captive market. For example, traditional clothing remained in great demand. Modern and monopolistic industries, selling at prices poor people cannot afford, paradoxically saved a traditional activity. Moreover, tourism—a result of international transport modernization—creates a privileged market for artisan production.

The modernization of a poor state brings about the expansion of a bureaucracy with its attendant low salaries. Late industrialization benefits from cheap manpower. Propulsive activities, like admin-istration and manufacturing, pay little, and salaries in "peripheral" jobs like services and trade are lower still. The income of the "periphery" depends on that of the "center" (Frankenhoff, 1971), but the labor cost is also a function of labor supply. The incessant flow of migrants swells the number of poor, inflates the labor market, and stabilizes the trend toward low salaries. The dramatic vicious circle is associated with the growth of the city, which partially modernizes its economy without creating a well-paid and large middle class, while poverty expands both qualitatively and quantitatively.

In no other Latin American country did society remain divided into two ethnic sectors, separated and almost antagonistic from a socioeconomic and cultural viewpoint, for so long a time period as in Peru. In some countries, export activities were more redistributive than in Peru, thus the number of effective consumers was greater. Everywhere the building of railways provoked economic and demographic concentration, but never as sharp as in Peru. In Brazil's coffee belt or Argentina's wheat belt, the homestead system helped railways entrepreneurs, town traders, and farmers. Big cities expanded, but local towns took advantage of the growing consumption.

In Venezuela, the proceeds of oil exploitation accelerated urbanization and allowed the development of costly public works and other equipment, but by the same token the middle class with relatively high salaries was considerably enlarged. In Argentina the railways network converges on Buenos Aires, but agricultural activity was from the beginning relatively redistributive. The European farmers already had modern consumption patterns when they came to South America. This has also been the case on the coffee plantations of Sao Paulo, where identical growth factors appeared together.

In Peru, the situation has been different. The time lag between the introduction of modern variables and the impact imposed by a particular model of demographic and economic growth has led to the unique situation that exists in Lima. As a growth pole, if we take in account the magnitudes of its economic, social, and political activity, Lima attracts an increasing flow of poor people from the countryside and other urban places. They join the multitude of urbanites having to fight each day for survival. This condition is typical of that in many Third World countries, where the economic pole houses an enormous social periphery.

REAPPRAISAL OF THE
POLE-PERIPHERY MODEL

The current notion of the pole-periphery presupposes perfect opposition, with wealth concentrated in the metropolis and peripheral lack of dynamism and growth. The schema, however, is not so rigid. The antinomic model is not valid. In many situations the "hierarchical filtering down" rule (Thompson, 1965; Berry, 1971, 1972; Berry and Prakasa, 1968; Ridell, 1970) does not fit with reality. In some areas the rural economy as well as the urban economy experience a quick growth (in terms of output) and a capital-deepening process. This is the result of a loosening of locational requirements in the establishment of units of multinational corporations. If the local infrastructure is supportive, industries can be established a great distance from the metropolis. Whether modern activities are located in already existent settlements or built-up new towns, they fail to create links with the immediate regional space; they are directly related to the national metropolis and dynamic regions and/or to foreign markets. If the country, however, is not highly industrialized, these technological enclaves have no choice but to develop external linkages. According to some economic reports, there is growth, but the difficulty is to determine where the surplus value migrates.

Modern agricultural or industrial activities settled in the hinterland have higher levels of functional needs than the small or middle-size cities can provide. But metropolises which have the required environment retain a part of the external-oriented capital flow.

Improved utilization of services, banking, trade, and transport organization, as well as information and politico-administrative methods, give primate cities a new economic impulsion. But middle-size and small cities are short-circuited and an economic and demographic imbalance ensues. Indeed, efforts to improve the situation of the intermediate cities lead to a greater metropolitan dynamism. The growth is upstream, an *ascendent* growth (Nichols, 1969).

When Hilhorst (1970: 41) considers regional growth theory as nonexistent, his sharp criticism is justified. Without a theory a workable regional plan cannot be formulated. The usual examination of regional inequalities or pole-periphery models are not sufficiently analytical and are often inconsistent. Mainly they fail in analyzing the nature of the phenomenon itself and the nature of space in the Third World.

Most of the studies are little more than a listing of statistical data which pay little attention to the intrinsic importance of key variables. In a dynamic process, variables change in value and even in nature, both in time and space, along with the process and as a result of the process itself. On the other hand, most of the authors still reason the existence of a universal law which governs spatial organization. They forget the specificities of Third World geographic patterns. The elements of space do not function in the same manner in developed and underdeveloped countries. Sophisticated quantitative analysis is only useful "to measure what exists and cannot be changed" (Armstrong, 1973: 115). At most, such elaborations can provide a demonstration, but a demonstration is not an explanation.

Distance and price are still the basis of many spatial theories and these variables have also influenced the development of the pole-periphery model. Friedmann acknowledges his debt to Schultz, when he states that "the further an area lies from an urban center, the less promising will be its outlook for development" (Friedmann, 1960: 12). "Distance comes to be considered as the major variable of study in the present context. It is the variable that alone may be said to account for the observed order in the space economy" (Friedmann, 1958: 251). Distance nor price do not provide explanatory power by themselves. A space so considered is simply a bidimensional space.

Actually, the space elements everywhere have specific *ages*. These elements are also structures and systems, which depends on the resolution level considered. If space is analyzed out of a holistic approach, the analytic categories are not real "values" but only isolated elements. Time and space structures are fundamental to transforming elements into *variables*. Otherwise, there is no way to evaluate the importance of relationships maintained in space on the whole and in each spatial sub-unit. Distance-time and distance-price, as well as price itself, change, but the pole-periphery notion fails to consider the historical perspective. The history of underdeveloped countries is often taken as a late version of Western history (see Perloff and Wingo, 1961). Terms like "pseudo-urbanization," "over-urbanization," "hyperurbanization," and the like are used to define primacy illustrate this erroneous interpretation. Such an approach leads one to adopt a growth model similar to those utilized in developed countries.

The argument comforts those who praise generalized technological modernization, when there is much evidence that technological modernization is against deconcentration. We could ask if the planning principles involved in the pole-periphery notion support increasing concentration rather than favor deconcentration. Lefever (1958), quoted by Friedmann (1963: 49), believes that "an optimum allocation is achieved when the relevant factors of production are so distributed that further shifts among uses and location are not possible without reducing the national product." It is in the present situation where no decentralization is possible. Berry (1971: 139) saw the problem differently when he observed that higher growth rates in big cities were "to worsen the way of life outside the metropolis." But the degradation of the way of life is *outside* as well as *inside* the metropolis. To have real meaning, according to social practice, the pole-periphery notion requires a more elaborate explanation. It is an illusion to think we are able to build normative theories before an adequate explanation of the problems involved.

After World War II primacy was aggravated and became generalized. It is a geographic consequence of the modern economy. The use of technologies permanently renewed, the increasing returns of scale and intensification of capital, the new institutional and organizational needs aggravate the concentration tendencies (Wingo, 1969: 121). Macrocephaly is both cause and effect of monopolistic structure (Santos, 1974 and forthcoming). Administered prices, bound to monopolies, handicap peripheral activities and populations. The buying power of individuals shrinks (see Johnson, 1970: 78-79) as does the relative number of jobs. Unemployment and underemployment become a general feature and mass poverty is present everywhere. Whether or not salaries go up, the value of manpower decreases because productivity is ever increasing. It helps to understand the phenomenon of relative pauperization (Salama, 1972: 69). To earn more does not mean to be less poor. We must consider the way the total product is distributed and the permanent changes in consumption structure. Poverty cannot be measured in absolute terms (Santos, 1974). New needs are created and imposed on consumers whatever their income might happen to be.

Moreover, poor and middle-class savings are syphoned off in the modern circuit through different channels (modern and conspicuous consumption, lotteries, housing programs, public expenditures in social and economic infrastructures required by modern activities).

This structural situation is universal in Third World countries but very scarcely considered in social and planning theories. Our problem lies in the confrontation between a growth model leading to demographic and geographic concentration and the will to decentralize activities and population and to reduce or eliminate poverty. If the metropolitan economic structure that evolves with technological changes is the ordering force upon the country's economic and spatial structure (Friedmann, 1970: 8), economic structure and spatial structure have to be defined and transformed as a whole.

Any undertaking to decentralize, if production structure remains unchanged, leads to a reinforcement of the center. Only marginality can be redistributed, not well-being (Santos, 1974). Theories like growth poles (Perroux), deliberated urbanization (Friedmann), or concentrate decentralization (Rodwin, 1961) must be reconsidered in this light.

NOTES

1. Primacy has been extensively studied. See, among others, Linsky (1965), Browning (1958), Davis (1962), Shaks (1965), Mehta (1969), Wingo (1969), Hoselitz (1957), and Clarke (1972).

2. The rank-size concept was developed by Zipf (1949). It was originally proposed by Auerbach.

3. The Peruvian case is representative of an internal colonial relationship established before industrialization. In Africa, Asia, and Latin America, many ports and capital cities have identical characteristics.

4. For a discussion of "internal colonialism," see, among others, Gonzalez Casanova (1969), Coleman (1960), Hoselitz (1962), Lean (1969), Fanon (1961), Mills (1965), Cotler (1967), Quijano (1965).

5. In contrast to dependency, autonomy implies essentially the capacity for self-direction, the possibility of mobilizing and utilizing resources to reach well-defined objectives (Friedmann, 1970: 9).

6. Other authors have the same opinion: McKee and Leahy (1970: 82), Robirosa et al. (1971: 57), Baer (1964: 269), Linsky (1965), Rivkin (1964), and Escudero et al. (1972: 13).

REFERENCES

Algunas Caracteristicas Socio-Economicas de la Educacion en el Peru (1971) Servicio del Empleo y Recursos Humanos. Lima: Centro de Estadistica y Mano de Obra, Ministerio de Trabajo (November).

ARMSTRONG, W. (1973) "Critica de la teoria de los polos de desarrollo." Revista Eure 3, 7 (April).

—— and T. G. McGEE (1968) "Revolutionary change and the Third World city." Civilisations 18, 3: 353-377.

ARNAO, J. (1972) "Estudio analitico de la industria del Distrito de La Victoria." Cuadernos de Geografia 1 (January): 12-46. Lima: Centro de Investigacion Geografica.

AUERBACH, F. (n.d.) "Der gesetz der bevölkerungskonzentration." Petermann's Mitteilungen 59: 74-76.

BAER, W. (1964) "Regional inequality and economic growth in Brazil." Economic Development and Cultural Change 12, 2 (January): 268-285.

BEGUIN, H. (1963) "Aspects geographiques de la polarisation." Revue Tiers Monde 16: 559-608.

BERRY, B. (1972) "Hierarchical diffusion, the basis of developmental and filtering and spread in a system of growth centers," pp. 340-359 in P. English and R. Mayfield (eds.) Man, Space and Environment. London: Oxford University Press.

––– (1971) "City size and economic development," in L. Jakobson and V. Prakash (eds.) Urbanization and National Development. Beverly Hills: Sage.

––– and R. PRAKASA (1968) Urban-Rural Duality in the Regional Structure of Andhra Pradesh, A Challenge to Regional Planning and Development. Wiesbaden: Franz Steiner Verlag.

BROWNING, H. L. (1958) "Recent trends in Latin America urbanization." Annals of the American Academy of Political and Social Science 316 (March): 111-120.

CLARKE, J. I. (1972) "Urban primacy in tropical Africa." La Croissance Urbaine en Afrique Noire et à Madagascar. Paris: CNRS.

COLEMAN, J. S. (1960) "The political system of the developing areas," in G. A. Almond and J. Coleman (eds.) The Politics of Developing Areas. Princeton: Princeton University Press.

Comision Multisectoral del Plan Nacional (1973) Diagnostico de la Situacion Alimentaria y Nutricional de la Poblacion Peruana. Lima.

COTLER, J. (1967) "The mechanics of internal domination and social change in Peru." Studies in Comparative International Development 3: 229-246.

DAVIS, K. (1962) "The role of class mobility in economic development." Population Review 6 (July): 67-73.

DELER, J. P. (1973) "Crecimiento acelerado y formas de subdesarrollo urbano en Lima." Buletin Geografico de Lima 2.

Desco (1967) Muestra 1967. Lima.

DOBB, M. (1965) Croissance Economique et Sous-Développement. Paris: F. Maspero. (French version of Economic Growth in Underdeveloped Countries [1963]. London: Lawrence & Wishart.)

EMMANUEL, A. (1969) L'Echange Inégal. Paris: Maspero.

ESCUDERO, J., A. GOMEZ et al. (1972) "Sintesis del estudio 'Region Central de Chile': Perspectivas de Desarrollo." Revista Eure 2, 6 (November): 9-30.

ESPADA, A. de (1971) Estudio de la Comercializacion Minorista en Lima Metropolitana. Informe No. 18. Lima: Ministerio de Agricultura.

FANON, F. (1961) Les Damnés de la Terre. Paris: F. Maspero.

FOX, R. (1972) Urban Population Growth in Peru. Urban Population, Series No. 3 (November). Washington, D.C.: Interamerican Development Bank.

FRANKENHOFF, C. (1971) "Economic activities," pp. 127-149 in Improvement of Slums and Uncontrolled Settlements. New York: United Nations.

FRIEDMANN, J. (1968) "The strategy of deliberate urbanization." Journal of the Institute of American Planners 24, 6 (November): 364-373.

––– (1966) Regional Development Policy, A Case Study of Venezuela. Cambridge: MIT Press.

––– (1963) "Regional economic policy for developing areas." Papers and Proceedings, the Regional Science Association (Vol. 10).

––– (1958) "Economy and space: a review article." Economic Development and Cultural Change 6, 3 (April): 249-255.

––– (1955) The Spatial Structure of Economic Development in the Tennessee Valley. Research Paper No. 1. Chicago: University of Chicago, Program for Education and Research in Planning.

––– E. COHEN, and E. BOURDON (1970) Polos de Desarrollo Social. Santiago: CIDU, Universidad Catolica de Chile.

FRIEDMANN, J. and W. STOHR (1967) "Planeamiento de politicas de Chile." Cuadernos de la Sociedad Venezolana de Planificacion 51 (May).

FUNES, J. C. [ed.] (1972) La Ciudad y la Region para el Desarrollo. Caracas.

GAUTHIER, H. L. (1971) Economic Growth and Polarized Space in Latin America: A Search for Geographic Theory. Conference of Latin Americanist Geographers (December), Syracuse, N.Y. (mimeo)

GIANELLA, J. (1970) Marginalidad en Lima Metropolitana (Una investigacion exploratoria). Lima: Desco.

GONZALEZ CASANOVA, P. (1965) "Internal colonialism and national development." Studies in Comparative International Development 1: 27-37.

HAGERSTRAND, T. (1967) Innovation Diffusion as a Spatial Process. Chicago: University of Chicago Press. (Translated from Innovations Forleppet ur Korologist Synpunkt [1953]. Lund: Gleerup.)

HARGOUS, S. (1973) Les Deracinés du Quart Monde. Paris: F. Maspero.

HILHORST, J. (1970) "Teoria del desarrollo regional, un intento de sintesis." Cuadernos de la Sociedad Venezolana de Planificacion 76-77 (June): 41-54.

HIRSCHMAN, A. C. (1964) Strategie du Developpement Economique. Economie et Humanisme. Paris: Les Editions Ouvrieres. (Translated from The Strategy of Economic Development [1958]. New Haven: Yale University Press.)

HOSELITZ, B. (1962) SOciological Aspects of Economic Growth. Glencoe, Ill.: Free Press.

——— (1957) "Urbanization and economic growth in Asia." Economic Development and Cultural Change 6, 1 (October): 42-54.

Instituto Nacional de Planificacion (1973) Modelo de Largo Plazo. Peru.

JALEE, P. (1969) L'Impérialisme en 1970. Paris: F. Maspero.

JEFFERSON, M. (1939) "The law of the primate city." Geographical Review 43, 3.

JOHNSON, E. A. (1970) The Organization of Space in Developing Countries. Cambridge: Harvard University Press.

LEAN, W. (1969) Economics of Land Use Planning: Urban and Regional. London: Estates Gazette.

LEFEBER, L. (1958) Allocation in Space-Production, Transport and Industrial Location. Amsterdam: North Holland Publishing.

LINSKY, A. (1965) "Some generalizations concerning primate cities." Annals of the Association of American Geographers 55, 3: 506-513.

MADDISON, A. (1970) Economic Progress and Policy in Developing Countries. London: Allen & Unwin.

MEHTA, S. K. (1969) "Some demographic and economic correlates of primate cities: a case for reevaluation," pp. 295-308 in G. Breese (ed.) The City in Developing Countries. Englewood Cliffs, N.J.: Prentice-Hall.

MEIER, G. M. and R. E. BALDWIN (1957) Economic Development: Theory, History, Policy. New York: John Wiley.

MERHAV, M. (1969) Technological Dependence, Monopoly and Growth. Oxford-London: Pergamon Press.

McGEE, T. G. (1974) The Persistence of the Proto-Proletariat: Occupational Structures and Planning of the Future World Cities. Melbourne: Australian National University, Research School of Pacific Studies, Department of Human Geography.

MILLS, C. W. (1965) "The problem of industrial development," p. 154 in I. L. Horowitz (ed.) Power, Politics and People. New York: Oxford University Press.

MURPHEY, R. (1969) "Colonialism in Asia and the role of port cities." The East Lakes Geographer 5 (December): 24-49.

MYRDAL, G. (1971) Economic Theory and Underdeveloped Regions. New York: Harper & Row. (First published as Rich Lands and Poor [1957]. New York: Harper & Row.)

——— (1973) "Peasants in cities: a paradox, a paradox, a most ingenious paradox." Human Organization 32, 2: 135-142.

——— (1971) The Urbanization Process in the Third World. London: Bell & Sons.

McKEE, D. and W. LEAHY (1970) "Dualism and disparities in regional economic development." Land Economics 56, 1 (February): 82-85.

NICHOLS, V. (1969) "Growth poles: an investigation of their potential as a tool for regional economic development." Discussion Paper Series, No. 30. Philadelphia: Regional Science Institute.

NORRO, L. (1972) "Urbanisation et développement économique dan les pays africains: théories et méthodes de recherche," in La Croissance Urbaine en Afrique Noire et à Madagascar. Paris: CNRS.

PATCH, R. W. (1973) La Parada, un estudio de clases y asimilacion. Lima: Mosca Azul Edit. (Also published [1967] as La Parada, Lima's Market, West Coast South America Series 14, 1-3. New York: American Universities Field Staff.)

PEDERSON, P. O. and W. STOHR (1969) "Economic integration and the spatial development of South America." American Behavioral Scientist 12, 5 (May-June).

PERLOFF, H. S. and L. WINGO, Jr. (1961) "Natural resources endowment and regional economic growth," in J. J. Spengler (ed.) Natural Resources and Economic Growth. Washington, D.C.: Resources for the Future.

PERROUX, F. (1970) "Note on the concept of 'growth pole'," pp. 93-103 in T. G. McKee, R. Dean and W. Leahy (eds.) Regional Economics. New York: Free Press.

——— (1963) "Consideraciones en torno a la nocion de 'polo de crecimiento'." Cuadernos de la Sociedad Venezolana de Planificacion 2, 3-4 (June-July): 1-10. (Translation of "Note sur la notion du Pole de Croissance" [1955] Economie Appliqué 1, 2.)

——— (1961) L'Economie du Vingtième Siecle. Paris: Presses Universitaires de France.

[Peru, Government of] Ministerio de Agricultura, Direccion General de Commercializacion, Sub-Direccion de Asistencia Tecnica y Economica (1972) Estudio de la Commercial-izacion Minorista de Productos Alimenticios. Informe No. 23. Lima.

[Peru, Government of] Ministerio del Trabajo (1971) Situacion Ocupacional del Peru. Informe 1971. Lima.

QUIJANO, A. (1965) "El movimiento campesino peruano y sus lideres." America Latina 8 (October): 189-211.

RATTNER, H. (1972) Regional Inequalities and Planning for Development. Cambridge: MIT. (mimeo)

RIDELL, J. B. (1970) The Spatial Dynamics of Modernization in Sierra Leone. Evanston: Northwestern University Press.

RIVKIN, M. D. (1964) Regional Development in Turkey. Cambridge: MIT Press.

ROBIROSA, M. A. ROFMAN, and O. MORENO (1971) Elementos para una Politica Regional an la Argentina. Instituto Torcuato Di Tella, Centro de Estudios Urbanos y Regionales.

RODWIN, L. (1961) "Metropolitan policy for developing areas," in W. Isard and J. H. Cumberland (eds.) Regional Economic Planning. Paris: OEEC.

ROFMAN, A. (1970) "Mercado comun lationo-americano y estructuras especiales." Cuadernos de la Sociedad Venezolana de Planificacion.

ROSSI, A. (1968) Aspectos de la Planificacion Urbano-Regional. Lima: IPL-PIAPUR.

SALAMA, P. (1972) Le Procès de Sous-Développement. Paris: F. Maspero.

SANTOS, M. (forthcoming) The Shared Space. London: Methuen.

——— (1974a) L'Espace Partagé: Les Deux Circuits de l'Economie Urbaine et Leurs Répercussions Spatiales dans le Tiers Monde. Paris: Editions M.-Th. Genin-Editions Techniques.

——— (1974b) "La marginalite urbaine en Amérique Latine." Paper prepared for International Labour Organisation, Geneva.

——— (1972a) "Dimension temporelle et systèmes spatiaux dans les pays du Tiers Monde." Revue Tiers Monde 13, 50 (April-June): 247-268.

——— (1972b) "Los dos circuitos de la economia urbana de los paises sub-desarrollados," pp. 67-99 in J. C. Funes (ed.) La Ciudad y la Region para el Desarrollo. Caracas: Comision de Administracion Publica de Venezuela.

——— (1971) Les Villes du Tiers Monde. Paris: Editions M.-Th. Genin-Librairies Techniques.

SCHULTZ, T. W. (1953) The Economic Organization of Agriculture. New York: McGraw-Hill.

——— (1971b) Encuesta de Comercio Ambulante (December). Lima.

SERH-CEMO (1971a) Encuesta de Establecimientos comerciales de menos de 20 trabaja-dores (October-November). Lima.

SHAKS, S. E. (1965) "Development, primacy and the structure of cities." Ph.D. dissertation. Harvard University.

SLATER, D. (1968) The Modernisation Process—Spatial Aspects and the Latin American Case. Discussion Paper No. 21 (July). London: London School of Economics, Graduate Geography Department.

SMITH, A. D. [ed.] (1969) Les Problèmes de la Politique des Salaires dans le Développement Economique. Cahiers de l'Institut International d'Etudes Sociales, No. 10. Paris: Librairie Sociale et Economique.

STOHR, W. B. (1971) "Regional planning as a necessary tool for the comprehensive development of a country." Paper presented at the United Nations Inter-Regional Symposium on Training of Planners for Comprehensive Regional Development (June), Warsaw, Poland.

THOMPSON, W. (1965) A Preface to Urban Economics. Baltimore: Johns Hopkins University Press.

WILKINSON, R. G. (1973) Poverty and Progress. London: Methuen.

WILLIAMSON, J. G. (1968) "Regional inequality and the process of national development: a description of the patterns," pp. 99-158 in L. Needleman (ed.) Regional Analysis. Baltimore: Penguin Books.

——— (1965) "Regional inequalities and the process of national development." Economic Development and Cultural Change 13 (July).

WINGO, L. (1969) "Latin American urbanization: plan or process?" pp. 115-146 in B. J. Frieden and W. Nash (eds.) Shaping an Urban Future. Cambridge: MIT Press.

WROBEL, A. (1971) "Theories and models of regional development." Paper presented to IGU Commission on Regional Aspects of Economic Development (April), Victoria, Brazil.

ZIPF, G. K. (1949) Human Behavior and the Principles of Least Effort. Cambridge, Mass.: Addison-Wesley.

11

Indicators of Labor Market Functioning and Urban Social Distress

ROBERT MIER
THOMAS VIETORISZ
JEAN-ELLEN GIBLIN

INTRODUCTION

□ INDIVIDUAL PROBLEMS ARISE IN A SOCIAL SETTING. It is generally accepted that important aspects of physical and mental health are associated with the stresses of poverty and unemployment. Other urban problems such as illegitimacy, homicide, or poor housing conditions are recognized as being closely related to poverty-level incomes. Much less clear, however, is the extent to which these phenomena originate not only in poverty itself, but in the way the social process of production is organized. As this process reproduces itself from day to day, from year to year, from generation to generation, the resulting structure of job oppor-

AUTHORS' NOTE: Support for this research was provided by Grant #MH-23615 from the Center for the Study of Metropolitan Problems, National Institute of Mental Health, to the Research Center for Economic Planning, New York, N.Y. Project monitor: Dr. Eliot Liebow, Metro Center. Principal investigator: Thomas Vietorisz. Opinions expressed are not necessarily those of the funding organization.

tunities—or lack of opportunities—is inseparable from other aspects of the reproduction process which are commonly referred to as "social pathologies."

A number of recent studies have established the relationship between socioeconomic factors and social or health phenomena. Shapiro et al. (1968), Chase (1970), U.S. Department of Health, Education and Welfare (1972), Smiley et al. (1972), and Wright (1972) all established relations between infant mortality and various socioeconomic indicators. Harvey Brenner (1973a), in his study on fetal, infant, and maternal mortality found significant changes associated with fluctuations in unemployment. Studies by Guerrin and Borgatta (1965) on tuberculosis incidence and Brenner (1971) dealing with heart disease mortality show strong relationships between health indicators and economic indicators. Studies by Brenner (1973b), Hurley (1959-1963), Srole et al. (1962), and Dunham (1965) indicated that poverty and mental illness are closely related.

Many studies of poverty areas have found that poverty is associated with instability in family structure (for example, Leighton, 1965; Lewis, 1966; and Moynihan, 1965). Fleisher (1966) has shown that juvenile delinquency may result from a search for alternative sources of income. Several recent studies on the economics of hard drugs have emphasized the relationship between urban unemployment and narcotics involvement (Helmer and Vietorisz, 1973; Schick et al., 1972; and Cutler, 1967). Finally, Muth (1969) and Leighton (1965) have established a relationship between poverty and substandard housing.

THE LABOR MARKET AND POVERTY

Poverty, in either a relative or absolute sense, is best understood as insufficient *earned* income. Projector, Weiss, and Thoresen (1969) have shown that wages and salaries in 1962 accounted for 75% of total income for the population as a whole. More importantly, they have found that wages and salaries comprise almost 85% of income for individuals in the $5,000-$15,000 per year income bracket. This proportion drops to 72% in the $3,000-$5,000 income bracket, owing to welfare payments.

Part of the income from sources other than wages and salaries represents retirement income received by the aged. In the lower bracket, such income is almost wholly Social Security. The high concentration of the aged in these lower brackets attests, first, that the past earnings of broad segments of the working class have not been adequate to finance private retirement benefits; and second, that Social Security benefits based on these past earnings are greatly inadequate to provide support at minimal levels of decency.

A theory of poverty must, therefore, largely be a theory of employment, either directly, via current earned incomes, or indirectly, via the welfare and retirement income systems. To observe close relationships between specific social distress phenomena and poverty also implies close relationships between those phenomena and the functioning of the labor market.

Economic theory recognizes a circular causation of poverty: low productivity leads to low wages, which imply low living standards with their associated conditions of poor nutrition, inadequate housing, poor health, improper sanitation, and so on, and these in turn reinforce low labor productivity. The standard policies advocated to break this vicious circle focus on the individual. They relate the observed low productivity to individual deficiencies of the workers such as low levels of education, skills, or motivation, and do not consider the technical and skill structure of available jobs, or the lack of available jobs.

The 1973 Comprehensive Employment and Training Act, which represents the culmination of five years of manpower reform efforts, provides for the usual package of employment counseling, supportive services, classroom education and occupational skills training, training on the job, and work experience. Yet it provides only for "transitional" public service employment to offset "temporary" unemployment (CETA, 1973). The history of such manpower efforts causes one to question if the focus has been misplaced—should it be on institutions and not individuals?

The basic hypothesis guiding this research is that urban social and economic institutions have developed in such a way that a persistent condition in urban labor markets is a broad demand for low-skill, low-wage, unstable labor. The labor market as a whole responds to the technical and social organization of the production process, and is structured so that the demand for each grade of labor is inelastic. The range of substitutability between different grades of labor, within socially tolerable limits of price variation, is thus slight. In other words, a decrease in the cost of carpenter's skills will result in little, if any, increase in the demand for carpentry work. Such a labor market can be viewed almost as a set of fixed job slots, since it does not readily respond to changes in the size and productive quality of the labor force. Some segment of the labor force—which, at one extreme may consist always of the same group of individuals, or at the other extreme may consist of a revolving pool of workers with considerable upward and downward mobility—will therefore always be involved in low-skill, low-wage work, no matter how much effort is expanded on training for higher skills.

The constant presence of this segment, with its substantial concentration of minority group members, leads to the institutionalization of poverty, and becomes a major cause of many of the social phenomena which contribute to a sense of crisis in modern urban society (Vietorisz, 1973; Freedman, 1973; Oppenheimer, 1974; Harrison, 1972; Gordon, 1972). Given such a frame of reference, conventional labor market indicators are seen as inadequate because they fail to show the full impact of labor market conditions on individual workers.

OBJECTIVE

It is the objective of this chapter to discuss a set of indicators of labor market functioning that will permit a more penetrating analysis of a variety of social phenomena by directly linking them to conditions in the labor market. In order to achieve this, it is necessary to clarify the precise relationship between poverty and labor market conditions.

'Poverty,' measured as the percentage of families with an income below a stated cut-off level, is an inadequate labor market indicator for two reasons. First, it lumps currently *earned* income with property income, retirement income, and transfers. Second, it aggregates and hence obscures such important labor market information as the number and family relationships of labor force participants and the nature of their fortunes in the labor market. 'Unemployment,' measured as the percentage of labor force participants currently seeking work, is, at the other extreme, also an inadequate measure of labor market conditions because it considers only one kind of adverse outcome—an immediate lack of employment—and does not capture other important dimensions of adverse social impact such as substandard, low-paying, unstable employment.

SOURCES OF INFORMATION

In developing a set of labor market indicators, extensive use will be made of information available from the 1970 *Census Employment Survey* (CES). The CES was a detailed sample survey of employment characteristics of individuals in each of 60 urban and 8 rural clusters. (The 60 urban clusters were located within 51 cities.) The areas were selected because there was evidence that they were more likely to contain high proportions of persons with low incomes. The urban survey clusters represented a population totaling an average of 15% of the respective SMSA population and 33.5% of the respective central city population. Much of the information from the individual enumeration areas has been aggregated for each survey area and summary results published in a special series (U.S.B.O.C., 1972). For the purposes of this research, the individual responses in five of the survey areas have been directly analyzed.[1]

INDICATORS OF
LABOR MARKET FUNCTIONING

UNEMPLOYMENT

A key purpose of developing labor market indicators is to associate urban social conditions with labor market conditions. The intervening variable between these two is, of course, poverty. Table 1 shows the results of regressions of eleven social indicators on either a "family poverty" index or an "employment" index. The statistical units are the 51 cities surveyed in the CES, and all information is for the year 1970. The "family poverty" index is defined as the proportion of families with income below the Bureau of Labor Statistics (BLS) "lower level" budget. This budget allows approximately $7,000 per year for an urban family of four.

The regression results show that seven of the eleven social indicators of crowding, family structure, health, and crime are related in a statistically significant degree to family poverty. In the seven cases in which family poverty was a significant independent variable, it explains over twice as much of the variation in the social indicators, as does unemployment. Unemployment is a better

TABLE 1
LOG/LOG LINEAR REGRESSIONS OF SOCIAL DISTRESS INDICATORS IN 51 C.E.S. CITIES AGAINST FAMILY POVERTY AND UNEMPLOYMENT[a]

Dependent Variable	Family Poverty R^2	Unemployment R^2
Households with 1.01 or more persons per room (%)	43.5[b]	34.5[b]
Families receiving A.D.C. (%)	1.5	10.9
Marriage/divorce rate	11.7[b]	2.6
Out of wedlock birth rate	32.0[b]	25.7[b]
Infant mortality rate	15.4[b]	2.2
Deaths due to complications of pregnancy	8.9	4.3
Deaths due to cirrhosis of the liver	2.5	0.0
Deaths due to syphilis	24.0[b]	4.5
Suicide rate	5.2	15.1[b]
Homicide rate	39.4[b]	19.9[b]
Burglary rate	10.4	1.3

a. The specification of all equations included logarithmic transformations of all variables and a variable to normalize for coverage. Family Poverty = percent of families with income below $7,000 per year; Unemployment = official unemployment rate. Table 1 shows log/log regression results of two individual explanatory variables, unemployment and family poverty, on various social indicators. A factor to normalize for sample coverage has been used in each case based on the reference area for the social indicator. The regressions show that family poverty is generally a better explanatory variable than unemployment. In explaining infant mortality, for example, family poverty is a significant variable while the unemployment rate is not.
b. Statistically significant at the .05 level.

SOURCES: U.S. Bureau of the Census, Census of Population: 1970 Employment Profiles of Selected Low-Income Areas. Final Report PHC(3). Government Printing Office, Washington, D.C., 1971.

U.S. Bureau of the Census, Census of Population and Housing: 1970. Government Printing Office, Washington, D.C., 1971.

U.S. Bureau of the Census, County and City Data Book, 1972 (A Statistical Abstract Supplement). Government Printing Office, Washington, D.C., 1973.

U.S. Department of Health, Education, and Welfare, Vital Statistics of the U.S., 1969. Government Printing Office, Washington, D.C., 1973.

explanatory variable only in relation to suicide, as might be expected since suicide has been shown to correlate strongly with the *cyclical* aspects of unemployment (Hammermesh and Soss, 1974). It is not surprising that unemployment is a weaker explanatory variable for selected social traits than poverty itself since unemployment is not a good predictor of poverty in any static, cross-sectional analysis (Lampman, 1971).

A regression of family poverty on unemployment in the 51 CES cities yields the following result:

$$\log \text{FP} = 3.85 + 0.05 \log \text{UE} \qquad R^2 = 0.01$$
$$\phantom{\log \text{FP} = 3.85 + 0.0}(.08)$$

where FP is the porportion of familes earning less than $7,000 per year, and UE is the official unemployment rate. The regression is not statistically significant; unemployment explains next to none of the variation in family poverty, and its coefficient is overwhelmed by statistical noise.

At another level of analysis, a detailed examination of the Detroit CES sample area showed that 48.5% of the 27,730 individuals estimated to be unemployed at the time of the survey were members of families whose combined income, after adjusting for family size, exceeded the equivalent of $7,000 per year for a family of four.[2] Conversely, 54,618 family heads or unrelated individuals received less than this amount, yet only 14%, or 7,757, of these were unemployed. From either the perspective of the individual or the family head, there is a weak relationship between current unemployment and family poverty.

In conclusion, an unemployment index is seen to be an unsatisfactory indicator of poverty-related social phenomena because unemployment is not a major *direct* contributor to the incidence of family poverty. This may leave out the lagged effects of past unemployment (which these data do not capture) as well as *indirect* effects of unemployment (e.g., via family break-up). Such lagged and indirect effects will require careful consideration in subsequent research, especially because (as shown below) only about half of total poverty can be statistically related to the current direct effects of various labor market indicators. Recalling that most income is derived from wages and salaries, and that, hence, poverty results from labor market conditions, it is necessary to go beyond current unemployment to be able to associate labor market conditions with poverty-related phenomena.

POVERTY INDICATORS

If unemployment is an inadequate indicator of labor market conditions, is it possible to introduce some basic modifications in the poverty index itself to achieve a more satisfactory indicator? The customary poverty index is a threshold measure that counts the proportion of families whose income falls beneath some specified threshold level. Often the Social Security Administration "official" poverty line (or 1.25 times that amount) is the threshold used, although recently the BLS "lower level" family budget is, for reasons to be demonstrated later, gaining wide acceptance. (See, for example, CETA, 1973: 38.) Using the BLS lower-level family budget, an analysis of the Detroit CES area reveals that 38% of the families are classified as poor.

This sort of a poverty indicator is too broad to capture a sense of current labor market conditions, because it includes income received by important groups of people from sources other than current earnings in the labor market. Such groups of people who are frequently poor include the elderly, those in poor health, and members of households with dependent children and only one adult member, usually a woman. Thus a poverty index measures the adequacy of retirement payments, disability benefits, welfare transfers, and current earnings as a single aggregate whose components cannot be distinguished.

It is possible to modify the standard poverty index to sharpen its *current* connections to the labor market. Such an indicator is derived by including in a threshold measure only those households headed by a labor force participant, with unrelated individuals treated as one-member households. The computation of such a measure is made more difficult by the male chauvinist practice of the Bureau of Census which identifies a male member of the household as the head even if a woman is the prime wage earner. In the Detroit CES sample area in 1970, there were an estimated 198,498 unrelated individuals and heads of families, and only 125,284 of these were labor force participants. Of the latter, 31,806 or 25%, were poor or in poor families by the BLS standard. Even in a static framework, three problems arise with this sort of a labor market indicator.

First, this modified family poverty indicator considers only those unrelated individuals or family heads who are currently in the official labor force. In the Detroit sample area, there were an additional 15,190 family heads or unrelated individuals in poor households who expressed a desire to work, and who must be considered as "discouraged workers." (For further discussion of the "discouraged worker" concept, see Flaim, 1971.) Inclusion of these persons raises a modified family poverty index to 32%. (The additional persons, of course, must be included both in the numerator and in the denominator of the threshold measure.)

Second, a modified family poverty index masks the *type* of labor market failure. Some unrelated individuals or heads of families in the labor market fail to break the grip of poverty because they are unemployed (17%); others, because they are "discouraged workers"

(32%), because they hold full-time, but low-paying jobs (45%), or because of involuntary part-time work (6%).

Third, a modified family poverty index does not include all individuals who are in demoralizing labor market circumstances. A number of family heads may individually be experiencing adverse labor market conditions, but may be members of families which generally—as a result of having multiple wage earners—are not in poverty. In the Detroit survey area, there were 28,380 such family heads. Other members of a family, especially wives, may also be trying, but not succeeding, to hold a job paying adequate wages. In the Detroit sample area, this included an additional 104,262 secondary family members.

In conclusion, neither the standard nor the adjusted poverty index serves as a good indicator of labor market conditions. A good set of labor market indicators are, nonetheless, important for policy purposes to monitor the level of performance of the economy and to assess the comparative needs of different local areas for federal grants, subsidies, or revenue sharing. A number of economists and manpower specialists following Willard Wirtz' 1966 initiative[3] have therefore begun to turn to developing an entirely new measure of labor market failure—the subemployment index (Spring, Harrison, and Vietorisz, 1972; Levitan and Taggart, 1973; Miller, 1973; Leggett and Cervinka, 1972).

SUBEMPLOYMENT INDICATORS

The purpose of a subemployment index is to capture in a single measure two major dimensions of labor market functioning that reproduce poverty: first, the lack of opportunity for work; and second, substandard wages. Therefore, the various subemployment indices which have been developed as generalizations of the unemployment rate comprise the following detailed categories of workers or work conditions:

The officially unemployed.

The discouraged jobless. This includes workers who have been unemployed for long periods and have given up searching for jobs they know are not there. The official unemployment statistics do not include them in the labor force. They are regularly identified only in the *Current Population Survey.*

Involuntary part-time workers. This estimate requires a separation of persons who work part-time because they are unable to find full-time work from persons working part-time by choice.

Workers earning substandard wages. The Bureau of Labor Statistics has defined a "lower level" family budget, essentially a watershed between family living wages and substandard wages. The national urban average "lower level" family budget for 1970 was $6,960 per year, or about $3.50 per hour, for a family of four; and for 1973 it was $8,181 per year, or about $4.00 per hour. The Social Security Administration has, on the other hand, defined for 1970 $4,200 as an adequate annual income for a family of four, and this is the basis of the "official poverty line" (Orshansky, 1965). Conventional economics would argue against using any wage criterion, stating that a person is paid his marginal product—he is paid all he is worth, hence he cannot be considered subemployed on wage criteria.[4]

Despite an overall agreement about these general categories, at least three mutually inconsistent definitions of subemployment have emerged. Spring, Harrison, and Vietorisz; Levitan and Taggart; and Miller have each developed different indicators because they disagree over such issues as whether the elderly or students should be considered as potential labor force participants, whether a secondary wage earner in a nonpoor family should be considered as subemployed, and at which level the income adequacy threshold should be set. The detailed definitions of these three subemployment indicators are given in Appendix A.

An application of the three indicator definitions to the Detroit Census Employment Survey area based on identical statistical data reveals a range of subemployment rates of 26% for the Miller Index to 66% for the Spring, Harrison, and Vietorisz index. It would appear on first examination that the major discrepancy between the three indicators involves the choice of an income adequacy threshold. In fact, this accounts for only a little more than half the variation among the three sets of indicators. Figure 1 is a graph of the three indicators with the income adequacy threshold treated as a parameter. This graph shows that even if there were agreement on the income adequacy threshold, there would still be wide disagreement about the subemployment rate.

Such lack of agreement should not be surprising when the construction of social indicators is viewed not as an objective, scientific process, but rather as an exercise in political economy. A proponent of one indicator explained in public that he could not

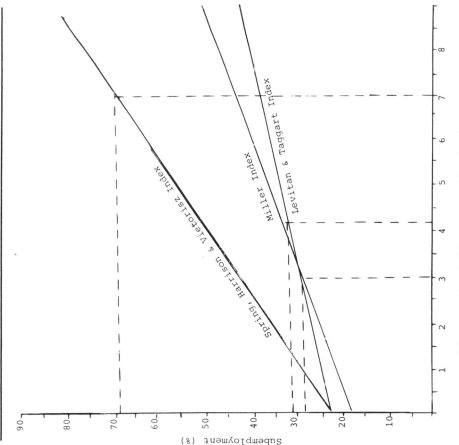

Income Adequacy threshold: Wages and Salaries ($1000 per year)

In the Levitan and Taggart Index, income is adjusted to be equivalent to that of a family of four. The subemployment rate derived without treating wages and salaries parametrically is denoted by dashed lines. The figure is derived from data in Appendix B.

Figure 1: COMPARISONS OF SUBEMPLOYMENT MEASURES IN DETROIT CES SAMPLE AREA

accept the political implications of 60-70% subemployment rates in geographical areas having "only" 8-12% unemployment; and therefore he set out to construct an indicator which would yield the lowest possible subemployment rates. He said that he was astounded to find that even his conservative definitions of the subemployment estimates exceeded 20%.

The debate over the development of a single subemployment

index is far from over and has become the subject of continuing research as a result of a formal mandate written into the 1973 Comprehensive Employment and Training Act (CETA, 1973: 23). At this point in the debate, it appears that much of the problem of differing indicators occurs because different concepts of subemployment are involved, not just different judgments of who should or should not be counted. At least three different concepts of subemployment have emerged:

(a) *Subemployment as the lack of individual opportunities for finding a decent job.* The Spring, Harrison, and Vietorisz subemployment calculations are a first approximation to this type of subemployment index. We have developed a new index called the Exclusion Index, which makes the measurement of subemployment under this definition more precise and consistent.

(b) *Subemployment as the shortfall of earned family incomes,* with the stress on family structure and the inability of family heads to support their families adequately. The subemployment indices by Levitan and Taggart and by Miller were aiming to capture this aspect of labor market functioning. We have developed a new index called the Inadequacy Index, which yields a more consistent test of the economic adequacy of earned family incomes.

(c) *Subemployment as the manpower waste inherent in labor market functioning.* This concept suggests a Manpower Underutilization Index by quantifying the gap between potential labor supply and actual demand. The Department of Health, Education and Welfare is currently considering research to estimate an important dimension of manpower underutilization—the difference between worker qualifications and the actual skill requirements of the job. We have not found it possible, within the limits of our data base, to estimate an Underutilization Index.

In Figure 2, the original three subemployment indicators are supplemented by our Exclusion and Inadequacy Indices. The detailed definitions used in computing all five indicators are included in Appendix A, and the estimates used in constructing the graphs are given in Appendix B. Vietorisz, Mier, and Giblin (1974) contains a more detailed discussion of the comparative properties of the indicators. Appendix B reveals important differences between the indicators in terms of the absolute numbers of individuals involved. Such differences are obscured by the use of ratios which are inherent in the computation of rates of subemployment.

In further work with these indicators in the present paper, the

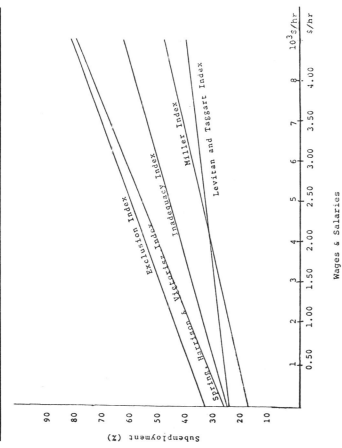

Wages & Salaries

In the Levitan and Taggart, and Inadequacy Indices, income is adjusted to be equivalent to that of a family of four. The figure is based on data from Table 1. In the Spring, Harrison and Vietorisz Index, and the Levitan and Taggart Index, annual wages and salaries are considered. In the other, hourly wages and salaries are considered.

Figure 2: COMPARISON OF SUBEMPLOYMENT MEASURES IN DETROIT CES SAMPLE AREA

income adequacy threshold will no longer be treated parametrically, but rather the hourly equivalent of the BLS "lower level" budget for 1970 will be used as a threshold. Rainwater (1973) has shown that the much lower official poverty line falls far beneath the perceived cutoff level for poverty as expressed by respondents in over thirty years of Gallup Poll surveys. These surveys sought to establish a perceived family income level which would enable a family of four to "get along" or physically survive and socially participate in a minimally adequate sense in urban society. For our reference period, 1970, the "get along" budget corresponds closely, at the national level, to the BLS "lower level" budget, which indicates that the BLS estimates of minimum adequate family budgets have been close to the perception of the population at large. Over the longer period, this correspondence cannot be expected to hold if the BLS budgets are

TABLE 2

COMPARISON OF FIVE SUBEMPLOYMENT INDICATORS FOR DETROIT, MICHIGAN C.E.S. SURVEY AREA IN 1970

Subemployment Category	Spring, Harrison & Vietorisz		Miller		Levitan & Taggart		Exclusion Index		Inadequacy Index	
	n	%	n	%	n	%	n	%	n	%
Unemployment	27,157	23.0	17,601	9.1	19,163	8.8	27,730	11.6	11,900	8.3
Involuntary part-time	9,069	4.4	5,122	3.0	6,373	2.9	9,231	3.9	5,365	3.7
Discouraged workers	11,872	5.9	9,651	5.9	24,895	11.4	39,686	16.7	17,783	12.4
% Earning less than $4,200 per year or $2.00 per hour	47,142	23.0	20,436	10.6	14,804	6.8	43,225	18.2	14,594	10.2
% Earning $4,200 to $7,000 per year or $2.00-$3.50 per hour	38,329	18.7	21,267	11.0	15,580	7.2	59,788	25.1	25,734	18.0
Total	133,569	65.2	74,677	38.7	80,815	37.0	179,660	75.5	75,377	52.7
Denominator (Potentially subemployed or labor supply)	204,979	—	193,108	—	218,008	—	238,080	—	143,067	—

SOURCE: See Appendix B

merely adjusted for price level changes. The work of Kilpatrick (1973) indicates that the perceived property level, in constant prices, increases with rising median real family incomes, even though at a somewhat slower rate; a doubling of the median raises the perceived poverty level by some 50%.[5] Unless the BLS budgets are periodically expanded in the future to reflect this popular perception, they will cease being an acceptable threshold for substandard earnings. The same critique applies a fortiori to the Social Security Administration's official poverty line.

Table 2 is a comparison of all five subemployment indicators for the Detroit CES sample area using $7,000 per year or its hourly equivalent, $3.50 per hour, as the income adequacy threshold. (In the Inadequacy Index and the Levitan and Taggart Index, the income threshold is adjusted for family size to make it equivalent to that required by a family of four.) With the operational development of these new indicators of subemployment it is possible to return to the original problem of examining the relationship between labor market functioning and urban social distress.

APPLICATION OF
SUBEMPLOYMENT INDICATORS

SUBEMPLOYMENT AND SOCIAL DISTRESS

The eleven social indicators examined earlier will now be reconsidered in light of the new set of indicators of labor market functioning. Table 3 is a presentation of the results of a series of separate regressions of the social indicators against the Exclusion Index and then the Inadequacy Index for the 51 cities in the CES. As a basis of comparison, the results from Table 1 are again included. In the six cases in which poverty is a significant independent variable, the Inadequacy Index and the Exclusion Index explain respectively 40% and 50% more of the intercity variation in the social indicators than does the unemployment indicator. The Exclusion Index is statistically significant in every case in which poverty is significant and in one additional case, deaths due to complications of pregnancy and childbirth. Subemployment is thus an improvement over unemployment as a measure of how the labor market affects a

TABLE 3

LOG/LOG LINEAR REGRESSIONS OF SOCIAL DISTRESS INDICATORS IN 51 C.E.S. CITIES AGAINST SELECTED LABOR MARKET INDICATORS[a]

Dependent Variable	Family Poverty R^2	Unemployment R^2	Inadequacy R^2	Exclusion R^2
Households with 1.01 or more persons per room (%)	43.5[b]	34.5[b]	40.7[b]	45.5[b]
Families receiving A.D.C. (%)	1.5	10.9	0.5	2.0
Marriage/divorce rate	11.7[b]	2.6	4.4	12.0[b]
Out of wedlock birth rate	32.0[b]	25.7[b]	35.5[b]	25.5[b]
Infant mortality rate	15.4[b]	2.2	14.3[b]	15.7[b]
Deaths due to complications of pregnancy	8.9	4.3	17.0[b]	10.4[b]
Deaths due to cirrhosis of the liver	2.5	0.0	0.0	0.5
Deaths due to syphilis	24.0[b]	4.5	7.1	12.7[b]
Suicide rate	5.2	15.1[b]	5.7	5.7
Homicide rate	39.4[b]	19.9[b]	31.8[b]	33.7[b]
Burglary rate	10.4	1.3	9.5	7.0

a. The specification of all equations included logarithmic transformations of all variables and a variable to normalize for coverage. Family poverty = % of families with income below $7,000; unemployment = official unemployment rate; inadequacy = inadequacy index; and exclusion = exclusion index.

b. Statistically significant at the .05 level.

SOURCE: Same as Table 1.

variety of social phenomena, especially those which are poverty-related.

Tables 4 and 5 compare the relationship between poverty and the alternative indicators of labor market functioning which have been considered, namely, the Exclusion Index and the Inadequacy Index. Table 4 shows 348,378 people in the Detroit survey area over the age of 16, and 149,987 of these, or 43%, are living in poverty circumstances. Of those who are poor, 76,032 or 51% are currently subemployed, while only 14,288 or 9% are unemployed. In looking at poverty-related phenomena, therefore, about half of the observed poverty can be related to subemployment conditions in the labor market, and only a tenth to official unemployment.

Looking at the Detroit survey area situation in another way, we see that 179,638 or 50% of the individuals are subemployed. This group divides about 40/60 between poor and not poor. Contrariwise, those in the labor supply and not subemployed—58,421 persons or

TABLE 4

**LABOR MARKET STATUS OF INDIVIDUALS
IN THE DETROIT C.E.S. AREAS[a]**

	Poor	Not Poor	Total n (%)
Not in labor supply	67,363	42,956	110,319 (31.7)
In labor supply, not subemployed	6,592	51,829	58,421 (16.8)
In labor supply, subemployed	76,032	103,606	179,638 (51.6)
(unemployed)	(142,881)	(13,442)	(27,730) (18.0)
Total n (%)	149,987 (43.1)	198,391 (56.9)	348,378

a. Subemployment is measured by the exclusion index. The labor supply consists of the official labor force plus discouraged long-term jobless.
SOURCE: See Appendix B.

17% of the total—are almost certain not to be poor. This group breaks into 11% poor and 89% nonpoor. Finally, those not in the labor supply—110,319 individuals or 32% of the total—are 55% more likely to be poor than to be nonpoor. This group breaks into 61% poor and 39% nonpoor.

While Table 4 tabulates labor market conditions for all individuals, and thus uses the Exclusion Index as the measure of subemployment,

TABLE 5

**LABOR MARKET STATUS OF FAMILY HEADS AND
UNRELATED INDIVIDUALS IN THE DETROIT C.E.S. AREA[a]**

	Poor	Not Poor	Total n (%)
Not in labor supply	43,277	12,154	55,431 (27.9)
In labor supply, not subemployed	7,621	60,070	67,691 (34.1)
In labor supply, subemployed (unemployed)	46,996 (7,757)	28,380 (4,144)	75,376 (38.0)
Total n (%)	97,893 (49.3)	100,604 (50.7)	198,498 (100)

a. Subemployment is measured by the inadequacy index. The concept of labor supply is the same as that used in Table 3.
SOURCE: See Appendix B.

Table 5 covers a narrower group. It deals only with family heads (including unrelated individuals) and thus uses the Inadequacy Index as the measure of subemployment. As against some 340,000 individuals (Table 4) there are less than 200,000 family heads in the Detroit CES area.

Comparing the cross-tabulations of Table 4 and 5, an important difference emerges from the row-by-row comparison. If an *individual* is subemployed, there is a 57% chance of escaping poverty (Table 4, row three). Conversely, if a *family head* is subemployed, his chances of escaping poverty drop to one-third (Table 5, last row; 28,380 nonpoor are 38% of the row total of 75,376). Likewise, if an *individual* is not in the labor supply, his chances of escaping poverty are 39%; if a *family head* is not in the labor supply, his chances of escaping poverty drop to 22% (Table 5). Moreover, when a family head is poor, other persons in the same household are also in poverty.

The regressions and cross-tabulations of this section are convenient for describing the structure of our data, but by themselves they explain nothing. They merely suggest that we are on the right track in translating our initial assumptions about the labor market and social pathologies into measures that capture some of the dimensions relevant for explanation. In other words, the statistical material presented is a halfway house between two analytical tasks. The first, summarized in the introduction, preceded and underlay our definition of labor market and social pathology indicators. The second, awaiting further work, must follow up the statistical description by detailed specification of the mechanisms in which the statistical regularities are rooted. The latter analytical task lies outside the limits of our present paper.

SUBEMPLOYMENT AND URBAN STRUCTURE

As stated in the introduction, a basic hypothesis underlying this research is the sensitivity of subemployment to urban economic and institutional structures.

To amplify this, subemployment is compared in Tables 6 and 7 for four cities: San Antonio, Cleveland, Houston, and San Francisco. In both Tables 6 and 7 there are considerable variations in the components of subemployment among the four cities. It may be observed, however, that the percentages of the unemployed and of discouraged workers tend to vary in a parallel fashion (also see Rosenbloom, 1974).[6] These two components jointly may be regarded as measuring the first major dimension of subemployment: the lack of opportunity for work. Discouraged workers, it is recalled, are persons who are unemployed but not looking for jobs they know are not there; excluding them from the official labor force is thus a statistical artifact that makes the official unemployment figures look lower. In the tables, the sum of the first two components is shown as a subtotal.

Similarly, the remaining components of subemployment tend to vary together. They may jointly be regarded as measuring the second major dimension of subemployment: substandard wages. It is not clear that involuntary part-time work belongs in the group—it appears to represent both a lack of work opportunity and a low-paying job.[7] In the absence of clearer evidence, it has arbitrarily been assigned, in the tables, to the low-wage cluster, and the sum of the last three components is shown as a separate subtotal.

The four cities were not selected haphazardly. San Antonio and Houston are relatively young cities whose economic structure has been laid down largely since World War I. Cleveland and San Francisco, on the other hand, are older cities that developed their basic economic structure in the nineteenth century and, as is common in such older cities, have remained unchanged in their essential features since they completed their formative burst of development. Thus San Francisco, despite its recent flurry of building activity, has remained virtually unchanged in spatial and economic organization. Commerce, industry, housing—the same sort of economic activity—is being built in the same sector of the city where that activity has always been located. Only the scale, not the mix, of activity has changed.

Tables 6 and 7 show that the newer cities differ significantly from the older ones in their labor market structure. The newer cities have better work opportunities, but a much higher incidence of substandard wages, than the older ones. Using the first subtotal in Tables 6 and 7 as a measure of the lack of work opportunities, the difference is particularly striking for family heads and unrelated individuals (Table 7): an average of 9.3% for the two new cities, versus 18.8% for the two old cities. In other words, the old cities have almost twoce as many unemployed or discouraged family heads than the new cities. For substandard pay, the situation is reversed. Using the second subtotal as a measure, the new cities show up with

TABLE 6

**COMPARISON OF SUBEMPLOYMENT RATES BY MEANS OF THE EXCLUSION INDEX
IN FOUR SELECTED CITIES (in percentages)**

C.E.S. Area Labor Supply	San Antonio 118,320	Houston 237,394	Cleveland 141,928	San Francisco 123,874	Weighted Average— New Cities	Weighted Average— Old Cities
(1) Unemployed	8.1	5.2	7.5	10.9	6.1	8.9
(2) Discouraged workers	16.4	11.9	15.0	12.2	13.4	13.7
Subtotal (1+2)	24.5	17.1	22.5	23.1	19.5	22.6
(3) Involuntary parttime workers	5.8	4.3	3.6	3.8	4.8	3.7
(4) Fulltime workers earning $0-$2.00 per hour	35.8	32.4	21.0	15.1	33.5	18.2
(5) Fulltime workers earning $2.00-$3.50 per hour	19.5	22.6	32.8	32.1	21.6	32.5
Subtotal (3+4+5)	61.1	59.3	57.4	51.0	59.9	54.4
Total % subemployed	85.6	76.4	79.9	74.1	79.4	76.7

SOURCE: Same as Appendix B.

TABLE 7
COMPARISON OF SUBEMPLOYMENT RATES BY MEANS OF THE INADEQUACY INDEX IN FOUR SELECTED CITIES (in percentages)

C.E.S. Areas Labor Supply	San Antonio 59,930	Houston 135,523	Cleveland 86,955	San Francisco 90,572	Weighted Average— New Cities	Weighted Average— Old Cities
(1) Unemployed	4.8	2.7	5.7	10.7	3.4	8.3
(2) Discouraged workers	7.8	5.0	10.6	10.0	5.9	10.2
Subtotal (1+2)	12.6	7.7	16.3	20.7	9.3	18.8
(3) Involuntary parttime workers	4.7	4.2	3.1	3.1	4.4	3.1
(4) Full-time workers earning $0-$2.00 per hour	31.5	20.6	11.5	8.5	24.0	10.0
(5) Full-time workers earning $2.00-$3.00 per hour	23.9	28.3	21.9	12.0	26.9	17.0
Subtotal (3+4+5)	60.1	53.1	36.5	23.6	55.3	30.1
Total % subemployed	72.7	60.8	52.8	44.3	64.6	48.9

SOURCE: Same as Appendix B.

half again as many substandard pay recipients as the old cities. This difference holds both for family heads and for all workers. (In Table 6 the ratio is 59.9/54.4 = 1.10; in Table 7 it is 55.3/30.1 = 1.84.)

The thesis behind an age categorization of cities is that the economic and spatial structure of a city affects conditions in its labor market. The older cities tend to have an earlier mode of industrial activity which is spatially concentrated, labor-intensive, more unstable, and lower in productivity. Such cities tend to have a well-developed blue-collar working class earning adequate wages, although wages not sufficient to permit a high standard of living. Preliminary evidence by Watkins (1973) suggests that these older cities tend to be more static and inert, with relatively few of the newer, more dynamic activities. As a result, labor market entry is relatively easy and subemployment is lower. Yet, as a consequence of economic instability, unemployment is high.

Newer cities, on the other hand, tend to capture the leading edge, high-skill, high-productivity activities. Owing to their expansion into new markets, these activities also tend to be more stable than the more mature activities. This, however, creates dynamic effects in the labor market. As a result of low unemployment rates and a consequent sense of job availability, migrants are attracted to the area (Lowry, 1966; Harrison, 1974). Entry into skilled, high-wage jobs is difficult; therefore, a pool of largely unorganized, low-skilled labor is formed which has limited bargaining power. This situation holds down wages in the low-productivity sector. Yet, as a result of the dynamics of the leading sector, economic activity in the low-productivity (often service) sector is stabilized. The combined effect is one of higher incidence of substandard wages simultaneously with lower unemployment. San Antonio and Houston, which were selected from a group of "newer" cities, and Cleveland and San Francisco, selected from a group of "older" cities, exhibit just these sorts of subemployment patterns.

CONCLUSION

We started this chapter with the assumption that the various symptoms of social distress found in urban areas were inherent in the social organization of the process of production, and in the way this process reproduces itself day after day, year after year, generation after generation. This view has led us to formulate measures of labor market functioning that go beyond the traditional unemployment concept and tie the labor market closely to the phenomenon of poverty.

Our measures are variants of the recently introduced subemployment concept. This concept unites two poverty-related dimensions of labor market functioning: a lack of job opportunities, and substandard wages. As such, it captures the plight of the working poor, yet maintains a close link with current earnings which a threshold measure of poverty—an aggregate of the effects of earnings, retirement and disability pay, welfare transfers, and other income —fails to provide.

Acceptable definitions of subemployment are far from a closed issue; the subemployment indicators presented here are subject to modification and improvement. They are constrained by the necessity of tying the social impact of jobs to a single quantitative aspect, wages, when a series of qualitative aspects, such as alienation, are important. The definitions of subemployment are also constrained by prevailing definitions of poverty, which themselves are subject to much debate. Finally, the definitions are very, very constrained by available data, which emerge from traditional economic viewpoints. A wealth of data are available by a variety of units of aggregation on the *individual*, but little is available on the *job*.

The static nature and the static uses of labor market indicators have been continually emphasized. Within such a static framework, it has been seen that the best labor market indicator can account for little more than half of the incidence of poverty. Before concluding that the other half is not rooted in the functioning of the labor market and thus not a concern of labor market policy, the analysis must be expanded to take account of lagged and indirect effects. The aged poor today are the victims both of current circumstances, such as adequate social security payments, and of earlier conditions in the labor market which prevented them from providing for their old age. A poor urban family, whose head deserted under the pressure of substandard earnings, is victimized indirectly by the labor market as well as directly by the welfare system. Further research, based in part on new approaches to data gathering, is required for a more complete clarification of such lagged and indirect effects between the labor market and poverty.

Further work is also required to confirm and expand the

descriptive statistical relationships involving subemployment that have been presented in this chapter. The work pertaining to various social distress phenomena requires much refinement, and the exploration of the way subemployment flows from urban history and economic geography must be extended to industrial structure.

No amount of empirical research can, however, lead to an understanding of the role these phenomena play in the social reproduction process. In fact, to be productive, such research must be rooted in a previous interpretation of society-wide forces. The persistent demand for low-skill, low-productivity work has thus been left unexplained by our empirical material. The reason for the low level of pay associated with such work has not been touched by the statistical research we have reported upon. The role of poverty in reproducing the low-skill, low-productivity, unstable labor force has likewise been a moot question throughout our presentation. These issues have been present only implicitly, determining our organization of the statistical material and our definition of concepts and measures.

The labor market is one of the system-defining institutions of our society. Empirical research can illustrate and document problems pertaining to it, but cannot begin to analyze its structure in depth. Our study of subemployment is intended to pose—not to answer— some fundamental questions of labor and its rewards in America. Hopefully, we have posed these questions in a way that rule out old homilies as acceptable answers. A detailed interpretation, placing the labor market in the framework of society as a whole, is the aim of our ongoing study.

COMPONENTS OF SUBEMPLOYMENT FUNCTIONS

	Spring, Harrison & Vietorisz Index	Levitan & Taggart Index	Miller Index	Exclusion Index	Inadequacy Index
Numerator	The numerator consists of the sum of all individuals who fall into one of the following categories.				
A. Unemployed	1. Officially unemployed	1. Officially unemployed 2. and *not* over 64; 3. and *not* a 16-21 year old student; 4. and *not* a resident of a household with above average family income in the previous year. (Hereafter defined to be the nation-wide mean family income for SMSAs or non-SMSAs, as appropriate.)	1. Officially unemployed 2. and *not* over 64; 3. and *not* a 16-21 year old student; 4. and *not* a resident of a household with above average family income in the previous year. (Hereafter defined as the mean family income for the SMSA in which the family resides.)	1. Officially unemployed	1. Officially unemployed 2. and a household head or an unrelated individual;
B. Discouraged worker	1. Not in the official labor force; 2. and *not* over 64; 3. and "inability to find work" is the primary or a secondary reason for not seeking work.	1. Not in the official labor force; 2. and *not* over 64; 3. and *not* a 16-21 year old student; 4. and *not* a resident of a household with an above average family income in the previous year;	1. Not in the official labor force; 2. and *not* over 64; 3. and *not* a 16-21 year old student; 4. and *not* a resident of a household with an above average family income in the previous year;	1. Not in the official labor force; 2. and desires work.	1. Not in the official labor force; 2. and a household head, or an unrelated individual 3. and desires work.

APPENDIX A (continued)

	Spring, Harrison & Vietorisz Index	Levitan & Taggart Index	Miller Index	Exclusion Index	Inadequacy Index
		5. and currently desiring work but not looking because they cannot find work for either job market or personal reasons. (Job market reasons are: "looked, but could not find," and "think they are too young," or they "lack education, skills or training," or they possess "other person-handicaps.")	5. and not looking for work, because they believe no work is available.		
C. Involuntary parttime	1. In official fulltime labor force; 2. and working less than 35 hours per week for economic reasons.	1. In official fulltime labor force; 2. and *not* over 64; 3. and *not* a 16-21 year old student; 4. and *not* a resident of a household with an above average family income in the previous year;	1. In official fulltime labor force; 2. and *not* over 64; 3. and *not* a 16-21 year old student; 4. and *not* a resident of a household with an above average family income in the previous year;	1. In official fulltime labor force; 2. and working less than 35 hours per week or less than 50 weeks per year for economic reasons.	1. In official fulltime labor force; 2. and a household head or an unrelated individual; 3. and working less than 35 hours per week or less than 50 weeks per year for economic reasons.

		5. and not a household head or an unrelated individual earning less than a "poverty" income (hereafter defined as the official Social Security Administration annual family poverty budget adjusted for family size.) 6. and working less than 35 hours per week for economic reasons.	5. and working less than 35 hours per week for economic reasons.			
D. Earnings	1. In official labor force; 2. and working more than 34 hours per week; 3. and earning less than an "adequate" individual income (defined either as the Bureau of Labor Statistics "lower level" annual family budget for a family of four, or as $4,000 per year, corresponding to	1. In official labor force; 2. and a household head or unrelated individual; 3. and *not* over 64; 4. and *not* a 16-21 year old student; 5. and *not* a resident of a household with above average family income in the previous year;	1. In official labor force; 2. and working more than 34 hours per week; 3. and a household head or an unrelated individual; 4. and *not* over 64; 5. and *not* a 16-21 year old student; 6. and *not* a resident of a household with an above average family income in the previous year;	1. In official labor force; 2. and working more than 34 hours per week; 3. and *not* previously counted as involuntary parttime; 4. and earning less than an adequate income in the previous year (defined parametrically).	1. In official labor force; 2. and working more than 34 hours per week; 3. and a household head or an unrelated individual; 4. and *not* previously counted as involuntary parttime;

APPENDIX A (continued)

	Spring, Harrison & Vietorisz Inedx	Levitan & Taggart Index	Miller Index	Exclusion Index	Inadequacy Index
	the proposed minimum wage level of $2.00 per hour debated by Congress in 1972).	6. and earning less than "poverty" income in the previous year adjusted for family size; 7. and not previously counted as unemployed, discouraged, or involuntary parttime.	7. and earning less than an adequate weekly income adjusted for family size (defined as the legal minimum wage unadjusted for family size).		5. and earning less than an adequate family income in the previous year adjusted for family size (defined parametrically).
Denominator	The denominator consists of the sum of all the individuals who fall into one of the following categories.				
	1. Official labor force; 2. plus discouraged workers.	1. Official labor force; 2. plus discouraged workers.	1. Official labor force; 2.	1. Official labor force; 2. plus discouraged workers.	1. Official labor force; 2. plus discouraged workers; 3. and a household head or an unrelated individual.
Subemployment rate	The subemployment rate is obtained by dividing the numerator by the denominator. The number of individuals subemployed equals the number of individuals included in the numerator.				

DEFINITIONS COMMON TO ALL INDICES:

1. *Officially unemployed:* No work during the survey week, tried to find a job in the last four weeks, and available for work during the survey week; *or* waiting to be recalled from a layoff; *or* waiting for a new wages and salaries job within thirty days; *and* over 16 years old, civilian, and non-institutionalized.

2. *Officially employed:* Performed any paid work, or more than fourteen hours unpaid work, on a family farm or in a family business during the survey week; *or* were with a job but not at work due to illness, bad weather, industrial dispute, vacation, or other personal reasons.

3. *Official labor force:* All those either officially employed or officially unemployed. Notably excluded are members of the armed forces; persons whose activity consisted of work around the house; persons doing volunteer work; persons unable to work because of long-term physical or mental illness; the voluntarily idle; and seasonal workers for whom the survey week fell in an "off" season and who were not reported as unemployed.

4. *Not in the official labor force:* Includes all civilians over 16 years old, and not institutionalized who are not in the previously defined "official labor force."

5. *Official fulltime labor force:* Persons working on a fulltime (over 34 hours per week) schedule, persons working parttime involuntarily because fulltime work is not available, and unemployed persons seeking fulltime jobs.

APPENDIX B
COMPARISON OF SUBEMPLOYMENT MEASURES IN DETROIT C.E.S. SAMPLE AREA*

	Spring, Harrison and Vietorisz		Miller	
	n	%	n	%
Base	204,979	—	193,108	—
Unemployed	27,157	13.2	17,601	9.1
Discouraged worker	11,872	5.9	9,651	5.0
Involuntary parttime	9,069	4.4	5,722	3.0
Subemployed sub-total	48,098	23.5	32,974	17.1
Earnings cutoff ($)**	Cumulative n	Cumulative %	Cumulative n	Cumulative %
500/yr. or $0.25/hr.	58,476	28.6	37,290	19.3
1,000 or 0.50	63,587	31.1	38,595	20.2
1,500 or 0.75	67,911	33.2	40,282	20.9
2,000 or 1.00	71,493	34.9	42,125	21.9
2,500 or 1.25	76,709	37.4	44,779	23.3
3,000 or 1.50	80,381	39.2	46,429	24.2
3,500 or 1.75	87,453	42.7	49,664	25.9
4,000 or 2.00	92,319	45.1	52,037	27.1
4,500 or 2.25	99,623	48.7	55,470	28.9
5,000 or 2.50	104,433	51.0	58,081	30.3
5,500 or 2.75	113,907	55.6	63,682	33.2
6,000 or 3.00	117,565	57.4	65,578	34.2
6,500 or 3.25	128,545	62.8	71,713	37.4
7,000 or 3.50	133,569	66.2	74,677	38.9
7,500 or 3.75	146,501	72.5	83,401	43.4
8,000 or 4.00	152,321	75.3	86,835	45.2
8,500 or 4.25	163,457	80.7	93,953	48.9
9,000 or 4.50	167,940	82.9	96,631	50.3

*See also Figure 1.

**The Spring, Harrison, and Vietorisz, and the Levitan and Taggart indices use an annual earnings cutoff criterion; the others use an hourly earnings cutoff criterion.

SOURCE: All data are derived from the 1970 Census Employment Survey of a selected low-income area in Detroit, Michigan. The survey was conducted by the Bureau of Census for the Bureau of Labor Statistics.

	Levitan and Taggart		Exclusion Index		Inadequacy Index	
	n	%	n	%	n	%
	218,008	—	238,080	—	143,067	—
	19,163	8.8	27,730	11.6	11,900	8.3
	24,895	11.4	39,686	16.7	17,783	12.4
	6,373	2.9	9,231	3.9	5,365	3.8
	50,431	23.1	76,647	32.2	35,048	24.5

Cumulative n	Cum. %	Cumulative n	Cumulative %	Cumulative n	Cumulative %
54,264	24.9	83,215	35.0	38,133	26.7
55,206	25.3	85,635	36.0	39,018	27.3
56,512	25.8	88,006	37.0	39,926	27.9
57,624	26.3	91,023	38.2	41,054	28.7
58,986	26.7	95,851	40.3	42,264	29.6
60,088	27.4	103,393	43.4	44,171	30.9
62,651	28.6	110,862	46.6	46,531	32.5
63,992	29.2	119,875	50.4	49,643	34.7
67,101	30.6	129,008	54.2	53,407	37.3
70,240	32.0	140,622	59.1	58,451	40.9
73,266	33.4	148,366	62.3	62,860	43.9
75,587	34.5	159,670	67.1	66,888	46.8
78,720	35.9	169,439	71.2	71,371	49.9
80,815	36.9	179,664	75.5	75,378	52.9
84,114	38.4	192,713	80.1	80,120	56.0
87,290	39.9	203,038	85.3	84,849	59.3
90,288	41.3	209,671	88.1	88,766	62.0
92,366	42.3	214,790	90.2	92,097	64.4

NOTES

1. In each designated area interviews were obtained from about 5,000 persons 16 years of age or older residing in about 2,750 households. The interviews were conducted between mid-November 1970 and early March 1971 and lasted from nine to twenty weeks in each area. Each person interviewed was asked approximately 250 questions on, first, a variety of personal and housing characteristics; and then, on a variety of employment characteristics such as employment status, training and educational experiences, job-seeking methods, job tenure, sources and levels of income, residential mobility, transportation to work, and many other factors. The individual interviews were coded and stored on computer tape. The 24 tapes which have been made available within disclosure limitations have been converted to BCD fixed record-length machine-readable format and are available, along with a detailed Codebook, from the Research Center for Economic Planning, New York, N.Y.

2. The Detroit CES sample area was chosen to permit comparisons in other research of ours with the findings of Gordon (1971). Unemployment is defined according to the Bureau of Labor Statistics convention: no work during the survey week, tried to find a job in the last four weeks, and available for work during the survey week; or waiting to be recalled from a layoff; or waiting for a new wages and salaries job within 30 days; and over 16 years old, civilian, and noninstitutionalized. The 27,730 unemployed individuals consisted of 11,901 family heads or unrelated individuals and 15,829 secondary family members.

3. The original subemployment index was developed to underpin President Lyndon Johnson's call for a war on poverty (Spring 1971).

4. A set of subemployment categories such as these are converted to a subemployment *rate* by examining some population of individuals and categorizing each as either subemployed by one of the criterion above or not subemployed. The subemployment rate is the number subemployed divided by the sum of those subemployed plus those not subemployed. The rate may be expressed as a fraction or as a percentage.

5. Technically, the elasticity of the perceived poverty level in constant prices, with regard to median real family income, is estimated as 0.6, which yields the numerical approximation cited in the text.

6. An analysis of partial correlations based on data from the 60 CES survey areas reveals that the discouraged worker component correlates significantly only with the unemployment component.

7. Again using data from the 60 CES areas, the involuntary part-time components correlates significantly with all other components except discouraged workers.

REFERENCES

BRENNER, M. (1973a) "Fetal, infant and maternal mortality during periods of economic instability." International Journal of Health Services 3, 3: 145-159.

——— (1973b) Mental Illness and the Economy. Cambridge: Harvard University Press.

——— (1971) "Economic changes and heart disease mortality." American Journal of Public Health 61, 3 (March).

CHASE, H. (1970) "A study of infant mortality from linked records: registration aspects." American Journal of Public Health 60, 11 (November): 2181-2195.

CETA (1973) "Comprehensive employment and training act of 1973." Public Law 93-203. 93rd Congress, S. 1559 (December 28).

CUTLER, R. (1967) "An assessment of the meaning of work to the male narcotic addict in a voluntary treatment center." Unpublished Ph.D. dissertation. New York: New York University.

DUNHAM, H. (1965) Community and Schizophrenia. Detroit: Wayne State University Press.

FLAIM, P. (1971) Employment in Perspective: Discouraged Workers and Recent Changes in Labor Force Growth (BLS Report 396). Washington, D.C.: Bureau of Labor Statistics.

FLEISHER, B. (1966) The Economics of Delinquency. Chicago: Quadrangle Books.

FREEDMAN, M. (1973) "Good jobs, bad jobs: the search for shelters." Document from Conservation of Human Resources. New York: Columbia University.

GORDON, D. (1972) Theories of Poverty and Underemployment. Lexington, Mass.: D. C. Heath.

——— (1971) Class, Productivity, and the Ghetto. Unpublished Ph.D. dissertation. Cambridge: Harvard University.

GUERRIN, R. and E. BORGATTA (1965) "Socio-economic and demographic correlates of tuberculosis incidence." Milbank Memorial Fund Quarterly 43 (July): 269-290.

HAMMERMESH, D. and N. SOSS (1974) "An economic theory of suicide." Journal of Political Economy 82, 1 (January/February): 83-98.

HARRISON, B. (1974) Urban Economic Development: Suburbanization, Minority Opportunity, and the Condition of the Central City. Washington, D.C.: Urban Institute.

——— (1972) Education, Training and the Urban Ghetto. Baltimore: Johns Hopkins University Press.

HELMER, J. and T. VIETORISZ (1973) "Drug use, the labor market and class conflict." Paper presented at the Annual Meeting of the American Sociological Association, New York (August 30).

HURLEY, R. (1969) Poverty and Mental Retardation. A Causal Relationship. New York: Vintage Books.

JACOBS, J. (1970) The Economy of Cities. New York: Vintage Books.

KILPATRICK, R. (1973) "Income elasticity of the poverty line." Review of Economics and Statistics 55, 3 (August): 327-332.

LAMPMAN, R. (1971) Ends and Means of Reducing Income Poverty. Chicago: Markham.

LEGGETT, J. and C. CERVINKA (1972) "Countdown: labor statistics revisited." Society 10 (November): 99-103.

LEIGHTON, A. (1965) "Poverty and social change." Scientific American 212, 5 (May): 21-27.

——— (1963) My Name is Legion. Vol. 1 of The Stirling County Study of Psychiatric Disorder and Sociocultural Environment. New York: Basic Books.

——— C. HUGHES, M. TREMBLAY, and R. RAPOPORT (1960) People of Cove and Woodlot, Vol. 2 of The Stirling County Study of Psychiatric Disorder and Sociocultural Environment. New York: Basic Books.

LEIGHTON, D., A. LEIGHTON, J. HARDING, D. MACKLIN and A. MacMILLAN (1963) The Character of Danger, Vol. 3 of The Stirling County Study of Psychiatric Disorder and Sociocultural Environment. New York: Basic Books.

LEVITAN, S. and R. TAGGART (1974) Employment and Earnings Inadequacy. Baltimore: Johns Hopkins University Press.

——— (1973) "Employment and earnings inadequacy: a measure of worker welfare." Monthly Labor Review (October): 19-27.

LEWIS, O. (1966) "The culture of poverty." Scientific American 215, 4 (October): 19-25.

LOWRY, I. (1966) Migration and Metropolitan Growth: Two Analytical Models. San Francisco: Chandler.

MILLER, H. (1973) "Measuring subemployment in poverty areas of large United States cities." Monthly Labor Review (October): 10-18.

MOYNIHAN, D. (1965) The Negro Family, The Case for National Action. Washington, D.C.: Government Printing Office.

MUTH, R. (1969) Cities and Housing: The Spatial Pattern for Urban Residential Land Use. Chicago: University of Chicago Press.

OPPENHEIMER, M. (1974) "The sub-proletariate: dark skins and dirty work." Insurgent Sociologist 4, 2 (Winter): 6-20.

ORSHANSKY, M. (1965) "Counting the poor: another look at the poverty profile." Social Security Bulletin (January): 7-9.

PROJECTOR, D., G. WEISS, and E. THORESEN (1969) "Composition of income as shown by the survey of financial characteristics of consumers," in L. Soltow (ed.) Six Papers on the Size Distribution of Wealth and Income, Studies in Income and Wealth, No. 33. New York: Columbia University Press.

RAINWATER, L. (1973) "Economic inequality and the credit income tax." Working Papers for a New Society 1, 1 (Spring): 50-61.

ROSENBLOOM, M. (1974) "Discouraged workers and unemployment: a review of recent evidence." 1974 Meetings of the N.Y. State Economics Association, Buffalo, N.Y.

SCHICK, R. et al. (1972) Strategic Factors in Urban Unemployment: An Analysis of Thirty Cities with Policy Implications. New York: Praeger.

SHAPIRO, S., E. SCHLESINGER, and R. NESBITT (1968) Infant, Perinatal, Maternal and Childhood Mortality in the United States. Cambridge, Mass.: Harvard University Press.

SMILEY, J., S. EYRES, and D. ROBERTS (1972) "Maternal and infant health and their associated factors in an inner city population." American Journal of Public Health 62, 4 (April): 476-482.

SPRING, W. (1971) "Underemployment: the measure we refuse to take." New Generation (Winter): 23.

——— B. HARRISON, and T. VIETORISZ (1972) "The crisis of the underemployed." New York Times Magazine (November 5): 42-60.

SROLE, L. T. LANGNER, et al. (1962) Mental Health in the Metropolis. Vol. 1 of The Midtown Manhattan Study, T. Rennee (ed.). New York: McGraw-Hill.

United States Bureau of the Census (1972) Employment Profiles of Selected Low Income Areas, Vols. 1-68. Washington, D.C.: Government Printing Office.

United States Department of Health, Education and Welfare (1972) Infant Mortality Rates: Socio-economic Factors. Washington, D.C.: Government Printing Office.

VIETORISZ, T. (1973) "We need a $3.50 minimum wage." Challenge 16, 2 (May/June): 49-62.

——— and B. HARRISON (1972) "Earned family incomes and the urban crisis." Project proposal funded by the National Institute of Mental Health.

VIETORISZ, T., R. MIER, J-E. GIBLIN (1974) "The concept and measurement of subemployment." Selected papers from the 1974 North American Conference on Labor Statistics. Washington, D.C.: U.S. Department of Labor.

WATKINS, A. (1973) "City age and the typology of subemployment." Working Paper No. 12. New York: Research Center for Economic Planning.

WRIGHT, N. (1972) "Some estimates of the potential reduction in the U.S. infant mortality rate by family planning." American Journal of Public Health 62, 8 (August): 1130-1134.

12

The Distribution of Coloured Immigrants in Britain

CERI PEACH
STUART WINCHESTER
ROBERT WOODS

□ FOR AN AMERICAN to visualize the position of black immigration in Britain, he would have to imagine the United States experiencing the Puerto Rican immigration as its first large-scale black migration. The movement has occurred in the last twenty years and is concentrated in the major urban centres. There have been black people in Great Britain since the seventeenth century (Shyllon, 1974), but their numbers have been small. Large numbers are a postwar phenomenon dating particularly from the late 1950s and 1960s.

As the numbers of immigrants have grown, so the literature has changed from that small, academic cottage industry dealing with small communities, mainly in dockside areas of the 1940s and early fifties (Little, 1948; Collins, 1957; Banton, 1955 and 1959) to part of what the press disparagingly refers to as the race-relations industry of the 1960s—a literature large enough to have a whole series of its

AUTHORS' NOTE: *The authors acknowledge the financial assistance of the United Kingdom Social Science Research Council. This paper forms part of their project HR 1774/1 on the Segregation of Coloured Immigrants in Britain.*

own internecine struggles (see particularly the surveys by Rose et al., 1969, and Deakin, 1970).

There are altogether perhaps 1.5 million black people living in Great Britain who form between 2% and 3% of the total population. They are thus not only more recent and less numerous than the black population of the United States, but form a much smaller portion of the total population. In their rapid rate of growth and their high degree of spatial concentration, however, they are distinct.

The largest groups are the West Indians, Indians, and Pakistanis, who differ as much from each other as from the rest of the population. Table 1 gives the number of persons born in various birthplaces, living in Great Britain in 1971.

The largest single immigrant group of all, however, is white. The Irish from Eire (the Irish Republic) number 721,000. While the West Indians, the Indians, and, until 1972, the Pakistanis belong to the British Commonwealth, they have been subject to immigration control since 1962. The Irish, however, though alien, are subject to no immigration control and may come and go to the country without restriction; they vote, and they benefit from the provisions of the welfare state as British citizens do. Migration statistics, of varying accuracy, have been kept by British government departments since 1955 for Commonwealth immigrants, but no comparable annual figure exists for the Irish (Jackson, 1963).

TABLE 1
ORIGIN OF POPULATION IN GREAT BRITAIN, 1971

Birthplace	
U.K.	50,514,820
Irish Republic	720,985
Old Commonwealth	145,250
New Commonwealth	1,157,170
India	322,670
Pakistan	139,445
West Indies	302,970
Cyprus	72,665
Africa	176,060
Other countries	143,355
Foreign and not stated	1,076,935
Not read by machine	211,215
Total	53,826,375

SOURCE: *Census Great Britain 1971, Advance Analysis* (London: H.M.S.O., 1972): Table 2, p. 103.

DATA PROBLEMS

British census data refers to birthplaces, not race, and as the two are imperfectly correlated there are pitfalls to the interpretation of the position of immigrants in Britain. The census distinguishes between "New" Commonwealth, and the "Old." The New Commonwealth countries include India, Pakistan, the West Indies, and Commonwealth countries in Africa; they also include Cyprus and Malta. The Old Commonwealth countries are Australia, Canada, and New Zealand. The population of the Old Commonwealth is predominantly white while that of the New is predominantly black. The correlation of newness and blackness is very imperfect, however, and the 300,000 persons born in India, living in Britain in 1971, are thought to include some 100,000 white British born in India (Peach and Winchester, 1974). Given this composition of the Indian-born, it becomes difficult to distinguish how much the higher socioeconomic position of that group compared with other immigrants is due to the presence of colonial-born British (see Table 2).

Additionally, children born to immigrants in Great Britain are listed as British-born, though additional information is available on those persons both of whose parents were born in the New Commonwealth. The 1971 census gives information on those parents' birthplaces. One estimate puts the number of coloured children born in Britain at 500,000 in 1971 (Kohler, 1973).

One of the difficulties associated with the census data is that of underenumeration. In 1971 it was calculated that the Pakistan-born population was underenumerated by over 30% while the West Indian population was underenumerated by about 15% (Peach and Winchester, 1974). It is not known to what extent local underenumeration differs from that at the national level.

TABLE 2

DISTRIBUTION BY TYPE OF JOB BETWEEN WHITE AND MINORITY EMPLOYEES (in percentages)

	All Men	Men from Minority Group	All Women	Women from Minority Group
Non-skilled manual	37	67	50	65
Skilled manual	21	19	6	Nil
Non-manual	42	14	44	35

SOURCE: Smith, 1974: p. 39 (based on a sub-sample of 263 plants).

FACTORS AFFECTING IMMIGRATION

AND DISTRIBUTION

The movement into Great Britain of black migrants in the postwar period was dominated by the demand for labour in that country. Analysis of the West Indian movement in the period before immigration restrictions were imposed in 1962 show a strong positive correlation between prior fluctuations in the number of unfilled job vacancies and fluctuations in the number of arrivals (Peach, 1965 and 1968). It also shows an inverse correlation with the GDP per capita in the sending territories (Davison, 1962: 43-44). Britain was not unique in its demand for labour in northwestern Europe, as all of the major industrial countries—Germany, France, the Netherlands and Switzerland—required substantial immigrant labour. These countries differed mainly in their source of supply. Germany received substantial amounts from East Germany, Italy, Greece, and Turkey; Switzerland from Italy; and France from North Africa, West Africa, and Portugal. Kindleberger (1967) has gone so far as to argue that migrant labour has been the critical factor in explaining the buoyancy of much of the West European economy in the postwar era.

The demand for labour in certain sectors of public employment such as hospitals and public transport led to the direct recruitment of labour from Barbados in the 1950s. This recruitment was conducted through the Barbados government-assisted schemes (Barbados, 1962). During World War II labour had been recruited in Jamaica to work in English munitions firms (Banton, 1955). Actions of this type led to the movement of a relatively small number of immigrants. But it did foster the development of a communications network between Great Britain and the West Indies which led to an acceleration of immigration during the more recent period.

The movement from India and Pakistan started later than the West Indian movement and operated without restriction for only a short time before the restrictions of the 1962 Commonwealth Immigrants Act were imposed. The most recent input to the black migratory movement is the East African Asians. About 27,000 were forced out of Uganda in 1972 by political pressure, and many of these refugees settled in Britain (Humphry and Ward, 1974). However, while the economic demand for labour was the dominant factor attracting immigrants to Britain, there was also at work a number of repelling factors. Among these were social and political pressures against immigration, as well as such factors as a shortage of housing and school crowding. This pressure culminated in a series of restrictive measures, the Commonwealth Immigrants Acts of 1962, 1965, and 1971.

Although the movement was dominated by a demand for labour, the type of employment available to coloured immigrants was selective. Smith (1974: 37) speaks of black labour as marginal labour, and thus argues that there is a tendency for it to be employed when there was a shortage of white labour. An earlier analysis also showed a similar tendency for West Indians to be concentrated in industries which were not attracting white labour (Peach, 1967). Smith (1974: 27) showed that there was a tendency for coloured labour to be concentrated in a small number of plants; 74% of the black employees were in 28% of the plants in his national sample. Firms which operated a night shift were particularly dependent on coloured labour (Smith, 1974: 74). Male coloured workers were strongly concentrated in non-skilled manual employment in the sample, though women had found access to nonmanual employment somewhat easier (Smith, 1974: 39) (see Table 3).

Within Great Britain the black migrants avoided regions of heavy unemployment at the periphery of the country; thus, they are found in small numbers in Scotland, Wales, northern England, and the northwest (see Table 2). Elsewhere in the country they appear to serve as a "replacement population" (Peach, 1968), moving to those

TABLE 3

GREAT BRITAIN REGIONAL RATES OF POPULATION GROWTH, 1961-1971, AND NEW COMMONWEALTH PERCENTAGE OF EACH REGIONAL POPULATION, 1971

	% New Commonwealth	1961-71 % Growth	% Indian, Pakistan and West Indian Born
Scotland	0.589	0.864	0.280
North	0.523	1.316	0.268
Yorks & Humberside	1.600	3.265	1.285
N.W.	1.216	2.426	0.859
E. Midlands	1.816	9.236	1.255
W. Midlands	3.123	7.294	2.658
E. Anglia	1.046	13.364	0.536
S.E.	3.799	5.297	2.357
Wales	0.504	3.009	0.264
S.W.	1.263	10.468	0.623

SOURCE: *Census Great Britain 1971, Advance Analysis* (London: H.M.S.O., 1973).

regions which, despite a demand for labour, had not been able to attract much white immigration. In 1971 this pattern was yet in evidence. The smallest percentage of New Commonwealth-born persons were found in those regions which had longstanding problems of unemployment and slow population growth, and those which had experienced high demands for labour and rapid growth of the total population (see Table 4). These data show that regions of unemployment were unattractive to black immigrants, but even within regions with demand for labour those subregions which were attractive to white migrants were less dependent on black immigration. Black migrants formed larger concentrations in those regions which, despite their demand for labour, did not show rapid rates of population growth.

Similar patterns were previously identified in the United States. Thomas (1954) demonstrated the inverse relationship of white immigration to the United States and the movement of blacks from the southern to northern states. Willcox (1931: 109) has likewise

TABLE 4
THE DIFFERENTIAL CONCENTRATION OF PERSONS BY AREAS OF BIRTH IN GREAT BRITAIN'S SEVEN CONURBATIONS, 1971

Conurbation	Total Population	Birthplace			
		New Commonwealth	India	Pakistan	West Indies
Greater London S.E.L.	7,392,915	476,535	106,925	30,135	166,970
(Manchester)	2,387,275	46,690	12,540	12,385	10,415
W. Midland (Birmingham)	2,368,585	120,725	44,795	23,170	39,535
W. Yorks (Leeds-Bradford)	1,725,040	55,485	16,020	21,845	10,285
Merseyside (Liverpool)	1,264,180	9,040	2,200	455	1,455
Tyneside (Newcastle)	804,455	5,010	1,850	855	280
Clydeside (Glasgow)	1,726,785	10,345	3,675	2,440	455
Total Conurb. Pop.	17,669,235	723,830	188,005	91,285	299,395
Total Br. Pop.	53,826,375	1,157,170	322,670	139,445	302,970
% of each population group living in these areas	32.83	62.55	58.27	65.46	75.72

SOURCE: *Census Great Britain 1971, Advance Analysis* (London: H.M.S.O., 1972):Table 2.

demonstrated the inverse spatial distribution of black population and foreign immigrants to the United States at the end of the era (1920) of unrestricted European immigration.

RESIDENTIAL PATTERN OF COLOURED IMMIGRANTS

The black population of Britain is overwhelmingly urban in its concentration, and it is concentrated disporportionately in the largest centres. While about one-third of the total British population lived in the seven census-defined conurbations, 58% of the Indian-born, 65% of the Pakistani-born, and 76% of the West Indian-born population lived in these areas (see Table 4). The concentration of the Indian, Pakistani, and West Indian-born populations was most pronounced in four (Greater London, West Midlands, West Yorkshire, and Southeast Lancashire) of the seven conurbations.

Over half of the West Indian population lived in Greater London (55.1% compared with 13.7% of the total population) and a further 13% lived in the West Midlands conurbation. The degree of metropolitan concentration was less for the Indian and Pakistani born (33.1% of Indians and 21.6% of Pakistanis).

These conurbations, with the exception of the West Yorkshire, were losing population. Without immigration from the Commonwealth, the West Yorkshire conurbation would also have shown a decrease between 1961 and 1971. Thus, the bulk of the black population was not only concentrated into the largest urban centres, but into urban centres which were losing population.

Within the cities in which large-scale settlement took place, the areas most affected were the edges of the inner central districts. In Greater London, for instance, the centre of the conurbation (south Kensington, Westminster, and Holborn) appears as a white crater, ringed by areas of West Indian settlement. These, in turn, are surrounded by the white suburbs. The ring of areas of West Indian settlement is broken into three sections. The southern part of the arc (from Wandsworth through central Lambeth, central Southwark, and into northern Lewisham) is separated from the northern part by the river Thames. The northern part is divided into two by a white cordon sanitaire extending north from Westminster and Marylebone into Hampstead. The western part of this northern arc extends northwestward from North Kensington and Paddington into Willesden (in Brent) and Ealing. The eastern sector extends northwards from north of the City (the London equivalent of Wall Street) from Hackney and into Haringey (Glass, 1960; Davison, 1963; Glass and Westergaard, 1965; Doherty, 1969; Lee, 1973a and 1973c).

In Birmingham the central area was similarly fringed, with notable concentrations in Handsworth/Soho/Aston to the north and Sparkbrook to the south (Jones, 1967 and 1970; Woods, 1971 and 1973). In Bradford, the areas of immigrant concentration are located much closer to the centre than is the case in Birmingham, where the centre is a focus of urban redevelopment. An almost complete zonal encirclement of the Bradford centre has been broken into a series of wedges by a series of linear industrial areas (Tommis, 1974). In Coventry, the main area of immigrant settlement was around the central core, but distorted out of a regular concentric shape by the pattern of railway lines. Collinson (1967) concluded from the lower indices of segregation of the amalgamated Asian population in Oxford in 1961 that it was a socially privileged elite. However, by 1971, the Asian population in Oxford had increased its segregation index to 43, while the West Indian level of segregation had decreased to 42.

If the data were treated at the finer scale of the enumeration district (roughly comparable to American blocks) the index of segregation would be much higher. In Bradford, the index of dissimilarity for the Asian population in 1961 at the enumeration district level was 71.0 and in 1971 it was 68.5 (Tommis, 1974: 19).

Again the areas in which concentration was taking place tended to have lost population. In 1961, 80% of the West Indian population living in the London Administrative County lived in boroughs which had lost white population, and in Birmingham 70% lived in wards which had lost population (Peach, 1968: 89). In 1971 87% of the West Indian-born, 92% of the Pakistani-born, and 94% of the West Indian-born lived in boroughs which had lost population between 1961 and 1971 (compared with 80% of the population as a whole).

Given a situation in which black migration had taken place on quite a large scale over a relatively short period of time, a degree of residential segregation of the black population was to be expected. The degree of segregation of the black population in Greater London (measured by wards), however, was much less than that for the black population of New York or Chicago when measured on a comparable scale (community areas or tracts).

If one takes a strict Parksian view of society, then the degree of social integration of groups is reflected in their degree of spatial mixing. Parks stated "human relations can always be measured ... with more or less accuracy in terms of geographical distance" (Park, 1926). Duncan and Duncan (1955), Lieberson (1963), Taeuber and Taeuber (1965) have subscribed to, and produced evidence for, this general hypothesis. The index employed by most of these authors to define the level of spatial segregation is the index of dissimilarity. Its use in such studies has become widespread. Since this index measures the percentage of a group which would have to shift its area of residence in order to reproduce the distribution of the group with which it is being compared, there should be a spectrum of degrees of integration/assimilation (however defined) from complete acceptance to complete rejection which corresponds closely to the range of value from zero residential dissimilarity to 100% dissimilarity.

The position of groups on such a scale should be a fairly good guide to the relative ranking of those groups in the wider society, and their scores might be employed as a surrogate for social distance between groups. However, the scores on such a scale represent the *net* balance between two different forces. On the one hand, there may be the negative discriminatory pressures of the dominant society leading to enforced segregation, while, on the other hand, there may be a positive desire by minority groups to preserve their identities through the process of self-segregation. Thus one may identify the intensity of spatial clustering among individual groups without being able to ascertain the extent to which clustering is externally imposed or internally desired.

The Cypriot-born population has the highest degree of residential dissimilarity of any birthplace group in Greater London, but colour is not a material factor in the case of Cypriot segregation. The Caribbean population ranks next in terms of residential dissimilarity, followed by Pakistan and India. The indices of dissimilarity of the West Indian and Indian population to the native white population (England and Wales) are notably different. One observer (Lee, 1973) has used this as an argument that social and spatial structures are not as congruent as the human ecologists tend to imply. Lee's argument is that the literature shows Indian and Pakistan society to be tightly knit. The geographical evidence (for India and Pakistan combined) is that the group is dispersed in London and not segregated. Therefore, he argues, it is possible to have tight social organization without close spatial proximity. The shortcoming of this view, however, is that the Indian and Pakistani populations differ not only from each other but contain two very distinct groups, the whites and blacks. When one examines the data for Bradford (or Coventry) where the white Indians are not present, the pattern conforms to the expected high degree of segregation (see Table 4).

The correspondence of social and physical distance is more marked in the Coventry case. Certainly the ranking of birthplace groups on the basis of the spatial evidence conforms to the expected situation derived from knowledge of their social organization (see Table 5).

Having identified the degree of segregation present, we will now attempt to explain it. Basically, we are interested in two main points. To what extent does discrimination account for the degree of segregation observed, and to what extent has segregation increased or decreased since the first data were available in 1961?

If a population is both poor and black, part of their segregation may be due to their poverty rather than their blackness. Taeuber and Taeuber (1964) attempted to measure the degree of segregation of the black population of Chicago which could be imputed to income differentials. The Taeubers demonstrated that income differentials explained about 14% of observed residential segregation in Chicago in 1950 between whites and nonwhites. By 1960 income differentials explained only 12% of the observed pattern. Thus, certainly in terms of income, very little (and a decreasing amount) of the segregation of whites and nonwhites was explained.

TABLE 5
RESIDENTIAL SEGREGATION OF BIRTHPLACE GROUPS IN
COVENTRY ON A WARD BASIS (1971)

Birthplace	Index of Segregation
Wales	10.3
Scotland	7.8
Northern Ireland	11.9
Irish Republic	14.1
Old Commonwealth	13.5
New Commonwealth	51.9
Africa	52.4
America (West Indies)	34.3
Ceylon	43.1
India	58.9
Pakistan	69.9
Far East	29.0
Remainder	12.1

A similar exercise was carried out for the various components of the black population of London, Birmingham, and Coventry. Instead of income differentials, the distributions were standardized by six socioeconomic groups.

Using such a standardization it is possible to "explain" 15% of the segregation of London boroughs for the West Indians, 13% for the Pakistanis, 10% for the Indians, and 22% for the Southern Irish. At the ward level one may account for 17% fo the West Indian, 9% of the Indian, 11% of the Pakistani, and 21% of the Southern Irish, again for 1971. For Birmingham in the same year, but at the tract level, 12% and 11% respectively can be explained of the West Indian and Asian segregation. This represents a decrease from 1961 when the two percentages were 14% and 15%. For Coventry at the ward level the amount of West Indian segregation which was explained fell from 40% to 15% and that for the Asian population from 26% to 10% between 1961 and 1971. Such a low level of explanation in terms of socioeconomic differences necessarily leads one to infer the existence of a high level of ethnic segregation over and above the level of class segregation.

A second method of estimating the causal factors in segregation was to measure the amount of segregation due to self-imposed spatial separation. This can be gauged from the amount of segregation that is present between groups that are culturally rather similar, or between groups where colour is not one of the distinguishing traits. The Cypriots have consistently the highest indices of dissimilarity. The Irish from Eire, despite having one of the lowest indices of dissimilarity, would still have to shift the ward of residence of over a quarter of its population in order to reproduce the distribution of those born in England and Wales. Similarly, if one breaks down the West Indian population into its component islands and territories (Table 6), it can be seen that the indices of dissimilarity between West Indians vary between 15 and 41 with a mean value of 30. Thus, it would seem that even between very similar populations, at this scale of investigation, there is a minimum index of dissimilarity of about 15 points which is due to self-segregation. In a situation in which the maximum values vary between 63 and 64, it seems that at least between 23% and 24% of this segregation may be self-imposed.

In Coventry the difference between the minimum and maximum levels of segregation (at the ward level) were even more acute. Only 8% of the Welsh, 12% of the Scottish-born, 15% of those born in Northern Ireland, and 16% for Southern Irish would have to change

TABLE 6
LONDON: INDICES OF DISSIMILARITY OF
SELECTED BIRTHPLACES, BY WARDS, 1971

	(1)	(2)	(3)	(4)	(5)	(6)	(7)
(1) Barbados	—						
(2) Guyana	27.83	—					
(3) Jamaica	26.41	31.91	—				
(4) Trinidad & Tobago	31.73	29.99	40.90	—			
(5) Other Caribbean	31.31	37.76	39.16	37.59	—		
(6) Total Caribbean	17.45	24.82	14.86	31.62	26.34	—	
(7) England & Wales	49.11	46.21	57.13	44.45	54.36	50.92	—

SOURCE: Special tabulations of 1971 Census supplied by Office of Population Censuses and Surveys.

their ward of residence to produce a distribution which would correspond to that of the population born in England. However, within the New Commonwealth-born category, West Indians were separated from Indians by 36% and from Pakistanis by 49%. In the case of West Indians these degrees of segregation from other New Commonwealth subgroups is higher than that from the population as a whole. For Birmingham at the tract level West Indians were separated from Indians and Pakistanis by some 44% in 1961 and 36% in 1971, but unlike Coventry this separation was not higher than from the population as a whole.

Thus, the different black immigrant groups which were of roughly similar socioeconomic levels were not only markedly segregated from the native-born population but from each other also. While there may be doubt as to the relative strengths of discrimination by the white population and self-imposed separation by the black population in explaining the degree of spatial segregation of black and white, the high levels of segregation of the Asians from the West Indian population indicates that voluntary, self-segregation must be a major factor.

BLACK IMMIGRANTS AND
THE HOUSING MARKET

Positive evidence of discrimination in the selling of houses to black people does exist (Peach, 1968: 90-91). The most notable and carefully researched examples were embraced in a report by the Political and Economic Planning Report (P.E.P., 1967). In a series of controlled experiments in which a series of subjects matched for socioeconomic characteristics but differentiated by ethnic background approached realtors to ask details of houses, the Englishman was given a much more extensive list than the West Indian (P.E.P. Report, 1967: 77 and Appendix 1: A13). Thus, while the percentage of West Indians owning houses is similar to that of the local white population, there may have been limitations in the choice of areas available to them over and above differences produced by income.

As significant from the point of view of the British situation is the distribution of municipally controlled housing. Municipally controlled housing accounts for over 30% of all tenures in Great Britain (G.H.S. Table 5.8: 92). In 1961 it accounted for a quarter. Immigrant population, however, is substantially underrepresented in this part of the housing stock. In the English conurbations for 1961, for which data are available (Commonwealth Immigrants in the Conurbations, Census of England and Wales, 1961), 5.3% of households headed by a person born in the West Indies, India, or Pakistan lived in accommodations rented from local authorities, while 23.0% of the total households lived in such accomodation.

This situation of underrepresentation in council housing was not unexpected for two reasons. The first is that the allocation of council housing operates largely through a points system in which length of residence in a borough is an important element. Newcomers would not initially score well on such a scheme. The second reason is that, for many immigrants, the Asians particularly, ownership and the maintenance of wealth within the community was a desired end. Paying money, in rent, to a local authority was regarded as undesirable (Dahya, 1974: 97). Indians and Pakistanis, for the greater part, did not want council housing (Burney, 1967: 35). Thus, not only were black immigrants underrepresented, but there was evidence that very few applied for such housing (Richmond, 1973: 136).

Nevertheless, there was evidence of discrimination by local authorities in the type of housing allocated to those immigrants whom they were forced to rehouse as a result of need or urban redevelopment (Burney, 1967: 65, 193-194; P.E.P. Report, 1967: 12). Burney indicated that many of that small number were housed in "patched" houses awaiting destruction and that such houses tended to be located in existing areas of immigrant settlement. They, therefore, tended to maintain the existing patterns rather than promote dispersal (Cullingworth Committee, 1969: 125). In Bristol, however, Richmond (1973: 140) found that all applicants for council housing were given impartial treatment.

It is difficult therefore to assess whether immigrants had excluded themselves, had been excluded through local authority discrimination, or through the working of the length of residence or points systems of allocation. The first reason would seem most applicable to the Asians, the third to the West Indians, and the second is "non-proven" (see Plant, 1971). Certainly some local authorities have been forced to rehouse coloured immigrants in rehousing schemes and after road developments.

Nevertheless, by 1971 the percentage of black households living in council accommodation had increased substantially. The G.H.S. reported (1973: 143) that 19% of black heads of households lived in local authority housing compared with 32% of U.K.-born heads of households. The black heads of households were mainly West Indian, and this factor may help to account for the decreasing degree of segregation of the London West Indian-born population in 1971.

While the degree of residential dissimilarity of the West Indian-born population has decreased between 1961 and 1971, that of the Indian-born has increased. Since a large number of Indians are white, this phenomenon is difficult to interpret. Supposing the white Indians and the black Indians had opposite kinds of distributions, the whites in the middle-class periphery and the blacks in central areas, then the overall distribution of both groups taken simultaneously would be similar to that of the population as a whole (that is, they would have low indices of dissimilarity). The indices of dissimilarity are indeed quite low. Any increase in the black Indian-born population without a corresponding rise in the white element would lead to an increasing index of dissimilarity with the total population, even if it replicated exactly the existing distribution of black Indians. Now while the indices of dissimilarity of West Indians who entered the country before and after 1960 is very low (about 2) and that for the Pakistanis is also low (about 5), that for the Indians is much higher (about 10). It seems unlikely that newcomers would have so radically different a distribution from established migrants, and the more likely explanation is that this large difference reflects the increasing size of the black proportion of the Indian-born, which thus accentuates the existing difference in the distribution of black and white Indians. The lack of appeal of council housing to the

TABLE 7

SEGREGATION OVER TIME

Location	West Indies	Indians, P & C
Coventry (ward basis)		
1961	51	60
1971	34	61
Birmingham (tract basis)		
1961	66	54
1971	56	69
Oxford (tract basis)		
1961	43	32
1971	42	43

Indian population would mean that they have been less subject to the forces of municipal dispersal than the West Indians. Merely maintaining their existing distribution, they would thus appear to become more segregated.

In London, Birmingham, Oxford, and Coventry the situation was the same. The segregation of the West Indian-born population decreased dramatically in London, Birmingham, and Coventry between 1961 and 1971 and was accompanied by an increase, albeit slight in Coventry's case, in the Asian-born levels of segregation. In Oxford Asian segregation rose as markedly as in Birmingham but the decline in West Indian was almost insignificant.

Not only are the general levels of the indices of dissimilarity of the black population in British towns lower than those commonly found in the United States, but the degree of concentration in areas in which they are concentrated is generally much lower. The highest percentage which persons born in the New Commonwealth formed of any London ward was 54%; in a Coventry enumeration district it was 68%, and in a Birmingham enumeration district it was 61%. But such figures can be very misleading since that same Birmingham enumeration district had 91% of its population born to parents both of whom were born in the New Commonwealth (i.e., the coloured children of immigrant parents are included in this percentage). This compares with situations of nearly 100% in the densest areas of black settlement in U.S. cities (see, for example, Kantrowitz, 1969b).

Similarly, the extent to which the various black populations are found in areas in which the group forms a high percentage of the population is much lower than is general in the U.S. Duncan and Duncan (1957) demonstrate that in Chicago in 1950, 80% of the Negro population lived in areas that were over 75% black. With the

Taeubers' (1965) demonstration that the median index of dissimilarity for the Negro population in 207 U.S. cities in 1960 was 86.7, such high degrees of concentration which the Duncans demonstrated must have been common. In English cities, however, the percentage of the black population which is clustered is generally much lower. West Indians, for example, show twin characteristics of clustering and dispersal (Glass, 1960). In London in 1961 less than 15% of West Indians were living in arbitrarily defined dense clusters (Glass and Westergaard, 1965). In Birmingham the concentration was higher (30.5%), but in both cases about a third of the West Indians were living in scattered distributions (Peach, 1968: 88). In both London and Birmingham in 1961 the Irish showed similar proportions of clustering and dispersal (ibid.).

In 1971 on the basis of enumeration districts for Coventry West Indians were absent from 41.4% of the areas, Indians from 33.2%, and Pakistanis from 74.9% of the districts.

Much of the interest in the 1971 census centres on an examination of the extent to which the black population has been able to penetrate the white periphery as opposed to the extent to which existing concentrations have increased. In the terms of Duncan and Duncan (1957) the interest centres on the balances between "penetration" and "piling-up."

It is at this stage that the contrasts between the contemporary American and British situations seem to diverge. While the major American cities seem to have reached a stage of black saturation segregation (so high that it is difficult to increase), the British situation, at least as far as West Indians are concerned, is one of decreasing segregation. The trend that was increasing from 1961 to 1966 (Jones, 1970) has been reversed both in terms of the 1966-1971 comparison and in terms of 1961-1971.

SOCIAL AND SPATIAL DISTANCES

Although, in terms of our social distance interpretation, this should represent a closer adjustment of West Indian and British societies, the sociological evidence is more ambivalent and the alienation of some British-born black youths is noticeable (Community Relations Commission, 1974).

Two of the most important areas in which differences between the black and the white populations might be expected are in job types and housing standards. However, before these can be examined in detail, it should be noted that the age structure of the populations differs very markedly. Since age is closely associated with many other variables, direct black-white comparisons are not of direct value in explaining how far colour accounts for variation.

The black population had a notably younger age distribution than the white. Forty-one percent of the black population were aged 14 or less, compared with 24% of the white population. In the age group 25 to 39 were another 26% of the black population, compared with 19% of the white. Only 4% of the black population was aged 60 or over, compared with 19% of the whites (G.H.S., 1973: 80).

In terms of household size and amenities there were marked differences between the black and white groups. More than half of the households described as "coloured" by the G.H.S. had four or more persons compared with less than a third of the white households. Not only were black households larger than average, they were more crowded. Using a system of norms, the G.H.S. estimated that 23% of black households had fewer bedrooms than the standard, compared with 6% of white households of U.K.-born persons. Thirty-four percent of black households had bedrooms in excess of the standard, compared with 60% of U.K.-born white headed households (G.H.S., 1973: 144). Similarly, only 23% of New Commonwealth black households not in shared accommodation were living at low densities (less than half a person per room), compared with 39% of the Irish (North and South) and 53% of the British-born heads. At the other end of the scale, 18% of the households headed by a black born in the New Commonwealth were living at a density of one person or more, compared with 14% of the Irish and 5% of the British (G.H.S., 1973: 145, Table 5.47). Twenty-one percent of black households shared a bathroom and 25% a lavatory, compared with 2% and 3% respectively for whites born in the U.K. (G.H.S., 1973: 145).

Although the black population has been treated as if it were homogeneous, the attitudes of the West Indians and Asians are sharply different. The West Indians of the first generation, at least, could be viewed as coming from an assimilationist background (Patterson, 1963: 15-16). They were reared with English history and patriotism as their own (see, for example, Glass, 1960: 3-4). This does not mean that there were not significant differences in their attitudes from those of the British, nor that they did not desire to return to the West Indies. Nevertheless, their degree of identification with British society was high. For the Indians and Pakistanis, however, England was not the central reference point. That point was their home society.

England represented a means to betterment at home, not a place in which to be comfortable. While both West Indians and Asians professed a desire to return, this determination seems, from external evidence, to be stronger in the case of the Asians than the West Indians. The West Indians seemed, from a very early date, to be prepared to bring their womenfolk and children to Britain. In 1971 the sex composition of the Asian population was still strongly male-dominated, particularly for Pakistanis, and all-male households were common:

India	175,305 males;	146,885 females
Pakistan	100,200 males;	39,005 females
West Indies	151,465 males;	150,870 females

Having the home society as the dominant reference point for the Asian population produced a number of spatial consequences. Regional conflicts in India and Pakistan were reflected in spatial separation of the groups from each other in British cities. The larger the groups became, the more easy it was for the individual groups to become self-sufficient as communities and thus to become spatially separate. The early Asian population of Birmingham during World War II was a mixture of Indians and (what were to become after 1947) Pakistanis, who lived in the same hostels and areas. As the group increased in size, the Muslim Pakistanis and the Indians separated.

The dominance of the home society as the reference point for the Pakistani population means that spending in Britain is kept to a minimum and the maximum of what is disbursed in Britain is spent within institutions which are part of the ethnic group. Ownership is preferred to renting—council housing is thus considered a waste of money (Dahya, 1974: 97). Cheap housing is preferred to expensive housing, and geographical concentration, which allows contact and facilitates the support of ethnic organizations and shops and reinforces ethnic values, is seen as a positive virtue (Dahya, 1974: 94-95, 97; for a contrary view, however, see Rowley and Tipple, 1974). Thus, the clustering of the Pakistani population, its high degree of underrepresentation in council housing, and the low standards of amenities enjoyed seem to have arisen through the positive desire of the immigrant group for these things and not through discrimination on the part of the white society (Dahya, 1974: 102-103). The cheapness of the West Riding housing is perhaps one of the reasons for the concentration of this group in that area (Burney, 1967: 35). It should be pointed out, however, that these observations apply only to the working-class Indians and Pakistanis. Middle-class Indians and Pakistanis, such as doctors, conformed to the housing and locational norms of the white middle classes (Dahya, 1974: 105).

SUMMARY AND CONCLUSIONS

Thus, the British situation differs at the moment very markedly from that in America. There is no ghetto in the American sense. The percentage of the black population living in areas where they form a majority of the population is low.

Although the black population is not yet found in every subunit of the major cities, about a third live in areas in which they form a small proportion of the total population. Thus, both concentrations and dispersal characterize the distribution of the black population in British cities, but the concentrations are not exclusively black and the ability of the population to spread is evident from the evidence of changes between the 1961 and 1971 censuses. From the bulk of the evidence it seems that concentration is the result of voluntary rather than enforced distributions.

The positions of the different elements in the black British population contrast in important ways. The Asian population—the Pakistanis in particular—are inward-looking and tightly-knit social groups, tending more to tight spatial clustering and focused on their institutions such as mosques, temples, shops, and cinemas which bind their communities together. They have achieved a high degree of residential dissimilarity which has remained high in some cities (such as Bradford) or increased in other places (such as London, Birmingham, and Coventry). However, as their sex ratio changes to a more balanced community, there may be changes in the spatial distribution also. Certainly there is evidence that the distribution of the highest socioeconomic members of these groups conform to the distribution pattern of the white members of their class rather than the spatial pattern of their own ethnic group.

The West Indians, on the other hand, show a higher degree of assimilation into the British patterns of life. They are, paradoxically, more segregated than the Indians and Pakistanis in London, but their degree of separation is decreasing. The changes in their distribution in relation to the centres of some British cities show remarkable shifts toward the suburban peripheries over time. However, there may be dangers that a more alienated second generation may not pursue this trend.

Perhaps the most critical way in which the British and American situations differ is in the availability of local authority housing. With one-third of the housing controlled by local municipalities and with a substantial section of the suburban peripheral housing development controlled by the council's allocation of points, rather than the money market and income, Britain possesses, in incipient form, the most powerful levers for the sociospatial engineering of society. This power brings with it a need to be clear about our understanding of the social significance of the spatial concentrations of minorities. The conclusion must be that ghettos are bad but that concentration of minorities is not necessarily so. The key distinction is that in ghettos there is an element of compulsion and prevention of dispersal. As long as concentrations represent the positive desire of individuals to cluster together and as long as there are no special impediments to their dispersal should they wish it, pressure for the break-up of such groupings would be counter-productive. The Cullingworth Committee Report (1969), which is the sanest and most humane of documents, nevertheless concluded that dispersal should be a *policy* but that there should not be coercion or special treatment for black minorities. The black population was more than proportionally concentrated into areas of multi-deprivation, but the help must be given to the *area*, not particularly to the black element of the disadvantaged population. Dispersal might be a desirable by-product of assistance, but not its aim.

REFERENCES

ABBOTT, S. [ed.] (1971) The Prevention of Racial Discrimination. London: Oxford University Press [for the Institute of Race Relations].

ALLEN, S. (1971) New Minorities, Old Conflicts: Asian and West Indian Migrants in Britain. New York: Random House.

BAGLEY, C. (1970) Social Structure and Prejudice in Five English Boroughs. London: Institute of Race Relations.

BANTON, M. (1959) White and Coloured. London: Jonathan Cape.

——— (1955) The Coloured Quarter: Negro Immigrants in an English City. London: Jonathan Cape.

Barbados, Government of (1962) Report for the Years 1960 and 1961. Bridgetown: Government Printing Office.

BAYLISS, F. J. and J. B. COATES (1965) "West Indians and work in Nottingham." Race 7, 2.

BHATNAGAR, T. (1970) Immigrants at School. London: Cornmarket.

BROWN, J. (1970) The Un-melting Pot. London: Macmillan.

BURNEY, E. (1967) Housing on Trial. London: Oxford University Press [for the Institute of Race Relations].

BUTTERWORTH, E. [ed.] (1967) Immigrants in West Yorkshire. London: Institute of Race Relations Special Paper.

CHEETHAM, J. (1972) "Immigration," ch. 14 in A. H. Halsey (ed.) Trends in British Society Since 1900. London: Macmillan.

COATES, B. E. (1968) The Distribution of the Overseas Born Population of the British Isles. Transactions of the Institute of British Geographers 43: 37-43.

COLLINS, C. (1971) "The distribution of New Commonwealth immigrants in Greater London." Ekistics 32: 12-21.

COLLINS, S. (1957) Coloured Minorities in Britain. London: Butterworths.

COLLISON, P. (1967) "Immigrants and residence." Sociology 1, 3: 276-293.

"Commonwealth Immigrants in the Conurbations." Census of England and Wales, 1961. London: H.M.S.O.

Community Development Council (1970) "Coventry" (1 in 6 household survey)—unpublished.

Community Relations Commission (1974) Unemployment and Homelessness: A Report. London: H.M.S.O.

CULLINGWORTH, J. B. [chairman] (1969) Council Housing: Purposes, Procedures and Priorities. Ninth Report of the Housing Management Sub-Committee of the Central Housing Advisory Committee, Ministry of Housing and Local Government. London: H.M.S.O.

DAHYA, B. (1974) "Pakistani ethnicity in industrial cities in Britain," in A. Cohen (ed.) Urban Ethnicity. London: Tavistock.

DALTON, M. and J. M. SEAMAN (1973) The Distribution of New Commonwealth Immigrants in the London Borough of Ealing 1961-1966. Transactions of the Institute of British Geographers 58: 21-39.

DANIEL, W. W. (1968) Racial Discrimination in Britain. London: Penguin.

DAVIES, J. G. and J. TAYLOR (1970) "Race, community and no conflict." New Society 406 (July 9): 67-69.

DAVIES, P. and K. NEWTON (1972) "The social patterns of immigrant areas." Race 14: 43-57.

DAVISON, R. B. (1966) Black British: Immigrants to England. London: Oxford University Press [for the Institute of Race Relations].

—— (1963) "The distribution of immigrant groups in London." Race 5, 2: 56-69.

—— (1962) West Indian Migrants. London: Oxford University Press [for the Institute of Race Relations].

DEAKIN, N. (1970) Colour Citizenship and British Society. London: Panther.

—— (1969) Race and Human Rights in the City. Urban Studies 3 (November Special Issue).

—— (1964) "Residential segregation in Britain: a comparative note." Race 6: 18-25.

—— and B. COHEN (1970) "Dispersal and choice: towards a strategy for ethnic minorities in Britain." Environment and Planning 2: 193-201.

DEAKIN, N. and C. UNGERSON (1973) "Beyond the ghetto: the illusion of choice," in London: Urban Patterns, Problems and Policies. London: Heinemann [for the Centre for Environmental Studies].

DESAI, R. (1963) Indian Immigrants in Britain. London: Oxford University Press [for the Institute of Race Relations].

DOHERTY, J. (1973) "Immigrants in London: a study of the relationship between spatial structure and social structure." Ph.D. thesis, University of London.

—— (1969) "The distribution and concentration of immigrants in London." Race Today 1, 8: 227-231.

DUKE, C. (1971) Colour and Rehousing: a Study of Redevelopment in Leeds. London: Institute of Race Relations Research Publications.

FOOT, P. (1965) Immigration and Race in British Politics. London: Penguin.

DUNCAN, O. D. and B. DUNCAN (1957) The Negro Population of Chicago. Chicago: University of Chicago Press.

GENERAL Household Survey (G.H.S.) (1973) London: H.M.S.O.

GEORGE, V. and G. MILLERSON (1967) "The Cypriot community in London." Race 8, 3: 277-293.

GLASS, R. (1960) Newcomers: the West Indians in London. London: Bell.

FITZHERBERT, K. (1967) West Indian Children in London. London: Bell.

GRIFFITH, J.A.G. et al. (1960) Coloured Immigrants in Britain. London: Oxford University Press [for the Institute of Race Relations].

—— and J. WESTERGAARD (1965) London's Housing Needs. Centre for Urban Studies, Report No. 5. London: University College.

GRIFFITHS, P. (1966) A Question of Colour? London: Leslie Frewin.

HADDON, R. (1970) "A minority in a welfare state: location of West Indians in the London housing market." New Atlantis 2, 1: 80-123.

HARRISON, P. (1974) "The patience of Southall." New Society (April 4): 7-11.

HEPPLE, B. (1968) Race, Jobs and the Law in Britain. London: Allen Lane-Penguin.

HILL, C. S. (1971) Immigration and Integration. Oxford: Pergamon.

HIRO, D. (1971) Black British, White British. London: Eyre & Spottiswoode.

HUMPHRY, D. (1972) Police Power and Black People. London: Panther.

—— and G. JOHN (1971) Because They're Black. London: Penguin.

HUMPHRY, D. and M. WARD (1974) Passports and Politics. London: Penguin.

JACKSON, J. A. (1963) The Irish in Britain. London: Penguin.

JEFFREY, P. (1972) "Pakistani families in Bristol." New Community 5 (Autumn).

JENKINS, S. (1971) Here to Live: a Study of Race Relations in an English Town. London: Runnymede Trust.

JOHN, A. (1970) Race in the Inner City. London: Runnymede Trust.

JOHN, De W. (1969) Indian Workers' Association in Britain. London: Oxford University Press [for the Institute of Race Relations].

JONES, E. and D. J. SINCLAIR (1968) Atlas of London. London: Oxford University Press.

JONES, H. R. and M. DAVENPORT (1972) The Pakistani Community in Dundee. Scottish Geographical Magazine 88, 2: 75-85.

JONES, K. and A. D. SMITH (1970) The Economic Impact of Commonwealth Immigration. Cambridge, Eng.: Cambridge University Press.

JONES, P. N. (1970) Some Aspects of the Changing Distribution of Coloured Immigrants in Birmingham 1961-1966. Transactions of the Institute of British Geographers 50: 199-219.

——— (1967) "The segregation of immigrant communities in the city of Birmingham, 1961." University of Hull, Occasional Papers in Geography, No. 7.

JONES, T. P. and D. McEVOY (1974) "Residential segregation of Asians in Huddersfield." Delivered at the annual conference of the Institute of British Geographers at Norwich.

KANTROWITZ, N. (1969a) "Ethnic and racial segregation in the New York Metropolis, 1960." American Journal of Sociology 74: 685-695.

——— (1969b) Negro and Puerto Rican Population of New York City in the Twentieth Century. New York: American Geographical Society.

KATZNELSON, I. (1973) Black Men, White Cities. London: Oxford University Press [for the Institute of Race Relations].

KINDLEBERGER, C. P. (1967) Europe's Postwar Economic Growth: The Role of Labour. Cambridge: Harvard University Press.

KING, J. (1972) "Immigrants in Leeds: an investigation into their changing spatial distribution." University of Leeds, Department of Geography, unpublished paper.

KOHLER, D. (1973) Facts and Figures About Commonwealth Immigrants. London: Community Relations Committee.

KRAUSZ, E. (1972) Ethnic Minorities in Britain. London: Paladin.

LAMBERT, J. R. (1970) Crime, Police and Race Relations. London: Oxford University Press [for the Institute of Race Relations].

——— and C. J. FILKIN (1971) "Race relations research: some issues of approach and application." Race 12, 3: 329-335.

LAWRENCE, D. (1974) Black Migrants—White Natives: A Study of Race Relations in Nottingham. Cambridge, Eng.: Cambridge University Press.

LEE, T. R. (1973a) "Concentration and dispersal: a study of West Indian residential patterns in London, 1961-1971. Ph.D. thesis, University of London.

——— (1973b) "Ethnic and social factors in residential segregation: some implications for dispersal." Environment and Planning 5: 477-490.

——— (1973c) "Immigrants in London: trends in distribution and concentration 1961-1971." New Community 2, 2: 145-158.

——— (1972) "Socio-economic considerations in the residential segregation of ethnic and racial groups." London School of Economics, Graduate School of Geography, Discussion Paper No. 43.

LEECH, K. (1966) "Migration and the British population." Race 7: 401-408.

LIEBERSON, S. (1963) Ethnic Patterns in American Cities. New York: Free Press.

LESTER, A. and C. BIRDMAN (1971) Race and Law. London: Penguin.

LITTLE, K. L. (1948) Negroes in Britain. London: Routledge and Kegan Paul.

NG K. C. (1968) The Chinese in London. London: Oxford University Press [for the Institute of Race Relations].

PARK, R. E. (1926) "The urban community as a spatial pattern and a moral order," in E. W. Burgess (ed.) The Urban Community. Chicago: University of Chicago Press.

PATTERSON, S. (1971) "Immigrants and minority groups in British Society," in S. Abbot (ed.) The Prevention of Racial Discrimination in Britain. London: Oxford University Press [for the Institute of Race Relations and the U.N. Institute for Training and Research].

—— (1969) Immigrants and Race Relations in Britain 1960-1967. London: Oxford University Press [for the Institute of Race Relations].

—— (1968) Immigrants in Industry. London: Oxford University Press [for the Institute of Race Relations].

PEACH, G.C.K. (1968) West Indian Migration to Britain: A Social Geography. London: Oxford University Press.

—— (1967) "West Indians as a replacement population in England and Wales." Social and Economic Studies 16, 3: 289-294.

—— (1966) "Factors affecting the distribution of West Indians in Great Britain." Transactions of the Institute of British Geographers 38: 151-163.

—— (1965) "West Indian migration to Britain: the economic factors." Race 7: 31-46.

—— and S.W.C. WINCHESTER (1974) "Birthplace, ethnicity and the underenumeration of West Indians, Indians and Pakistanis in the censuses of 1966 and 1971." Oxford University School of Geography, unpublished research paper.

PLANT, M. A. (1971) "The attitudes of coloured immigrants in two areas of Britain to the concept of dispersal." Race 12, 3: 323-328.

Political and Economic Planning (P.E.P.) (1967) Report on Racial Discrimination. London: Political and Economic Planning.

REUBENS, E. P. (1971) "Our urban ghettoes in British perspective." Urban Affairs Quarterly 6, 3: 319-340.

REX, J. (1968) "The social segregation of immigrants in British cities." Political Quarterly 39.

—— and R. MOORE (1967) Race, Community and Conflict: A Study of Sparkbrook. London: Oxford University Press [for the Institute of Race Relations].

RICHMOND, A. H. (1973) Migration and Race Relations in an English City: A Study of Bristol. London: Oxford University Press [for the Institute of Race Relations].

—— (1970) "Housing and racial attitudes in Bristol." Race 12, 1: 49-58.

ROSE, E.J.B. et al. (1969) Colour and Citizenship: A Report on British Race Relations. London: Oxford University Press [for the Institute of Race Relations].

ROWLEY, G. and G. TIPPLE (1974) "Coloured immigrants within the city: an analysis of housing and travel preferences." Urban Studies 11: 81-89.

Select Committee on Race Relations and Immigration (1971) Housing I: Report. London: H.M.S.O.

SHYLLON, F. O. (1974) Black Slaves in Britain. London: Oxford University Press [for the Institute of Race Relations].

SMITH, D. (1974) Racial Disadvantage in Employment. London: Political and Economic Planning and the Social Science Institute.

TAEUBER, K. E. and A. F. TAEUBER (1965) Negroes in Cities. Chicago: Aldine.

—— (1964) "The Negro as an immigrant group: recent trends in racial and ethnic segregation in Chicago." American Journal of Sociology 69: 374-382.

THOMAS, B. (1954) Migration and Economic Growth. Cambridge, Eng.: Cambridge University Press.

THOMAS, C. J. (1970) "Projections of the growth of the coloured immigrant population of England and Wales." Journal of Biosocial Science: 265-281.

TOMMIS, S. (1974) "Urban residential segregation and an immigrant community." M.A. thesis, University of Dundee, Department of Geography.

WARD, C. (1971) Coloured Families in Council Houses: Progress and Prospects in Manchester. Manchester, Eng.: Manchester Council for Community Relations.

WATERHOUSE, J.A.H. and D. H. BRABBAN (1964) "Inquiry into the fertility of immigrants: preliminary report." Eugenics Review 56.

WILKINSON, R. K. and S. GULLIVER (1971) "The impact of non-whites on house prices." Race 13: 21-36.

WILLCOX, W. F. (1931) International Migrations 2. New York: National Bureau of Economic Research.

WINCHESTER, S.W.C. (1973) "Immigration and the immigrant in Coventry: a study in segregation." Presented at the conference of the Institute of British Geographers.

WOODS, R. I. (1973) "The role of simulation in the modelling of immigrant spatial sub-systems: an application to Birmingham." Presented at the conference of the Institute of British Geographers.

——— (1971) "Coloured immigrants and the 'zone of transition' in Birmingham." B.A. dissertation, Cambridge University.

WRIGHT, P. (1968) The coloured workers in British industry. London: Oxford University Press [for the Institute of Race Relations].

Part V

POVERTY AND RESOURCES: ANALYSIS AND MOBILIZATION

Introduction

□ THIS SECTION IS CONCERNED with the relation between the distribution of resources and the conditions of urban poverty—a topic foreshadowed in the previous section on distressed urban populations. What is the relationship between the distribution of economic progress and the manifestations of economic marginality? Another major question of urban concern is whether a rise in the absolute material standard of living in low-income communities is likely to be more "effective" than an attempt to deal realistically with the conditions which prevail in that community (short of any major commitment to eradicate poverty). Or is it more desirable to focus on a narrowing of the boundaries of economic inequality within a metropolian area by focusing on redistribution mechanisms of several kinds, both local and federal (Rein and Miller, 1974)?

Although it is obviously important to raise the absolute material standard of living for those who live in malnourished and under-sheltered conditions, it is the problems of relative economic inequality and deprivation which present a much more perplexing series of questions for urban social and economic policy. After "floors" have been raised, and after more "doors" to opportunity have been opened, how much inequality will persist and what will it mean? What happens to those left behind? Does not being an economic "winner" condemn one to being a social "loser" (Gappert, 1973)?

There are those who suggest that there is a social underclass whose habits and behavior contribute to their economic failure. Others have sketched out a notion of a "culture of poverty" containing self-perpetuating, psychological elements which create a social system effectively isolated from the larger external socioeconomic system (Leacock, 1971).

Others have developed more of an ecological model of urban deprivation and behavior. This approach suggests that poverty is a consequence of factors within the national urban system in which poor people are "encapsulated" and is not due to factors endogenous to the poor. This approach seeks to explain the adaptation which a population makes to its environment by an analysis of the interrelationships, interaction, and feedback between the variables which influence and regulate adaption (Kurtz, 1975). As Rainwater (1969) has written:

> one can hope that as a result of social science efforts to date, thinking people will stop deluding themselves that the underclass is other than the product of an economic system so designed that it generates a destructive amount of income inequality, and face the fact that the only solution of the problem of underclass is to change that economic system accordingly.

If it is accepted that the resources which the poor are able to marshal are severely restricted, it then is clear that the conditions associated with such mobilization becomes the necessary focus for both analysis and policy development. Associated with this focus on resource mobilization is a concern with a group's perceived needs and expectations and the controls over the distribution of resources and access to them.

The existence of the social and cultural attributes which have been more or less associated with poverty (such as the matrifocal household, patterns of shopping, child labor, consensual marriage, and so on) represent adaptions to the paucity of resources and the objective external circumstances which restrict the ability of the community to mobilize any additional resources.

In Table 1, a colleague has provided a system perspective on so-called poverty traits. He distinguishes between the "effectors" of poverty (controlled externally) and the internal adaptations and

TABLE 1
THE SYSTEMIC NATURE OF POVERTY TRAITS

Effectors	Adaptations	Manifestations
Unemployment and underemployment	Child labor	*Sociocultural*
Low wages	Working women	Absence of food reserves in home
Miscellany of unskilled occupations	Patterns of frequently buying small quantities as need arises	No bank credit
Low level of education	Pawning	Lack of privacy, gregariousness
	Informal credit devices	No doctors, hospitals
	Free unions	Partial integation into national institutions
	Consensual marriages	Absence of savings
	Trend to matrifocal household	Chronic shortage of cash
	Reliance on home remedies, curers	
		Psychological
		Martyr complex of women
		Critical attitude toward values and institutions of dominant class
		Mistrust of government and those in high positions
		Cynicism which extends even to church
		Feelings of marginality, helplessness, alienation, etc.
		Feelings that existing institutions do not serve their interests and needs

SOURCE: Kurtz (1975).

manifestations. "Adaptations" to poverty are self-regulating mechanisms by which the poor maintain a viable relationship with their habitat and the larger national-urban system. "Manifestations" are the social, cultural, and psychological consequences of the adaptations. These manifestations are often the source of the demands voiced from the poverty habitat across the boundary to the external metropolitan system.

In this section are assembled four different perspectives on the problems of analyzing and responding to the manifestations associated with the conditions of economic deprivation within an urban society. The conceptual ways by which we observe the "problem" become an important determinant of how resources are mobilized and allocated in the quest for a "solution."

The Greers in their paper explore the significance of an important question: what is the future of social class in a society with an egalitarian ideology? Their evolutionary analysis of the emergence of occupation-based class structure is related to changes in organizational size and the shift to a larger scale of the networks of functional social interdependence. Although they view class as a multidimensional phenomenon, it is suggested that the role-system in a metropolitan society is increasingly governed by bureaucratic considerations. The need for impersonal screening by credentials lowers a "diploma curtain" that creates both the underemployed and unemployed. Real achievement and competence becomes less important than other attributes such as sex, group membership, interpersonal skills, and conformity-seeking behavior.

Their analysis provides a perspective that suggests that urban, postindustrial society will be characterized by (a) relative deprivation among much of the college educated, white-collar occupation groups; (b) attempts by the national societal leaders to convince their electorate either that the "system" is roughly fair or that minor reforms can make it so; and (c) the proliferation of occupational counter-cultures in which the existential commitment to "playing the game"–a game in which there are many small winners and losers—becomes a substitute for higher class status or income position. At the same time, as the national economy enters a period of economic decline, there will be relative deprivation for many social categories and the prospect of a rising social bitterness generally, with a more rapid turnover in governing elites.

Lowenthal provides a different perspective. He indicates that the struggle for survival by working-class families has always been a difficult one. He explores this question: how do people with inadequate incomes meet their needs and manage to survive? His analysis argues that many critical needs are provided for, not through money income, but by the social networks in which people participate. Social support systems within working-class communities are just as important to survival as family incomes and the availability of governmental services.

This analysis suggests that governmental services, both local and federal, need to be assessed as to whether they tend to reinforce family and community networks or whether they tend to undermine those necessary activities characteristic of the informal helping systems.

The central importance of locally based social relationships must be recognized by the makers of urban policies and the writers of eligibility requirements. Equally significant in the analysis is the realization that the organization and evolution of the urban industrial economy has depended upon the existence of the family and the community and the economic functions they perform. The evolution of these socioeconomic functions will be equally critical to the development of the postindustrial metropolitan economy. Lowenthal points out that "cohesive stable working-class communities have generally developed intricate and complex systems of reciprocal arrangements which can be effective as redistribution mechanisms within the community." It may well be that transitory, white-collar suburbs will, in a time of general decline of real income, be the most in need of such helping networks. Class and occupational differences in social network behavior have been rarely investigated and even less seldom used in the planning of housing, community facilities, and transportation routes. The recreating of networks of obligation, cooperation, and shared experiences among those of the middle class who will be subjected to the vicissitudes of economic dislocation may well be one of the organizing issues of the 1970s and 1980s.

A different perspective is provided by Blair, Gappert, and Warner. They examine the idea of the "grants economy" as a construct by which urban problems and policy can be rethought and redirected. The notion of "granting," of one-way transfers of income or wealth, was introduced to economics by Kenneth Boulding. He has suggested that any complex social system represents a combination of different organizing relationships. Social granting has arisen in the modern Western economy to offset some of the disruptive effects of "free market" exchange relationships.

In the paper written for this section, the notions of the grants economy are placed within an urban context. The analysis of organizational transfers by major urban institutions is discussed. An ideal political-economic theory of grants would postulate the regulation of the urban market economy by a set of transfers; a complimentary social-economic theory of the grants economy would postulate the support of a series of low-income urban lifestyles by a different set of directed transfers. "Granting," out of either love or fear, might be a more efficacious, if not humane, way of dealing with the persistence of the urban subeconomy. The authors suggest that a well-functioning city or metropolitan system attempts to integrate itself by ameliorating the problems of its subeconomies through a series of grants, both private and public.

One of the more publicized urban granting efforts has been the involvement of the nation's major corporations with urban projects over the last decade. Jules Cohn (1971) has reviewed these trends, and the more significant of these developments have been summarized in his article for this volume. Cohn's analysis of the several directions taken by corporate programs emphasizes their paradoxical nature: if the programs are too successful, their very success cause conflict with other objectives; if the programs are not successful, the primary objective of corporations—to make a profit for their stockholders—comes under increased attack.

Even if we accept that the cities shall have the poor with them always, it should be possible to project policies which could make the circumstances of a low-income life a less punishing experience. Additional questions, however, remain unanswered.

(1) From a gini coefficient of .56 for the SMSA of Newark to a coefficient of .29 for the Cedar Rapids SMSA, how important is the extent and nature of income inequality to the quality of urban life in the different metropolitan areas? Can the urban community ameliorate the manifestations of the unequitable distribution of resources within its own midst?

(2) Would a change in the "rules" of the national socioeconomic system (i.e., tax reform) have a more dramatic effect on the condition within cities? What are the separate but related responsibilities of the national and regional cultures in dealing with social and economic injustice?

(3) Where should the emphasis be—on the services to enable low-income people to compete more effectively for those opportunities which do exist, or on services to those who must continue to occupy the communities with insufficient resources? And who should control the decisions about the nature and distribution of those services?

REFERENCES

COHN, J. (1971) The Conscience of the Corporation: Business and Urban Affairs. Baltimore: Johns Hopkins University Press.

GAPPERT, G. (1973) "The future of economic inequality and the planning of urban service." Journal of the American Institute of Planners (May).

KURTZ, D. V. (1975) "The ecology of poverty." (unpublished manuscript)

LEACOCK, E. B. [ed.] (1971) The Culture of Poverty: A Critique. New York: Simon & Schuster.

RAINWATER, L. (1969) "The American underclass." Trans-action 6, 4.

REIN, M. and S. M. MILLER (1974) "Standards of income distribution." Challenge (July/August).

13

Urban Work in a Changing Economy

SCOTT GREER
ANN LENNARSON GREER

☐ A HUMAN SOCIETY may be conceived of as a man-machine system in an environment. As such, it exists reciprocally with the environment, drawing energy from it and returning energy to it, using up a quantum of energy in the process through the upkeep costs and wastage. The internal structure of the society turns energy to human purposes, however defined. The Aztec emperors used their relatively vast controls of energy for the purpose of ripping out the hearts of thousands of sacrificial victims annually, believing that this prevented the doom of the world. Today we stockpile thermonuclear devices under some similar illusion. Thus a simple "functional prerequisite" view of society will not do; we must keep in mind the values created and the actions permitted by the basic economic structure of the society. For the basic prerequisite becomes dependent upon actions and products that are clearly not prerequisites for survival; millions of new automobiles built and sold every year are not a functional prerequisite of our society, yet as Detroit slows production the economy slumps abruptly. What the man-machine system does is to provide options, and limits, for the exercise of human purpose.

In the last few centuries, there has been an enormous increase in the scale of the man-machine systems. The number of people involved in the networks of interdependence has grown at staggering rates in most societies—in the United States, from three to well over two hundred millions in less than two hundred years. The amount of energy processed has increased at an even more spectacular rate. Whether measured by the number of people involved in the same system, or the amount of energy processed, the United States today is among the largest in scale of any society humanity has ever organized.

This increase in scale has been due to changing technology, social organization, and labor force in the man-machine system. The direction of technological change has been unilinear—the increase in the amount of non-human energy made accessible to humanity through *energy converters*. Thus we have moved from chlorophyll and animal muscles as converters to wind, in the age of the commercial empires; coal, in the early years of the industrial revolution; and finally oil and gas in the period of "mature capitalism." Each step has greatly increased the amount of energy available for organized human communities to use in pursuing their purposes (Cottrell, 1955).

In order to exploit the changing technology, we have organized our work in new and different ways. Wilbur Moore (1946) has noted the very old prototypes of the factory system, the gradual evolution of industrial society from roots as old as urban man. What is new, however, is the sheer magnitude of the change. While there are hundreds of thousands of enterprises in the United States today, a thousand of them control an overwhelming majority of the market.[1] The social organization of our society is characterized by giantism. There are corporations such as General Motors, unions such as the United Auto Workers, which with their dependents would make respectable-sized members of the United Nations, even as the State of California has a gross national product twice that of the entire Indian nation. California is, predictably, organized in much larger networks of interdependence than is the predominantly peasant society of the subcontinent.

Concurrent with these shifts in *scale* is a change in the *nature of* the *labor* market. There is no such thing as a "labor market" in a peasant society; work in such societies is organized by status—age, sex, lineage, or what-not. The labor market emerged with the industrial revolution. However, treating labor as a commodity was cognate with the social organization of large-scale societies. Marx deplored it in *Das Kapital*, Adam Smith legitimated it in the *Wealth of Nations*; the key point is that it happened. The man-machine-in-environment systems called corporations were geared to market thinking; input, process, output to the market, and surplus—or profit. So instead of inheriting work-roles as appropriate to one's sex, age, and lineage, humanity went on the market to sell its labor. That market was extended, with growing demand for workers, to include the young, the old, the ethnic minorities and the second sex. Indeed, citizenship in the fullest sense tended to become associated with participating in the labor force and the market (Form, 1968).

Both of these changes (organizational size and the creation of the labor market) associated with increasing scale had the effect of changing the distribution of power, rewards, and social honor—in short, creating "social class." For we follow Weber in seeing class as a multi-dimensional variable, dealing with political privilege, economic privilege, and honorific privilege. Sometimes in human history the three have been so highly intercorrelated that they could be considered unitary; at other times one has to deal with them separately and leave their relationships problematic, to be determined by empirical research. We suspect that, on the whole, our era is such a time (Weber, 1946).

The remainder of this paper will deal with (1) the importance of occupation in large-scale society; (2) the effects of increasing scale on economic and political status; (3) short-run and long-run antici-pations of the future; (4) the politics of a highly differentiated but large-scale society; and (5) some speculations about desirable futures. What is the future of social class in a society with an egalitarian ideology? What could, and what should, be done about it?

I. THE IMPORTANCE OF OCCUPATION

As we have noted, the increase in overall societal scale is both cause and consequence of increasingly large-scale sub-organizations. These include massive economic, political, religious, leisure, and other bureaucracies: more behavior is channelled through fewer social groups. In the same way, large-scale society is pervasively *urban*. More people live in fewer, very large, metropolitan areas.

Furthermore, through mass media, the mass market, and mass tourism, the urban world permeates the small towns and open country neighborhoods. The approximate one-third of the United States population living in the latter give a deceptive impression of the persistence of the old pre-metropolitan society; in fact, many of these smaller cities are pseudopods of dominant metropolises. They may be resort towns, university towns, state capitals, or other specialized centers of the national grid of cities. Aspen, Colorado; Carbondale, Illinois; Olympia, Washington—these share many attributes with the specialized sub-areas of the metropolis. They just happen to be farther away from the center of density.

And even when the smaller centers are "traditional" small towns, engaged in providing the market between primary producers and the larger system, they are hardly free of that system. They are deeply implicated through the national commodities and money markets, the national government and its giant bureaucratic agencies, and perhaps most important, the national communications flow. Messages of the mass media, presented across the board to almost all citizens of the society, originate in metropolitan areas and are controlled by the interests and preferences of those at the center. Small-town America learns from Hollywood, New York, and lesser centers of mass culture how to view itself, its occupations, and achievements; the increasingly uniform culture of the United States is urban in origin, direction, and interpretation.

We have emphasized these points in order to make clear to the reader our use of societal scale to explore "the changing economic structure and social class system of the city." "The city," or rather, a national network of cities, is most of the contemporary United States. Further, so much of the concept of class implies a cultural definition that the homogenization of our culture through the media is an important aspect of class today. That culture has a fluidity, a volatility, unknown in the past: custom becomes fad; the given changes; our shared world of meanings and values is in constant flux. Keeping this in mind we shall examine what has happened to the occupational structure as society increases in scale.

A person's occupation is a role. It determines and limits rights and duties, belief systems, and normative systems. It is in turn dependent upon access to the job and ability to perform at whatever is the minimal level required to keep the job. As a role it is part of a role-system, a way of ordering an aggregate of persons so as to produce a finished product that has survival value for the role-system and the role performer.

The dominant type of role-system in large-scale economic structures is some form of the bureaucratic. In their efforts to control behavior, the managers try to separate the position from the individual occupant so that performance may be objectively evaluated. Management also tries to define clear lines of authority so that each actor may be accurately evaluated, each success or failure attributed to the correct actor. Finally, records are kept and balances are struck periodically, to assess the overall success of the enterprise.

Entry into such organizations is usually controlled formally and informally. Formally, certain previous experience documented in the dossier is usually a prerequisite for being considered for the job. In the early stages of a career, education is perhaps the most powerful selective mechanism, as measured by the certificates of formal schooling which the candidate can claim.

In fact, the requirement of certificates is as much a function of the state of the labor market and the organization of privilege as it is of competence. It is said that the requirement of a high school diploma began during the Great Depression of the 1930s when dozens of workers were equally able to do a single job; requiring high school graduates simplified the task of selection by eliminating many (and insured the hiring of younger workers). But certificates are also powerful ways of protecting the jobs of the incumbents; apprenticeship, educational degrees, and professional degrees all have the effect of excluding other candidates who might do as well or better. Thus the "diploma curtain," as Peter Drucker calls it (in a personal communication to the authors), is a major barrier to jobs for many and protects the jobs of others.

To be sure, exposure to formal training is critical in some occupations—one thinks of such highly responsible and demanding tasks as piloting commercial aircraft, performing brain surgery, and employing nuclear physics technology. Yet a great many jobs are defined in an *over*-selective fashion and the underemployed, as well as the unemployed, are created by the diploma curtain. At the same time, the bureaucratic management has a vested interest in keeping a stable labor force while the employees have their vested interest in keeping the job. The result is a tendency toward stasis, with achievement no more important than sex, ethnic group, or ability to "get along" with fellow workers. It is frequently less important.

One can then play the occupational game by "mini-max" strategy. This means minimizing the maximum risk; the classical example is in government "classification" of official documents—one is punished for *failure* to classify, but not for wrongly classifying what should be public. The result is apparent in the vast archives (containing mostly dead letters) to be found in national capitals. Unclassifiers face the same problem. Minimizing the maximum risk leads inevitably to the development of group work norms. In almost all occupations sociologists have identified three items of occupational culture: the "bogey," or the accepted rate of work; the role of "make out artists," or those who do less than the bogey; and the role of "rate buster," or "company man," for those who are conspicuously more productive than the bogey would require (Caplow, 1964).

Such a culture of work applies to the majority of the labor force. However, there remain roles which call for exertion, ingenuity, and commitment to the task at hand. In public and private life the entrepreneurial roles remain, whether we are considering an independent long-lines truck driver who owns his own vehicle, a free professional, an elected official, or the creator of a private economic empire such as Henry Kaiser. The farmer, shopkeeper, free craftsman linger on, but the agribusiness conglomerates, the supermarket chains, and the craft unions increasingly call the tune.

There remain the emerging careers which C. Wright Mills (1951) first called to our attention more than twenty years ago. In a bureaucratic society one may still be an entrepreneur, for much is required beyond the "minimum job description" to make such societies work. Mills was fascinated with the role of the "operator," the person who facilitates inter- and intra-organizational combinations. A go-between, he maximizes the interests of various parties, including himself. A good mayor may perform the same role (Greer, 1974a). Indeed, if we expand the concept to include all those who are social facilitators and coordinators within and among bureaucracies, this is one of the major types of job in contemporary society. There are indications that effective urban governments are in cities with more of these managerial types in the electorate. Those who create such economic empires as I.T.T. are clearly some kind of entrepreneur. It is not so obvious that city managers, city planners, the directors of local public authorities for Urban Renewal programs, are also entrepreneurs. Yet they also combine a variety of enterprises in pursuing what they usually conceive to be the public interest. These are some of the ways of going from rags to riches in large-scale society. Most people do not succeed at such movements, in the United States or any other society. (There seems to be little reason to believe that the American occupational mobility rate today is much higher than that of any large-scale society.) Instead, workers either take a job with little possibility of upward mobility or they progress in a fairly orderly fashion through a few rungs in a bureaucracy—from box boy to grocery clerk to department supervision in a supermarket, from assembly line worker to foreman in an automobile assembly plant. (The latter is limited by actual or anticipated sanctions from fellow workers.)

Yet Gallup poll results (Gallup Opinion Index, October, 1973) comparing various large-scale societies show workers in the United States to be among the most contented (if not *the* most contented) with their work in the world. They are much less alienated than workers in Japan, for example. It may be that they have simply had a longer history of dealing with bureaucratic work structures as against traditional, hierarchical organizations of work. How do they make their peace with the economic structure?

Some do so through pursuing the entrepreneurial career. Whether they win or lose—and there are many small winners and losers—the job is at least interesting (Terkel, 1974). So Studs Terkel finds in interviewing an interstate truckdriver:

When you're in that truck you're not Frank Decker, factory worker. You're Frank Decker, truck owner and professional driver. Even if you can't make enough money to eat, it gives you something . . .

There's a joke going around with the truckdrivers. 'Did you hear the one about the hauler that inherited a million dollars? What did he do with it? He went out and bought a new Pete.* Well, what did he do then? He kept running until his money ran out.' Everybody knows in the business you can't make no money. Owning that big Pete, with the chrome stacks, the padded dashboards, and stereo radio, and shifting thirty-two gears and chromed wheels, that's heaven. And in the joke, he was using up his inheritance to keep the thing on the road.

*That's a Peterbilt, the Cadillac of trucks. It's a great big, long-nosed outfit. The tractor alone costs 30,000 dollars.

Others are satisfied that, compared with those they know, or with their fathers, or with their own origins, they have won in the bureaucratic role-system. Others still consider the job in the traditional manner, as "earning a living by the sweat of the brow," and after the payment of the pound of flesh retire to after-hours, the weekend, the vacation, and hopes for the children. Then there is the large residue, perhaps a third of the labor force, which is unsatisfied with the occupation and alienated from the economic structure. For them there is nothing resembling a "career," an orderly procession through given positions of responsibility and rewards—and they do not like what they have to do.

The reward system in urban society is, *in principle*, based upon achievement. That is, it does not stem from status. In fact it varies enormously with states; whether we are considering French Canadians in Quebec or women in the United States, we find that when we control for educational achievement they are substantially underpaid compared with English Canadians or men. Atypical populations, whether so because of ethnicity, sex, or age, have difficulty in achieving access to jobs, in progressing through the career trajectory, in achieving equal pay for equal work, in protecting their jobs in the event of a work shortage, and in keeping their jobs in general. In consequence, they not only have a lower level of job and a lesser reward, but also a less predictable income. This, in turn, inhibits participation in the community at large; a likely consequence is less ability to protect their privileges (Form, 1968).

Some authors have suggested further that lack of access to the urban job market may foster among excluded groups a rejection of urban skills and values. In their own minds cultural inversion creates greater value for the undervalued roles these persons occupy but moves them further from competitive entry into the control networks of urban society. Hughes (1943: 210) tells us that French Canadian "counter-culture" values the literary and artistic rather than the scientific and the valued intellectual is the "savant" rather than the expert. A similar argument might explain the preference of 1950s housewives for the large families of pre-urban America. Female workers of the wartime years reported a desire to retain their jobs after the war (Grey, 1972: 236). Eased out of those jobs by returning soldiers, they retreated to the values of house and family and rejected work values.

One of the most important rewards of the occupation is, simply, information. Some trades make a rational attack on the bureaucratic system much easier, others provide no information on the problem. Thus accountants have a much better knowledge of the tax structure and how to protect themselves from overpayment than most; the same is seldom true of unskilled laborers. This may explain why at every income level the federal income tax appears to be about the same—15% of gross annual income. To be sure, an executive or professional or entrepreneur may not know what the accountants know, but he knows enough to hire the knowledge. In the same way, given occupations lead to associated opportunities: the lawyer is inevitably involved with business interests or labor interests, as well as local and state, even national, politics. The college professor has easy access to publishing (as witness this volume). And indeed, upholsterers in the furniture industry have the basic "nut" for becoming furniture manufacturers; cutters in the clothing industry move easily into entrepreneurship. The key point is that one of the *tools* of the trade is, with luck, one of the rewards.

The combination of (1) advantages based upon certification, (2) privilege based upon position, and (3) specialized knowledge resulting from this position, tends to influence, powerfully, one's world-view. Third-generation union craftsmen see the world from the point of view of their own; so do third-generation inheritors of corporate wealth. The Republican Party, with its key constituency of entrepreneurs, managers, free professionals, and *rentiers*, tends to oppose redistribution of income. The Democratic Party, however, with its working-class and ethnic constituencies, tends to favor redistribution of wealth; after all, they have less. (There is, however, the vested interest of the craft union in the present distribution of wealth.)

There are subtler and fascinating aspects of occupational specialty. Speer (1970) saw the Third Reich of Hitler through the eyes of an ambitious architect and monument builder. Henry Thoreau, an itinerant man of letters, could afford to see New England society through the eyes of a philosophical anarchist. In short, our vision, our very way of abstracting from the flow of experience, is profoundly affected by our occupation. The economist looks at the world through eyes which emphasize markets, factors of production, exchange; poets see other things. Unskilled laborers, living in small-scale niches of the world, see mostly persons, routines, petty penalties and rewards. Their job is not much of a learning environment. They become the true "urban villagers." The academic professional, free to follow his intellectual nose where it leads, with easy access to information and a basically secure job, tends to exaggerate the freedom available to workaday man.

When, however, the professor's device for protecting his privileges, his tenure, is called into question as it has been recently, we find the same response as the union man to "union busting" activity. Whether the professor's freedom of inquiry demands tenure or not, it is clear that *professors* do so (Greer, 1974b). Those with the most universalistic frames of reference become very particularistic when their own privilege is in question. Occupation is a powerful limit and determinant of frames of reference and therefore—world-views.

II. INCREASING SCALE AND SOCIAL CLASS

Most industrial societies evolved from feudalism. The sharp divisions of political and economic power and of social honor in the feudality were the basis for authority in such industrial societies as nineteenth-century Britain and Germany. Yet the feudality had been based upon a reciprocal master-man relationship; the market society of the industrial revolution had no place for such a bond. The very basis for justice was eroded, leaving nothing in its place. The result was the creation of a class society (Thompson, 1964).

As industrial society developed into a system where people were divided by classes, unlike feudal society divided by estates, it depended for its legitimacy upon the belief that the higher classes deserved their positions because of their *achievement*. This is much harder to define and defend than the feudal argument from lineage. It was challenged from the day it was promulgated, for the thrust of the industrial society was toward the destruction of the rights of the working population (e.g., manual workers). They responded in kind, from the weavers who followed "General Lud" to the political opponents of "Old Corruption," that superstructure of placemen and pensioners which was so costly to the public budget. The middle classes, as we now call them, created the working class as an inevitable response to new privilege that had essentially been vested from them.

Thus, we see the birth of the "two nations," the rich and the poor. Each developed its own self-definition vis-à-vis the other, while in between were the "strainers and strivers," those who were middle-class in aspiration but poor in worldly goods. Each class developed a version of history, past and future, to suit its interests. Each had its heroes and its martyrs, its sages and prophets (Thompson, 1964).

Such a class system did not develop easily in the transplanted cultures of Northwestern Europe. In the overseas colonies, particularly in the United States, the absence of a feudal tradition and the abundance of opportunity for new starts as farmer and entrepreneur blurred the class structure from the start. It was strongest where there was a clearly defined underclass, as in the plantation South; it was weakest on the frontier. That frontier was to become the bulk of the present United States (Elazer, 1970).

Nevertheless, the basic structure of emerging industrial society pitted workers against management, farmers against financiers, buyers against sellers, and the result was a modified version of the English and European class systems. The chief differences were (1) an ebullient optimism, generated by the growth of settlement and industry, (2) accompanied by a utopian faith in the perfectability of society and, through society, mankind. Thus from the organization of the Knights of Labor to the International Workers of the World, militant working-class movements were inclusive and egalitarian. As their enemies were blurred, so were the boundaries of the movements (Wiley, 1967: 529 ff.).

There were, however, sharp gradations in economic power and (indirectly) political power, by occupation. And occupations were so evaluated by those who had the most money and did not work with their hands. Despite ritual obeisance to the virtue of "labor," it was clear where status honor was conferred—to those entrepreneurs and free professionals who served under nobody and who got a good profit from their position. Thus the class system of industrial America, circa 1914 (Allen, 1952; also Lynd and Lynd, 1937). It was less harsh than in the systems derived from a feudal society, where the underclass had the worst of both feudal and capitalist worlds; it was far from the egalitarian world imagined by Eugene Debs, W.E.B. DuBois, and Walt Whitman. Under the glaze of egalitarian rhetoric there were enormous disparities in wealth, power, and honor, as these and other working-class intellectuals knew.

The very ecological layout of the nineteenth-century city, surviving in the central areas of our older metropolitan areas, tells the story.

Near the place where the railroads met the river or the sea, the heavy industry and warehouses clustered. Movement of the heaviest loads was reduced to a minimum. Near the workplace grew the row houses and tenement buildings that housed the workers, so that they could reach their jobs by walking. The shops and services that relied upon these populations and activities for trade, as well as those serving the entire city, were near the center. Other shops supplying necessities to the households were scattered through the city within walking distance of the various neighborhoods. Farther away from the noise and the unsightly workplaces came the neighborhoods of average housing, tall structures built close together with little yard space. Only the rich could afford so expensive a commodity as free space. With the ability to pay for space or transport, they built tall town houses in the center, or magniloquent testimonials to their wealth in the outlying districts [Greer, 1962: 118].

Lewis Mumford calls such a city the conjunction of the factory and the slum. There is some justification for such a characterization in most industrial cities, and from the beginnings, American entrepreneurs tended to locate their dwellings as far as possible from either factory or slum.

It was a city characterized by propinquity in public space —factory, shop, warehouse—and segregation in private space. Thus the white Protestant entrepreneur might rub shoulders with immigrant workers and black sweepers in the workplace, but at night he returned to the cool haven of the Main Line or the North Shore; the workers returned to the parish of Bridgeport or the neighborhood of Dago Hill. The Blacks returned to Harlem or the South Side.

The various class-worlds and ethnic-worlds communicated through the media of the political "machine"—a loose confederation of ethnic wards—and the public print. All shared a version of the "news" through the papers and magazines. In a sense, they got along together because they hardly knew one another. When the "Jazz Singer" left the Jewish community to become a star instead of a cantor it was news; when Abie met his Irish rose it was a cause celebre. Segregation made a functional integration possible.

In the very seed of the industrial society lay its supercession by the post-industrial. For the demiurge of industrialization was the substitution of fossil fuels for muscle and chlorophyll as the energy base, and use of the machine as actor. At first the revolution simply released men from agrarian bondage to work in factories; however, as the process continued, the factories required fewer and fewer workers to supply the commodities the society needed. As labor was freed from industrial work, it moved into white-collar jobs. Today a majority of the labor force is made up of white-collar workers.

Those who had been an elite, their white collars symbolizing their freedom from manual labor, had become the statistical mode. The great gap between the small number of educated and privileged middle-class and the mass of blue-collar workers had been filled by a whole new structure of jobs, larger than either of the old antagonists. At the same time the meaning of the occupation for social class came unglued, for many of the white-collar jobs were simply tedious drudgery in (relatively) clean surroundings. While most of this work is with people and symbols, rather than with things, it is often as stupefying as that of the assembly-line workers (Greer, 1962). Studs Terkel is listening to a receptionist at a large business firm in the midwest:

You come in at nine, you open the door, you look at the piece of machinery, you plug in the headpiece. That's how my day begins. You tremble when you hear the first ring. After that, it's sort of downhill—unless there's somebody on the phone who's either kind or nasty. The rest of the people are just non, they don't exist. They're just voices. You answer calls, you connect them to others, and that's it. . . . The machine dictates, this crummy little machine with buttons on it—you've got to be there to answer it. You can walk away from it and pretend you don't hear it, but it pulls you. You know you're not doing a hell of a lot for anyone. Your job doesn't mean anything. A monkey could do what I do. It's really unfair to ask someone to do that [Terkel, 1974: 29-30].

Then, too, when work is so routine that almost anyone can do it, it is hard to bargain for good salaries. The white collar simply means clean semi-skilled labor.

Meanwhile the blue-collar workers are sharply divided into those who maintain strong, disciplined unions, on the one hand, and the unorganized, on the other. The former often earn much more money and, through union-organized political power, have in the aggregate more power than white-collar workers. Of the perquisites of higher class, many white-collar workers have only a questionable claim to greater social honor. For these reasons there is a growing trend to union organization among those who had been considered "unorganizable"—clerks, teachers, musicians, athletes, and others. Indeed, the most rapidly growing segment of organized labor today consists of unions organizing public service workers at federal, state, and local levels.

The unorganized blue-collar workers are in the worst position of all. They have neither money, political power, nor status honor. Their work is usually simple and they are easily replaceable. Having no technological monopoly and, usually, no work group large enough to force an employer to bargain (as is the case in mass manufacturing) they are vulnerable to those better equipped and organized. Their best bet is to be included in a "conglomerate" union, where the sympathy strike power of union allies will be respected. Otherwise, they are apt to be those to whom the minimum wage law represents the maximum they can expect.

As non-human energy sources and machines freed labor first from agriculture, then from industry, so today it is freeing labor from the clerical drudgery necessary for large-scale society. The computer is to white-collar work what the steam engine was to manual labor. In three decades an entirely new industry has emerged from the primitive efforts of Dr. Watson at I.B.M. We are only beginning to explore the possibilities of such tools. They can already do the work of clerks and accountants, draftsmen and such; much of the work of "middle management" who execute but do not make policy is also computer-manageable, while computer simulation is increasingly at least an adjunct tool for executive training and high-level decision-making. Key urban occupations shifted from machine tender to clerk to computer programmer.

Perhaps some of the malaise of occupational life these days is due to two facts. First, people know, or believe, that their work need not be this way; second, they cannot count on its remaining even the way it is. The relatively serene acceptance of work possible in a traditional society, even the American society of 1914, is difficult then. As William Form (1968: 253) noted some years ago, "It is now problematic whether any occupation has its destinies in its own hands."

Social class has also become blurred in its consummatory effects. In the near-global increase of affluence (at least among the privileged nations) since World War II the luxuries of other epochs have become the necessities of all. Mass higher education and access to the great, at least indirectly through mass media, have reshaped our view of the world. Mass higher education may be partly spurious, a matter of lowering standards, and access to celebrities via television may be a shoddy substitute for true presence, but together they have given the modal population a view of the world which only the very well-placed and well-informed would have had in, say, 1914.

In both public education and the media, the dogma of egalitarianism is pervasive. Nevertheless, there are sharp differences in the rating of various jobs by status honor. In an old but wise summary of the findings, Theodore Caplow reports two basic attributes which seem to account for that variation. First, the higher the rank of occupations (banker, college professor, physician, clergyman, lawyer) the greater one's freedom and ability to control one's relationship to others. As he writes, "... it is worthy of note that the position of the subject with respect to the control of other people's behavior, and their control of his, appears to conform rather well to what is reported as occupational prestige" (Caplow, 1964: 55). A second attribute of high-ranking occupations is the likelihood of further vertical mobility—is it a dead end job or an opportunity? This is why college students working in the summer do not really understand the work situation: they have a probable way out.

Control also seems to be a key variable in the occupational status of women. As work shifts from "big muscle," to "small muscle," the physical disadvantages of women in the labor force diminish. At the same time, the technology of fertility gives them more control over pregnancy. Consequently, 45% of adult women now work–altogether some 40% of the United States labor force (Montly Labor Review, April, 1974). Yet their positions relative to equally qualified men are low. The reasons seem to be, again referring to Caplow, the hesitancy of the upperclass (men) to have women in positions of superordination where workers are male, and consequent "gender typing" of jobs. In business, industry, and politics, there are specified "women's jobs." These may be, respectively, in personnel records (business), as inspectors (industry), and as Secretary of State. Sometimes what is women's work in one industry is men's work in another, with a consequent adjustment in the rewards it confers. In no case do they have equal opportunity and access to the occupational groupings which yield the highest economic gain, political power, and status honor.

Yet these women are largely from the same social class as their fathers and husbands, their ethnic identity is apt to be the same, and they have gone through school and college with their competitors. We must ask *why* they suffer such a status handicap in jobs and, therefore, in social class. Their jobs are more likely to be marginal and they are apt to be paid three-fifths of the wages of men in the same level of jobs, yet they have come from the same background as their male competitors (and frequently a "better" one), have the same intelligence test scores and the same formal schooling (though usually higher grades).

III. THE DILEMMA OF CLASS
IN A MASS DEMOCRACY

As we have noted earlier, once the estate system based upon birth is overturned, one is left with the problem of legitimation of privilege through achievement. Yet whoever controls access to the appropriate certification processes and jobs controls the opportunity to achieve. Women, visible ethnic minorities, the young uncertificated worker (or dropout), and the old are discriminated against in a systematic fashion. Consequently, we do not know what their achievements would be if the competition were fair.

But the competition is never fair. The problem for the state, then, is to convince the citizens that it is *roughly* fair; otherwise, how can one justify the existing distribution of wealth, political power, and status honor? To do so one must make every effort (or show of effort) at extending equal opportunity to all. At the same time it is useful to emphasize the rhetoric of equality. Through dress, manner, speech, and the like, the society should become homogenized when possible, but tolerant of deviance when necessary. (If we seem to be anthropomorphizing the state, we should add that the average individual of "good will" tends to do the same, facing the same dilemma.) Thus we may maintain the fiction of equality at public places, at most a "division of labor," while enjoying extreme inequality in private. Politicians of great wealth eat salami and kiss babies in public: in private they collect art and pilot yachts.

Whether this is the recipe for "la dolce vita" or not, it seems to be the American resolution of the dilemma. All persons are not created equal; class and ethnic distinctions start in the womb deriving from the mother's status and emerge at birth. Class *consciousness*, however, is unevenly distributed; this is a measure of the success of the perhaps unconscious strategy of those who direct the mass educational system and the mass media. Such successful persuasion is sometimes the result of past achievements, but it is often the exercise of privilege to protect privilege.

In metropolitan society the moral bases of social class and the physical means of repression are greatly weakened. The various classes and ethnic divisions have a kind of immediate knowledge of each other through the mass media, yet the poor are segregated in neighborhoods near the centers of transportation and communication. The bases of invidious comparison through the media, on the one hand, and sabotage of the urban structure at the center, on the other, are obvious. Culturally, the metropolis blurs and blends: the young Blacks see the rewards of our society on television, and greet "the hard sell" with an expletive. The rich psychiatrist in Beverly Hills resonates to his disgust. Where is the legitimation necessary for effective repression?

Politically, our tradition has been to find cleavages and exploit them. But what social classes are clearly enough defined to provide leverage for an ambitious politician? And if he cannot find "his people," nor they him, how are coalitions possible? And without coalitions how can such a differentiated society survive? The city is the center of our society; it is here among the anonymous crowds that the loss of earlier hierarchy is most evident, and here that the problem must be solved. Lake Forest and the South Side in Chicago, Watts and Beverly Hills in Los Angeles, must be included in a new polity, allowing a new expression of fluctuating class and ethnic interests. The alternative is some repressive system which, in its demands on the society, staggers the imagination.

IV. SOME SPECULATIONS CONCERNING FUTURES

We return to a question asked earlier in this paper: "What is the future of social class in a society with an egalitarian ideology?" More specifically, what are the possibilities as the society becomes primarily urban in nature due to the increase in societal scale? How have these problems been handled in the past, and what are the possibilities for the future of the United States and other large-scale societies of the present?

These are not new questions. The conflict between tribal identity, loyalty to the polis, on the one hand, and increasing economic inequality, on the other, was, in the views of many scholars, the reason for the collapse of Athens at the height of its glory. The conflict between the chosen people of Israel and the increasing social class differences among the chosen people can be seen as contributory to the defeat and extinction of that state. In each case, increasing scale and an urban society, with its riches and complexities, eroded the traditional basis of legitimacy (Turner, 1941; also Kitto, 1956: esp. chapter 9).

That legitimacy, based upon common language and culture and the myth of common origin, was doomed as the society increased in scale. For it excluded those who were not born to the polis or the tribe; it limited the extension of the societal networks of control, particularly the political structure. Rome evolved an answer, of a sort: the extension of citizenship in the City of Rome to achieving, worthy, or potentially dangerous foreigners and foreign cities. Indeed, by the end of the Empire, practically all the inhabitants were citizens with the exception of slaves (though by then citizenship was not worth much). In its last days Rome developed a concept central to the contemporary state—citizenship by territorial membership.

The contemporary term "nation-state" symbolizes the effort of contemporary large-scale societies to fuse the concept of territorial citizenship to the concept of tribalism. It is a myth, in the sense that none of the large-scale societies is ethnically homogeneous. White Protestant Americans make up a bare majority of the population in the United States; the so-called Great Russians are a minority in the U.S.S.R., and the United Kingdom includes the Scottish and the Welsh (both with active independence movements).

The myth, in essence, says that because we have all been in the same territory and under the same political system for quite a while, we are all one people. It does not solve the problem of gross inequity (social class and ethnic discrimination) within the people. There have been a number of efforts to solve this problem.

In a context of "coerced consensus," the Soviet Union has promulgated the ideal of the "New Soviet Man." This is a communally oriented, other-directed person whose primary concern is the state, which symbolizes other human beings. (The Chinese version is not very different.) All are doing their best for the good of the whole; whatever their station, their work is useful to the state. This myth is a version of the old Thomist myth—all work is good, and each worker deserves those means necessary to achieving his job. Commissars deserve limousines.

In the United States and Great Britain the general myth is still that of the self-regulating society: "You get what you deserve." It is held that all have equal opportunity, and achievement, evaluated by the marketplace, determines rewards in wealth, power and honor. Of course, opportunity is not equal, and those with greater advantages determine the relative value of different achievements. Millionaires deserve it.

In both patterns, the real structure of control is one of

bureaucratic hierarchies modified by plebiscite and/or party *apparat* maneuvering. In general one can say that when the economy goes well, the ordinary citizen seems to accommodate to the demands of the state. The real test of the legitimacy of that entity occurs when the society faces depression, rebellion, or reversals in war with other states.

Translated into the class politics of the city, this set of propositions reflects, once more, the volatility of modern urban society. When all goes well economically, all goes well. But when there is rapid change in economic fortunes the result is a general anomie, or a sense of normlessness, and a general anger with the governing structure (Durkheim, 1951). This portends alienation among *all* social classes in the city, all ethnic categories. The English lost their empire and worked out their coal mines; the class war becomes increasingly naked. (When acorns are scarce the squirrels fight.) When the economy of the city is depressed, there is relative deprivation for each social category and it is expressed through rejection of the governing elite.

These considerations lead us to pose a number of research leads which we think might clarify the nature of the futures available to cities in large-scale societies. We list them with no bias as to priority:

(1) What is the relationship between *relative* mobility and political consensus? How does one's achievement (as a mature citizen) compared to one's parents, one's own past, or one's peers, relate to general affiliation with the governing structure?

(2) Are we, in large-scale society, making a middle class? If so, what are its cultural components—belief systems, norms, general assumptions? Does class override ethnicity? Some studies indicate this to be the case.

(3) What are the developing infrastructures of the modern city? We know that local area organizations, mostly voluntary, are growing in numbers, and are developing federal characters—within the local areas of the city and within "regions" of the city. What is the significance of this as symptom and as probable cause of change in the distribution of wealth, power, and prestige in the city?

(4) What are the political effects of inter-city variance: within a society; by the societal scale of the society; by the relative scale of a single city within that society (for example, societies with the "cabaza de Goliath"—head of a giant on the body of a pygmy—versus multi-centered societies); by the specialized nature of a city within an urban network? That is, what are the consequences for occu-

pational differentiation, its interpretation as social class, and the consequent organization of privilege—economic, political, honorific?

(5) What are the effects of expanding societal scale upon the nature of political opportunities and political craft? That is, as societies expand in scale, do they inevitably become bureaucratic, "managed" societies? Are they inevitably expressions of the efforts of specialists to contain change and maintain order? Or do we, as human beings, have other alternatives?

NOTE

1. See any annual "Fortune 500" and "Fortune Second 500" in Fortune magazine. The literature on concentration of economic control is vast and persuasive; see Mason (1968).

REFERENCES

ALLEN, F. L. (1952) The Big Change. New York: Harper.

CAPLOW, T. (1964) The Sociology of Work. New York: McGraw-Hill.

COTTRELL, F. (1955) Energy and Society: The Relation Between Energy, Social Change, and Economic Development. New York: McGraw-Hill.

DURKHEIM, E. (1951) Suicide. New York: Free Press.

ELAZER, D. J. (1970) Cities of the Prairie. New York: Basic Books.

FORM, W. (1968) "Occupations and careers," in The International Encyclopedia of the Social Sciences 2. New York: Macmillan and Free Press.

GREER, A. L. (1974a) The Mayor's Mandate. Cambridge: Schenkman.

——— (1974b) "Faculty Attitudes on Tenure." Milwaukee: University of Wisconsin. (unpublished)

GREER, S. (1962) Governing the Metropolis. New York: Wiley.

GREY, B. M. (1972) "Sex bias and cyclical employment," in N. Malbin-Glaser and H. Y. Waehrer (eds.) Woman in a Man-Made World. Chicago: Rand McNally.

HUGHES, E. (1943) French Canada in Transition. Chicago: University of Chicago Press.

KITTO, H.D.F. (1956) The Greeks. London: Penguin.

LYND, R. and H. LYND (1937) Middletown. New York: Harcourt-Brace.

MASON, E. S. (1968) "Corporation," in The International Encyclopedia of the Social Sciences. Vol. 3. New York: Macmillan.

MILLS, C. W. (1951) White Collar. New York: Oxford University Press.

MOORE, W. (1946) Industrial Relations and the Social Order. New York: Macmillan.

SPEER, A. (1970) Inside the Third Reich. New York: Macmillan.

TERKEL, S. (1974) Working. New York: Pantheon.

THOMPSON, E. P. (1964) The Making of the English Working Class. New York: Pantheon.

TURNER, R. (1941) The Great Cultural Traditions. New York: McGraw-Hill.

WEBER, M. (1946) From Max Weber, ed. H. W. Gerth and C. W. Mills. New York: Oxford University Press.

WILEY, N. (1967) "America's unique class politics." American Sociological Review 32.

14

The Social Economy in Urban Working-Class Communities

MARTIN D. LOWENTHAL

□ THE STRUGGLE FOR SURVIVAL by working-class families in America has always been a difficult one. The problems of inadequate income from employment have been noted by many commentators on the American political economic scene. The estimates of the number of persons with incomes below the poverty line range between 25 million and 40 million people, depending upon the standard which is used. For urban working-class families, the more realistic measure of poverty is probably the one developed by the Bureau of Labor Statistics, which looks at the actual living cost for an urban family of four living under specified conditions. The Bureau computes three levels or standards of living—lower, intermediate and higher—and the lowest standard for a family of four is now in excess of $7,000 per year. This would indicate that approximately 20% of the population of the country is near or below the governmental income measures for economic survival.

I. PROBLEMS OF SURVIVAL WITH LIMITED ECONOMIC RESOURCES

The question that arises at this point is, how do people with inadequate incomes meet their needs and manage to survive?

[447]

Although many needs may go unmet and thus give rise to such problems as poor health and substandard housing, it is the argument of this essay that many needs are provided for, not through money income but by the social networks that people participate in.

Many of the current public debates on welfare, social services, housing programs, and health services often overlook aspects of the helping networks which people use and depend on. In many communities where networks provide care in times of illness, the patient may feel more comfortable receiving care from kin, friends, and neighbors than from a county hospital. In other communities, however, these networks may be weak or undeveloped and so no alternative to public institutions exists. The entire question of economic vulnerability is one which must be examined in the context of social support systems in addition to family income and the availability of governmental services.

These questions are important not simply in terms of the alternatives which are available to populations which are potentially dependent upon governmental services. They are also important in terms of what services tend to reinforce the family and community networks of people and which tend to undermine those activities. These social activities may be undermined by being overburdened or relegated to inappropriate times and places. For example, by requiring a mother on welfare to take employment on a full-time basis, the other work roles which she performs such as providing care and protection for her children, maintaining her household, and maintaining the network of other relationships on which she may be dependent must be completed after employment hours or may not be done at all. This may put increased burdens upon the mother, the children, and the network of other people who depend upon her as a participant. In addition, welfare mothers who cannot manage the added work load may become more vulnerable to stress, such as physical illness and emotional tension, and their children may be further delimited in their parent-child relationship. The family may also become increasingly dependent upon governmental services for child care and protection and for dealing with crises if provisional social networks are not maintained.

The problem of limited economic resources is most obvious in the cases of dependent populations such as children, the elderly, and the disabled. Without sufficient income to meet their own needs and without the capability or opportunity to participate in income-producing employment, these populations must depend upon social relationships and governmentally financed institutions to meet both their basic needs as people and special needs, particularly for care and protection, as dependent populations. Historically, there have been few and generally inadequate public institutions which are designed to meet the entire range of needs for children, the elderly, and the disabled. The burden of provision, care, and protection has generally fallen upon the nuclear or extended family. In the case of children, the responsibility has fallen primarily upon women.

Some of the research on lower-income working-class communities has identified many of the informal helping systems which are crucial for the survival of families in those communities and for the survival of particular dependent populations. In her study, *Blue Collar Marriage*, Mirra Komarovsky found that among the families included in her study the extended family played a crucially important role. In addition to the part played by relatives in the socialization of their children, in providing emotional support, and in being companions in recreation, parents were also economically important to the survival of the young family. Although they were usually in no position to help their children to buy a home or establish a business, some financial aid was frequently provided in emergencies. Komarovsky went on to note that the economic arrangements among kin were often of the form "reciprocal aid":

Thus, a widowed father shares his home with a married son who pays no rent but is responsible for household expenses; a widowed mother residing with her daughter works as a waitress, paying rent and her share of the grocery bill; a widow and her bachelor brother inherited the parental home and rented rooms to a married daughter who is the homemaker for the whole group and expenses are shared.

Komarovsky goes on:

Apart from such more-or-less permanent arrangements, relatives frequently exchange services which among wealthier families are purchased from specialists—such as housepainting, carpentry, repair, laying linoleum, building partitions, and help in moving [Komarovsky, 1967: 237].

In their work, *Family and Kinship in East London*, Young and Willmott also discussed the exchange of services between members of working-class families. Among the most important services that were noted, the ones provided by mothers for their daughters received particular emphasis. In most cases the mother would look after the home in cases where her daughter was in bed preparing to give birth to a child. Usually there was at least one other child as well as a husband to provide for. The child had to be looked after, food bought, meals cooked, clothes washed, housecleaning to be done, arrangements and schedules to be coordinated, and preparations made for the care of the new baby. They note that "the wife's dependence for help on her kin, and especially upon her mother, does not end with the confinement and its aftermath (at times of childbirth). Whatever the emergency, and whether her need is big or small, the wife looks to her mother for advice and for aid" (Young and Willmott, 1957: 54).

In his study of widows and their families, Marris (1958) examined the manner in which social relationships were continued, altered, or eliminated after the death of the husband. Most of the women in the study were young and had children. He found that immediately after the death of the husband, parents, brothers, sisters, and children all helped to provide companionship to distract the widow from her loss and, at the appropriate time, to insist that she had grieved long enough. During this period, relatives helped in protecting her from harrassing practical problems such as the funeral arrangements and often helped to defray the cost. Money was tight in many cases, particularly if the husband had been ill for a length of time. Some help in money came from relatives and sometimes from neighborhood groups and the workmates of the husband. The extended family tended to be the most important in providing for widows with children by giving meals, clothes for children, and financial assistance.

Marris found that most of the women worked after the death of their husbands. Many had worked before they became widowed in order to supplement the family income; after the death many worked not only for the income but also for the companionship of the work situation and the status derived from being financially independent. As most of the women became less dependent on their relatives for money, they became more dependent on them for services. The mother and other female relatives helped with such tasks as housekeeping and looking after the children. Many of the male relatives performed odd jobs around the house for the widow such as painting, wallpapering, mending furniture, and plumbing. The brother of the widow often had the responsibility for advising her on the education of the children and of handling the unresolved financial affairs of the husband (Marris, 1958).

The issue of the importance of the kin-group for older people was the focus of Townsend's (1957) study in a small working-class borough near London. He found that the primary self-image of old people was as a member of a family and secondarily as an individual old in years. Forty-five percent of his sample lived with relatives and only 21% lived by themselves. Most of his population had children who lived in the same household or nearby. Townsend indicated that many activities and transactions were involved in the kin relations of older people. For instance, if a widow lived with her married daughter and her children, they would share many of the chores, with the grandmother doing much of the cooking and babysitting in return for her washing and shopping being done. This pattern was evident in most of the cases; the older people were helped with shopping, cleaning, washing, and transportation, and they assisted with cooking and watching after grandchildren.

The older man tended to play a relatively small part in the relationship with kin, possibly owing to his limited domestic role. The strongest ties were formed between grandmother-daughter-grandchildren since the grandmother was able to offer more useful services and because there had traditionally been strong ties between mother and daughter. The availability of care was sometimes limited by the ability of the people concerned to reciprocate services. Older people were more reluctant to ask for help and at times relatives were not as willing to offer help to them if they were unable to give help in return.

Townsend points out that the major problems of old age and retirement centered around the fall in income and subsequent drop in the standard of living at a time when needs tended to increase. New expenses grew out of the need for medicines, additional fuel, and devices of home entertainment such as radios and televisions. Many received help from their children in meeting their expenses; however, others who were the most isolated from their families were socially and financially the poorest. Townsend concludes by postulating that those who are socially isolated in old age, particularly those with fewest contacts with relatives, tend to make greater claims on hospitals and other health and social services and to die earlier.

Fried, in his study of the West End in Boston, notes that the finding "which emerges most consistently from studies of the working class is the central importance of locally based social relationships." He characterizes the pattern of the social relationships as being "close knit networks." These close knit networks may include family, extended kin, neighbors, community workers, local shopkeepers, and people who work and live in the community. "What is common to the various close knit networks is not their bases in kinship or common enterprises but rather the ready availability of members and the binding set of expectations for a mutual dependence and dependability in emergencies and in daily encounters, in sorrow and in joy, in routine contact and in the presence of conflict" (Fried, 1965: 132-133).

One of the most revealing studies of how social relationships and networks operate to meet economic needs was conducted by Carol Stack and reported in her book, *All Our Kin: Strategies for Survival in a Black Community*. Stack lived in and studied a community suffering from severe economic depression and found that extensive networks of kin and friends supported and reinforced each other in devising schemes for self-help and strategies for survival. Lacking the money income resources to meet family needs at a subsistence level, the people of the "Flats" immersed themselves in a domestic circle of kinfolk and friends who would participate in a system of mutual help. Stack notes "to maintain a stable number of people who share reciprocal obligations at appropriate stages in the life cycle, people establish socially recognized kin ties":

People in the Flats borrow and trade with each other in order to obtain daily necessities. The most important form of distribution and exchange of the limited resources available to the poor in the Flats is by means of trading, or what people usually call "swapping." As people swap, the limited supply of finished material goods in the community is perpetually redistributed among networks of kinsmen and throughout the community [Stack, 1974: 29, 33].

In the Flats the idiom of kinship is used to describe the important social relations. Friends are considered kinsmen when they assume recognized responsibilities of kin and are people one can "count on." The extension of kin relationships to friends allows for the creation of mutual aid domestic networks which are not limited by genealogical distance and genealogical criteria. A more determining factor is the practical requirement that people in the kin network live near one another, since the physical proximity permits the high level of interaction and activity between kin which is essential to the survival of the network.

The responsibility for providing food, care, clothing, and shelter and for socializing children is often spread over several households within domestic networks. Although residence changes are common, they usually take place within a narrowly defined geographical area so that kin relationships and domestic networks are maintained. The daily exchanges and dependencies of people engaged in common activities and relations are generally not affected by household changes, which are frequent owing to the poor housing, overcrowding, unemployment, and poverty. Stack concludes that without the help of kin and friends in the domestic networks, fluctuations in the limited flow of available income and goods could easily destroy the ability of a family to survive.

Although these and other studies touch upon the material importance of social relationships in working-class communities, there has been no formal development of a theory which includes or incorporates the economic functions which these helping systems and social relationships perform. There has been no systematic effort to identify the principles upon which the economic aspects of social relationships operate nor the infrastructures upon which these networks of activities and transactions depend. In analyzing modern industrial economic systems, formal economics makes no attempt to include or understand the non-monetized activities and transactions which play such an important role in the survival of working-class families and communities.

II. ECONOMIC FORMULATIONS

In searching for theories and models to assist in an understanding of helping and support networks, the field of economics, as it is articulated at this time, offers very little. When referring to the "economy," economists usually mean the sphere in which goods and services are produced for sale, are sold, and are purchased. This formulation derives out of the basic characteristics of the market economy. A market economy is based on the exchange of goods and services in the market for equivalent value, usually based upon a standardized monetary system. Market exchange depends upon measured payments in the buying and selling of goods and services and is the organizing principle for transactions involving material products, labor, and natural resources. This means that, in general, people derive their livelihood and meet their needs from selling something in the market; for most people this is their labor.

It has been recognized by most economists that the payment of wages as an essential aspect of the organization of the productive process is a comparatively recent development. In almost every period of recorded history there existed transactions which could be described as the hiring of labor for a contractual payment. However, such transactions were only typical for a small sector of the economic process. Generally work was done and livelihoods gained through systems of social relationships which defined the rewards and responsibilities of the participants. It is only with the breakdown of the feudal restrictions and the replacement of the domestic worker by the factory system that the basis for a general wage system arises. In addition, the gradual monetization of an increasing number of economic transactions provides a further basis. Under these conditions, the "free" but property-less worker must offer his only possession, his labor power, in order to maintain himself and his family, while the owner of the tools or land, the employer, can obtain the necessary labor force only by inducing people to work for him by offering them a wage.

Labor is thus considered a particular kind of commodity subject to the forces of supply and demand in the marketplace. This formulation of labor power in the industrialized societies of the West has become the sole preoccupation of most economists to the exclusion of other perspectives.

The concern of economists since the nineteenth century has been the development of economic models which attempt to deal with the question: what are the factors which determine the prices of labor, natural resources, and products in a national, market economy? Activities and transactions not subject to pricing mechanisms have generally been excluded from formal analysis, and thus social scientists and social planners at all levels, by accepting this traditional understanding of the economy, have tended to disregard the significance of women, the family, and the community from their economic analyses and their economic policies.

Other social scientists have also tended to accept these formulations of the nature of the modern economy. For example, Smelser (1963) examines only those activities in the market place as they are affected by or affect sociological variables in terms of status, political position, attitudes, and the family. The discussions of roles, function, authority, status, and change focus on wage labor and business within the context of a market economic system. While recognizing the importance of social dimensions and noneconomic elements in the market system and the impact of the market system on social arrangements, Smelser ignores the economic dimensions of social arrangements which are not necessarily in the market economy.

Within the classical and neo-classical framework of economic thought, a housewife cooking a meal is not performing economic activity, whereas if she were hired to cook a similar meal in a restaurant she would be. A so-called "retired" person who looks after grandchildren during the day is not performing apparent economic activity, although if the same person were hired to care for children by a day-care center it would be economic activity. If a daughter nurses and cares for a disabled or ill parent, she is not considered to be engaged in economic activity; however, the same work performed in a nursing home or in a hospital would be considered as part of the economic sphere. This conception of "economic" excludes activities within the family and the community, and an analysis of the economic dynamics of society based on this conception tends to exclude many economic actors, particularly women, except in their roles as wage earners.

To the extent that households are mentioned by traditional economic theory, they are treated as a collection of individuals who are engaged in the process of production or consumption of goods in the market place. There is no account taken of goods which are produced by a family for its own use—what the Greeks call "householding,"—or of those transactions which transpire as part of the social and kinship relationships which make up major portions of people's lives.

This exclusion is most readily seen in the case of women's work, which for a variety of reasons has generally been excluded from the market economy. The responsibilities of women in relation to household management include the jobs of cleaning, maintenance, purchasing of household and family goods and supplies, cooking, laundering, and household planning. In addition to these tasks, the woman is generally responsible for the care, early education, and protection of children, not to mention the labor of bearing children. Most of these services and goods can be purchased in the market; however, most families are not in financial positions to do so. In addition, the role of the woman, the family, and social relations in times of crisis—during illness of a member of the family or the sudden absence of a wage earner, for example—is crucial to the survival of the family and the maintenance of stability in communities.

If the origins of the term "economics" are traced back to its Greek roots, a basic concern becomes evident. The Greeks gave us the term "oikonomos" which is a compound of the word "oikos," meaning "house, and a derivative of "nemein," meaning to distribute or to manage. Thus the word meant household management. The Latin word "oeconomis" meant specifically household management. When economics is approached from the question of how households are managed and maintained, the limitations of the market economy in our modern industrial society become obvious. For example, the poor know that the market system provides them with limited amounts of income through their wages and that this is often insufficient to meet their normal needs for goods and services and provides little protection in times of crisis. They know that only high-income persons can purchase many of the services they provide for themselves in the maintenance of their households, such as child care and household management. They know the importance of relatives, friends, and neighbors in time of illness, family problems, and loss of the job, for the survival of the people in the family and the maintenance of the household over time.

In searching for alternative economic theories to explain the nature, the extent, and the significance of non-market, non-governmental economic transactions, a clue is suggested by Karl Marx. Marx at one point made reference to a larger conception of the economy in the preface to the *Critique of Political Economy*. The economic structure, he wrote, was "the total ensemble of social relations entered into in the social production of existence." However, from this broad starting point, Marx narrows his concern to the study and critique of market capitalist economies. The bulk of his theory and concerns revolves around the class structure as it is derived from the operations of capitalism.

This same limitation appears to apply to a lesser extent to the work of Engels (1942). In his *Origin of the Family, Private Property, and the State*, Engels attempted to write a history of the family from a materialist and economic viewpoint. He saw production and reproduction of "immediate life" as the determining factor in history. By this he meant the production of the goods and services by which people subsist as well as the production of "human beings themselves." By reproduction he meant more than simply physical reproduction but also socialization and care and protection. Social institutions are in turn conditioned by both these kinds of production—by the stage of development and organization of labor for subsistence and by the organization of the family for reproduction. Engels described the family as being an important determinant of the historical development of societies in that it determined much of what occurred in the process of reproduction.

For Engels, the ties of sex within the family play a more important role in the division of labor in societies in early stages of development; however, as production processes are developed, the family system becomes dominated by the forces of the property system in which class distinctions evolve around the ownership of the means of production and class conflict develops. According to Engels, the role of women was always domestic. In early hunting societies the domestic role of the woman is strong and primary vis-à-vis the male hunter and warrior. With the development of cultivation and domestication of animals, the pastoral shepherd and the farmer with their own property become the dominant force and the calims of property dominate the social relations between the sexes and within the family. Later anthropologists have questioned the view that the social relations within the family were and are always consistent with particular property relations and that the stages of development of production and reproduction have followed universal patterns.

III. ANTHROPOLOGICAL FORMULATIONS

The field which has contributed most to the description of non-market exchange systems is anthropology. Anthropologists tend to examine the economic significance of all forms of social relationships. They also describe the units of social relations in terms of the factors which make them cohesive and in terms of how the units themselves are bound economically, socially, and politically into larger systems.

One of the few community studies that examines the mechanisms and significance of non-market economic activities is the investigation by Arensberg and Kimball (1968) of rural communities in Ireland in the 1930s. Production activities in these rural communities are limited to the production of farm goods and goods produced in the home for use by the family. The family is the primary economic unit with labor divided on the basis of sex and age. The work done by women is considered essential and an integral component of the total production scheme. The work done in the home, the care of the children, the care of chickens and smaller farm animals, and lighter farm work. The father is in charge of the farm as a whole and does all of the heavy farm labor. The children are required to begin working early in life and assist the parent of their own sex in work on the farm.

The economic nature of the family is apparent in all aspects of daily life. A major criterion for the choice of a marriage partner is the potential economic contribution of the individual. Arensberg and Kimball note that the immediate family of husband, wife, and children carry on a kind of "corporate economy" in which the economic aspects are part and parcel of life of this basic social unit.

The secondary economic unit in the social economic system in rural Ireland is the community. The "community" refers to the families whose homes and farms are in the vicinity surrounding any particular farm. The communities will often consist of one or more extended families which are bound by an intricate network of kinship ties. The ties between community members were reported to have strong economic dimensions. The word "cooring" is used by people in these communities to describe all types of non-monetary economic cooperation among neighbors. The word is derived from the Irish "comhair" which is used similarly. The authors observed that this non-monetary economic cooperation between families takes the following forms: lending tools, assisting with labor, lending a member of the family to assist in farm work, pooling goods such as butter, lending a girl when extra help is needed in the household, helping in times of distress or crisis, communal harvesting, and obligations surrounding rites of passage and ceremonies.

Other anthropological works, particularly those concerned with tribal societies, contain many observations and descriptions about the operation of social economic systems which are primarily non-monetized. The works of Richards (1932), Watson (1958), Epstein (1962), Ishwaran (1966), Bohanan and Dalton (1962) are examples of such studies. However, there have been few attempts to treat the economies of these societies outside the framework of classical Western economics. Polanyi (1968) and Dalton (1971) have noted the limitations of the analytic power of traditional economic theory in anthropology and have suggested some alternative categories for a better comparative analysis of economic systems. On the basis of empirical studies, Polanyi suggests that economic activities fall into three main patterns. The first he calls "reciprocity" which is illustrated by the ritualized gift-giving among families, clans, and tribes—as can be seen in the works of Malinowski and Mauss, for example. Another illustration of reciprocity is the cooperation among farming families to assist each other in the building of barns, known as "barn raising," and the mutual assistance they lend to each other at harvest times. In patterns of reciprocity goods and services are given because of bonds which mutually obligate the parties involved. The rights and obligations of each party are usually determined according to some traditional concept of how they are supposed to relate socially.

The second pattern noted by Polanyi was "redistribution." This involves the gathering of economic goods and services to some form of central place—usually controlled by governmental or religious agents—and then redistributing the goods and services throughout the populace. Polanyi notes that many Asian societies and African tribes utilize this economic pattern of behavior and that, like reciprocity, redistributive patterns are characterized by the absence of equivalency calculations and price mechanisms. The principle of calculation in redistribution patterns seem to be one of a kind of "justice" —namely determining what each class in the population traditionally deserves.

The third pattern noted by Polanyi is that of "exchange," by which he means the exchange of economic goods and services within some form of market context. Under exchange, prices and distribution are not determined on the basis of tradition but they are the result of bargaining mechanisms which adjust and match supply and demand.

Although Polanyi and his associates and followers applied these categories to the study of primitive and non-industrial systems, this type of analysis was not extended into the investigation of modern industrial society. Polanyi argues, particularly in his work *The Great Transformation* (1957), that the non-market patterns of economic integration which prevailed in archaic economies were supplanted in the nineteenth century by the growth of the market economy under capitalism. Polanyi argues that the market economy transformed the whole of society and harnessed the economic and productive dimensions entirely to the institutional mechanism of the market. "The rise of the market to a ruling force in the economy can be traced by noting the extent by which land and food were mobilized through (market) exchange, and labor was turned into a commodity free to be purchased in the market." Polanyi argues that in the nineteenth century the organization of production under a market economy—when needs for food, clothing, and shelter had to be met through the purchase of commodities—hunger and gain became purely "economic." Thus, the meeting of basic needs was linked with the production system through the need of "earning an income." In order to get an income, a person needed to sell some goods in the market. For workers this meant that unless they had land to rent or own ownership of a capital means of production, they had to convert their labor power into a commodity and sell it in the market place for wages. "No propertyless person could satisfy his craving for food without first selling his labor in the market." For Polanyi the rise of capitalism in the market economy transformed all resources, goods, services, and labor into commodities which were bought and sold in the market place and thus the whole of society was transformed.

Polanyi, like Marx, overstated the extent to which the dynamics of the capitalist market organized the economic activities of Western societies. There is growing historical, sociological, and anthropological evidence that a large portion of goods and services are produced and distributed through mechanisms other than the market and governmental intervention. It has become clear that low-wage working-class populations are particularly dependent upon non-monetized, non-market systems of production and exchange. The operations of these systems permit working-class families to survive and to reproduce, and in this sense they have also played a crucial historical role in the development of industrial capitalist societies. If survival had in fact been totally dependent upon the income derived from selling of labor in the market place, many workers would have perished or would have had to receive significantly higher wages than they did. In either case, the growth and development of large-scale capitalist enterprises would have been severely limited. In this way, the organization of production and the development of capitalism in Western society has depended upon the existence of other economic units such as the family and the community and the economic functions which they perform. For example, the wage-labor system is sustained by the economically and socially necessary labor of housewives and mothers as they perform such functions as child rearing, cleaning, food preparation, property maintenance, health care, household purchasing, and reproduction; these are all necessary to maintaining life.

IV. TOWARD A THEORY OF
THE SOCIAL ECONOMY

The thesis of this paper is that many goods and services are provided through an economic system which is based on the network of social relationships. I call this economic system "the social economy," since the economic transactions which are being considered are imbedded in and based on the network of social relationships which people maintain over time.[1] These transactions have characteristics which suggest that they be treated as a system. They are a structured set of arrangements for providing material goods and services. In addition, they are governed by certain rules which integrate the transactions and interdependencies and assure the continued cooperation of those involved in the provision of goods and services.[2]

The economic aspects of social relationships in the social economy are usually not primary; the relationships are based on other principles of organization which give them their primary significance and meaning, such as kinship, tradition, religion, friendship, community, or neighborhood. This is not to say that the economic dimension is not important, for it may play a vital role in maintaining the ties between the participants. The rules for the initiation and maintenance of relationships in the social economy derive out of the cultural values and social norms for the relationship in which the participant is engaged. For example, among some ethnic groups it is expected that grandparents will assist in the care and socialization of grandchildren. In many communities neighbors who are friendly with the corner grocer or pharmacist may purchase items on credit without interest, collateral, and without a credit check. Neighbors may borrow food or share appliances. Parents may provide a newlywed couple with some funds to begin a household, and sons and daughters may provide money to help their parents in their old age.

One of the operating principles of the social economy is that of reciprocity. Reciprocity is the mutual recognition of rights, responsibilities, and privileges. This means that a party in a relationship has certain rights to goods and services from others in meeting his/her needs. The others in the relationship have the responsibility to respond to that need. In return the party is obligated to respond to certain needs of the others in the relationship.

In complex situations, the patterns of reciprocity may not be so obviously two-way. For example, members of a particular kin group need not reciprocate with one another but may do so with the corresponding members of a third kin group toward which they stand in an analogous relationship. Among the Trobriand islanders, the man's responsibility is toward his sister's family, but he himself is not necessarily assisted by his sister's husband. If he is married, he is assisted by his own wife's brother—a member of a third family. The point is that the rights and the corresponding responsibilities apply to all participants in the network of social relationships according to mutually accepted criteria.

In communities in which resources and commodities are very scarce, reciprocal arrangements tend to operate as a redistribution mechanism. Scarce goods are thereby spread among members and to those in need. One of the important reasons for this sharing and giving is that those who are poor realize how dependent they are upon familial, neighborhood, and community networks for their day-to-day survival and for assistance in times of crisis. Realizing this dependency, a family will tend to share its resources and services with those in need, with the expectation of having its own needs met at the appropriate time by others within the network.

The principle of reciprocity involves another principle for effective operation—the principle of adequacy of response. Unlike market-exchange transactions in which mathematical equivalencies are computed and in which the value of the good or service you give in an exchange is theoretically equivalent to the good or service you receive and is standardized in money terms, the principle of adequacy of response requires that those responding to a need do so as fully as they are able even though the person in need may not have responded to others to the same extent, owing to his own limitations. For example, a family may pass on outgrown children's clothing to relatives with a newly born baby and the recipient family, in responding to relatives with a time of illness of the mother for a few days, may do the cooking for the donor family. In another instance a grandmother may care for the children of her daughter during the day while the daughter works and in turn may be able to turn to the daughter for extra cash when the rent is due, or utility bills must be paid, or when medical bills arise.

Cohesive stable working-class communities have generally developed intricate and complex systems of reciprocal arrangements which can be effective as redistribution mechanisms within the community. These reciprocal arrangements are also important for the survival and integration of the community itself over time and help cement the social relationships themselves. The redistributional aspect of the social economy is particularly important in integrating groups in communities and assuring the permanence of the arrangement. The participants in the relationships are able to derive a measure of security within a larger society in which they are considered marginal by wage-market standards.

Another allocation principle which operates in the social economy in many communities can be called "command-subordinate." This allocation mechanism is based in hierarchical relationships in which a person in a superordinate position can command another person in a subordinate position to produce and deliver goods and services to him. The superordinate person usually has some form of recognized authority within the relationship or has the coercive resources to insure the hierarchical direction of transactions. This kind of relationship must be distinguished from those legal and governmental relationships where a person is acting as an agent of a legally constituted and recognized government.

Within command-subordinate relations and transactions there may be elements of reciprocity involved. A subordinate, in return for certain services, may be entitled to expect protection and sometimes assistance in times of need. However, the subordinate is not an equal partner in the relationship and does not have the authority or power to determine when and how much the person in command will deliver. These forms of relationships can be seen within certain family structures, within hierarchical kinship systems, within gangs, and within communities which have hierarchically organized patronage systems.

In many cases the superordinate in a command-subordinate set of relationships has many functions. If the person is a political boss of some sort, he may serve as the broker between his community constituency and the larger society. In this instance he is an advocate for the community in obtaining resources from the larger society and an advocate for the larger society in obtaining the community support for outside interests. In some instances the superordinate has the role of keeping order within the society and of maintaining certain community structures. The superordinate may also be an important mechanism by which redistribution of resources within the community takes place. By being able to command goods and services from people, he can also distribute them to others within a social network or community. This form of redistribution resembles the kind of central redistribution principles discussed by Polanyi.

Another allocation principle which operates in many traditional working-class communities is what might be called "ascriptive-prescription." In this case people with certain ascribed characteristics or statuses receive goods and services from others in the network according to certain prescribed rules. Gifts and various forms of assistance when a person has reached a certain age often fall within this category. A person who is considered incapable either physically or mentally may often be entitled to assistance from members of his social network because of his incapacity according to certain customs and traditions. An elder in a community may receive goods and services from people because of his age and status. A widow may be entitled to certain forms of assistance from kin and from people in the community until she remarries. A person who works within a religious order, such as a rabbi, a minister, or a priest, may be entitled to services from the community because of his religious calling and status.

This allocation mechanism can also play an important role in the redistribution of resources within a community. In many instances it serves to provide goods and services to those who are considered to be legitimately incapacitated or exempt from having to engage in wage labor because of some highly valued calling or purpose. The rules which govern this form of allocation are usually established by custom and tradition and these are exhibited most strongly in kin networks, communities, and societies which have resisted many of the values and orientations of modern industrial society.

V. RESEARCH AND POLICY ISSUES

Unfortunately, the nature of non-market allocation mechanisms and the extent to which industrialized societies depend upon such mechanisms have received little attention in research and in the formulation of public policy.

It is clear from existing work on social economic networks that a great deal more research needs to be undertaken. Questions about what social economic relationships and networks promote neighborhood cohesion and stability need to be answered. The importance of such networks for the success of community business ventures also requires assessment. The class differences in network behavior are rarely investigated and utilized in the planning of housing, community facilities, and transportation routes.

Research of the type required cannot be readily accomplished by going to existing statistical data sources. Most of the data has been collected because there were "market economic" reasons for using the information, such as the census of manufacturers and various labor surveys of occupations and unemployment, or because the activities being surveyed were easily reduced to some quantitative measure. Hopefully social economic relationships and activities will be measured in such ways in the future, but at this time we lack the conceptual and empirical basis for finding and developing these kinds of indicators. This should not be interpreted to mean that planners should wait until the field of study of social economic networks has reached an advanced stage of statistical sophistication. The methods exist for qualitative observation and analysis which can tell most of what planners need to know with the richness of detail that can never be possible from quantitative sources such as the census or surveys of housing conditions.

The reliance of planning agencies on existing data sources and on the skills of more statistically oriented and design-oriented professionals also tends to aggravate the economic class bias already built into political structures. With the concern for the viability of the market economy in terms of successful business location and activity and the accessibility of an appropriate labor force, little attention is given to the requirements of the physical and social infrastructure which facilitates social economic activities. Even with better information and understanding of such activities in working-class communities, planning efforts in a social economic direction will have little impact unless the political process is changed so that working-class communities can exert some influence over planning decisions. One of the goals of various efforts aimed at community control and citizen participation was the use of agency programs to maintain and create community activities, instead of the subversion of community structures through the imposition of projects which served outside interests and responded to the priorities of market economic decision-makers.

As the field of social economy becomes the subject of greater investigation, a number of crucial issues need to be examined and confronted. For example, among the most important issues suggested by social economic analysis is the role, status, and significance of the work performed by women outside of the labor market. This means that economic analysis cannot be limited to the factory or office, or to paid work, or to the process of production management (excluding the management of household consumption). Simply putting a wage value on the hours or tasks performed by women in the home is not satisfactory because such activities have generally low market value and because the activities are thereby assessed outside of the context from which they receive their social value.

The issue of work activities by women is closely related to the question of the importance of neighborhood viability and organization. We have little precise information on the process of neighboring and the extent of dependence in working-class communities on these activities, but the work of Stack and the piecemeal data from other projects suggest the need for more extensive investigation of factors of propinquity, intensity, frequency, and infrastructure. For example, one assumption in most literature on cities is that in highly urbanized areas the priority, frequency, formality, and intensity of neighbor relations decrease in favor of relations with the immediate family and personally selected friends. It will probably be found that the validity of this proposition varies with the social economic class of families and with the cultural context in which families operate. As family income rises and members relax traditional ties, the loosening of the networks of obligation, cooperation, and shared experience may result in the declining role of kin and neighbors as significant economic participants in the life of the family and as economic and social intermediaries between the larger society and the family members.

These issues have great import for the success of policy attempts to integrate communities economically so that low- and moderate-income families can live in higher-income communities. Many of the housing programs being proposed tend to overlook the significant differences in the economic activities and dependencies of the populations involved. The dependence upon neighbors, particularly in times of crisis, is greater for lower-income groups than high; the modes of transportation differ because lower-income populations rely more on walking and public transportation; the kinds of recreation that are utilized will depend upon income and the social functions which may be served.

The social economic activities which exist in a community also affect the extent of social mobility of the members. For families to move out of the community often means that they are cut off from the support systems which provide security in times of crisis. As families take the risk of moving and beginning to use income for their own mobility, they become more and more dependent upon their job or the government to provide the kinds of support which the community provided in the past. This is particularly true in terms of the care of children and during times of unemployment and illness. This dilemma means that as long as people remain in the community, they are often obligated to others when their income increases, but that if they move out, they may become more dependent upon governmental and private agency services, unless they can achieve high enough levels of income to provide financial security.

One of the important implications of the state of the social economy is the extent to which needs must be met by alternative mechanisms. As low-income populations are dispersed or women who head households are required to take jobs, the networks within the social economy may be weakened. Support systems begin to disappear and people must turn to private agencies and government for assistance. In many instances, their income is not sufficient to purchase the required services directly, such as day care, and subsidization from the public sector is essential. Currently, there is little work being done to trace the relationship between the operation of social economic networks and the need for governmental services. Indices of the levels and viability of social economic activities need to be developed for use in the planning of urban settlements and in the development of social policies which are more responsive to the needs and relationships of working-class populations.

This does not necessarily mean that policies should be oriented toward reinforcing existing traditional relationships. Many of these relationships are oppressive, constraining, and ultimately unworkable. The point is that even undesirable relationships cannot simply be destroyed. New institutions must be created to meet needs and alternative forms of relationships must be available. At this point, limited attention has been given to kinds of institutions which are required to assist people in providing for their material well-being. Even less work has been completed on the creation of alternative social relationships and social system forms which can be emotionally and economically supportive as well as liberating for the individual members. Some of the experiments in new communities and with communes signify fragmentary efforts in this direction, but they are beginnings at best. More systematic and sensitive research on families and communities, more imaginative experimentation in programs, and greater daring and compassion in the creation of supportive social forms are all required if the social economic issues in the lives of working people are to be resolved.

NOTES

1. For analytical purposes those systems in which economic transactions are primarily based in social relationships are social economies. It should also be noted that there are social economic dimensions in all economies and that the study of these dimensions is the study of the field of "social economy."

2. The literature on social networks is useful in tracing the social economic arrangements and activities discussed in this paper. Of particular value is Mitchell (1969).

REFERENCES

ARENSBERG, C. and S. KIMBALL (1968) Family and Community in Ireland. Cambridge: Harvard University Press.

BOHANNAN, P. and G. DALTON [eds.] (1962) Markets in Africa. Evanston: Northwestern University Press.

DALTON, G. (1971) Economic Anthropology and Development. New York: Basic Books.

ENGELS, F. (1942) The Origin of the Family, Private Property, and the State. New York: International.

EPSTEIN, S. (1962) Economic Development and Social Change in South India. Manchester: Manchester University Press.

FRIED, M. (1965) "Transitional functions of working-class communities," in M. Kantor (ed.) Mobility and Mental Health. Springfield, Ill.: Charles Thomas.

ISHWARAN, K. (1966) Tradition and Economy in Village India. London: Routledge & Kegan Paul.

KOMAROVSKY, M. (1967) Blue-Collar Marriage. New York: Random House.

MARRIS, P. (1958) Widows and Their Families. London: Routledge & Kegan Paul.

MARX, K. (1967) Critique of Political Economy. New York: International.

MITCHELL, J. C. (1969) Social Networks in Urban Situations. Manchester: Manchester University Press.

POLANYI, K. (1968) Primitive, Archaic and Modern Economics: Essays of Karl Polanyi. G. Dalton (ed.). Boston: Beacon.

RICHARDS, A. (1932) Hunger and Work in a Savage Tribe. New York: Meridian.

SMELSER, N. J. (1963) The Sociology of Economic Life. Englewood Cliffs, N.J.: Prentice-Hall.

STACK, C. B. (1974) All Our Kin: Strategies for Survival in a Black Community. New York: Harper & Row.

TOWNSEND, P. (1957) Family Life of Old People. London: Routledge & Kegan Paul.

WATSON, W. (1958) Tribal Cohesion in a Money Economy. Manchester: Manchester University Press.

YOUNG, M. and P. WILLMOTT (1957) Family and Kinship in East London. London: Penguin.

15

Rethinking Urban Problems:
Inequality and the Grants Economy

JOHN P. BLAIR
GARY GAPPERT
DAVID C. WARNER

Progress . . . always passes into retrogression as inequality is developed.
Henry George, 1880

INTRODUCTION:

ECONOMIC KNOWLEDGE AND

URBAN DISTRESS

□ A FUNDAMENTAL CHARACTERISTIC of the urban condition in contemporary America is the existence of both relative and absolute deprivation in the midst of affluence. Yet serious discussion by economists about such questions as income distribution has generally been inhibited by the reward structure of the profession.

A large number of academic economists insist that any analysis must be rigorous to be of value. In practice, rigor often implies applying fairly standard mathematical techniques to trivial or even plainly wrong assumptions about society in order to discover some

AUTHORS' NOTE: *The authors are indebted for comments, critique, and encouragement from Victor Goldberg, Jonas Horvath, Patricia Burke Horvath, and David Porter. The traditional pledge of immunity should perhaps be reiterated in this case.*

"optimum arrangement."[1] Just as a high percentage of mainstream economists close their articles with statements "proving" the efficiency of the price systems and competitive markets, radical economists predictably close by suggesting a conspiracy of the "haves." This dichotomy has seriously inhibited the development of techniques for social policy analysis, particularly concerning questions of income distribution. As Bronfenbrenner (1971) wrote: "For the present, however, distribution issues affect policy decisions mainly by indirection. Their indirect influence typically takes the form of policy paralysis."

Although the debate between radical and conventional economists will continue to flourish, the field of grants economics promises to bridge the gap between the two schools. By providing a forum that gives credence to both approaches, the debate on issues of income distribution and market failure may advance beyond the stage where radicals accuse the mainstream economists of being heartless mathematicians, and radical economists are considered sociologists.

Improved knowledge of the grants economy will have particular relevance for urban analysts because many urban problems center around issues of nonmarket transfers. It is the purpose of this paper to survey the operation of the grants economy as it bears on urban society. Our contention is that the concepts of grants economics, inchoate though they may be, provide a perspective from which urban issues may be reconsidered. If systems of transfers are integral to the urban economy, then improved knowledge of the granting sectors will be an important element in the generation of solutions to many urban problems.

First, we explore the basic concepts of the grants economy in the context of urban systems. Second, recent work is reviewed. Equity and efficiency considerations are emphasized. Next, specific studies of the urban grants economy are discussed, and finally some principles central to a granting approach to urban analysis are suggested.

FUNDAMENTAL CONCEPTS OF
THE GRANTS ECONOMY

The notion that grants economics ought to be a separate field of study was introduced by Kenneth Boulding (1969) in his presidential address to the American Economic Association. The central concept of grants economics is that the study of exchange, whereby A gives something to B and B gives something of equal value to A, is unnecessarily limiting. Grants or one-way transfers (A gives something exchangeable to B, B gives nothing exchangeable to A) are a large and significant part of the economy.

Government expenditures at the federal, state, and local levels amount to more than $300 billion per year. Government economic activity is probably best seen as two sets of one-way transfers. Governments at all levels impose taxes and receive funds from taxpayers without directly providing services on any quid pro quo basis. At the same time, government expenditures amount to one-way transfers to those whom they benefit or harm.[2] Quasi-public activities such as health and hospital insurance allocate more than $40 billion a year from premium payers to institutions and individuals who provide care to selected insurees. Regulated industries also operate partly in the grants economy. Other one-way transfers include intrafamily grants and private philanthropy estimated for the year 1970 at $313.2 billion and $20 billion, respectively (Baerweldt and Morgan, 1971). Finally, even in exchange-dominated sectors there are substantial grants elements within the firm. In regard to the budgetary process, workman A does not transfer to Department Y because the market has mandated it, but rather because his superior has ordered him there (Coase, 1952).

Grants are transfers of goods, services, or assets which could be exchanged. Grants economics encompasses those one-way transactions that are in the economic sphere. However, a pristine conception of a grants economy isolated from the exchange system is untenable. Just as the subsidies and costs of the grants economy determine the terms of exchange, the technical facts and consumer needs underlying the exchange system limit and determine the rules by which the grants economy functions.

Many operations of the grants economy have been unstudied or considered phenomena thought best left to political scientists or sociologists. As the field has increased in importance, a number of economists have tried to incorporate the study of one-way transfers into the rubric of mainstream economics. Such approaches emphasize selfish reasons for transfers such as securing future payoffs, increasing status, or gaining grace. Symbolically an individual's motivation may be expressed by the utility function:

$$(1) \qquad Ua = f(Ya)$$

where $Ua = A$'s utility function and $Ya = A$'s income or vector of goods.

For instance, Mancur Olson (1965) developed an elaborate model which explains most collective action as motivated by selfish interests. Similar extensions of traditional economic analysis are offered by several writers in *Public Expenditures and Policy Analysis* (Haveman and Margolis, 1970). The exchange model has also been adopted by other social scientists and applied to social (Homans, 1961) and political exchange (Easton, 1965; Ilchman and Uphoff, 1969).

Boulding's formulations clearly deviated from narrowly conceived self-interest theories. His work on the grants economy grew out of involvement with the Society for General Systems Research and the *Journal of Conflict Resolution*. This involvement led to his synthesis of conflict theory and general systems theory (Boulding, 1968).

Discussing the development of his thinking, Boulding (1973a) wrote:

A critical question, however, remained unsolved—what elements of the social system make some conflict creative and fruitful and some conflicts destructively damaging to all parties? I concluded that the main problem lay in what I have come to call the "integrative system"—that is, that aspect of society that deals with status, identity, community, legitimacy, loyalty, benevolence and so on, and of course the appropriate opposites.

Granting is presumed to result from utility functions that include the welfare of others such as:

$$(2) \qquad Ua = f(Ya, Yb...Yn)$$

where the definitions are the same as (1) and $1/b...1/n = $ others' income or vectors of goods. Weights assigned to the welfare of others are determined by community and social relationships as well as other sociopsychological factors. Likewise, granting could result from a social consensus or theory of justice which would mandate transfers.

Boulding concluded that the study of the "grants matrix" (that is, who gives what to whom) is an important approach to social integration. Grants are products of integrative relationships and the integration system as well as of threats and the threat system. The grants economy is an important link between economics and the other social sciences. At the integrative level it moves toward sociology and anthropology; at the threat level it moves toward political science and psychology.

SOME REPRESENTATIONS

Granting may be analyzed from both micro and macro perspectives. Figure 1 represents a situation in which transfers from A to B might result from A's altruistic feelings (Boulding, 1962).

Consumption is measured on the horizontal axes. Since the total amount of goods is fixed at 100, as A's consumption increases, B's consumption declines. In reality, the consumption of any person is at the expense of someone else since the consumer may transfer goods to potential beneficiaries if he so chooses. The model can be extended to n-person communities.

The point could be reached where further increases in A's consumption decrease his total satisfaction because he sees B with so little.

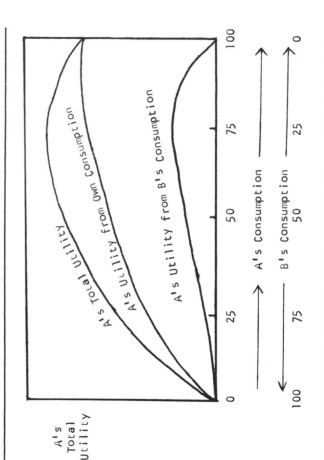

Figure 1: A MICRO REPRESENTATION OF GRANTING

A's utility actually declines, in spite of the fact that he is gaining more goods, because the direct gain to himself is outweighed by pity which he feels for B's miserable poverty [Boulding, 1962: 69].

If the initial distribution was such that A had more than 75, voluntary granting from A to B would be observed. Likewise, if A had more than 75, he might support public redistribution programs which would help individuals similar to B at the expense of those in his own income category.

It is important to distinguish between transfers represented by Figure 1 and satiety in the sense of eating too many candy bars. In the case represented by Boulding, A would gain satisfaction from extra consumption except for the fact that B has so little. It is the interdependence of welfare, illustrated by equation (2), that induces granting.

At the macro level, Boulding has suggested a social triangle as one way of showing economic relationships. Point T in Figure 2 represents a social system organized 100% around the concept of threat, E represents a complete exchange system, and L represents total organization of a social system around the integrative concept of "love." In reality, any particular social system is located at some point within the triangle representing a combination of different organizing relationships. For instance, M represents the "mean-spirited" city of early industrialization. Most exchanges were motivated by market exchange or just plain fear and threat. Point G represents the "good spirited" city of Paul Goodman's *Communitas*

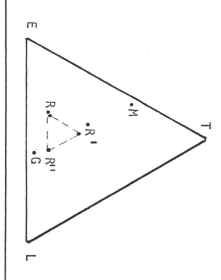

Figure 2: THE SOCIAL TRIANGLE OF THE GRANTS ECONOMY

where both market efficiency and social togetherness dominate exchange relationships. Points R, R', and R'' represent contemporary urban society. Threat and market exchanges are continually being tempered by a concern for social integration or "love."

Figure 3 portrays the institutional relationships and complexities of the urban grants economy. The model is similar to the traditional circular flow diagram of the national exchange economy. The essential difference is that the feedback does not represent a quid pro quo reward for the transfers, but rather represents perceptions used in evaluating grant's efficacy. At each stage, perceived results are compared to desired outcomes—both intermediate and ultimate. Future flows are adjusted accordingly. A and B can even be the same person.

We know much more about the goods, services, and money flows from A to B than the feedback and behavioral aspects of the model. A businessman might make a large contribution to the Boy Scouts because he thinks he likes the organization's values. What happens when he discovers that the modern Boy Scout hardly resembles Horatio Alger? It is in understanding the less quantifiable aspects of the system that urban grants economics merges with other social sciences and where the need for interdisciplinary research is paramount.

Figure 3 is useful in distinguishing between intrafamily and other private grants which are likely to avoid intermediate systems, as opposed to government grants which involve a multiplicity of institutions and are generally coercive. As the effectiveness of one kind of transfer—such as intrafamily support for the aged—diminishes, we can then study alternative approaches to filling gaps.

Intermediate systems increase in importance as urbanization increases since direct granting becomes more difficult due to information and other transactions costs. Furthermore, the same institution may have several roles in the grants economy. For examples, churches not only serve as intermediate granting and recipient organizations but also influence A's motives and B's response. The multiplicity of roles played by government is obvious. Indeed, an interesting aspect of the grants economy is the divergent goals of bureaus of the same organization.

The interaction of granting systems is particularly complex in the public grants sector where the goals of the component parts are diverse. The legislature, through tax agencies, constitutes an intermediate grantor system and such varied organizations as welfare,

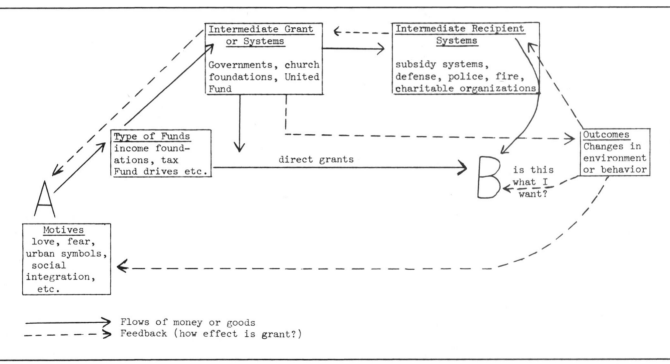

Figure 3: A SIMPLE MODEL OF THE URBAN GRANTS ECONOMY

defense, police, and fire departments receive appropriated funds with which to finance their transfer activities.

Private, government-subsidized granting includes permanently endowed charitable organizations such as foundations and the benevolence of private individuals. Foundations, originally created to avoid death duties and prevent dissolution of control over major firms and industries, now provide support for charitable and cultural organizations and finance innovation in sectors which have little economic incentive to innovate. Individuals give to charitable, religious, health, and education-related organizations and institutions. These gifts seem to be motivated by feelings of benevolence, social pressure, and a desire to invest in social insurance. Much of this expenditure is for club-like institutions constituted for the purpose of joint production of some services which neither government nor the private market adequately provide (Buchanan, 1955; Garn and Springer, 1975). Individual grants may also be attributed to private extorters.

Intrafamily grants incorporate a great deal of inadequately studied behavior. Increasingly, social institutions evolve to provide goods and services that were traditionally provided through the family. Changing roles and entitlements of family members will lead to quite different intrafamily decision processes and resource allocation.

TRANSFER SYSTEMS AND TECHNIQUES

The pathologies of the urban social economy are familiar —poverty, inflation, pollution, congestion, and so forth. If granting is to be an effective technique for reorganizing the social economy to alleviate such problems, an analysis of transfer systems is necessary. Without attempting to develop a comprehensive approach to the evaluation of granting processes, several considerations are central.

First, the grants economy is not isolated from the exchange economy. They overlap and dovetail in complex ways. Thus, a grants delivery system which ignores market forces is unlikely to be successful. Horvath (1971) contended that the pathologies of the grants economy outweigh the efficiency and equity benefits in rural America. A transfer policy designed to improve the housing conditions of the urban poor may only increase housing prices; the intended expansion in the quantity supplied may be negligible, particularly in the short run. Even over longer periods, a large percentage of the transfer expenditures might accrue to landowners, since the supply of land is fixed, rather than increasing the supply of low-income housing.

Since supplies are less elastic in the short run, grantors may become discouraged and discontinue an otherwise viable program when policies fail to show immediate results. Short-run inelasticities account, at least partially, for the boom-and-bust characteristics of many private foundation programs.

Leveson (1972) suggested that cost increases caused by grantings is largest when the expansion of funds is very large; the expanding sector is comprehensive so that similar resources will be less likely to be attracted from other fields; and the resources are highly specialized and have long gestation periods.

The interaction of the grants and exchange economies creates a situation such that exchange pathologies spill over into the grants economy. Policy dissent concerning failure of public grants may be viewed as a failing of the exchange system. For example, are abuses in the food stamp program a result of the grants or exchange economy?

The notion of a sacrifice trap is an important granting concept. Can the grantor cut losses or terminate a grant if it goes awry, or is the organization locked into long-term commitments? Some businessmen, in responding to the urban urgency of the 1960s, funded the right cause with inappropriate methods. Yet attempts to cut programs result in bad press. Publicity costs of shifting or terminating grants were high.

In order to avoid an interminable program, the complex of intermediate grantor and recipient systems is useful. Intermediate agencies, for example, cushion ultimate benefactors if they decide to discontinue support. However, the desire to avert locked-in effect may bias the grants system toward short-term projects. Yet, many urban solutions require long-term commitments. Insufficient longitudinal research problems may be a result of this weakness of granting systems.

"Serial reciprocity" (Boulding, 1973b: 27) is characteristic of many public and private grants delivery systems. Grants from A to B might be analyzed and evaluated according to whether they eventually elicit a grant from C. An epidemiological model might suggest itself where the original grant giver becomes the source of an infection of benevolence. Or perhaps the grant from A might be evaluated as to whether it also elicits grants to B from D and E. This seed money dimension is an obvious granting strategy for many agencies and foundations.[3]

Serial reciprocity is a technique for avoiding sacrifice traps as well as increasing total grants. If A's giving is linked to transfers of others, no one party needs to take responsibility for an abortive program. For example, when children fail to provide support for their parent, each may claim that they would have given their share, but the reciprocity techniques were inadequate; the others did not do their share.

Compulsion, an extension of the attitude that A's contribution be contingent upon the contribution of others, is an extreme form of serial reciprocity. Coercion is an increasing characteristic of urban delivery systems.

Urbanization, coupled with the public goods characteristics of grants, explains increased governmental involvement in the grants economy. To the extent that private contributions are motivated by the desire to see others better off or to eliminate the manifestations of poverty, an individual's benefits from his own contribution will be negligible; rather, benefits will depend upon what others contribute. Therefore, each individual will have an incentive to withhold his own share. The free-rider phenomenon is particularly an urban problem. In small communities, one individual's contribution will have a more visible impact and may induce others to chip in. Of course, other factors such as increased anomie, social disintegration, and invisibility of poverty may also increase the need for coercion in the urban granting process. Poverty may also make the urban grants economy more dependent upon the force of government.

Hochman and Rodgers (1973) empirically tested the relationship between private transfers through charity and urban size for a sample of U.S. metropolitan areas. They found a negative but statistically insignificant correlation between size and per-capita charitable contributions. Thus, although the theoretically expected sign resulted from their statistical work, further research is indicated.

Systems relying upon private contributions may also fail to provide a significant degree of income redistribution. Most private transfers are either intrafamily or religious, neither of which has redistribution of income among social classes as a paramount goal. Furthermore, private granting may itself be regressive. From a study of deductions on federal income tax returns, Vickery (1962: 48) concluded that there is a steady decrease in the proportion of after-tax income given to charity as income increases. Private, voluntary grants were themselves regressive.

Will the professionalization of the grants economy lead to a decline in both voluntarism and in related forms of giving? The desire to expand services causes granting groups to hire professional staff and to make quasi-market evaluations of programs as they adopt misguided economic efficiency. As United Funds have grown in various cities, areas of overlap and duplication of services are often ruthlessly eliminated. This not only affects certain granting groups or agencies, but also seriously impairs relationships between service vendors and significant client populations.

Newhouse (1970) suggested that charitable organizations will increase fund-raising activities to the point where the marginal cost of raising a dollar equals a dollar of benefits to the manager. Such a model, although incomplete, may serve as a basis for analyzing the dynamics of granting organizations. Newhouse implied that the goals of utility maximizing managers must be analyzed at least as carefully as a charitable organization's manifest goals. The analysis of grants delivery systems will benefit from recent theories concerning behavior of bureaucratic organizations.

Additional analogies between exchange and granting institutions provide further insights into grants delivery systems. If urbanization and professionalization tend to impersonalize giving, gifts to various charities may become nearly perfect substitutes. The United Fund fills a role similar to a cartel since vigorous promotional campaigns by one organization might be harmful to the charity industry as a whole just as price cuts by one oligopolist hurts the industry as a whole.

The dynamics of institutions in the grants economy must be explored in order to understand what causes some nonprofit organizations or agencies to grow while others to remain static or decline. The National Polio Foundation's decline is undoubtedly due to the success of polio vaccinations. As it declined, the American Heart Association experienced growth. The changes are probably related, but we do not understand the affects on the entire charity market. The relationship between growth, innovation, and efficiency in a grants environment seems quite different from the relationship that exists in a market context.

URBANIZATION, TRANSFERS, AND RELATIVE POVERTY

Although absolute poverty continues to be a major urban concern, distribution theory is increasingly viewed in relative terms. Veblen (1961: 392) pointed to the importance of relative position:

The existing system has not, and does not tend to make, the industrious poor poorer as measured absolutely in means of livelihood, but it does tend to make them relatively poorer, in their own eyes, as measured in terms of comparative economic importance, and curious as it may seem at first sight, that is what seems to count.

When the focus shifts from absolute to relative poverty, the need for interdisciplinary research becomes paramount. "Relative position" is a subjective concept and exact quantification becomes extremely hazardous. If the concept of relative position, however defined, is relevant to the urban grants economy, three questions should be considered: (1) how important is relative as opposed to absolute position; (2) how does urban living affect perceived relative position; and (3) how do the grants systems affect relative deprivation? Although such questions are on the frontiers of interdisciplinary research, tentative answers are possible.

Richard Easterlin (1973) analyzed 30 surveys concerning the relationship between income and happiness for several different countries over a period of twenty years. In spite of the variations in his samples, the results were consistent. Happiness increased with income in cross-sectional studies, yet the general level of happiness has not increased over time. Such results are what one would expect if happiness were a function of relative rather than absolute income. Furthermore, Vickery (1962), reporting on laboratory research undertaken by RAND, stated that in situations where rivalry exists, "there is a natural tendency on the part of individuals to maximize their relative rather than absolute position."

Of course, relative economic position is a subjective concept. A worker will have a different perception of his status depending upon whether he lives in a wealthy suburb or a black ghetto. Urbanization probably increases the perception of relative poverty among the urban poor; the low-income neighborhood may be but a few blocks from fashionable shopping areas. The hierarchical nature of urban society undoubtedly affects relative deprivation.

It seems paradoxical that concern over urban poverty has increased at the same time observers cite urbanization as a factor in increasing absolute income. One explanation is the importance of relative status in an urban society.

Harrington's (1963) conclusion that urban poverty is invisible because the wealthy seldom see the slum areas and because the poor, when they are seen, do not look obviously deprived (due to mass-produced, cheap clothing, for example) is true in terms of absolute poverty. It is less true if the concern is on relative poverty since subtle and not so subtle ways of showing relative income levels are constantly being devised.

Does the grants economy respond to relative economic position? Since, at any given time, transfers increase both relative and absolute income, empirical observation is difficult. Schwartz's (1970) study of personal philanthropy showed that the elasticity of charity was more responsive to income using cross-sectional than time series data. This is consistent with the hypothesis that relative position rather than absolute income is the more important determinant of voluntary transfers. On the other hand, much public granting, particularly transfers in kind such as food stamps and public housing, seem designed to eliminate absolute poverty or at least the offensive manifestations of poverty. Can granting eliminate relative poverty? The subjective nature of much relative poverty implies that market-created inequalities may not be as responsive to grants as absolute poverty. The stigma of welfare, for example, may in some cases be as much a source of unhappiness as absolute poverty.

EQUITY, EFFICIENCY, AND THE GRANTS ECONOMY

Much of the concern by economists analyzing the grants economy has been focused around two questions: (1) do granting processes lead to more or less equality, and (2) do granting processes contribute to economic and social efficiency? The next two sections summarize the leading arguments relating to these two issues.

EQUITY

Although the need for a more equitable distribution of income is widely accepted, at least seldom explicitly challenged, economists have no consensus as to what the income distribution ought to be. Hochman and Rodgers (1969) claim that income redistribution can be approved on Pareto grounds (a change is good if no single person is made worse off). However, this approach has been rejected by many on the grounds that its rigorous mathematical formulation has little operational content (Mishan, 1971).

In an early attempt at a normative theory of distribution, Lerner (1944) developed the idea that since the marginal utility of income declines as income increases and since we do not know which individuals get the greatest utility from income, any increase in equality will most likely increase total welfare. Transfers from the rich will probably cause smaller loss in utility than the corresponding gain from the increased income received by a poorer individual. Nevertheless, the consensus is that questions of optimal distribution are beyond the scope of legitimate inquiry by economists. In the absence of a generally accepted, normative theory of distribution, the issue has been eschewed. Principle issues relating to equity which have been raised and studied by economists working in the grants area are: (1) developing alternative methods for moving most of the poor above the poverty line; (2) building better analytical techniques for measuring the distributional impact of policies or programs; and (3) discussing the impact of particular antipoverty programs on income distribution.

Alternative Methods for Ameliorating Poverty

With the creation of the War on Poverty it was necessary to determine who was poor. The Social Security Administration defined a poverty line determined by income and family size. Lampman (1972a) found that a little more than a third of the poor were lifted above the poverty line in 1966 while 10 million remained below after transfers of $39.2 billion. Okner (1972b), using 1969 data, found that transfers raised the poor's share of income from 3% before taxes and transfers to 10% afterwards. He noted that

the twelve percent average poverty rate obscures the appallingly high incidence of poverty among certain subgroups even after receipt of such payments. Thirty-three percent of all nonwhite families remained poor even after transfers, whereas the incidence of poverty among white families is only ten percent . . . for nonwhite families headed by a female, the chances of remaining poor are better than six out of ten even after receipt of transfer income.

His optimistic conclusion that "redistributing income from the nonpoor to the poor is feasible and need not be too painful" is an economic and not a political judgment.

Improperly designed income-maintenance programs can induce perverse behavioral changes. Green and Tella (1972), using existing census data, investigated the effect of nonemployment income on hours and weeks worked. They found that the income and substitution of an income transfer plan which raised family income from $600 to $1,000 annually and reduced the marginal wage rate 25% to 50% would lead to a significant diminution of work effort. Michael Taussing (1972) argued that automatic, impersonal approaches to income maintenance should avoid isolating the poor into permanently separate, nonparticipating, dependent subpopulations as has happened in part with the aged. The Social Security system has discouraged work and limited intergenerational responsibility. Also AFDC programs have built-in incentives which break up family units by strict enforcement of different subsidies to differently constituted family units. Although grants may have useful integrative functions, Taussing warns that grants could just as easily create a very objectionable socioeconomic class structure. These studies and Taussing's caveats are important beginnings in the continuing discussion that must take place concerning serious attempts to eliminate poverty.

Development of Analytical Techniques

Analytical techniques for measuring the distributional consequences of particular projects or programs are also important. Cost-benefit analysis is too often performed independent of the distributional outcomes. Equity questions are assumed to be settled by the "public choice" process (Harberger, 1971).

Several articles proposed explicit consideration of the unemployment figures of a state when measuring the benefits of federal grants (Bahl and Warford, 1972). Furthermore, since poor communities and poor people are generally less efficient at producing goods publicly, equity considerations should be explicitly taken into account by weighting benefit-cost ratios by a welfare index when engaged in project selection (Aaron and McGuire, 1972). Approaches to equalizing resource allocation criteria are interesting, but a difficulty persists. Although equity criteria would suggest subsidizing projects in poor areas with high unemployment, there is no guarantee that the poor and unemployed in these poor areas actually receive the benefits of the projects. For instance, it would be possible to argue that certain kinds of irrigation projects subsidize technologies in agriculture which have pushed the poor off the land. This is probably not an isolated example. Area-oriented equity projects may not touch the disadvantaged population at all, particularly if the target population lacks the traits necessary to benefit from the creation of stable, primary labor forces.

Equity of Existing Programs

Benefits of existing government programs do not necessarily accrue inversely to income. Stigler (1970) and Tullock (1971) recently argued that the effects of government policies in Social Security, urban renewal, public education, and the farm program is that the poor and the rich are subsidized by the middle class. Victor Goldberg (1974c) took issue with their methodology, noting that "if these programs are redistributive, what then of enforcement of private contracts, right to work laws, restrictions on class section suits, or the common law protection of private property?"

Aaron and McGuire (1972) showed that previous studies of the distributional impact of public expenditure have implicitly assumed utility functions which are not applicable. For example, allocation of public goods benefits by family-income class amounts to the assumption that the marginal utility of income is constant as income rises. They argue that studies focusing on income brackets alone obscure the more significant distributional questions, such as the differences between the impact of public budgets on the working poor and the impact of such budgets on the dependent population. The need for a general approach to the fundamental question of whether or not the poor are made better off by government action should not rule out the careful analysis of the distributional impact of current arrangements in particular sectors.

The grants aspects of the tax system have been inadequately studied, although many assertions are regularly heard about the progressivity or regressivity of particular taxes. Pfaff and Pfaff (1972) looked at three sources of "implicit public grants" in the income tax system: tax deductions, tax credits, and the taxing of taxpayers, such as married couples, by a different schedule than others. They found that the implicit public grants were large ($63.9 billion in 1965) and regressive. However, progressivity and regressivity depend upon some norm of equity. Talking only about part of the whole economic system is tenuous. Everyone in the system benefits from some subsidy so that net subsidies between groups are not easily measured. In addition, Gabriel Rudney (1972) analyzed the system of tax aids accounting developed by the U.S. Treasury Department over the last ten years to measure the implicit grants from exemptions and subsidies in the tax system. This definition of an implicit grant is more restrictive than the Pfaff and Pfaff definition for the whole tax system, since Rudney estimated that tax aids amount to less than $50 billion. The alternative definitions results not only from difficulties in identifying subsidized flows, but from "assumptions about the normative income tax structure."

Microeconomic policy has major impacts on income distribution. Hollister and Palmer (1972a, b) claimed that a poor people's price index would show that the expenditure effects of inflation for families below the poverty line have not been deleterious to the poor since World War II. Rather, if inflation is the price the poor pay for expansionary periods, then their incomes would be enhanced more than proportionally since their fortunes are so closely tied to the unemployment rate. Hollister and Palmer have also assembled evidence implying that most poor, even the aged poor, do not receive substantial portions of their income from fixed value assets vulnerable to inflation. They concluded: "Surely if middle or upper income people were asked if *they* themselves were willing to bear the tax of unemployment in order to remove the tax of inflation, they would answer resoundingly no!" Thus, while it may be true that the poor are hurt by inflation, the alternative of high unemployment is even less desirable. Unfortunately, more recent experiences of recession and inflation occurring simultaneously make this analysis somewhat less conclusive.

A number of articles dealt with particular sectors or programs. Charles Schultze (1972) pointed out that most farm subsidies, which amounted to more than $5 billion in direct payments and perhaps $7 billion in higher prices to consumers annually between 1969 and 1970, seem to be proportional to the output and inverse to the income of particular farmers. Such subsidies become capitalized into the value of the land and newcomers buy in at that level. After a long enough period, subsidies become frozen into asset values and their removal causes hardships. Stuart (1972) showed that the Medicaid program has distributed real income from poor to wealthy states. The reasons for Stuart's findings are complex, but two explanations are that (1) the fiscal effort required by the poor states in spite of an 87% federal matching provision as opposed to 50% in the richer states, and (2) the relatively high political power of medical schools, hospitals, and other corporate medicine in industrial states compared to poor rural physicians who would benefit from more comprehensive Medicaid benefits in poorer states (Warner, 1975b).

Hansen and Weisbrod (1972) estimated beneficiaries of the higher education system in California by income class. Hartman (1972) summarized the raw facts assembled by Hansen and Weisbrod and concluded that (1) poor people pay taxes and very few of them use public higher education. Those who do, gain thereby; those who don't, don't. (2) Middle-income people are heavy users of the system, and their taxes do not cover the costs. (3) A very few rich people use the system and gain handsomely thereby. The rest of the rich pay substantial taxes and get no direct return. There is no clear way to reduce these facts to a statement of who gains and who loses. Hansen and Weisbrod have since taken this lesson to heart (Hansen and Weisbrod, 1972).

The recent work on equity has not evolved to the point where definitive principles can be put forth. Since no single theory or technique is clearly superior, we can expect continued controversy over equity issues. Improved conceptual and statistical tools will be forged from such a debate. The articles in the Boulding and Pfaff (1972) volume are very useful because a number of different approaches to income distribution are under one roof.

There may be no consensus on what a "just" society is (see Rawls, 1971), but that does not mean that we should not study the impact of nonprofit activity more closely than we have heretofore. Indeed, much more analysis is required to reduce the unfortunate implications of Robert Goldfarb's rule that policy is made by coalitions between people who think they will be better off and people who know they will be (Goldberg, 1974b).

EFFICIENCY AND THE GRANTS ECONOMY

In addition to responding to impulses of benevolence and appreciation of threat, the grants economy is intended to compensate for the market's failure to achieve a preferred arrangement.

Arrow (1974) took an extreme view when, in his presidential address to the American Economic Association, he said that welfare economics should sometimes be considered an empirical discipline. He believes that if an opportunity for a Pareto improvement exists, there will be an effort to achieve it by developing a market if that is practical. If a market is impractical, Arrow predicted that some other social device will develop. Such a position is consistent with the sociology of Talcott Parsons (1937), for whom all institutions of society are integrative and functional. Without being quite so sanguine about the good effects of all activity,[4] we can still see that grants often help the economic and social system work more efficiently. The principal problem is identifying those circumstances under which grants are more efficient than exchange at improving social conditions.

Various sources of market failure have been identified by many writers (Bator, 1958; Buchanan, 1968; and Burkhead and Miner, 1972). The ones most often mentioned with regard to efficiency are: (1) positive or negative externalities (spillover effects) engendered by private activity which are not captured by the economic agent in the first case and are not included in the cost he bears in the second; (2) public goods which will be consumed by many and from which the consumer cannot be excluded; and (3) the failure of certain markets to function because of high information and transactions costs.[5]

The case of externalities has been treated very comprehensively. An example of "positive" externalities would be the case in which a firm engages in general training of its employees (teaching some to be accountants, for example). The employee can appropriate the benefits of the general training and go to work for another firm. For this reason firms do not usually provide general training, leaving it to educational institutions.

Negative externalities also have been widely discussed. Indeed, five of the articles in the Boulding and Pfaff volume (1973) deal with the way in which transfers can function as instruments of urban ecological policy. Freeman (1973) noted that three approaches to pollution control have been effluent or user charges, judicial sanctions, or grants to dischargers who satisfy certain specified pollution-control activities. He preferred the first solution, with the provision that when harm is done to groups of employees or localities the government will arrange a method of compensation. Much of the distributional impact of the ecological revolution will be to enhance the quality of life and value of the property of the rich.

Thomas Havrilsky (1973) picked up the theme and in addition suggested that the prices of necessities are likely to increase more than the price of luxuries. He predicted that idealistic technicians and better-functioning subsidized markets for information will lead to more responsible public policy regarding the environmental impact of activity in the exchange economy.

Schmid (1973) contended that property rights seem to fall outside traditional exchange or grants analysis. American property law favors use of property such as irrigation or waste disposal to the detriment of public goods use such as fishing or scenic enjoyment, even when the latter constitute prior uses. Daly and Giertz (1973) showed that if there is substantial joint consumption of pollution, any individual will have some "free-rider" motivations and not pay as much as he would be willing to pay in order to abate pollution. Further, they point out the high transactions cost of bargaining solutions which some writers have proposed as alternatives.

In allocating public goods, the granting problem is that the exchange economy does not always generate efficient solutions and therefore some mechanism has to be developed that determines how payments for the service are levied, who will receive what, and how efficiently the service will be provided. These "public choice" problems are extremely complex and are increasingly being studied by economists, political scientists, and others interested in policy. The determination of an efficient solution becomes even more difficult when one realizes that for many citizens participation in communal and political activity is an end in itself and not just a means to more goods and services for themselves and their loved ones (Hirschman, 1971).

Finally, there are substantial public and collective activities which are undertaken because of the high information and transactions costs associated with market alternatives. Kenneth Arrow (1963) pointed out that voluntary and charitable hospitals can be explained as an attempt to reduce risk in a world in which the market for health insurance functions imperfectly. Clifford Geertz (cited in Coleman, 1973) described credit associations in Southeast Asia where one person invites his neighbors (somewhere between 10 and 30 in number) to participate in a continuing set of social occasions carried out weekly or monthly. The total of the contribution at each session goes to one person. These associations are used to make capital investments which could not be made either because of the absence of saving habits or because of the absence of any kind of lending institutions at a reasonable rate of interest (Coleman, 1973). In most primitive societies there are extremely complex forms of gift exchange and reciprocity which serve to confer status among the givers and to reduce uncertainty in the social system (Belshaw, 1965).

Nonmarket activities often evolve as the most feasible solutions to people's needs. This does not mean that such activities are automatically carried out as efficiently as possible. A burgeoning literature has developed concerning techniques for improving the efficiency of the grants economy. Nearly all are agreed that simply adopting the jargon of efficiency and its appurtenances as was done with PPBS is insufficient.[6]

URBAN SYSTEMS AND THE GRANTS ECONOMY

A comprehensive, explicit approach to issues of urban granting processes is long overdue. After all, the pathological consequences of the failure of national grants and market economies to provide for adequate social and economic integration are most pronounced in the cities. Even in the best of urban worlds, market forces operate so as to create an underclass, "the lower city."

Improving the operation of markets may not reduce the size of the economic underclass, especially if relative income is important. Alternatively, "special markets" may be encouraged to create parallel forms of socioeconomic participation for the members of the lower city. Educational vouchers, food, heat, and transportation stamps as well as housing allowances, are devices that create special access to markets. Such attempts are perhaps more desirable than providing special services such as those associated with case work. At the same time, reliance upon special markets that are also subject to additional exchange pathologies might represent an undeserved veneration of the market economy.

We need to rethink urban problems (and solutions) around the notion of the grants economy. An ideal political-economic theory of grants would postulate the ideal regulation of the market economic by a set of transfers. In addition, a complementary social-economic theory of the grants economy would postulate the ideal support of a series of urban lifestyles (along a spectrum of income levels) by an additional set of directed transfers. Granting out of love and/or fear might be a more efficacious if not humane way of dealing with the persistence of the urban subeconomy. Perhaps the city should be viewed as the unfortunate poorhouse of the affluent metropolitan nation.

Neither a comprehensive theory of the grants economy nor an integrative model of the changing urban system has been developed. Most of the work on the urban grants economy has been less ambitious.

Buchanan (1950) concluded that horizontal equity—equal treatment of equals—required that the higher governmental unit compare the fiscal position of individuals within its subdivisions. Unless all citizens receive equal treatment in terms of payments and benefits from government activity, some citizens will be treated unfairly. "Fiscal residuum" should be equalized regardless of residence. Therefore, intergovernmental transfers should flow from wealthy suburbs to the cities. Alternative definitions of intergovernmental equity have been proposed. However, Scott (1952) suggested that indifference between locational choices of individuals of equal income is a more valid criteria for determining the distribution of federal grants.

Leveson (1973) took an alternative theoretical approach by starting with considerations of the objective function of a granting strategy. Building upon the new consumer theory of Lancaster (1973), Leveson developed a lifestyle approach to solving urban poverty. If we wish to improve the well-being of the urban underclass, income (and national granting systems of income maintenance) is not enough. It is also necessary to shift families into "more productive lifestyles," and improve the efficiency of resources used for consumption. He suggested that these improvements could be advanced by a number of changes in the financing and production of service components as well as by the integration of services at the neighborhood level.

Additionally, Leveson demonstrated that the efficiency-seeking approach to the provision of social services leads to more people being served but at a reduction in quality: "it appears that the desire for equity is frequently pushed to the point where services produced are of such low quality that they lead to very little if any improvement in well-being." Efficiency criteria, such as persons served per dollar, are inadequate since quality is not held constant.

An improvement in quality usually requires more resources per person served. A coordination of granting processes at the neighborhood or community level could lead to such quality improvements. This suggests that the urban grants economy which exists in the metropolitan centers of the United States must begin to evolve different evaluative criteria than those provided by the market economy model.

Although grants economists continue to be concerned with theoretical models the focus has shifted to empirical attempts to identify the distributional pattern of tax incidence and public expenditures.

Pfaff (1973) has looked at the distributive and poverty-reducing effects of social welfare payments in the twelve largest metropolitan areas. (These include Baltimore, Chicago, Cleveland, Detroit, Houston, Los Angeles, New York, Philadelphia, Pittsburgh, St. Louis, San Francisco, and Washington.) Her results are summarized in Table 1. Using figures for five different welfare programs in 1966, she showed that the poor areas received only about 7.4% of some $32 billion in welfare expenditures. Less than $2.5 billion went to the large urban ghettos of the twelve major American metropolitan areas. American society, contrary to political myth, has not been pumping large amounts of money into these areas.

A different approach is represented by the studies of Neenan and Mellor. Neenan (1973) analyzed data from the Detroit area to see if suburban communities adequately compensated the central city for expenses that it undertakes on their behalf. Mellor (1973) analyzed the Shaw-Cardoza area in Washington to determine if a low-income area is self-sufficient regarding its needs in the areas of safety, education, and welfare.

TABLE 1

DISTRIBUTION OF BENEFITS FROM SELECTED WELFARE PROGRAMS
IN THE TWELVE LARGEST SMAs, 1966

| Program | U.S. Total (in $ billions) | Percentage Allocations | | | | |
		12 SMSAs	Fringe Areas	Central Cities	Poor Area
Social Security	20.0	38	18	20	5.4
Unemployment Insurance	1.8	50	25	25	9.3
Workmen's Compensation	1.2	48	30	18	4.5
Public Assistance	4.3	42	19	32	20.3
Veterans Pension	4.4	34	10	14	4.3
Total All Programs	32.1	39	18	21	7.4

SOURCE: Derived from Pfaff, 1973: 95.

Neenan concluded that if benefits are measured by a "willingness to pay" criteria, suburban municipalities receive a welfare gain in various degrees at the expense of Detroit. For a family of four, this welfare gain was estimated to range from $7.00 to over $50.00 a year depending upon the frequency of contact with the facilities of Detroit.

Interjurisdictional public sector benefit flows were measured by a theoretically constructed willingness-to-pay multiplier applied to the basic cost of service for the following categories:

(1) Benefits provided directly by public sector to users
  a. Library
  b. Art museum
  c. Zoo
  d. Recreation and parks
  e. Streets and traffic control
(2) Costs of poverty in the metropolitan area
(3) Benefits provided indirectly by public subsidies
  a. Water department
  b. Tax exemption for hospitals, universities, limited access highways

A whole series of elaborate empirical determinations were then made by Neenan to demonstrate that the disproportionate (and involuntary) assumption of metropolitan responsibilities by the central city of Detroit constitute a welfare grant to the six suburban communities studied.

Mellor estimated that the Shaw-Cardoza poverty area in the District of Columbia pays taxes to all levels amounting to at least $40 million and possibly as much as $50 million. At the same time the benefits received from all levels of government were estimated at between $34 and $64 million. The wide variation in his estimates illustrates the fragility and incompleteness of the data available. However, as with the Neenan study, the development of a number of theoretical constructs as a guide to detailed empirical interpretation of explicit and implicit granting processes is an important pioneering contribution.

Table 2 shows estimated outflows to government bodies from the neighborhood for the fiscal year 1967 and "feasibly probable" estimates for benefits received. (The neighborhood had an estimated population of 83,700 in 1966 and a median income for families of $5,600.) The result is a small net outflow of $0.8 million.

TABLE 2

OUTFLOW/BENEFIT: SHAW-CARDOZA = 1967

Outflows (in $ millions)		Benefits (in $ millions)	Feasible
Contributions for social insurance	9.72	Education (incl. library)	13.06
Federal income tax	14.95	Police	2.14
D.C. income tax	2.65	Courts and corp. counsel	0.56
Property taxes	2.95	Corrections	0.47
D.C. sales tax	2.10	Fire	1.02
Burden of business income taxes	7.71	Other public safety	0.23
Tobacco and alcohol excise	1.73	Welfare department	8.19
Telephone excise	0.15	Health and hospitals	8.15
Auto and gasoline	2.67	Vocational rehabilitation	0.25
		Recreation department	0.90
		National Park Service and zoo	0.22
		Highways and traffic	0.60
		Sanitation	0.50
		Urban renewal	0.00
		Overhead	1.50
		Unemployment compensation	1.33
		OASDHI	4.72
Totals	$44.63		$43.84

SOURCE: Derived from Mueller, 1973: 55-56.

These two studies are useful, not necessarily for their conclusions but because they provide an empirical beginning to the examination of the outcomes of intraurban granting processes in metropolitan areas.

Mueller (1973) also studied the effect of several categories of implicit public grants received through land use and property tax provisions in two suburban counties in Maryland. He concluded that the substantial loss of tax revenue from these implicit granting provisions results in additional inequity to the county tax structure.

A more thorough analysis of the urban grants economy would require a similar but expanded examination of both public nongovernmental granting processes and private inter- and intrafamily granting processes. Before more empirical work can proceed, perhaps more explicit theoretical and institutional frameworks will have to be developed.

NEW DIRECTIONS IN

URBAN ORGANIZATIONAL ANALYSIS

Although exchange relationships will continue to dominate the attention of economists, the study of granting provides an alternative and necessary approach to understanding the organization of the urban social economy. Granting plays an integral part in urban development. As Boulding (1973a: 42) claimed, the nature of grants "is critical in determining whether there will be deterioration of cities as a direct result of the failure of the grants economy." Based upon the emerging concepts of grants economics, we suggest several observations around which urban problems may be reexamined.

First, *the same markets which create wealth also create poverty (the Henry George Rule)*. The duality of wealth and poverty is particularly apparent with respect to relative economic position. As my neighbor improves his position, mine deteriorates. In absolute terms, city markets often do not function in isolation from one another and from social conditions as in economic textbooks. Anthony Downs (1970) referred to a "frontier of growth" and a "frontier of deterioration." In the land and housing markets of the last two decades, the forces of speculation and subsidies created at least modest fortunes as the impetus of suburbanization led to rapid increases in the value of suburban property at the same time the central city deteriorated. Similarly one might refer to the dual labor market theorists (Doeringer and Piore, 1971) and suggest that the same phenomenon of wealth creation and economic deterioration also occurs in the urban labor market. Some have argued that the combined effects of structural changes in the American economy have been to "twist" the demand for labor: to push up the demand for highly skilled, highly educated labor and to push down the demand for lower skilled workers (Killingsworth, 1971).

Second, *wealth, particularly in urban areas, is created and destroyed by involuntary, one-way transfers*. This characteristic of the grants sector is widely recognized. The behavior of many actors in the urban milieu may be analyzed in terms of strategies for capturing grants and thereby increasing personal wealth. The controversies surrounding zoning changes are local examples. Wealth is also created by coercive transfers outside the urban area by state or national authorities. The disproportionate allocation of the federal defense budget to politically advantaged states such as Florida,

Texas, California, and Washington are examples. Declines in wealth have resulted from enforcement of environmental standards on taxpayers in industrial cities rather than through consumer charges. Our second proposition is a grants economy counterpart to the Henry George rule: grants create poverty as well as wealth, even though no exchange takes place.

Third, *a well-functioning city or urban system ameliorates problems of its subeconomies and attempts to integrate its social system through a series of grants.* Even a well-balanced, economically efficient city (however defined) will generate subeconomies and economic underclasses that are associated with various problems. Indeed, much of the advantage of urban areas rests upon the subeconomies. Short of the utopian elimination of the urban underclasses by a national incomes policy, the local social economy will develop an integrative system of transfers.

One of the authors concluded that "the evolving aspects of the regional economy" must be explored "so as to obtain a better perspective on the potential for achieving social objectives through the forces of economic growth. The leadership (both grassroots and establishment) supported by appropriate research could then link economic growth to social objectives on a continuing basis" (Gappert, 1973). A system of automatic integrators can be as valuable as the automatic stabilizers that are part of aggregate fiscal policy. Unfortunately, such local approaches to granting are often swamped by pervasive national forces such as inflation, tight money, induced recession, and so forth.

Fourth, *transfers tend to become permanent property rights.* Television licenses, for example, were granted for a period of years and renewal hearings were required for an extension. Currently, however, licenses are considered private property, the value of which is capitalized into the value of the station. The ownership has evolved to the point where government action to reclaim the license is viewed as imposing a hardship upon station owners. Likewise, rules of regulatory agencies create wealth but, once enacted, often take on characteristics of property rights. The concept of parity certainly implies an entitlement.

Not all grants evolve into property rights. Welfare and other transfer payments which are subject to legislative review may be easier to decapitalize. On the other hand, grants which appreciate the value of capital such as transportation licenses and farm subsidies seem to be viewed as entitlements.

One reason for the permanency of capital-appreciating grants is that beneficiaries will theoretically spend up to the net appreciation in market value caused by the grant (that is, the capitalized value) in order to retain it. Since the amounts are frequently large, very effective lobbying efforts result.

A principle of taxation is "an old tax is a good tax." The idea is that once the market has adjusted to a tax, changes will result in readjustment costs which could be just as harmful as distortions caused by a tax. An example is the undesirable effects which would result if the interest deduction of homeowners for federal income taxes were abolished. The market has adjusted so that a change will cause additional distortions. Since the homeowner subsidy is also a transfer, perhaps some of the rigidities in the grants economy may be justified by reference to an analogous fifth principle—*an old grant is a good grant*. Thus, the fact that spot zoning may have been a grant which creates unjustifiable wealth does not imply that a repeal of the variance will correct the original effects of the grant.

Sixth, *urbanization, by virtue of the breakup of the extended family and the small geographically-bounded community, has accelerated the replacement of private giving and helping with public granting* (either through tax-supported expenditures or through tax-exempt provisions for private charity). It is interesting to note, for instance, that the annual United Fund budget of about $10-12 million in Milwaukee is now equaled and exceeded by the $12-14 million budget of one local poverty agency, the Social Development Commission. It receives a growing mix of city-county-state and federal funds. During the same period the top twenty or so foundations in Milwaukee (those grants over $100,000) provided an annual total of over $7 million in grants in 1970.

Nationally, in 1971, United Way's nonwealth-related fund drives raised $740 million for charitable activities, as compared to federal grants to state and local governments of $6.5 billion for social programs other than welfare and $9.6 billion for welfare (U.S. Bureau of the Census, 1973).

Finally, the behavior of granting systems is dependent upon the behavior of bureaucratic organizations. Niskanen (1971) suggests that bureaucrats maximize budgets and output and therefore produce as much as possible. An alternative view is that bureaucrats are utility enhancers and that they often receive satisfaction for producing less than the budget-maximizing quantity if they thereby generate more resources which will be available for them to spend as they like (Migue and Belanger, 1974).

A more general approach states that individuals and organizations mobilize resources from multiple sources. A resource mobilization approach to studying the behavior of cities, state, and school boards in a system of fiscal federalism would appear to be especially appropriate. Earlier research found that school systems facing different mixes of local and federal resources behaved quite differently in their organization and administration of the school system (Porter and Warner, 1973). It was found that school districts and indeed most organizations mobilize resources from the margin- ally most productive source and then practice multiple-pocket budgeting, spending the most restricted funds first, in order to achieve their own objectives. Other studies seem to show very different outcomes from expenditure for similar services under different fiscal arrangements (Campbell and Sacks, 1967). As federal money poured over the cities during the past decade there was only a limited response in the bureaucracy that delivers the services. Edward Hamilton likes to quote Mayor Washington as saying, "Our tax base increases only 3 or 4 percent per year while our expenditures have increased at a rate of 10 percent." This only means that there has been a lot of federal aid in the past, not necessarily that needs have been increasing as rapidly or as slowly. Indeed, what is required is a thorough study of how to make the grants economy more effective at achieving its ends not only in a world where there is productive inefficiency but also in one where large organizations and complex social arrangements lead to quite different outcomes than the initiators of the programs intended.

The principles discussed above are not intended as definitive axioms from which the entire grants economy can be understood. Indeed, it is because knowledge of the grants economy is incomplete that recent research in this area has been so provocative. It is clear, however, that sole reliance on exchange relationships will fail to provide a complete model of the urban social economy.

As the long overdue debate on economic inequality and social justice accelerates, we must not lose sight of the interaction between the exchange and the grants economy. Granting will play an increasingly important role in the evolving urban economy. With the advent of revenue sharing as a special impetus, the elaboration and analysis of the grants economy should become a rather urgent interdisciplinary research priority. Further analysis will enable us to better understand who gets what from the complex system of grants. Whether such research is actually undertaken, however, depends to a large degree upon the grants economy itself.

NOTES

1. For example, Schall's (1972) conclusion that a condition for a Pareto Optimum is

$$\frac{\partial U_a/\partial X_a}{\partial U_a/\partial Y_a} \quad \frac{\partial U_a/\partial X_b}{\partial U_a/\partial Y_b} = \frac{\partial U_b/\partial X_b}{\partial U_b/\partial Y_b} \quad \frac{\partial U_b/\partial X_a}{\partial U_b/\partial Y_a}$$

where U represents a utility function, X and Y represent goods and a, b are subscripts denoting individuals.

2. Porter (1975) makes the point that most legislatures recognize the different clienteles and strategies used in tax policy and expenditure policy by having separate committees on Ways and Means (taxation) and Appropriations (expenditures).

3. For interesting discoveries of how such funds or projects can be co-opted see Selznik (1949) and Dresch (1971).

4. Demsetz (1969) calls the view that government activity is called for and will work when market failure exists the "nirvana approach."

5. Goldberg (1974) and Roumasset (1974) both discuss a number of the putatively positive approaches to public expenditure.

6. An excellent early discussion of the development of government budgeting and the techniques involved is Burkhead (1956); for a discussion of the political element left out of most PPBSEC, see Gross (1969); and for a review of the government's experiment with PPB, see Schick (1973).

7. Personal communication with the authors.

REFERENCES

AARON, H. and M. C. McGUIRE (1972) "Benefits and burdens of government expenditures," pp. 41-56 in K. E. Boulding and M. Pfaff (eds.) Redistribution to the Rich and the Poor. Belmont, Calif.: Wadsworth.

AKERLOF, G. (1969) "The market for lemons: quality uncertainty and the market mechanism." Quarterly Journal of Economics (August): 353-374.

ARROW, K. (1974) "Limited knowledge and economic analysis." American Economic Review 64 (March): 1-10.

——— (1963) "The welfare economics of medical care." American Economic Review 53 (December): 941-973.

BAERWALDT, N. and J. MORGAN (1971) "Trends in intrafamily transfers." Delivered at the annual meeting of the American Economics Association, December.

BAHL, R. W. and J. T. WARFORD (1972) "Real and monetary dimensions of federal aid to states," pp. 116-130 in K. E. Boulding and M. Pfaff (eds.) Redistribution to the Rich and the Poor. Belmont, Calif.: Wadsworth.

BATOR, F. (1958) "The anatomy of market failure." Quarterly Journal of Economics 72 (August): 351-379.

BELSHAW, C. (1965) Traditional Exchange and Modern Markets. Englewood Cliffs, N.J.: Prentice-Hall.

BOULDING, K. E. (1973a) The Economy of Love and Fear: A Preface to Grants Economics. Belmont, Calif.: Wadsworth.

——— (1973b) "Association for the study of grants economy in retrospect." A.S.G.E. Newsletter 5 (December): 2.

——— (1969) "Economics as a moral science." American Economic Review 59 (March): 1-12.

——— (1968) Conflict and Defense: A General Theory. New York: Harper & Row.

——— (1962) "Notes on a theory of philanthropy," pp. 67-71 in F. G. Dickinson (ed.) Philanthropy and Public Policy. New York: National Bureau of Economic Research.

——— and M. PFAFF [eds.] (1972) Redistribution to the Rich and the Poor. Belmont, Calif.: Wadsworth.

BOULDING, K. E., A. PFAFF, and M. PFAFF [eds.] (1973) Transfers in an Urbanized Economy. Belmont, Calif.: Wadsworth.

BOULDING, K. et al. (1972) "Grants economics: a simple introduction." American Economists VI (Spring).

BROWN, B. (1973) "State grants and inequality of opportunity in education," pp. 208-225 in K. E. Boulding, A. Pfaff, and M. Pfaff (eds.) Transfers in an Urbanized Economy. Belmont, Calif.: Wadsworth.

BRONFENBRENNER, M. (1971) Income Distribution Theory. Chicago: Aldine.

BUCHANAN, J. (1968) The Supply and Demand for Public Goods. Chicago: Rand McNally.

——— (1955) "An economic theory of clubs." Economica 32: 1-14.

——— (1955) "Federalism and fiscal equity." American Economic Review 40 (September): 583-599.

BURKHEAD, J. (1956) Government Budgeting. New York: John Wiley.

——— and J. MINER (1971) Public Expenditures. Chicago: Aldine.

CAMPBELL, A. K. and S. S. SACKS (1967) Metropolitan America: Fiscal Patterns and Governmental Systems. New York: Free Press.

COASE, R. (1952) "The nature of the firm," pp. 331-351 in G. Stigler and K. E. Boulding (eds.) Readings in Price Theory. Chicago: Richard Irwin.

COLEMAN, J. S. (1973) "The social basis of markets and governments," in K. Arrow et al. Urban Processes. Washington, D.C.: Urban Institute.

COMMONS, J. (1935-1936) "The place of economics in social philosophy." Journal of Social Philosophy I: 7-22.

DALY, G. and F. GIERTZ (1973) "Pollution abatement, Pareto optimality and the market mechanism," pp. 350-357 in K. E. Boulding, A. Pfaff, and M. Pfaff (eds.) Transfers in an Urbanized Economy. Belmont, Calif.: Wadsworth.

DAVID, M. and J. LEUTHOLD (1972) "Formulas for income maintenance: their distributional impact," pp. 312-341 in K. E. Boulding and M. Pfaff (eds.) Redistribution to the Rich and the Poor. Belmont, Calif.: Wadsworth.

DEMSETZ, H. (1969) "Information and efficiency: another viewpoint." Journal of Law and Economics 12 (April): 1-22.

DOERINGER, P. B. and M. J. PIORE (1971) Internal Labor Markets and Manpower Analysis. Lexington, Mass.: D. C. Heath.

DOWNS, A. (1970) Urban Problems and Prospects. Chicago: Markham.

DRESCH, S. (1971) "Testimony before the Committee on Ways and Means," pp. 381-386 in General Revenue Sharing. Washington, D.C.: U.S. Government Printing Office.

EASTERLIN, R. A. (1973) "Does money buy happiness?" Public Interest (Winter): 3-10.

EASTON, D. (1965) A Framework for Political Analysis. Englewood Cliffs, N.J.: Prentice-Hall.

ELESH, D. et al. (1973) "The New Jersey-Pennsylvania Experiment: a field study in negative taxation," pp. 181-202 in K. E. Boulding, A. Pfaff, and M. Pfaff (eds.) Transfers in an Urbanized Economy. Belmont, Calif.: Wadsworth.

FREEMAN, A. M. (1973) "Grants and environmental policy," pp. 309-316 in K. E. Boulding, A. Pfaff, and M. Pfaff (eds.) Transfers in an Urbanized Economy. Belmont, Calif.: Wadsworth.

FUSFELD, D. (1973) "Transfer payments and the ghetto economy," pp. 78-92 in K. E. Boulding, A. Pfaff, and M. Pfaff (eds.) Transfers in an Urbanized Economy. Belmont, Calif.: Wadsworth.

GAPPERT, G. (1973) "A regional perspective on urban development." Milwaukee: University of Wisconsin, Center for Leadership Development. (mimeo)

GARN, H. and M. SPRINGER (1975) "Formulating urban growth policies: dynamic interactions among people, places and clubs." Publius (Winter).

GEORGE, H. (1880) Progress and Poverty. New York: Robert Schalkenback Foundation.

GOLDBERG, V. (1974a) "Public choice-property rights." Journal of Economic Issues 8 (September): 555-579.

——— (1974b) "On Pareto optimal redistribution." University of California, Davis. (mimeo)

——— (1974c) "On positive theories of redistribution." University of California, Davis. (mimeo)

GOLDFARB, R. (1974) "Learning in government programs and the usefulness of cost-benefit analysis: lessons from manpower and urban renewal history." George Washington University. (mimeo)

GOODMAN, P. and P. GOODMAN (1947) Communitas: Means of Livelihood and Ways of Life. Chicago: University of Chicago Press.

GREEN, C. and A. TELLA (1973) "Effects of nonemployment income and wage rates on work incentives of the poor," in K. E. Boulding, A. Pfaff, and M. Pfaff (eds.) Transfers in an Urbanized Economy. Belmont, Calif.: Wadsworth.

GROSS, B. (1969) "The new systems budgeting." Public Administration Review 29 (March/April): 113-137.

HANSEN, W. L. and B. WEISBROD (1972) "Distributional effects of public expenditure programs." Public Finance: 414-420.

HARBERGER, A. (1971) "Three basic postulates for applied welfare economics: an interpretive essay." Journal of Economic Literature 60, 3 (September).

HARRINGTON, M. (1963) The Other America. Baltimore: Penguin.

HARTMAN, R. (1972) "A comment on the Pechman-Hansen-Weisbrod controversy," in K. E. Boulding and M. Pfaff (eds.) Redistribution to the Rich and the Poor. Belmont, Calif.: Wadsworth.

HAVEMAN, R. H. and J. MARGOLIS [eds.] (1970) Public Expenditures and Policy Analysis. Chicago: Markham.

HAVRILESKY, T. (1973) "Technological innovativeness, the grants economy, and the ecological crisis," in K. E. Boulding, A. Pfaff, and M. Pfaff (eds.) Transfers in an Urbanized Economy. Belmont, Calif.: Wadsworth.

HIRSCHMAN, A. (1971) "Political economies and possibilism," in A Base for Hope. New Haven, Conn.: Yale University Press.

HOCHMAN, H. M. and J. D. RODGERS (1973) "Utility interdependency and income transfers through charity," in K. E. Boulding, A. Pfaff, and M. Pfaff (eds.) Transfers in an Urbanized Economy. Belmont, Calif.: Wadsworth.

——— (1969) "Pareto optimal redistribution." American Economic Review 59 (September): 542-557.

HOLLISTER, R. G. and J. L. PALMER (1972a) "The impact of inflation on the poor," pp. 240-269 in K. E. Boulding and M. Pfaff (eds.) Redistribution to the Rich and the Poor. Belmont, Calif.: Wadsworth.

——— (1972b) "The implicit tax of inflation and unemployment: some policy implications," pp. 369-374 in K. E. Boulding and M. Pfaff (eds.) Redistribution to the Rich and the Poor. Belmont, Calif.: Wadsworth.

HOMANS, G. C. (1961) Social Behavior: Its Elementary Forms. New York: Harcourt-Brace-World.

HORVATH, J. (1971) "Rural America and the grants economy." American Journal of Agricultural Economics 53 (December): 740-749.

HORVATH, P. B. and J. BURKE (1972) "Federal research and development funding: the grants component and its distribution," pp. 131-148 in K. E. Boulding and M. Pfaff (eds.) Redistribution to the Rich and the Poor. Belmont, Calif.: Wadsworth.

JOHNSON, H. (1974) "Review of The Economy of Love and Fear." Journal of Political Economy 82 (September/October): 1055-1056.

ILCHMAN, W. and N. T. UPHOFF (1969) The Political Economy of Change. Berkeley: University of California Press.

KILLINGSWORTH, C. (1971) "A structural view of unemployment," in A. Brown and E. N. Euberger (eds.) Perspectives in Economics. New York: McGraw-Hill.

KNIGHT, F. H. (1971) Risk, Uncertainty, and Profit. Chicago: University of Chicago Press.

——— (1965) Economic Organization. New York: Harper & Row.

KRISTOL, I. (1973) "Social return: gains and losses." Wall Street Journal (April 16).

LAMPMAN, R. (1972a) "Public and private transfers as social process," pp. 15-40 in K. E. Boulding and M. Pfaff (eds.) Redistribution to the Rich and the Poor. Belmont, Calif.: Wadsworth.

——— (1972b) "Discussion of new transfer plans," pp. 342-347 in K. E. Boulding and M. Pfaff (eds.) Redistribution to the Rich and the Poor. Belmont, Calif.: Wadsworth.

LANCASTER, K. (1966) "A new approach to consumer theory." Journal of Political Economy (April): 132-157.

LERNER, A. P. (1944) The Economics of Control. New York: Macmillan.

LEVESON, I. (1973) "Strategies against urban poverty," pp. 130-160 in K. E. Boulding, A. Pfaff, and M. Pfaff (eds.) Transfers in an Urbanized Economy. Belmont, Calif.: Wadsworth.

McGUIRE, M. and H. GARN (1972) "The integration of equity and efficiency criteria in public project selection," pp. 357-368 in K. E. Boulding and M. Pfaff (eds.) Redistribution to the Rich and the Poor. Belmont, Calif.: Wadsworth.

MEADE, J. E. (1964) Efficiency, Equality, and the Ownership of Property. Cambridge: Harvard University Press.

MELLOR, E. (1973) "A case study: costs and benefits of public goods and expenditures for a ghetto," pp. 38-58 in K. E. Boulding, A. Pfaff, and M. Pfaff (eds.) Transfers in an Urbanized Economy. Belmont, Calif.: Wadsworth.

MIGUE, J. L. and G. BELANGER (1974) "Toward a general theory of managerial discretion." Public Choice (Spring).

MISHAN, E. J. (1972) "The futility of Pareto-efficient distributions." American Economic Review 62 (December): 961-976.

MUELLER, T. (1973) "Income redistribution impact of state grants to public schools: a case study of Delaware," pp. 226-245 in K. E. Boulding, A. Pfaff, and M. Pfaff (eds.) Transfers in an Urbanized Economy. Belmont, Calif.: Wadsworth.

MUSGRAVE, R. A. (1959) The Theory of Public Finance: A Study in Public Economy. New York: McGraw-Hill.

NEENAN, W. (1973) "Suburban central city exploitation thesis: one city's tale," pp. 10-37 in K. E. Boulding, A. Pfaff, and M. Pfaff (eds.) Transfers in an Urbanized Economy. Belmont, Calif.: Wadsworth.

NELSON, R. R. (1972) "Issues and suggestions for the study of individual organization in a regime of rapid technical change," in V. Fuchs (ed.) Policy Issues and Research Opportunities in Individual Organization. New York: National Bureau of Economic Research.

NEWHOUSE, J. (1970) "Towards a theory of nonprofit institutions: an economic model of a hospital." American Economic Review 60 (March).

NISBET, R. A. (1953) The Quest for Community. New York: Oxford University Press.

NISKANEN, W. A. (1971) Bureaucracy and Representative Government. Chicago: Aldine.

OKNER, B. (1972a) "Alternatives for transfering income to the poor: the family assistance plan and universal income supplements," in K. E. Boulding and M. Pfaff (eds.) Redistribution to the Rich and the Poor. Belmont, Calif.: Wadsworth.

——— (1972b) "Transfer payments: their distribution and role in reducing poverty," pp. 62-76 in K. E. Boulding and M. Pfaff (eds.) Redistribution to the Rich and the Poor. Belmont, Calif.: Wadsworth.

OLSON, M. (1965) The Logic of Collective Action. Cambridge: Harvard University Press.

PARSONS, T. (1949) The Structure of Social Action. Glencoe, Ill.: Free Press.

PECHMAN, J. A. (1970) "The distributional effects of public higher education in California." Journal of Human Resources 5 (Summer): 361-370.

——— and B. A. OKNER (1974) Who Bears the Tax Burden? Washington, D.C.: Brookings.

PFAFF, A. (1972) "Transfer payments to large metropolitan poverty areas: their distributive and poverty-reducing effects," pp. 93-129 in K. E. Boulding and M. Pfaff (eds.) Redistribution to the Rich and the Poor. Belmont, Calif.: Wadsworth.

PFAFF, M. and A. PFAFF (1972) "How equitable are implicit public grants? The case of the individual income tax," pp. 181-203 in K. E. Boulding and M. Pfaff (eds.) Redistribution to the Rich and the Poor. Belmont, Calif.: Wadsworth.

PHARES, D. (1973) "Impact of spatial tax flows as implicit grants on state-local tax incidence: with reference to the financing of education," pp. 258-275 in K. E. Boulding, A. Pfaff, and M. Pfaff (eds.) Transfers in an Urbanized Economy. Belmont, Calif.: Wadsworth.

PORTER, D. O. (1975) "Responsiveness of citizen consumers in a federal system." Publius (Winter).

——— and D. WARNER (1973) "How effective are grantor controls? The case of federal aid to education," pp. 276-304 in K. E. Boulding, A. Pfaff, and M. Pfaff (eds.) Transfers in an Urbanized Economy. Belmont, Calif.: Wadsworth.

——— and T. PORTER (1973) The Politics of Budgeting Federal Aid: Resource Mobilization by Local School Districts. Beverly Hills, Calif.: Sage Professional Papers in Administrative and Policy Studies.

RAWLS, J. (1974) "Some reasons for the maximin criterion." Papers and Proceedings of the American Economic Association (May).

——— (1971) A Theory of Justice. Cambridge: Harvard University Press.

ROUMEASSET, J. (1974) "Institutions, social contracts, and second best Pareto optimality." University of California, Davis. (mimeo)

RUDNEY, G. (1972) "Implicit public grants under the tax system: some implications of federal tax aids accounting," pp. 175-180 in K. E. Boulding and M. Pfaff (eds.) Redistribution to the Rich and the Poor. Belmont, Calif.: Wadsworth.

SCHALL, L. D. (1972) "Interdependent utilities and Pareto optimality." Quarterly Journal of Economics 86 (February): 19-24.

SCHICK, A. (1973) "A death in the bureaucracy: the demise of the federal PPB." Public Administration Review 33 (March/April): 146-156.

SCHULTZE, C. (1972) "The distribution of farm subsidies," pp. 94-116 in K. E. Boulding and M. Pfaff (eds.) Redistribution to the Rich and the Poor. Belmont, Calif.: Wadsworth.

SCHWARTZ, R. A. (1970) "Personal philanthropic contributions." Journal of Political Economy 78 (November/December): 1264-1291.

SCOTT, A. D. (1952) "Federal grants and resource allocation." Journal of Political Economy 60 (December): 534-536.

SELZNICK, P. (1949) TVA and the Grass Roots. Berkeley: University of California Press.

SHMID, A. (1973) "The role of grants, exchange, and property rights in environmental policy," pp. 340-349 in K. E. Boulding, A. Pfaff, and M. Pfaff (eds.) Transfers in an Urbanized Economy. Belmont, Calif.: Wadsworth.

SJAASTED, L. and R. HANSEN (1972) "The distributive effect of conscription: implicit taxes and transfers under the draft system," pp. 285-308 in K. E. Boulding and M. Pfaff (eds.) Redistribution to the Rich and the Poor. Belmont, Calif.: Wadsworth.

SMITH, J. and S. FRANKLIN (1974) "The constitution of personal wealth, 1922-69." Papers and Proceedings of the American Economic Association (May): 162-167.

STIGLER, G. J. (1970) "Director's law of public income redistribution." Journal of Law and Economics (April): 1-10.

STROTZ, R. and C. WRIGHT (1973) "Externalities, welfare economics, and environmental problems," pp. 359-373 in K. E. Boulding, A. Pfaff, and M. Pfaff (eds.) Transfers in an Urbanized Economy. Belmont, Calif.: Wadsworth.

STUART, B. (1972) "The impact of Medicaid on interstate income differentials," pp. 149-168 in K. E. Boulding and M. Pfaff (eds.) Redistribution to the Rich and the Poor. Belmont, Calif.: Wadsworth.

SURREY, S. (1973) Pathways to Tax Reform. Cambridge: Harvard University Press.

TAUSSING, M. K. (1972) "Long run consequences of income maintenance reform," pp. 376-388 in K. E. Boulding and M. Pfaff (eds.) Redistribution to the Rich and the Poor. Belmont, Calif.: Wadsworth.

TULLOCK, G. (1971) "The charity of the uncharitable." Western Economic Journal (December).

U.S. Bureau of the Census (1973) Statistical Abstract of the United States. Washington, D.C.: Government Printing Office.

VEBLEN, T. (1961) The Place of Science in Modern Civilization. New York: Russell & Russell.

VICKERY, W. S. (1962) "One economist's view of philanthropy," pp. 38-45 in F. G. Dickinson (ed.) Philanthropy and Public Policy. New York: National Bureau of Economic Research.

WARNER, D. (1975a) "Fiscal barriers to employment." Annals of the American Academy of Political and Social Science (March).

——— (1975b) "Fiscal federalism and health care." Publius (Winter).

16

Recent Trends in
Corporate Urban Development Programs

JULES COHN

☐ IN THE PAST DECADE, the country's major corporations all became involved with urban affairs projects. Blacks and other racial minorities and women were hired, or promoted into jobs for which they previously were not considered, in numbers ranging from token to substantial. Special training programs were created for the disadvantaged. Programs to control or eliminate environmental pollution were intensified. Businessmen who once had only a passing acquaintanceship with the kinds of issues faced in the ordinary course of their work by the nation's mayors acknowledged the severity of some urban problems, and agreed to tool up to lend their energies to solve them. In the face of pressures for positive action from civil rights, consumers, and environmental groups, as well as from government agencies such as the U.S. Equal Employment Opportunities Commission, the Environmental Protection Administration, and the Justice Department, board chairmen and company presidents delivered a host of promises.

How does business's performance look when compared to the pledges made during the years of pressure and demands from outside interest groups? How have ten years of exposure to the cities' needs affected business leaders? And what, if anything, have corporations been able to contribute to today's urban development efforts? From the corporate standpoint, what are some of the obstacles and challenges encountered by businessmen who go forth to do battle against urban problems? Looking to the future, what can we expect of corporate urban affairs efforts in the next ten years?

AN OVERVIEW

A review of the nature of corporate urban affairs efforts provides grounds for a mixed view of what happens when forces of social change are brought to bear against the private corporation. The willingness, even the eagerness of businessmen to accept a share of responsibility for the problems of the cities, and to commit themselves to positive programs focused on these problems, is a highlight of the last decade of corporate history. Whatever their underlying motives, whether they were frightened for the future of corporate life, stimulated by the public relations possibilities inherent in the new programs, impelled by the hope that their social critics would eventually fade away, or that the urban crisis might become a profit-making opportunity, businessmen came through. They began to make changes in their companies, and in their ways of doing business.

On the whole, however, the evidence suggests that though businessmen made many good starts, the number of solid achievements with lasting value is very limited indeed. Even the most powerful and resourceful companies in the country came up against difficulties. In the first place, the economic recession raised the financial costs of programs, and raises, too, the likelihood that such improvements as changes in the racial and ethnic composition of the workforce may be erased. Second, businessmen, who know much more now about the complexity of urban problems, have grown less bold in defining what they can do, and more humble in the face of the odds which they now realize are against them. Corporate executives feel bruised by some of their encounters with community groups, for example, and accordingly have chosen to withdraw from certain kinds of programs in order to avoid the possibility of further conflict. When they concentrate their efforts on programs wholly within the company, such as training and upgrading, they feel in control of the environment, and more confident of producing good results.

A few developments arising out of corporate involvement in urban affairs programs were probably never anticipated either by businessmen or their critics. Nevertheless, these developments have been significant from the standpoint of urban development. Thus companies have uncovered new and potentially influential approaches to personnel relations and the management and definition of work. These approaches and their significance are discussed below, as are some other by-products of recent urban affairs efforts—the movement toward the corporate social audit, and the economic opportunities that businessmen have found in some of the new programs.

Major Developments in Corporate Urban Affairs Programs. The rest of this article will discuss the following major developments in business's urban affairs programs:

—the institutionalization of social responsibility;

—the tapering-off of man- (and woman-) power programs;

—the growth of human relations and work enrichment programs;

—the unanticipated consequences of affirmative action programs;

—the increasing importance of corporate environmental programs; and

—the conundrum of the corporate social audit.

THE INSTITUTIONALIZATION OF SOCIAL RESPONSIBILITY

A convergence of social, political, and even economic forces has had the result of transforming corporate urban affairs activities from the realm of the esoteric to the accepted way of doing business. Programs that would have been defined as pioneering ten years ago are taken for granted today. What once seemed bold now is accepted as commonplace. Corporate social responsibility has become institutionalized.

Federal, state and local laws regulating employment and environmental protection practices have placed many private sector activities outside the sphere of voluntary action. Since publication of the

Kerner Commission Report (1968), and passage of Title VII of the Civil Rights Act (1964), most major corporations are vulnerable to scrutiny by more than one federal agency as well as state and local commissions. The establishment in 1972 of the Equal Employment Opportunities Commission as a conciliation agency, with the power to sue, and promulgation of Executive Order 11246, enforced by the Office of Federal Contract Compliance of the Department of Labor, added to the beleaguerment of corporate officers by government officials, and the consequent bureaucratization of urban affairs efforts.

It is standard practice nowadays for large companies to have urban affairs departments charged with developing and/or administering monitoring and information-gathering efforts, as well as conducting programs. Most of these departments are run by former personnel officials, but some are staffed by recent business-school graduates, company lawyers, and public relations personnel. In the American Telephone & Telegraph Company, for example, a sizable new unit was created in the main office, and staffed with people transferred from affiliated Bell System companies, in order to collect and organize data about Bell employment practices on a nationwide basis.

The routinization of urban affairs within the corporate world has had both positive and negative consequences. On the credit side, businessmen have finally abandoned their old argument that their only proper business is business itself. In sharp contrast to classical economic views of their role, managers at all levels, as well as top executives, now acknowledge that they cannot justify their existence simply by attempting to maximize profits. The public takes it for granted that they will take an active part in dealing with urban problems, and they agree. Moreover, they have come to see that it is very much in their own interests, and in keeping with their profit-making objectives, to assume some responsibility for city problems. In their actions, as well as in their speeches and writings for public occasions, corporate chiefs aver that the long-range growth and welfare of their companies rests on their ability to foster policies and actions reflecting a sense of social responsibility.

In addition to managers and top executives, members of the work force, whether white- or blue-collar, have also come to accept the social changes that caused strains and stresses in some plants, factories, and offices in the early 1960s. Heterogeneity in race and ethnicity is now a fact of everyday life, and there is little sign of the backlash and bigotry that companies once had to contend with among their own employees at the outset of urban affairs efforts. To be sure, there is evidence that the struggle for equal opportunity for employment and advancement for women is nowhere near as favorably resolved as the struggle against racial discrimination in the business world.

On the negative side, the institutionalization of urban affairs activities has deprived them of the top priority they received from executives who, for what now seems a brief interlude in the mid-1960s, once devoted a lot of energy to them. When the urban crisis was headline news, when riots were disturbing the peace of city-dwellers and challenging the complacency of the inhabitants of executive suites, businessmen dramatically reordered their agendas and placed urban affairs in the forefront. They assigned some of their most talented and ambitious aides to the new programs, and set aside resources and funds to pay for them. Accordingly, some innovative and exciting programs were developed. Moreover, since some of the most admired people in major companies seemed to be working in these new areas, junior executives and recent business-school graduates soon aspired to similar roles.

Now, of course, the trend has shifted. For many business-school graduates, always on the alert to spot the newest directions on the ladder to success in the corporate world, urban affairs is a thing of the past, a has-been rung on the ladder. Ten years is a long time in organizational history, and the urban affairs box on the corporate organization chart is no longer the attraction it once was. Consequently, it is more and more the domain of company bureaucrats and time-servers. Fewer creative ideas are being developed; less and less prestige is attached to service in urban programs.

THE TAPERING-OFF OF
MAN- (AND WOMAN-) POWER PROGRAMS

From the beginning, the main emphasis of business's urban affairs program was on removing barriers to the employment and advancement of minority group members, particularly the disadvantaged. Thus most of the important programs created in the 1960s centered on special recruitment and training efforts. To augment and assist these efforts, the U.S. Labor Department, in cooperation with the National Alliance of Businessmen, a group founded with the express purpose of encouraging corporate social responsibility, established the JOBS program. Under the direction of John Gardner, a national Urban Coalition was also formed, with 48 local urban coalitions in major cities.

But the JOBS program and the Urban Coalition have foundered. The ambitious numerical goals of the former were never achieved, and only a few years after its heady beginnings it had nearly come to a standstill by the fall of 1974. The Urban Coalition no longer exists as a separate entity, and many of its local units have disbanded. And most of the less well-publicized programs for employment of blacks, browns, and Spanish-surnamed Americans that were initiated by businessmen in the 1960s, in the days when the economy was flush, there was an atmosphere of enthusiastic goodwill, and conditions seemed right for the emergence of a new corporate conscience, have been allowed either to die or to considerably contract.

The shift from the massive efforts of less than ten years ago to the present contraction can perhaps best be explained not only by changing national economic conditions but also by a growing awareness on the part of businessmen of how complicated and difficult urban problems are. Businessmen, supposedly hard-headed, practical men, were naive in their approaches to the new man- and woman-power training programs. In devising and setting up special projects, they made at least one major mistake. Instead of believing in their time-tested techniques for training new employees, they allowed themselves to be persuaded that entirely new—and costlier—methods were needed when bringing minority group members into their companies.

Perhaps because they felt guilty about years of neglecting the disadvantaged, businessmen let their attempts to provide training for these new workers be influenced by outsiders, including some of their own critics—social workers, poverty politicians, psychologists, and assorted bureaucrats. Many of these people know a good deal about urban problems. Unfortunately, however, experience proved that they knew very little about how to train the disadvantaged for jobs in industry.

Because they meant well, businessmen let themselves be led down the garden path. They followed advice based on the dangerous view that the disadvantaged are profoundly different from the rest of us. According to the rhetoric of the mid-sixties, the disadvantaged were said to come from different cultures, have different views of the world, and different attitudes toward work. Businessmen were told, therefore, that new kinds of training would be needed to turn them into willing and productive workers. An armada of "supportive services" would have to be established and brought into play, it was said, in order to adjust the disadvantaged to life in the corporate world. Thus some of the country's largest banks, insurance companies, and public utilities busied themselves with telling the disadvantaged how to use deodorants, comb their hair, and shop for frozen foods. With the best intentions, these and other companies sought to advise the newly hired workers about raising their children, selecting a lawyer, managing their finances, and resolving family tensions.

Providing the new supportive services drained a substantial amount of money as well as other corporate energy from the new employee training efforts. Training departments were reorganized, personnel departments expanded, special orientation materials written, and sometimes even new training facilities built or leased. Moreover, demands were made on the valuable time of top-level executives, who were often required to step away from their managerial responsibilities to tend to problems that arose in the new programs.

Aside from the monetary expense associated with training the disadvantaged, and the drain on management energies, there were also morale problems among members of the regular work force. Co-workers and supervisors of the specially treated new employees were antagonized by the new services being dispensed to others. They knew full well that they themselves never received what appeared to be the velvet-glove treatment. And they questioned whether an employee's personal habits should be given as much attention as his or her job performance.

Most large companies grew disillusioned about their ability to run effective programs for the disadvantaged, after the frustrating experiences arising from the problems detailed above. A few years after these new programs had been set up, they were being cut back as companies decided to concentrate their energies elsewhere. Businessmen never got back on the track, even though some of them probably genuinely wanted to do something to reduce the number of the so-called "hard-core" unemployed.

THE GROWTH OF HUMAN RELATIONS AND

WORK ENRICHMENT PROGRAMS

Such large American companies as American Telephone & Telegraph, General Foods, Texas Instruments, Corning Glass, Polaroid, General Motors, and Chrysler have been experimenting in the past decade with new ways to organize work. Partly as a result of their attempts to develop programs that would successfully bring the disadvantaged into the work force, and partly, too, as a result of the influence of similar experiments in Norway, Sweden, and to a lesser extent in other European countries, businessmen are increasingly interested in finding new ways to motivate employees by refashioning the work environment and by redefining work itself.

Programs for the disadvantaged, albeit never wholly successful, had the effect of acquainting businessmen with the growing body of literature on the subject of the morale problems of workers. The movement in industry to meet the challenges posed by these problems shows signs of outlasting programs specifically aimed at minority group members, and of having a beneficial effect on everyone who works for a large company, and accordingly on the quality of life in the cities. Inasmuch as large companies depend on the cities for their labor supply, therefore, the movement to reform work may prove to be of major significance.

The growing movement, focusing on the workplace—probably one of the most important institutions in the life of city-dwellers—is developing alternative approaches to how work is defined and assigned in large organizations, how employees are supervised, and how the office, plant, or factory is designed. The Department of Health, Education and Welfare, in a report published in 1972, went so far as to argue that instituting changes in the above areas might have a direct and positive effect on the physical and mental health of workers, family stability, community involvement, community cohesiveness, and sociopolitical attitudes, and possibly, too, could help reduce drug and alcohol addiction, aggressive behavior, and delinquency. Granted that HEW's *Work in America* report exaggerated the influence of work on urban problems, there still can be no doubt about the validity of its general point that the complaints of blue-collar workers are often linked to their dissatisfactions with their jobs. For that matter, the grievances of white-collar and management employees may also have their roots in the nature of work in large companies. Many of these grievances first came to the surface when new programs for the disadvantaged were being established, and as old-line employees observed that the latter group was being given better treatment—what they often called "human treatment"—on the job.

Discontent among workers has given the reformers of work a growing influence on top executives. Proponents of the movement point to studies that show that even workers at the lowest levels become happier and more productive when they are given some say over the pace at which they work and the surroundings in which they perform their duties. Employees have been expected to be passive on these issues, it is argued, as machines and technology, or top management working in isolation, determine the conditions of the workplace.

In general, the new human relations approaches stress the need for non-material incentives for the employee, arguing that sociological and psychological changes can have a positive effect on worker productivity. Those who have to be convinced of the validity of the reformers' arguments are the managers. Their jobs, of course, depend on increasing productivity and profits. Unfortunately, however, reformers are not yet able to show clear-cut evidence or hard data to back up their claims. In the meantime, as the movement grows, its main effects will probably benefit workers, if only by drawing more and more attention to their "human" problems.

THE UNANTICIPATED CONSEQUENCES OF AFFIRMATIVE ACTION PROGRAMS

Affirmative action programs were supposed to help solve problems caused by discrimination in employment. Yet only a few years after their inception, they are a subject of controversy and have presented employers with unanticipated and serious problems.

Members of groups *not* protected by the new programs complain that they are now suffering from inequities; they demand attention as well as redress of their grievances. They want corporate employers to worry about their careers, and provide advancement opportunities for them as well as for members of the protected minority groups. Affirmative action backlash looms as a challenge to corporate employers who mistakenly assumed that the issue of discrimination would be put to rest once the government-mandated programs were underway.

How can we account for hostile reactions to affirmative action among employees? What challenges does the backlash pose for personnel and urban affairs officers? And what are some of the other consequences of affirmative action?

White ethnic employees—Jews, Italian-Americans, and other non-protected groups—are worried about the prospect of compensatory discrimination. While they support corporate efforts to end inequitable practices of the past, they are understandably unwilling to be penalized themselves so that others can benefit. They want to have the same opportunities to advance and transfer to desirable jobs in their companies that affirmative action goals and timetables now guarantee to others.

Several major corporations have made major affirmative action commitments which may well have substantial side effects. Officials of American Telephone & Telegraph, for example, pledged special efforts to ease the movement of qualified women, blacks, and Spanish-surnamed Americans into better jobs, and to institute new promotion-pay policies for them, as well as special programs to assess their interests and qualifications for management jobs. Bell even agreed, as a result of an action brought against the corporation by the Equal Employment Opportunities Commission, to make one-time payments, totaling approximately $15 millions, to 15,000 female and minority-group employees deemed qualified for certain kinds of jobs, and held to have been delayed in obtaining them, or, in the case of women, paid less than male employees when they did obtain them.

By late 1974 there are indications that some Bell employees who are not members of the protected groups feel that their view of the law, and their conceptions of social justice, call for efforts in their behalf. They assert that they want to be dealt in on the newly established transfer and promotion policies. In response, company officials have asserted that they will attempt to give due consideration to these new sets of grievances.

The AT&T case is a portent for all corporate employers who, as they develop their affirmative action programs, should remember that the implications as well as the specifics of goals and timetables will be scrutinized by the work force, more and more of whose members are on the lookout for signs of discrimination against them.

For now that politicians, government officials, and the mass media have rediscovered the white ethnic groups, the self-consciousness of these groups and their willingness to assert their own interests have increased. Affirmative action has served to increase this new consciousness.

Businessmen worry about the implications of the law, and whether or not the government itself is, as their employees insist, encouraging reverse discrimination and quotas. These issues associated with affirmative action will soon have to be argued in the federal courts, for at least two cases have been brought by employees laid off or denied opportunity as a result, in their view, of corporate goals and timetables established at the behest of the E.E.O.C. So far the government has been insisting that goals are not the same thing as quotas. It argues that the former are nothing more than realistic and flexible objectives, to be differentiated from quotas.

Inside the corporate organization, it is usually the personnel officer who must meet the challenges raised by affirmation action. His (or her) greatest problem arises from the failure on the part of social policy-makers to resolve the ambiguities and conflicts inherent in this new instrument. Though the ideal of affirmative action is unassailable—an open-door policy for all minority group candidates for jobs and advancement—the conceptual underpinnings are obviously weak. What should be the proper balance between the claims of individual merit, on the one hand, and the facts of group membership, on the other, when a job or transfer applicant is being considered?

The challenges of affirmative action have been a cause of the bureaucratization of social responsibility in the business organization. Programs that will be acceptable to government regulators must be developed, and lengthy negotiations with federal agencies must be undertaken. The complex details of federal rules must be mastered, and intricate record-keeping must be maintained. Since the affirmative action concept is based, in part, on assumptions about the available labor market, employers need to assign staff members to compile data reflecting conditions not only in their own companies but throughout the regions where they do business. If nothing else, the advent of affirmative action has given personnel officers a new importance in their companies, and new responsibilities. They are asked to be the agents of social change in the corporate environment, and thereby are the objects of interest on the part of outsiders who monitor company progress, as well as those inside the company who have a more immediate interest in how things are going.

Thus the corporate personnel function, which was readily dismissed by most students of the corporate pecking order in the days before urban affairs programs were inaugurated, has taken an important place alongside other staff functions. By now, in most companies, new units have been created to conduct the new programs. An effective affirmative action officer can be a welcome help to a chief executive interested in ingratiating himself with government regulators. A sound and documented defense of existing policies can save a company significant amounts of court costs arising from equal-employment cases.

A few years ago, many companies created ombudsman posts to provide a special voice and special representation for minority-group employees who felt that they were being passed over in favor of members of so-called privileged groups. These same companies now might well consider creating similar roles to represent the interests of the new minorities—the white ethnic employees, particularly men, for whom there are no affirmative action goals and timetables.

In the final analysis, the major organizational (and social) consequences of affirmative action will be reflected in the composition of the work force. On the positive side, there is the promise of a more heterogeneous group of employees and executives, and a transfer and advancement process more open than ever before. By whittling away at social, racial, and ethnic factors as barriers to success in the corporate environment, affirmative action may eventually insure that merit and achievement will be the main routes to rewards. But opening those routes will also require satisfactory resolution of some of the conceptual and social problems engendered by the attempt to set goals that are tied to group membership.

THE INCREASING IMPORTANCE OF ENVIRONMENTAL PROGRAMS

Businessmen's views about their responsibility for environmental pollution range from reluctance to accept any obligation at all, to readiness to commit themselves to positive action programs. As environmental problems have moved to the forefront of public attention in recent years, major companies have become the targets of attack by conservation and consumer groups. Water pollution, a continuous and increasing source of danger to public health, is often attributed to industrial practices. Also well documented by now is the seriousness of air pollution as a public health problem, and the pollutant effects of industrially produced carbon monoxide, sulfur oxides, and hydrocarbons, as well as the public health problems caused by solid wastes from manufacturing and chemical pesticides.

As a result of new federal and local laws, most companies presently participate in some form of community or industry-wide antipollution program. But effective and uniform participation is hampered by the lack of explicit government standards implemented on an industry-wide and nationwide basis. Businessmen, continually badgered by community groups and consumer organizations, complain about finding themselves at a competitive disadvantage with companies that are able, in the absence of uniform national standards, to abdicate responsibility entirely for some kinds of programs.

Tax credits are discussed as possible incentives for businessmen who invest in the technology that is necessary to conduct certain kinds of antipollution programs, or to correct existing machinery. Federal, local, or state grants to help pay for the costs of new equipment are also talked about but not yet provided. Tax incentives, of course, are indirect government expenditures, and are not necessarily linked to government scrutiny and inspection. Their costs are buried in tax returns, and critics argue that the device enables companies to obscure real costs, sometimes by mingling them with other expenses. Governments could also contract with business for the development of new technology to reduce pollution, or for the initiate of experiments designed to identify new ways to combat environmental problems.

In order to wage an effective campaign against environmental problems, it is clear that business will have to do much more than it is doing. Doubtless the political and social ground rules within which business is conducted will have to be drastically amended. Changes will have to be made in the complex economic calculations that have had the effect of skewing business decision in destructive directions. In other words, business will have to change if the environment is to improve. And this change is unlikely to occur soon. As in the case or corporate programs for minority-group members, the evidence is that social pressure, and even government pleas, are not enough. Even when businessmen are persuaded to set up new programs, or to undertake new approaches, they will eventually evaluate the desirability of continuing such programs in economic terms. The bottom line will have to show a profit. Thus the only real hope for permanent shifts in the course of industry's approach to the environment rests on whether or not socially beneficial programs and approaches can justify themselves in the marketplace.

Although some companies pledged efforts to provide improved housing for low-income families burdened by inadequate shelter or deteriorating neighborhoods, only a few corporate housing programs were ever initiated as urban affairs projects. Aside from the activities of companies already in the housing construction business, a consortium of insurance companies agreed to undertake special efforts to set aside mortgage money for low-cost homes, but their project failed to achieve its goals even before the economic recession brought it to a halt. By far the major investment in the planning and construction of housing for the poor, as well as schools, hospitals, and other institutional facilities for the central cities, has been made with public funds. The fact is that private enterprise has been unable to find a profitable way to provide housing for low-income groups. Entrepreneurs have rarely built housing for poor people, and therefore it is probably true that an expanded federal program of subsidized construction and rents is the only workable solution. Unfortunately, the provision of decent new housing for families who cannot afford it is an expensive matter, involving heavy construction, financing, and operating costs.

A Note on Urban Management Consultants. Increased interest in urban problems on the part of businessmen in the 1960s had, among other by-products, the effect of introducing management consultants to city politics. Systems analysts and other management experts sometimes offered their services on a volunteer basis to help mayors master the growing complexities of urban administration, but more often were retained on a fee basis. Business consultants, who formerly limited their activities to industrial organizations, became well known in many city halls, including those in New York, Los Angeles, Chicago, Cleveland, Boston, and Cincinnati. They designed information systems for mayors, command and control techniques for police departments, and helped reorganize fire-fighting, sanitation, and personnel administration, as well as approaches to city planning, educational administration, and more general policy areas.

It is likely that the management consultants will outlast many of the other experts hired by cities in the heyday of the urban crisis. Specialists in training the disadvantaged, for example, have already found their welcome waning, as man- and woman-power programs have been cut back or discontinued. Their expertise is limited to a substantive area and a specific problem, whereas the management advisers claim a more generalized skill. Thus the advice industry will probably continue to grow in the cities, ready to offer answers to political patrons, in the form of newly designed organization charts, management kits, or decision-making diagrams. Describing themselves as detached, nonpolitical, and cost-conscious, these contemporary courtiers of political executives readily win contracts and retainers. In return, they offer hope that the city's problems can be solved simply by following the right flow chart and leaning on the right consultants.

While there can be no argument about the cities' need for help in managing programs, the assumption that a private firm's advice to a political executive can be objective, nonpartisan, and detached is misleading, if not downright fallacious. To whom are these technical advisers accountable? What interests do they represent? And are the solutions they propose adequate to the problem or designed merely to buttress the client's political position, perhaps by helping him win the next election? Moreover, what is good for the mayor may not necessarily be useful to the city. An appraoch to cost-cutting that is appropriate in the telephone company could be harmful in a public agency, where economic calculations must be balanced against such other considerations as social policy priorities.

Finally, the rise in influence of the technical adviser to mayors, with backgrounds in management consulting, leads to another danger. Management consultants are committed to the belief that there are solutions to all problems, that they can find them, and that these solutions generally lie in structural arrangements. In city government, unfortunately, many problems cannot be solved, only lived with, adjusted to, worked around. Short-term remedies are more common than long-lived solutions, and these remedies rest on political know-how, a skill for which mayors hardly need management consultants.

THE CONUNDRUM OF THE
CORPORATE SOCIAL AUDIT

The corporate social audit has been proposed by many public interest groups as a device by which companies can institutionalize their urban affairs programs and at the same time monitor and evaluate them. In essence, the idea is to permit firms to report their performance on issues of social concern in the same manner and with the same regularity that they report financial performance. "Social portfolios" and "social audits" are increasingly discussed, even as actual social programs wane, wither, and are allowed to die. The device has obvious appeal to the positivist tradition, and to those who gain reassurance from the belief, even if vain or merely nearly unrealizable, that all things can be quantified.

One approach to the social audit is simply to collect information about whether or not a company is doing "good" or "harm" by its activities. Who would collect this evidence, and how it would be weighed—that is, how the categories for judgment would be defined—is unclear. Among the public interest groups, the Council on Economic Priorities, a gadfly group formed in the late 1960s, and self-appointed (though partly foundation-financed) to the task of monitoring corporate activities, specializes in assessing corporate performance in selected areas. Its interests include air and water pollution, but also corporate involvement in military or armaments programs. Other groups have moved toward more comprehensive methods of measurement, and actually rate businesses in terms of their participation in community affairs (how many executives participate? how much time is volunteered?), opportunities provided for minority-group members and women (how many women have been advanced to management jobs? how many Mexican-Americans have been hired? how many disadvantaged blacks have been trained?), contributions (what socially valuable projects have been supported by the company? how many dollars donated to black social action groups? how many dollars to neighborhood self-renewal organizations?), consumerism, environmental protection efforts, and so on. The relevance of each of the rated dimensions will, of course, vary from industry to industry. Banks, for example, might be rated on their lending policies with regard to minority-group members or women, or their lending policies with regard to polluting industries. It is clear that any rating of industries will have to depend on a variety of sources—government reports, executive statements and speeches, annual reports, financial news reports, and the reports of monitoring agencies such as pollution control authorities, health departments, and sanitation inspectors, as well as equal employment commissions. Yet even if comprehensive data can be gathered from each of these sources on a company-by-company basis by outside evaluators, there will still be gaps in the record. For surely such other areas as morale inside the company, community and neighborhood attitudes about company policies, and even the nature and substance of company expense-account spending would bear upon a social audit. Furthermore, the indirect social effects of employee compensation scales and employee fringe benefits would need to be taken into consideration by an objective evaluator. What sort of health insurance does the company provide for employees? Are there educational benefits, recreation programs, day-care services, food services? And what about the quality of working life within the company—working conditions, work space, air quality, noise levels, and even the visual aesthetics of the plant or office? Surely each of these factors could well be justified in preparing an overall assessment of a company's contributions (or debts/injuries) to society. Yet it is virtually impossible to quantify them.

Though interest and curiosity about the corporate social unit are increasing, it seems unlikely that serious efforts will be conducted. A handful of companies claim to have undertaken self-audits to determine their social impact, but even these companies admit to proceeding on an experimental and temporary basis. Thus the rhetoric level far exceeds the actual activity in the field. For the most part, it would appear, attempts at a social audit will be undertaken only when a company's chief executive officer is bitten by the corporate social-responsibility bug. For it takes top management to spearhead such audit efforts, and to motivate line executives who would naturally see a social audit as an unwelcome diversion from their day-to-day activities. And even if its top executive decides to order a social audit, and the company is able to develop an instrument that will help, it will do well to pause before leaping into action. For what will the audit show? It might, after all, reveal a very bad record.

REFERENCES

ABT, C. (1972) "Managing to save money while doing good." Innovation 27 (January).

BANFIELD, E. (1974) The Unheavenly City Revisited. Boston: Little, Brown.

COHN, J. (1973-1974) "Coping with affirmative action backlash." Business and Society Review (Winter).

MILLER, S. M. and R. D. CORWIN (1972) "U.S. employment policy from the 1960's to the 70's." New Generation 54, 2 (Spring).

STARR, R. (1974) "America's housing challenge." New Leader 67, 19 (September 30): 3-31.

STEINER, G. (1971) Business and Society. New York: Random House.

U.S. Department of Health, Education and Welfare (1972) Work in America. Washington, D.C.: Government Printing Office.

U.S. Department of Labor, Office of Research and Development, Manpower Administration (1973) Productive Employment of the Disadvantaged: Guidelines for Action. Los Angeles: Human Interaction Research Institute.

ZIMPEL, L. and D. PANGER (1970) Business and the Hard-Core Unemployed. New York: Frederick Fell.

Part VI

THREE PERSPECTIVES ON EVALUATION

Introduction

□ THE EVALUATION PROBLEMS of the urban social economy begin with efforts to implement innovations or interventions within urban organizational systems but wind up in the murky domain of culture and institutions wherein individuals acquire social belief and purpose. At one end of the scale is the notion of the "effective organization," with the implication of accepted goals and the provision of reliable means for achieving them. At the other end of the scale are the simultaneous problems of people developing new social goals and inventing the means of achieving them with the implication of the need to develop new goals in accord with new perceptions of changing reality (Greenfield, 1973).

The tension between improving old organization roles and developing new ones is compounded by conflicting claims over the allocation of scarce resources, and by conflicting beliefs as to the need to provide for the restructuring of the social economy itself through a redistribution of power over those resources. The capacity of evaluation techniques to appraise the efficacy of public intervention to alleviate the problems generated by the urban social economy is not in question. The willingness to commit resources to the deliberate social invention of new laws, organizations, and procedures (Conger, 1973) that change the way in which the social economy operates raises questions of social purpose over which the evaluator must inevitably stumble. The papers in this section do not review the techniques of evaluation but are oriented to the broader issues of social criticism associated with the expanding use of the analytical technology of evaluation research.

THE THREE LEVELS

There is only partial agreement in the field of evaluation studies about the proper role, value, and use of evaluation outcomes. There is, however, agreement on the need to distinguish between different levels of evaluation research. The Office of Economic Opportunity has distinguished between what they regard as three quite separable types of evaluation research. Type I, called "summative" evaluation, is research to assess a national program as a whole with the variation in implementation averaged during the summing-up process. This is research essentially at the societal level to make a determination of the overall consequences of a program or policy for the national society.

Type II, "formative" evaluation, is concerned with the effects of different types of programs on the desired outcomes of a particular policy. This kind of research can be concerned with the significance of the outcomes of a newly designed program or with trade-offs between different program approaches to the same policy question. This kind of research seeks to provide an answer to the famous budgetary question put by Verne Lewis (1952): On what basis shall it be decided to allocate X dollars to Activity A instead of allocating them to Activity B, or instead of allowing the taxpayer to retain the money for his individual purposes?

Type III, or "monitoring" evaluation, is evaluation geared to the interests of program managers. This provides periodic review and feedback of results and problems to those who have responsibility for programmatic implementation. Programs managers obviously have a different interest in evaluation outcomes, which is the enhancement of their own managerial abilities and responsibilities.

The chapters in this section roughly correspond to this crude division of evaluation types. Flax, Springer, and Taylor review the social indicators movement from a societal perspective. Their paper

PART VI: INTRODUCTION [527]

examines the progress that has been made in the development of social indicators and discusses how these efforts relate to the development and assessment of public policy. They interpret the social indicators movement as the newest attempt (following PPBS and "management by objectives") to develop at the scale of the national socioeconomic system a set of tools, techniques, and analytical procedures that can assist national policy-makers in dealing with decisions that focus on the direction of the national system. The social indicators movement symbolizes a shift in preoccupation with the means of policy to a struggle to provide a better specification of the ends of public action in terms of human welfare, social purpose, and quality of life. Indeed, it is not improper to suggest that the inadequacies of the economic technology represented by program-planning-budgeting systems were related to the fact that economic technicians, having forgotten the origins of their discipline as a branch of moral philosophy, were unable to assist policy-making in the determination of appropriate goals. The social indicators movement provides a vehicle whereby social critics and social philosophers can, through the provision of technical support, participate in processes of policy formulation.

The second paper by Garn reflects a concern with the level of type II evaluation: How should policy-makers assess the evaluation of a new, multiple objective program mechanism (Community Development Corporations) as a way by which policy objectives (the development of inner-city communities) can be achieved? As Garn indicates in this chapter:

Both the wide variety of institutions which call themselves CDCs and the relatively short time in which they have operated make it extremely difficult to tell what criteria should be used in assessing them and, hence, whether the CDC experiment has succeeded or failed.

In his paper he develops a conceptual framework which will help improve understanding of what kind of institution the CDC is, how such an institution should be evaluated, and how policy issues associated with its operation and existence can be understood. At this level the basic problem is an extension of the cost-benefit framework to cover all potential classes of potential benefits relevant to the purpose of the policy. Heretofore, in traditional kinds of cost-benefit analysis, whole classes of benefits have gone unrecognized since the program or policy was inadequately conceptualized. Garn avoids this problem by building a conceptual model of what a CDC represents as a new community institution.

In dealing with the concept of community benefits he critiques the problems of benefit valuation when market prices do not exist. He goes on to explore a severity-of-need approach in order to adjust for the social environmental context in which a CDC operates. His model leads to a concern with two further considerations: (1) the problem of output identification and production, and (2) issues of resource mobilization as juxtaposed with issues of resource allocation. In the case of the former he points out that in most cases of the production and delivery of services, the interaction between the ostensible producer and the client-consumer is central to the valuation of output as well as being part of the production process. That is, the client's gain in welfare from the service is affected by both the process of production and the consumption of the eventual outcome in which he jointly participated with the service producer. In analyzing, therefore, a particular social program, policy-makers must concern themselves with the resources which the client-consumer contributes to the joint production of the service. For example, a public school is not a production unit which controls the amount and kind of resources which a student brings to a learning situation in the same way that an automobile producer controls his use of capital and labor. Thus a CDC may deviate from other service delivery systems in that it may concern itself with improving the consumption-production abilities of its membership.

In the latter case, Garn points out that the CDC, as a development institution, faces a dilemma between strategies of resource mobilization involving relations with the "outside" environment, and strategies of resource allocation in which choices between community services and economic self-sufficiency are prominent. As he writes: "If programs are to be valued, they must be responsive to the needs of clients. If programs are to be maintained the necessary resources to pay for them must be mobilized." Garn's chapter illustrates the reality that simple, mechanistic techniques are incomplete in providing policy-makers with a basis for choosing where to allocate scarce resources. Broader concerns including issues of social purpose, value, and participation must be factored into the analysis—not as an afterthought, but in the formulation of an initial conceptual framework of objectives.

Barndt's chapter addresses itself to type III evaluation and the concerns of local program managers. After suggesting that the rationality concerns of evaluation research play only a limited role in guiding local allocation choices in the face of the prevailing political culture and institutional histories, Barndt identifies and considers from the perspective of program managers, who face a real need to design and redesign "better" programs to service their client-customers, the limits that evaluation tools encounter because of their inherent assumptions and the organizational environment within which they are applied.

He details a number of assumptions inherent in the scientific paradigm from which evaluation methods are derived. He demonstrates that for some organizations these assumptions are not merely limited but are actually contrary to actual organization purpose. This analysis is then followed by a discussion of organization forms for which the assumptions are entirely inappropriate. He suggests that for certain organizations the traditional methods of evaluation have little relevance for program managers. These include those organizations which see themselves as (1) catalysts of change, (2) as counter-institutional, (3) as advocates for particular unpopular positions, (4) as requiring a substantial, ad hoc selection of diverse means and ends strategies, and (5) as participants in non-institutionalized forms of systems.

Barndt goes on to examine the potential for improving the usefulness of evaluation methods. He identifies six areas of potential improvement but also discusses the limits of these improvements in dealing with the genuine complexities associated with social values. He suggests that a series of more simple models be employed to discuss program results rather than investing in a more exact, high-technology type of model. As he writes: "The product of any evaluation process must, in the final analysis, improve the relevance of a design when it can and put up with the limitations of the design when little can be done about them."

Each of these chapters contributes to the conclusion that the field of evaluation research, if it is to contribute to the development of new programs and processes which will improve the conditions of the social economy of cities, must be rooted more substantially in those two social science paradigms associated with "experiential discovery" and "goal-directed, organizational development" rather than with those paradigms associated with "statistical significance" and "logico-deductive" methods.

RELATED ISSUES

There are several other related issues which can be discussed. First, one of the more important concerns which is important to the study of the urban social economy is the deliberate invention of regulations, organizations, or procedures which would help to move a socioeconomic system from a less desired state to a more desired set of social and economic outcomes. This is the concern of what Adolph Lowe calls "instrumental economics." It introduces concerns of social design. It also makes the study of innovation and diffusion processes of central importance. At the one hand a distinction between types of innovations in different systems is required: Does the innovation lead to (1) greater efficiency, (2) greater participation, (3) more responsiveness, or (4) more equality? On the other hand, is the innovation which is successful within the context of its own environment portable to the context of other socioeconomic systems?

Second, there is a basic dichotomy between paradigms of applied social science research. This dichotomy has been explored by Parlett and Hamilton (1972). Dominant in the field, according to their interpretation, is the classical or "agricultural-botany" paradigm which uses deductive methods rooted in experimental design traditions. They advocate instead "illuminative" evaluation guided by an anthropological paradigm drawing from social anthropology, psychiatry, and participant observation. Its primary concern is with description and interpretation rather than with measurement and prediction. The aim of illuminative evaluation as they define it is to study an innovative program, show how it operates, how it is influenced by the various contexts in which it is applied, how it is regarded by those directly concerned, and what are its most significant features, recurring concomitants, and critical processes. It is concerned with understanding the environment of implementation and diffusion.

This environment is described as a learning milieu, representing a nexus of cultural, social, institutional, and psychological variables. In most cases these variables cannot be controlled successfully so as to permit a more rigorous experimental design to be tested. In other cases, control groups are incomparable with treatment groups especially in innovation-orientation environments.

Third, there is, however, ample scope for more rigorous experimentation designs to be employed in evaluation in those program efforts where the objective is clear, easy to measure, and well understood, and where there is a widely accepted model which "explains" the relationship between variables. Rivlin (1974) has recently reviewed some of the conditions under which planned experiments lend themselves to cost-effective research. (She considers that the measurement of the response of individuals or households to a change in economic incentives is the neatest case. The HUD experiment with housing allowances is such an example.) Her colleague at Brookings, however, has concluded that if there is enough interest in some problem to support a major social experiment, then the interest will be so great that no one will be willing to wait for the conclusion of the experiment before passing legislation to implement a program. On the other hand, if there is not broad concern over the problem, then there will not be enough interest to support the funding of an experiment (Timpane, 1970). This, then, will lead to no experiment, or, in the Great Society case, to eventual accusations that the Democratic Administration "launched ineffective programs, wasted money and squandered public confidence" (Gorham, 1972).

Campbell (1969) has proposed a different orientation in distinguishing between "trapped" administrators and "experimental" administrators. He thinks that there should be a shift in posture away from the advocacy of specific reform to the advocacy of the seriousness of the problem, and hence to the advocacy of persistence in applying alternative policy and program efforts. An effort should be applied on an experimental basis until its success or failure is established. If it fails, a second effort should begin.

Fourth, other evaluation models and methods have been suggested (Guttentag). There is the notion of a legal or adversary model in which there are claims and counterclaims, arguments and counter-arguments, and each or every side is advanced by an advocate who attempts to make the best possible case for his interpretation or program or policy results. Issues of evidence and the inferences which can be drawn from multiple sources of evidence become important in this model. Another model could deal more explicitly with the personal probabilities and preferences of different types of decision-makers. The political, social, economic, and psychological needs of the various decision-makers become a major factor in the design and analysis of programs. This model could take on special significance when the diffusion of urban innovations into different municipal systems is considered. Other models or methods might attempt to deal more forthrightly with situational analysis and the study of behavior as influenced by spatial-design characteristics of particular sites as influences on program outcomes.

Fifth, there appears to be a need to develop a generalized system model which can be used to analyze, on a comparative basis, the usefulness and reliability of evaluation measures and methods. Such a generalized system model would postulate that a system of any type is the specified and organized conditions for the relations between nine basic elements—functions, inputs, outputs, sequences, environ-ments, feedback, physical catalysts, residuals, and unanticipated consequences—involved in a socioeconomic sub-system (Gappert, 1974). Insofar as one of the outcomes of evaluation research itself is a determination of intervention strategies into the production process of public goods and services, attention must be directed during the research to the identification of those elements most susceptible to change.

Without attempting to be definitive in this brief discussion of evaluation issues, several series of unresolved questions suggest themselves:

(1) How can we best understand the institutional obstacles to the use of evaluation research in different urban systems? How does evaluation take its place within different urban organizational structures?

(2) What kind of relationships between program administrators, evalu-ators, and practitioners either facilitate or retard the evaluation process? Who has a pro-active and who has a re-active role in developing social indicators and other normative criteria in the evaluation of the success of urban programs?

(3) How do you best prepare people to do evaluation research at both the local and national level? Are the assessment-of-need and the formu-lation-of-policy processes significantly different at the national and local levels that at least two different evaluation philosophies need to be applied?

(4) If the society remains unclear about the purposes of its urban and social policy, what then is the role of evaluation research? Is evaluation research likely to continue to exhibit a tension between scientific technique and social criticism?

REFERENCES

CAMPBELL, D. (1969) "Reforms as experiments." American Psychologist 24, 4: 409-429.

CONGER, D. (1973) "Social inventions." The Futurist 7, 4: 149-158.

GAPPERT, G. (1974) "Assessment and acceptability of urban futures." Futures (February): 42-58.

GORHAM, W. (1972) "What federal money didn't buy." Wall Street Journal (June 8).

GREENFIELD, T. B. (1973) "Organizations as social inventions." Journal of Applied Behavior Science 9, 5: 551-574.

GUTTENTAG, M. (1970) "Models and methods in evaluation research." Journal of the Theory of Social Behavior 1, 1.

LEWIS, V. (1952) "Toward a theory of budgeting." Public Administration Review 12, 1: 42-54.

LOWE, A. (1965) On Economic Knowledge. New York: Harper & Row.

PARLETT, M. and D. HAMILTON (1972) "Evaluation as illumination: a new approach to the study of innovating programs." Center for Research in Education Science, Edinburgh (October).

RIVLIN, A. (1974) "How can experiments be more useful?" American Economic Review, Peoples and Proceedings (May): 346-354.

TIMPANE, M. (1970) "Educational experimentation in national social policy." Harvard Educational Review 40: 547 566.

17

Social Indicators and Society: Some Key Dimensions

MICHAEL J. FLAX
MICHAEL SPRINGER
JEREMY B. TAYLOR

INTRODUCTION: POLICY NEEDS AND RESEARCH OBJECTIVES

☐ RESEARCH ON SOCIAL INDICATORS is one of the most ambitious undertakings of contemporary social science, offering the promise that scholars and researchers can deal with the broader aspects of social policy.[1] This paper examines the progress that has been made in the development of social indicators and indicates how these efforts might relate to the development and assessment of public policy. It is significant that the impetus for the social indicator movement came from the problems encountered in formulating social programs of the 1960s. Therefore, it would be

AUTHORS' NOTE: *This analysis was supported by funds from the National Science Foundation's Special Projects Program under Grant No. GS-38613. Opinions expressed are those of the authors and do not necessarily represent the views of The Urban Institute. We wish to thank our colleague, Harvey A. Garn, for his comments and suggestions, and Rhonda Truet and Jacqueline Swingle for their help in preparing the manuscript.*

well to first explore the character of this impetus and then outline the concepts we employ to assess how current social indicator research relates back to these social and political processes. Implicit in our approach is that critical questions concerning any large-scale research effort focus upon the two-way interactions between society and the scientific enterprise.

DEMAND FOR TECHNOLOGIES OF HUMAN ENDS

During the past decade and a half we have witnessed a fundamental shift in the orientation of public policy—a shift from preoccupations with means to a struggle to specify the ends of public action in terms of human welfare. In the years prior to 1960, public policy emphasized the tangible inputs to problem solving—if people lived in deteriorating neighborhoods, then a housing project was constructed; if they were victimized by crime, additional police officers were hired; if they were sick, a new wing was added to a hospital. Such responses, by themselves, are no longer sufficient. The primary focus of health policy has shifted from medical facilities to improving the actual health of the population. Crime is not simply a problem requiring more police manpower, but entails lowering the chance that citizens will be victimized by criminal acts. Solving the problems of deteriorating areas no longer means the construction of housing projects, but rather of somehow improving all aspects of neighborhood life. Deciding on whether education funds are justified no longer depends only on the number of buildings to be constructed or staff specialists to be hired. Rather, these decisions now involve whether skill levels might be increased or social norms inculcated. Such shifts are reflected in the thinking of political activists in almost every ideological camp. While most fully developed in federal policy arenas, these shifts are now taking hold at the state and local level.

This new orientation of public policy has clearly not been matched by understandings of how to guide policies that will affect the quality of people's lives. As a consequence, there has been a variety of undertakings to develop the tools, techniques, and analytic procedures that could assist policy-makers in coping with decisions that focus on the outcomes of public action. What is required, if you will, is a technology of human ends capable of identifying the social outcomes of public policies and programs.[2] Some concern for such outcomes is reflected in the Johnson Administration's Program-

Planning-Budgeting System, and to a lesser degree in the current administration's use of management by objective procedures. It is reflected, also, in the massive commitments being made to program evaluation and for the development of large-scale policy experiments. Perhaps the most ambitious of these undertakings has been a broadly-based intellectual movement concerned with the development of social indicators and social accounts.

The "social indicator movement" began with much promise, publicity, and confusion, reflecting the ebullient and optimistic intellectual atmosphere of the 1960s. Social indicators were to do many things—measure the benefits of social programs, estimate the consequences of technological innovation, evaluate the quality of life in cities, and monitor social change—all apparently in short order. Much energy and considerable resources were directed toward these objectives. But quite clearly the social indicator movement has not fulfilled its early promise. This is not to suggest that the social indicator movement is a failure. Rather, its early proponents and supporters tended to foster the belief that energy, enthusiasm, and adequate funding could quickly solve thorny intellectual problems. Real progress, however, in terms of improved understandings of social problems and policy outcomes, is a very slow and painstaking task.

Most of the developers of social indicators and those who support their activities now recognize this fact, and those who make careers following current intellectual fads have moved on. Sensible, coherent, and potentially productive plans of research and development have emerged. Yet the work on social indicators remains a large-scale and diverse enterprise. It is being seriously pursued not only in North America but also in Western Europe and Japan, and draws upon almost every conceivable academic and professional discipline. Efforts range from the development of elegant and sophisticated theoretical constructs to helping achieve solid and workmanlike improvements in the collection of official government statistics.

THE SOCIAL DIMENSIONS OF RESEARCH

At this time, it would be worthwhile to attempt to take stock of how social indicator research has evolved and where it might proceed in the future. There are obviously several alternative approaches to such an assessment. One could describe, in historical terms, the major figures and activities of the movement or could conduct an intensive content analysis of a well-developed approach to social indicators.[3] Such approaches are clearly instructive. Our own understandings of indicator research, however, were sharpened considerably by viewing the social indicator movement in terms of its own social character-istics—the roles, perspectives, and reference groups that structure interactions among the researchers and between them and a broader society. It is our view that not only is social knowledge shaped by political and social processes, but also that ideas about society have their impacts on those processes that gave them impetus. Given that the very notion of social indicators remains in the process of development, it would be premature to try to somehow predict the political and social consequences of indicator research. Rather, we aim simply to describe in a social context some of the key problems, methodologies, and theoretical approaches that characterize various research programs and delineate their potential social and political implications. Such considerations are particularly germane to a research enterprise that was conceived as and largely remains an effort to directly address critical problems in devising social policies.

We found it instructive to describe the social context of indicator research in terms of a series of starkly drawn dichotomies that identify important continua of social relationships. This paper focuses on the following four key dimensions of the social indicator movement.[4]

Basic or Applied Research. One can distinguish two broad orientations within the social indicator movement. One group is engaged in a basic research enterprise; that is, their key questions originate within academic communities and the primary audience for their efforts is other scholars. The second group is attempting to develop indicators that provide more useful approaches for the formulation and assessment of public policy and programs. These researchers are engaged in an applied enterprise; that is, their presumed clients are public officials and citizens whose problems and concerns help shape their key research questions.

Supply or Demand Emphasis. Some indicator researchers tend to focus their technical expertise on questions related to problems associated with the supply of social information for either basic or applied purposes. Their efforts, therefore, are directed toward methodological problems defined within various research or technical communities. The efforts of other indicator researchers tend to be demand-oriented. Their research activities are shaped by developments external to research communities—some problem or phenomenon perceived within society or the needs, concerns, and capacities of public officials or citizens.

Concern with Production or Consumption. Indicator research can emphasize the problems of production or consumption. Emphasis on production issues means that indicators would be particularly useful to decision-makers within production units, such as firms or government bureaus, while an emphasis on consumption would mean that indicators would be useful also to organizations representing consumer interests broadly conceived, such as trade unions, block clubs, clients of particular social programs, or ethnic associations as well as the general public.

Social Criticism or Technical Support. Researchers and intellectuals adopt various political postures. Most social indicator researchers see themselves as providing technical support for political or organizational decision-makers. A much smaller group sees itself engaged in the role of social critic, either representing counter-elites or the powerless. A tension between social criticism and technical support, however, marks much social indicator research.

The following sections examine the social indicator movement in terms of these dimensions. Our coverage of specific studies or programs is not intended to be complete or definitive. We endeavor simply to provide a general appraisal of the field of research with a discussion of selected approaches. It should be emphasized that these dimensions are only heuristic devices that help to structure observations of a set of highly complex and subtle relations. Even at this level of conceptualization our list of potential dimensions is by no means complete. Additionally, a particular study or program might combine several aspects of the same dimension. One research program, for example, might focus on problems associated with the supply as well as the demand for social information, while another might shift on this dimension over time. We conclude our discussion with some observations about where we believe the most productive avenues of future research might lie, given current emphases within the field.

BASIC AND APPLIED RESEARCH:
A QUESTION OF INSTITUTIONAL LOYALTIES

The idea of social indicators was originally put forth by a relatively small group of men engaged in an essentially political operation.[5] They were attempting to help policy-makers crystallize support for a wide range of social objectives that were emerging in the 1960s and to mobilize intellectual resources so that the pursuit of these objectives would be buttressed by the necessary analytic underpinnings. Ideas were purposely presented in a provocative and vague fashion so that the widest possible support could be secured. They were remarkably successful in their efforts to solicit the support of officials at the highest levels of the federal government and to engage the interest of large numbers of highly competent social scientists.

Once social scientists began the serious business of social indicator research, the character of their institutional associations and loyalties came to the fore. Demographers saw social indicators as an extension of the analysis of population trends and characteristics, while economists saw them as an extension of cost/benefit analysis or national income accounting. Since there was no well-developed theory or methodology for social indicators, it is not surprising that researchers initially addressed the problems of indicator development in terms of well-understood approaches. Additionally, there are significant institutionally based incentives for such behaviors. Serious researchers belong to particular academic and professional disciplines and are employed by particular kinds of organizations such as universities, research institutes, or units of government. We observed that it was this network of institutional relationships, rather than either the early proponents of social indicators or the policy-makers who supported them, that controlled the primarily personal incentives of most social indicator researchers. Thus, initially, social indicator research tended to be undertaken if compatible with previously established objectives such as those of an academic discipline, a statistical agency, or a private research organization.

These institutional loyalties shape the most significant dimension of the social indicator movement, i.e., whether the researchers are engaged in a basic or applied enterprise. The primary loyalties and orientations of the basic researcher lie with specific academic or scientific communities. Key research questions tend to flow from these communities which in turn are the primary audience for the resulting research products. Applied researchers, on the other hand, are primarily motivated by questions or problems that flow from the specific concerns of elements of society and their ultimate clients are firms, policy-makers, or citizens. There are obviously complex interactions between basic and applied research, and some of these are discussed below. Nonetheless, their fairly stark dichotomy does distinguish an important cleavage with the social indicator movement.

The cleavage emerged early in the development of the movement. Bertram Gross, Daniel Bell, and Raymond Bauer, in providing the initial articulation of the need for social indicators, were clearly calling for applied work. They observed the difficulties of policy-makers who were attempting to deal effectively with the second-order impacts of technological innovations, or to express social goals beyond full employment and economic growth, or to develop programmatic responses to the array of social concerns that emerged in the 1960s.[6] Social indicators were intended to monitor social consequences of technological innovation. They would provide a focus and help define a broad array of social goals, provide a framework for national social planning, and help order the needs of policy-making to academic and scientific communities.

Such concerns are quite consistent with the backgrounds of all three men. While quite different persons in many respects, Gross, Bell, and Bauer made academic reputations by ignoring the boundaries and strictures of academic disciplines. Each was a broad-gauged generalist, prone to focus on the interplay of political, economic, and social processes. Additionally, they had been active in political life and were keen observers of politics. As a consequence, each had the kinds of value orientations and intellectual perspectives consistent with a scheme that ventured to mobilize significant intellectual resources in the support of ambitious public policy objectives. Because of their backgrounds, they tended to underplay the technical and conceptual problems that would have to be addressed if social sciences were to provide tools to address large-scale social planning.

A response to this underemphasis occurred with little delay. One of the most vigorous and well-organized approaches to the idea of social indicators was undertaken by a group of demographers and sociologists who argued strongly that social science was unprepared for the demands of policy-making. The leading advocate of this faction of the social indicator movement was Eleanor B. Sheldon of the Russell Sage Foundation. She and others argued that since there is no macroscopic theory of social relationships or social change, the development of social indicators that would have the same kind of use in policy-making as economic indicators is not possible at this time. (See Sheldon and Freeman, 1970.) She further maintained that until our knowledge of social change becomes more complete, we must be limited to general descriptive reporting on the state of society. Therefore, work on social indicators must initially be a basic research effort that focuses on the development of fundamental understandings of social change. As will be noted in the following section, Sheldon and others have gone on to identify the kinds of studies that might be undertaken to provide these understandings. Not surprisingly, these consist primarily of data collection operations and analyses that are fully consistent with the long-term interests of university-based demographers and empirical sociologists.

For this faction of the social indicator movement, the linkage from fundamental knowledge to policy application is outlined by Kenneth C. Land (1972). Land assumes that applied science flows only from prior understandings developed by basic science. Once the empirically verified and nonnormative models of society are developed, all that remains is to identify desired states and policy-manipulatable variables in order to establish a system of social indicators. Given the current status of social knowledge, this conception of the development of policy-relevant knowledge argues for the absolute centrality of basic research for social indicators.

The history of the social sciences suggests that the interplay between basic and applied work is more complex than that outlined by Land. Certainly policy applications of basic studies are constantly being developed. Many policy applications as well as more fundamental understanding of social phenomena, however, have flowed from applied orientations. Our understandings of monetary behavior, attitude measurement, roles and reference groups, cost/benefit analysis, informal organization, macroeconomics, the unconscious, and so forth, have largely emerged from applied research. Applied concerns often lead researchers to identify limitations of the existing base of fundamental knowledge.[7] Moreover, attempts to solve applied problems provide criteria for assessing the validity of new approaches. This tends to be supported by systematic analysis of the history of contemporary social science. A report on major advances in the social sciences concludes:

our analysis indicates that practical demands or conflicts *stimulated* [author's emphasis] about three-fourths of all contributions between 1900 and 1965. In fact, as the years went on, their share rose from two-thirds before 1930 to more than four-fifths thereafter. The contributions of "ivory tower" social scientists in the future seem to be minor indeed.[8]

This conception of how applied knowledge develops is implicit in much social indicator research. Countless studies were undertaken to organize and present descriptive indicators so that the educated layman could make more informed judgments about the social condition of areas or groups of people.[9] Using well-understood and fairly simplistic conceptual approaches along with readily available statistics, these studies represent the initial applied approach to indicator development. For many researchers, such approaches represented the limits of indicator research. For some, these efforts helped focus attention on the limitations of bodies of fundamental knowledge. They helped pinpoint the conceptual and methodological problems of a long-term and durable character that constrained the development of more useful indicators.

Much applied indicator research is currently focused on constraints identified by efforts to develop more useful tools for the consideration of public policy. Mancur Olson's experience of extending the logic of program budgeting into the first national social report led him to identify, subsequently, gaps in the theory of public goods. Our own descriptive studies led to the problem of relating input/output relationships associated with the generation of welfare to structural arrangements of bureaucratic and political institutions. The key point here is that these problems within bodies of fundamental knowledge were revealed by attempts to address applied problems. The validity of new concepts and paradigms being developed will depend on whether they lead to the development of more useful tools for formulating and evaluating public policy.

New and more productive interaction between basic and applied work on indicators could eventually emerge from the role of the Special Projects Program of the National Science Foundation (the major source of funds for indicator research). Initial grants made for indicator research were distributed to a wide variety of projects of both a basic and applied character. A number of the projects simply extended the intellectual trajectories of existing academic disciplines. Some of the projects, however, appeared to be marking off new intellectual ground by identifying and pursuing unique sets of problems or issues. The current funding strategy of the Special Projects Program (NSF) is to help extend and provide for the further development of these efforts. Whether research on social indicators emerges as a coherent and fully defined field of study remains uncertain. However, the subtle and difficult role being performed by NSF—by first allowing indicator researchers to formulate issues and questions and then emphasizing common analytic and methodological concerns—is important and pertinent. In effect, this inductive process might lead to the articulation of a somewhat coherent field of study that could provide a focus for the institutionalization of indicator research having both basic and applied aspects.

In its early stages the social indicator movement was characterized by heated debates as to the proper strategy for social indicator research. We would expect such discussions to continue in that the institutional loyalties and orientations of basic and applied researchers remain quite different. However, there does appear to be some convergence of intellectual interests, generally focusing on theoretical issues which have largely sprung from applied approaches. The National Science Foundation's Special Projects Program is working to encourage these convergences.

SUPPLY OR DEMAND EMPHASIS

Basic and applied social indicator researchers confront two broad classes of problems—those associated with the supply of or with the demand for social information. Supply problems involve the scope and quality of available data sets. Demand problems deal with the articulation of the issues or relationships for which data might be sought to clarify. While much work is required on both classes of problems, individual researchers or particular research groups are inclined to emphasize either supply or demand problems.[10]

To a considerable degree, emphasis on a supply or demand approach to indicator research reflects the institutional associations and previous experience of the researchers themselves. Otis Dudley Duncan, along with many demographers and social statisticians, has chosen a supply approach which emphasizes the analysis of social trends, the conduct of replication studies, and the further standardization of measurement systems such as the census of population (Duncan, 1969). These approaches are extensions into the field of social indicator research of long-term concerns of demography and certain kinds of empirical sociology. Our own research can similarly be viewed as the extension of demand issues associated with the analysis of public policies and programs. These concerns reflect both the Urban Institute's commitment to policy research and the sensitivity to conceptual problems that stem from the academic associations of members of our research team.

Sheldon and Duncan, in particular, have made a series of strong arguments for a supply orientation and have written at length to define the proper scope and content of social indicator research. They suggest that, given the present state of social knowledge, there is little possibility of directly linking social measures with policy decisions. As a consequence, emphasis should be placed upon improving social information for purposes of furthering knowledge about the functioning of society, measuring social change and trends, and enhancing social prediction and forecasting. "Through the development and analysis of descriptive time series and the modeling of social processes, we will be able to describe the state of the society and its dynamics and thus improve immensely our ability to state problems in a productive fashion, obtain clues as to promising lines of endeavor, and ask good questions" (Sheldon and Parke, forthcoming). Otis Dudley Duncan makes the case for improving the measurement of social change even more forcefully: "The value of improved measures of social change . . . is not that they necessarily resolve theoretical issues concerning social dynamics or settle pragmatic issues of social policy, but that they may permit those issues to be argued more productively" (Duncan, 1975). These researchers appear in substantial agreement with the position that improving the measurement and description of social change must occur prior to the development of policy-relevant models which depict the relationships among such measures.

While we cannot fully accept all of these arguments, nonetheless we do recognize that the vigorous presentation of this position results in considerable benefit both to the social indicator movement and to the improvement of policy-relevant knowledge. Supply concerns facilitated a greatly expanded coalition of interests within the social indicator movement. It allows countless statisticians and data analysts, particularly in the employ of governmental units, to join the movement and more significantly to broaden their perceptions of the potential role of social statistics in the formulation of public policy. Supply concerns also encompass a variety of important efforts to improve the quality of policy-relevant data bases, such as the National Crime Survey and the National Assessment of Educational Progress. Additionally, the work of Duncan and others on replication studies and improving data bases will most certainly add to improved understandings of certain kinds of social phenomena. These will ultimately lead to improved policy knowledge.

Both supply and demand concerns often converge in efforts to develop descriptive indicator reports. Our early studies attempted to assess the quality of life in cities, to respond to questions being asked by public officials, and to identify deficiencies in available data sets. Perhaps the most significant of such descriptive indicator projects was the publication by the Office of Management and Budget (OMB) of *Social Indicators 1973*. Conceived as the first of a regular series, this document seeks to provide a foundation of "well organized and carefully selected social statistics appearing periodically and open to revision and change" (Tunstall, 1970: 3). OMB envisions two main purposes for such a publication, first, as a tool in decision-making, and second (and perhaps more importantly from a supply perspective), as a means for highlighting the gaps in available data, and thus as a guide for developing better social statistics. Similar functions are performed by many city and state indicator reports, particularly those developed by the Urban Observatories.[11]

Expressing demand issues in a disciplined and structured fashion is far more difficult than those associated with the supply of information. This is illustrated by the problems encountered when government-based information and statistical specialists try to develop systems of social indicators and descriptive social reports. By the character of their institutional support, these analysts must address supply as well as demand concerns. They can draw up fairly well-developed and agreed-upon statistical tests to assess the quality of available data. However, the selection and presentation of sets of data items that are useful to decision-makers tend to be far more ambiguous. For many, experience and intuition are the only available guides for anticipating possible uses and users of social indicators.

There are a good number of indicator researchers who are endeavoring to address demand issues in a structured fashion. Such work is predicated on two kinds of observations. First, the collection of data not addressed to satisfying the needs of specific users generally results in the development of large-scale and little-used data files. Second, the usefulness of sets of data depends on whether they are compatible with the implicit or explicit inference structures and associated modes of choice commonly employed by the analysts or decision-makers for whom they were designed.[12] Therefore, the demand problem in developing social indicators is to define appropriate inference structures so that data items provide useful insights in the consideration of policy issues.

Illustrative of such efforts is the work of two economists, Mancur Olson and Nestor Terleckyj, and the more recent efforts of our own team. Each of these is an effort to provide a structure to address what are viewed as central intellectual problems of policy formulation. Olson begins with the assumption that there are reasonably well-developed approaches to assist individuals making allocation decisions affecting their own self-interest. Approaches to identify the important elements of public decisions affecting collectively consumed goods and services are far less developed. Regarding social indicators, Olson (1969) initially saw the key problem as identifying the outputs of government actions in terms of outcomes that have clear and direct normative interest. Presently, Olson (1973) views the problem in terms of accounting for the spillover and external effects of policy decisions.

Terleckyj begins with a less abstract problem—how to view the allocation of resources in terms of multiple sets of social objectives and interactions among these objectives. Constrained by the level of available knowledge, he provides decision-makers with a framework that views expenditures and potential impacts of various levels of expenditures on areas such as health, housing, police services, nutritional programs, recreation, and income assistance. Additionally, this work attempts to demonstrate potential interactions among such expenditures. For example, how much improved health is secured by various levels of expenditures on nutritional and housing programs?

Our more recent work focuses on two kinds of issues. One is the consideration of what outcomes are important to different citizens and how such outcomes are generated. In investigating these questions, we are exploring the values and roles of consumers, particularly in relation to their utilization of publicly provided goods and services. Another issue is that it is important to try to relate these input/output relationships, which are associated with the generation of welfare, with how such relationships are influenced by the structure of bureaucratic and political institutions. While the emphasis in our work is on consumer activities, concern with these two issues enables us to explore how accountability on the part of either consumers or producers can be established for variations in outcomes. Like Olson and Tereleckyj, we are endeavoring to define the demand for indicators in terms of the needs and concerns of our presumed clients—citizens and public officials.

A strong emphasis on supply concerns characterized the movement initially. Problems associated with improving the technical quality of available data sets were well-defined and remedies well-understood. Therefore, initial efforts gravitated toward these manageable problems. On the other hand, it took considerably longer to state demand-side problems in a rigorous fashion. Subsequent efforts to address demand issues tended to reflect divergent intellectual perspectives. There are presently a number of well-elaborated approaches to these issues which require verification through empirical analysis and/or application in actual policy-making situations.

CONCERN WITH
CONSUMPTION OR PRODUCTION

As indicated above, much of the impetus for the social indicator movement stemmed from a fundamental shift in the orientations of public policy, i.e., from concerns with inputs and objects of expenditure to concerns with the outcomes of public policy in terms of the quality of people's lives. This shift was not only manifested in how analysts might formulate problems or how politicians might present issues, but also in significant alterations in the substructure of political life. The interests of consumers are becoming sufficiently mobilized so that they seriously challenge producer interests in many political struggles. This development is, of course, reflected in the vigor of national and local consumer lobbies as well as the environmental movement. It is reflected in concerns for client advocacy within many professional groups and the increasing willingness of ordinary citizens to challenge the action of producer organizations, whether retail stores or primary schools. It is reflected also in the growing skepticism that problems can be solved solely through the production activities of government. These political developments have resulted in significant policy innovations ranging from increased citizen participation in formal decision-making processes to allowing greater citizen choice in public goods and services through such devices as housing allowances and educational vouchers.

From this essentially political perspective, we view social indicators as an approach to structure consumer-oriented values and concerns into public policy considerations. Thus, social indicators are by no means value-free or apolitical; rather, we view them as potentially supportive of the values and concerns of consumer interests.[13] From our perspective, however, if one examines the bulk of social indicator studies and reports, an apparently paradoxical impression emerges—they appear to be far more useful to the producers of goods and services than to consumers. For example, they shed more light on problems of allocating funds to alternative housing construction programs than providing ways for consumers to assess the benefits provided by different kinds of housing units or neighborhoods. They are more responsive to problems associated with medical manpower than to those associated with the problems of citizens evaluating their own health or the quality of available medical services. How is it that the potential for consumer-oriented indicators has not been realized?

Several obvious explanations of the producer bias of much indicator work are possible. First, much available social information (e.g., data sets on crime, health, and learning) is the by-product of the administrative processes of producer organizations. Therefore, it tends to be bounded by the activities and concerns of such organizations. Second, independent data collection agencies, such as the Bureau of the Census, have been influenced far more significantly by producer than consumer interests, i.e., they are apt to collect information of use to industrial sectors rather than information useful to the consumers of their products. It is a truism of U.S. political life that producer organizations are far better able to organize and express their demands, and governmental statistical agencies must respond to these demands. Third, the costs of consumer-oriented data collection are high. The costs of the most obvious instrument for collecting consumer-oriented data, the sample survey, is rapidly becoming prohibitive for many purposes. Expansion of bureaucratic records, so that they include more information concerning clients or service recipients, is also costly. Finally, almost every strategy to expand consumer-oriented data bases is clouded by complex and heated privacy and confidentiality issues.

All of these explanations, however, do not sufficiently account for the producer bias of much indicator research. Most of the support for indicator work was motivated by a sincere concern for consumer interests by officials in producer organizations, that is, how do public policies and programs actually affect the quality of life of ordinary citizens? Despite the costs involved, there has been a significant growth in consumer-oriented information through sample surveys and the expansion of bureaucratic records. Imaginative ways are being worked out to cope with the problems of confidentiality and privacy in the collection and use of large-scale data bases.

The development of our own research revealed that the problem of producer bias in large measure stems from the very paradigms and concepts employed by researchers themselves. Key bodies of policy-relevant social theory emerged when the focal concerns of society centered on problems of economic theory in which the treatment of the processes of consumption is so truncated that it tends to be difficult, if not impossible, to structure questions related to important consumer values. (See, for example, Alonso, 1970.) Large bodies of political and administrative theory have little capacity to address questions involving the interactions between formal institutions and the clients or citizens these institutions serve (Ostrom, 1974). It should not be surprising, therefore, if several of the earliest theoretical efforts to define the demand for indicators were conceived in terms of fairly narrowly bounded production problems. *Toward A Social Report*, primarily the work of Mancur Olson, envisioned a series of grand social production functions which could provide guidance to federal executives (U.S. Department of Health, Education and Welfare, 1969). Bertram Gross (1966) couched the problem in terms of the development of a planning system directly analogous to those that might be employed by a firm or public bureaucracy. Nestor Terleckyj's work (forthcoming) on national goals is designed to provide decision-makers with a set of interrelated production functions to better manage resource allocation decisions.

Fulfillment of the promise of social indicators will, in large measure, depend on the development of inference structures capable of encompassing consumer values in a structured fashion and in relating these values to the production processes, the primary source of policy development. Considerable indicator research has been directed at this general problem. Such efforts fall into three categories: the work being done to expand notions of personal and national income; the exploration of social-psychological satisfactions as the key measures of human welfare; and our own work on expanding the view of the role of consumers in the generation of welfare.

For most economists, the most important measure of welfare is the level of family or national income. Monetary income, however, is a far too narrow measure to resolve the problems addressed by social indicators. A number of economists are attempting to expand the notion of income in order to provide a framework for social indicators. Richard Ruggles and his colleagues at the National Bureau of Economic Research are expanding national income accounting into a general framework for assessing social as well as economic performance (Ruggles and Ruggles, 1973). They are developing procedures to discount increases in economic output by estimating the cost of environmental damage as well as developing shadow prices for various nonmarket household activities. These efforts attempt to make a range of social costs and benefits commensurable in terms of monetary units. An economist, Karl Fox (1974), has developed an approach to social indicators using a total income concept in which costs and benefits are made commensurable through fundamental psychological processes.

For many sociologists and psychologists, human welfare must ultimately be reflected in levels of human satisfactions. Much indicator research has been directed at exploring this obviously consumer-oriented phenomenon. The Institute for Social Research of the University of Michigan is a leading center for this kind of indicator work. One group under the direction of Frank M. Andrews and Stephen B. Withey (1973) is working to identify patterns of social-psychological perceptions and satisfaction, while another group under the direction of Angus Campbell, Phillip Converse, and Stephen B. Withey is trying to develop methodologies to test perceptions of the quality of life (Campbell and Converse, 1972). A rather different approach to subjective indicators is being developed by Robert Newbrough (1973) at the Center for Community Studies of George Peabody College. He and his colleagues are exploring associations among indicators of psychological mood and neighborhood characteristics and public occurrences.

We believe that a focus on people's perceptions and satisfactions reflects a useful approach to develop consumer-oriented indicators. Several key problems, however, plague this area of research. One is identifying systematic relationships between subjective and objective measures of welfare—between, for example, feelings of well-being and levels of disability. Another is the relationship between objective and subjective measures of welfare and the outputs of particular service systems, between crime rates or fear of crime and the activities of police departments. Our own work in expanding understandings of consumer roles in the generation of welfare attempts to address these issues (Garn, Flax, Springer, and Taylor, 1975). Our key notion is that it is useful to view the generation of welfare in terms of two structurally independent sets of activities, actors and technologies—the first involving the transformation of resources into goods and services, and the second transforming the characteristics of these goods and services into welfare outcomes such as levels of learning and various satisfactions.

Thus, the producer/consumer dimension discussed here is less crystallized, less recognized, and in many ways more difficult to comprehend than those described previously. Production activities occur in stable and well-organized social units which are able to articulate their information requirements, mobilize resources necessary for data collection, and have the technical capacity to use such information. Furthermore, one can observe a rough division of labor among academic disciplines with some (such as economists and operations research analysts) historically more oriented toward producer questions, while others (such as sociologists and survey researchers) are more likely to focus on phenomena related to consumption concerns. The recognition and eventual synthesis of these intellectual streams is a difficult but essential task for social indicator research.

TENSIONS BETWEEN SOCIAL CRITICISM AND TECHNICAL SUPPORT

Any analysis or commentary on social relationships has potential political and ideological implications. In much social science, particularly of a basic character, such implications are neither paramount nor obvious; however, in social indicators research these implications are present but not always clear. Social indicators provide pictures of our society and its problems, as well as delineate key actions undertaken to address these problems. It is not our intent here to provide a full political critique of social indicator research, but we do wish to indicate in general terms some of its major political implications. Social indicator research can provide technical support to existing political or organizational decision-makers, or it can provide support to counter-elites or the powerless by systematically introducing new values and new concerns into the processes of policy formulation. Few social indicator efforts are unambiguously committed to either of these functions. What is interesting is that a tension between these objectives marks much social indicator research.

Early expressions of social indicator ideas by Gross, Bauer, and Bell were clearly efforts to provide decision-making tools for high-level policy-makers. In his early work on social indicators, Gross (1966) was attempting to develop tools that would assist in the management of social phenomena in a manner analogous to the way economic indicators and analysis were being employed in the management of the economy. Bauer (1967 and 1966) tended to view social indicators in terms of an information system analogy; social indicators were to be key data elements of a strategic planning information system. Bell (1969 and 1966) saw social indicators as a comprehensive cost/benefit analysis of key social issues. Therefore, the manifest purpose of social indicators for all three was to provide technical support to high-level policy-makers. Yet each had another set of purposes in mind. Gross saw social indicators as a way to stabilize the new political commitment of the 1960s. Bauer saw his information systems as anticipating new problems and concerns. Bell viewed his macroscopic cost/benefit analysis as introducing new values and concerns into the policy process. In effect they envisaged planning tools that would not only assist decision-makers but press for the sustained consideration of a set of social problems. It is of interest that *Toward A Social Report* provides a framework to accelerate the process of political change by systematically introducing new values and interests into the processes of policy formulation.

This tension between technical support and social criticism was present in the development of *Toward A Social Report*, the first federal attempt at social reporting. It was conceived as both a mechanism to state the social policies of the federal executive and to press for the sustained consideration of a set of social problems. It is of interest that *Toward A Social Report* provides a framework to formulate questions rather than offering solutions. It establishes two sets of questions. First, how can we define consensual social objectives in a clear and rigorous fashion? Secondly, what actions, private or public, facilitate these objectives? As a consequence, the report's approach had the capacity to reveal new problems, issues, and relationships.

The second federal social reporting effort, *Toward Balancing Growth: Quantity with Quality* (1970), was a radically different document. More journalistic than analytic in approach, it tried to justify a particular set of social policies rather than to provide a framework to plan or assess them. It clearly did not provide an analytic framework capable of providing surprises for anyone.

The development of an analytic framework which served to specify relationships, reveal new and unanticipated problems, as well as support a range of social values, has been central to our indicator research. By exploring the role of consumers in the generation of welfare, we developed the notion that consumers employ differing technologies and are subject to constraints that are independent of those associated with production processes. More liberal members of our group initially viewed these notions as a way to assign blame for policy failures to producer-determined constraints on consumers, while our more conservative colleagues viewed these notions as a way to identify the limits of social policy in terms of the inherent incapacities of the consumer. Aspects of our current work can be conceived as an effort to empirically sort out these ideological concerns.

Unfortunately, in much supply-oriented indicator research, there tends to be little or none of this tension between the functions of social criticism and technical support. This comes from the great stress placed on the standardization of definitions and data collection procedures. Often one has the impression that many supply-oriented indicator researchers are saying, in effect, that we should "measure something simply for the sake of an accurate and consistent measurement." Statistical systems, whether developed for health, education, or housing, are artifacts that should reflect ongoing changes in values and technologies. Excessive and exclusive concern with statistical niceties can limit the systematic introduction of such new perspectives into indicator systems.

This technically based conservatism of many indicator researchers led Bertram Gross and other more radical analysts to view the entire social indicator movement as largely an effort to provide technical support to an increasingly repressive and manipulative government (Gross and Straussman, 1974). While Gross perhaps overstates his case, there is much room in the social indicator movement for a greater emphasis on social criticism. Unfortunately, the political avant-garde has yet to propose an approach to social indicators that is geared to the systematic introduction of social criticism into political processes. Gross' preliminary reformulation of the concept of unemployment suggests that such an approach to indicator research would be both feasible and intellectually productive (Gross and Moses, 1973).

There has long been a tension between technical support and social criticism within the social indicator movement. The function of technical support has tended to dominate the enterprise, emphasizing supply rather than demand and production rather than consumption questions. Exclusive emphasis on supply issues tends to have a built-in technical conservatism, while a narrow production emphasis tends to be constrained by the short-run concerns and problems of established bureaucratic and political institutions. An excessive concentration on the technical support function limits the anticipatory capacity of any system of social indicators. We believe an increased concern with demand and consumption questions would potentially introduce a broader range of values into social indicator approaches.

SOME PRODUCTIVE FUTURE DIRECTIONS

We have attempted to provide an overview of the current status of the social indicator movement in terms of some of its key social dimensions. From this perspective we believe it premature and potentially counterproductive at this stage to define in a bounded fashion what should be the proper scope and content of indicator research. The magnitude of the intellectual and political problems to be addressed by systems of social indicators requires the serious pursuit of a variety of alternative approaches. While there appears to be some convergence of concerns within the movement, its vitality is to be reflected by the frequency and vigor of debates over significant theoretical and methodological options.

Obviously, we do not believe that it makes sense to argue that one particular approach to indicator research is correct or proper. We can conclude, however, with some observations about which general research strategies might have higher probabilities of success than others and which general problems require additional effort.

AN APPLIED EMPHASIS AS MOST PRODUCTIVE

Social indicator research was conceived and should remain a largely applied research enterprise. We are not arguing for an applied orientation that either demands or promises immediate practical results or entirely excludes basic research on indicator problems. We do, however, believe that applied concerns provide the intellectual problems that can force out new conceptual tools and disciplinary syntheses needed for indicator development. Additionally, an applied approach provides the requisite institutional constraints to keep the research effort on track, that is, indicator researchers will be presented with a mix of performance incentives controlled not only by colleagues in academic disciplines who allocate such things as promotions and access to refereed journals, but also incentives controlled by policy-makers and persons acting as their agents in the allocations of research funds.

STRESS ON DEMAND PROBLEMS

Defining the demand for social indicators in a structured and disciplined fashion remains the most important intellectual problem facing the indicator movement. While the pursuit of supply issues draws upon well-developed and agreed-upon concepts and methods, the selection of inference structures that can order and provide policy-relevant meaning to sets of social information is not buttressed by well-developed understandings. For applied researchers in particular, emphasis on demand issues provides a mechanism to articulate the needs of their presumed clients—citizens and public officials. Within the social indicator movement several alternative approaches to demand problems are emerging. These require further conceptual development and verification through both empirical analysis and policy application.

A FULLER UNDERSTANDING OF CONSUMER ROLES

A fuller understanding of consumer roles and values should continue as an important priority of social indicator research. Particularly difficult problems in this area of concern rest with the identification of linkages between subjective and objective measures of welfare and the relationship between the activities and values of consumers and the outputs of production systems. For example, what is the relationship between people's feelings of personal security and the objective risks presented by crime? What kinds of physical housing units and neighborhood amenities are associated with high levels of residential satisfactions? Aspects of questions such as these can usefully be addressed by basic social indicator research.

MAINTENANCE OF THE POLITICAL TENSION WITHIN THE SOCIAL INDICATOR MOVEMENT

Tension between the functions of social criticism and technical support have contributed to the intellectual vitality of the social indicator movement. From an applied perspective, this tension is necessary for the development of indicator approaches that can serve decision-makers by anticipating new political concerns and values. On balance, the efforts of the movement tend to be weighted too heavily toward the function of technical support. It would be a contribution to the field as a whole if its radical critics turn to the development of alternative indicator approaches geared toward the introduction of alternative values and concerns into the political process.

In summary, we believe productive avenues for future indicator research tend to lie with applied orientations that stress demand issues, including those reflecting consumer values and leavened with a concern for social criticism. We are not suggesting abandonment of alternative orientations. Our suggestions regarding future directions reflect questions of degree rather than of kind.

Our observations reflect a particular set of analytical and political orientations. Obviously, there can be valid differences with our point of view. We believe, however, in a far more categorical fashion, that debates concerning how to approach the development of social indicators should stress how research activities relate to important

NOTES

social and political processes. If these debates become enmeshed in narrow and technical discussions of methodological and conceptual issues, then this research enterprise could proceed unguided and undisciplined by the critical needs of policy-makers and concerned citizens. For social indicator research to fulfill its early promise, it must be undertaken on the basis of a clear understanding of its potential consequences outside the research community.

1. For an extensive bibliographic review of the social indicator field, see Wilcox et al. (1972).

2. In analytic terms we view the problem as establishing relationships between the bounded and structured activities of institutions producing goods and services or making authoritative decisions and the quality of life people experience as individuals or as members of households and communities.

3. For examples of such treatments, see Lear (1972) and Klages (1973).

4. Our understanding of these dimensions emerged from an ongoing self-assessment of our research agenda. We have been particularly concerned with the linkages between the theoretical orientations and potential uses of our and others' efforts by policy-makers and citizens.

5. Used loosely as synonyms for social indicators are social accounting, monitoring social change, social reporting, and assessing the quality of life.

6. For an analysis of this period of the social indicator movement, see Springer (1970).

7. Such limitations can be rooted in either conceptual or empirical problems.

8. Deutsch et al. (1971: 458); the detailed analysis of this can be found in Deutsch, Platt, and Senghaas (1970).

9. Illustrative of such work are Flax (1972 and 1971).

10. In a related discussion, Otis Dudley Duncan (1969) caricatures two fundamental approaches to indicator research: the empirical-inductive and the theoretical-deductive. The former emphasizes the development of standardized and reliable measurements of some social phenomenon, analysis of which will reveal conceptual propositions, while the latter primarily involves the development of deductive conceptual frameworks to structure observations and the analysis of data. In contrast to Duncan, who focuses solely on the analytic content of indicator research, our supply/demand distinction emphasizes the social interaction implicit in the research enterprise.

11. Illustrative of such work is Ontell (1972).

12. These notions are more fully explored in Garn, Flax, Springer, and Taylor (1975).

13. In our work, we have attempted to develop the information requirements of both producers and consumers.

REFERENCES

ALONSO, W. (1970) "The economics of consumption, daily life, and urban form." University of California, Berkeley, Department of City and Regional Planning, Working Paper 139 (December).

ANDREWS, F. M. and S. B. WITHEY (1973) "Developing measures of perceived life quality: results from several national surveys." Presented at the Conference of Urban and Regional Systems Association, Atlantic City, N.J., August.

BAUER, R. A. (1967) "Societal feedback." The Annals 373 (September): 180-192.

——— (1966) Social Indicators. Cambridge: MIT Press.

BELL, D. (1969) "The idea of a social report." Public Interest 15 (Spring): 72-84.

——— (1966) "The adequacy of our concepts," pp. 127-161 in B. M. Gross (ed.) A Great Society? New York: Basic Books.

CAMPBELL, A. and P. CONVERSE (1972) The Human Meaning of Social Change. New York: Basic Books.

DEUTSCH, K. W., J. PLATT, and D. SENGHAAS (1971) "Conditions favoring major advances in social science." Science 171 (February 5): 450-459.

——— (1970) "Conditions favoring major advances in social science." Ann Arbor: University of Michigan Mental Health Research Institute, MHRI Communications 273 (May).

DUNCAN, O. D. (1975) "Measuring social change via replication of surveys," ch. 5 in K. C. Land and S. Spilerman (eds.) Social Indicator Models. New York: Russell Sage Foundation.

——— (1969) "Toward social reporting: next steps." New York: Russell Sage Paper 2 in Social Science Frontier Series.

FLAX, M. J. (1972) "A study in comparative urban indicators: conditions in 18 large metropolitan areas." Washington, D.C.: Urban Institute Paper 1206-4 (February).

——— (1971) "Blacks and whites: an experiment in racial indicators." Washington, D.C.: Urban Institute Report 85-136-5.

FOX, K. A. (1974) Social Indicators and Social Theory: Elements of an Operational System. New York: Wiley-Interscience.

GARN, H. A., M. J. FLAX, M. SPRINGER, and J. B. TAYLOR (1975) "Models for indicator development: a framework for policy analysis." Washington, D.C.: Urban Institute Paper URI-89000.

GROSS, B. M. (1966) The State of the Nations: Social Systems Accounting. New York: Tavistock.

——— and S. MOSES (1973) "Measuring the real work force: 25 million unemployed." Social Policy 3 (September/October): 5-10.

GROSS, B. M. and J. D. STRAUSSMAN (1974) "The social indicator movement." Social Policy 53 (September/October): 43-54.

KLAGES, H. (1973) "Assessment of an attempt at a system of social indicators." Policy Sciences 4, 3 (September): 249-261.

LAND, K. C. (1972) "Social indicator models: an overview." Presented at the annual meeting of the American Association for the Advancement of Science, Washington, D.C., December.

LEAR, J. (1972) "Where is society going? The search for landmarks." Saturday Review (April 15): 34-39.

National Goals Research Staff (1970) Toward Balanced Growth: Quantity with Quality. Washington, D.C.: Government Printing Office, July 4.

NEWBROUGH, J. R., R. M. CHRISTENFELD, F. E. BURKE, M. HERTEL, H. BREGMAN, and C. G. SIMPKINS (1973) "Community mental health epidemiology: Nashville."

Nashville: George Peabody College for Teachers, John F. Kennedy Center for Research on Education and Human Development, August.

Office of Management and Budget (1973) Social Indicators, 1973. Washington, D.C.: Government Printing Office.

OLSON, M., Jr. (1973) "Evaluating performance in the public sector," pp. 355-410 in M. Moss (ed.) The Measurement of Social and Economic Research. New York: National Bureau of Economic Research.

——— (1969) "Toward a social report: its plan and purpose." Public Interest 15 (Spring): 85-97.

ONTELL, R. (1972) Toward a Social Report for the City of San Diego. San Diego: Urban Observatory of San Diego.

OSTROM, V. (1974) The Intellectual Crisis in American Public Administration. University, Ala.: University of Alabama Press.

RUGGLES, N. and R. RUGGLES (1973) "A proposal for a system of economic and social accounts," pp. 111-162 in M. Moss (ed.) The Measurement of Social and Economic Performance. New York: National Bureau of Economic Research.

SHELDON, E. B. and H. E. FREEMAN (1970) "Notes on social indicators: promises and potential." Policy Sciences 1, 1 (April): 97-111.

SHELDON, E. B. and R. PARKE (forthcoming) "Social indicators." Science.

SPRINGER, M. (1970) "Social indicators, reports and accounts: toward the management of society." The Annals 388 (March): 1-13.

TERLECKYJ, N. E. (forthcoming) Estimates of Possibilities for Improvement in the Quality of Life in the United States, 1973-1983. Washington, D.C.: National Planning Association.

TUNSTALL, D. B. (1970) "Developing a social statistics publication." Presented at the annual meeting of the American Statistical Association, Detroit, December 27.

U.S. Department of Health, Education and Welfare (1969) Toward a Social Report. Washington, D.C.: Government Printing Office.

WILCOX, L. D., R. M. BROOKS, G. M. BEAL, and G. E. KLONGLAN (1972) Social Indicators and Societal Monitoring: An Annotated Bibliography. San Francisco: Jossey-Bass/Elsevier.

18

Program Evaluation and Policy Analysis of Community Development Corporations

HARVEY A. GARN

□ COMMUNITY DEVELOPMENT CORPORATIONS (CDCs) are a relatively new invention. "The oldest of the CDCs, Progress Enterprises in Philadelphia, was organized in 1962; few existed prior to 1967" (Faux, 1971a: 30). From 1968 to 1973, CDCs were the primary recipients of funding from the Special Impact Program of the Office of Economic Opportunity. In 1974, the Administration requested a transfer of responsibility for funding CDCs to the Office of Minority Business Enterprise (OMBE) in the U.S. Department of Commerce. By 1973, OEO had allocated $68.5 million to 23 urban CDCs and $29.8 million to 40 rural CDCs. Table 1 shows the annual expenditures for CDCs through the Special Impact Program from 1968 to 1974 and those planned for OMBE. The table does not include funds for CDC activities from other federal sources, other governments, or private funds. The Ford Foundation has become the

AUTHOR'S NOTE: *This analysis was supported by funds from the Ford Foundation. Opinions expressed are those of the author and do not necessarily represent the views of the sponsor or the Urban Institute. The discussion which follows draws heavily on work currently under way at the Urban Institute for the Ford Foundation in designing and implementing an evaluation of Foundation-supported CDCs. My colleagues, Herrington J. Bryce, Carl E. Snead, and*

[561]

TABLE 1

OEO AND OMBE EXPENDITURES FOR CDCs BY FISCAL YEAR

(in millions of dollars)*

1968	1969	1970	1971	1972	1973	1974	1975
$1.6	10.9	30.0	31.3	24.4	36.6	38.1 (est.)	13.9 (est.) 11.0 (est.)

*Figures for fiscal years 1968-1972 are drawn from the March 1973 Newsletter of the Center for Community Economic Development. Fiscal year 1973-1975 figures are from the Federal Budget. The figures for fiscal 1975 are $13.9 from the OEO budget and $11.0 from the OMBE budget.

major private funding source for CDCs. Mitchell Sviridoff, Vice President of the Ford Foundation, wrote in December 1972: "The Foundation has so far spent $25 million for CDCs and if present hopes are sustained by experience, it may spend an additional $75 million over the next five years."[1] Any estimate of total funds used by CDCs up to 1974 is speculative, but it probably is in the range of $150 to $175 million.

Estimates of the total number of CDCs vary—anywhere from 70 to 100—depending upon the definition used. While the growth from only one CDC in 1962 to even the lower estimate is large, the growth in funds available has been neither spectacular nor assured.[2] In fact, the movement of funding responsibility for CDCs from the Special Impact Program, with its emphasis on both economic and social development, to OMBE, with its emphasis on minority entrepreneurship and provision of managerial and technical assistance, could mark a significant downturn in support, if not the demise, of this new institution. At the present time, continued federal support is required.

Both the wide variety of institutions which call themselves CDCs and the relatively short time in which they have operated make it extremely difficult to tell what criteria should be used in assessing them and, hence, whether the CDC experiment has succeeded or failed. Examples that might be viewed as successes and failures of

Nancy L. Tevis, will see many of their ideas reflected here. Similarly, the concern with methodology and approach expressed here reflects ideas generated in the Social and Urban Indicator Project at the Urban Institute. My colleagues in that work, Michael J. Flax, Michael Springer, and Jeremy B. Taylor, are continually pressing to discover and develop frameworks useful for policy and evaluation purposes. Some residual responsibility is, of course, mine. Jacqueline Swingle shepherded the paper through several drafts and prepared the final manuscript. My thanks to her, also.

particular activities supported within the CDC framework can be readily found. But such ad hoc examples are not particularly helpful in painting a complete picture.

Although there has been a considerable body of literature developed which discusses CDCs, a sufficiently broad and simultaneously detailed framework has not yet emerged which will permit conclusive assessments to be drawn about CDC performance. This paper attempts to develop a perspective which will help improve understanding of what kind of an institution the CDC is (both in terms of its structure and in comparison with other institutions), how such an institution might be evaluated, and how key policy issues associated with its operation and existence can be addressed. This kind of a framework is required in order to make assessments about appropriate policies and in judging whether or not the performance of a particular CDC is reasonable or acceptable. This paper, therefore, is a discussion of requirements for the analysis of community development corporations rather than a report on their performance to date.[3] Also, in the course of this discussion, an attempt will be made to draw out some of the generalizable implications of the analytic framework discussed for program evaluation and policy analysis. It should be mentioned here, however, that CDC performance (like that of many new institutions) is mixed. Leadership is being trained, and new forms of activities are being created. Some mistakes are being made, while at the same time there are considerable signs of promise. At the present time, it is highly unlikely that many CDCs will be self-sufficient over the near term, if self-sufficiency is defined as the ability to achieve current activity levels (or higher) and current mixes of economic and social programs without outside funds. Some activities generate profits which can be used elsewhere, but these profits are neither frequent enough nor large enough to sustain the institution in the next few years.

A useful way to get into the broader issues is to develop an analytical understanding of what a CDC is, trace the implications of this for program evaluation, and then describe some additional considerations for policy analysis.

HOW TO CONCEIVE OF A CDC

The conception of a CDC developed here requires for its explanation something more than a definition. There are so many variants in structural form, ownership, and program structure among CDCs that it seems fairly pointless to become overly enamored of a simple definition. Some elements of a definition, of course, spring forth from the name. CDCs usually are corporations—as opposed to governmental bodies or more loosely knit business enterprises. They usually have an identifiable set of links with a particular community, although the nature of the linkage varies. And they usually are concerned with an array of programs that have development —political, social, and economic—as their stated purpose.

This mix of programs—some with private market analogues and some with public (governmental) analogues—is characteristic of CDCs, however they are organized. Typically CDCs are engaged in operating some commercial and business enterprises; providing assistance to other entrepreneurs; providing manpower training; building, rehabilitating, and managing housing; providing access to welfare and related services; and dealing with publicly funded agencies (such as the police, fire department, and schools) on behalf of community residents. In short, they are hybrid organizations, quasi-private and quasi-public. As a consequence, those who have examined CDCs have chosen a variety of approaches in assessing them.

Milton Kotler, who helped create an early CDC, defines the primary function of CDCs as political. He sees a CDC as the only kind of embryonic neighborhood government feasible for minority communities in the short run and his aim is the creation of neighborhood government.[4] In 1971 he wrote:

Faced with the new economic policy of OEO, the community organizers agreed to write grant proposals promoting entrepreneurial development based on private enterprise or community ownership. But they also understood that tactically as much money as possible under these economic programs would be applied to the political organizing in a community development framework. The result would be that the political organization could continue while attention was paid to developing a local economy.

Further, the priority of economic enterprise would accord with the requirements of advancing political organization [Kotler, 1971: 8].

Kotler's view could not be more sharply challenged than it is by those speaking for the National Advisory Council on Economic Opportunity in their sixth annual report in June 1973. They write:

Further, economic development programs should be designed to establish profitable enterprises; federal agencies administering economic development programs should not impose social goals on business development programs.[5]

The Advisory Council recommends, therefore, that the broad social development aspects of CDCs be subordinated to the objectives of establishing self-sufficient, successful business enterprises capable of attracting financial, technical, and other business development resources and providing employment for target-area residents. The Advisory Council recommends that CDCs channel *all funds* into economic ventures and programs for their support.

The Council further urges that *no money* granted to community development corporations be specified for social development activities; social development programs should be administered by more appropriate agencies.[6]

Bennett Harrison (1974: 174) has noticed the conflict implicit in these two views of the essential character of a CDC. He writes:

A fundamental issue in the community development movement—and its technical literature—concerns the trade-off between commercial profits and community benefits. No single subject more clearly distinguishes this approach to inner city development . . . than the avowed commitment of most CDC leaders to give high priority to such "external" outputs as manpower training, consumer education, improvement of the ghetto environment, reduction of anomie (especially among young black men), and consolidation of the political "clout" in the community's relations with city hall, the state house, the federal bureaucracy, and the private sector.

We have here, then, three distinct views: (1) the CDC is an economic and social institution designed to pursue political objectives of neighborhoods; (2) the CDC is an economic institution with solely economic objectives; and (3) the CDC is an institution which must make trade-offs between economic gains and "external" outputs and to pursue "political clout."

It is difficult to be as sanguine about the ability of neighborhoods to resolve their problems solely through political means as Kotler suggests, given the pervasive (and frequently adverse) effects of private market forces at work in these communities. Even the larger urban governments find it difficult to apply sufficient political leverage to control private market forces in the interests of the citizens of their jurisdictions. Nor can one agree with the view of the National Advisory Council on Economic Opportunity (1973: 3) that

The Council has concluded that the most crucial element in the long-range success of economic development programs is an immediate and intense effort to educate the disadvantaged to understand and use the economic opportunities that exist in the free enterprise system.

More is required than minority education and the generation of individual entrepreneurs in communities served by CDCs.

Harrison is much closer to the mark with his emphasis on the importance of trade-offs between commercial profits and community benefits. His characterization, however, seems to treat commercial profits and community benefits as if they were objectives on the same plane. Analytically, his approach is clearly an advance over the view that maximizing profit in CDC ventures is the same thing as maximizing community benefits. He has not gone quite far enough, however. One needs to ask a further question about the appropriate standards for making trade-offs among alternative outputs. He implies that sometimes one should choose commercial profits and sometimes community benefits, but does not indicate a basis for making this choice.

I would argue that the major justification for a CDC is that it should be an organization that weighs all outputs in terms of their contribution to community welfare (or community benefits). In these terms, it is indeed possible that CDCs will not seek or obtain maximum profits from their individual ventures (Garn, Bryce, and Tevis, 1973b: 8). If they do not, however, it should be because they have decided that other actions make a greater contribution to community benefits. The point here is that profitability as well as other outcomes should be judged from the standpoint of community benefits. Profits should be considered intermediate goods along with other outcomes; all of which need to be judged in terms of their contribution to community benefits.

This point is further developed in Garn, Bryce, and Tevis (1973b).[7] There my colleagues and I state that

We depart from many other rationales for CDCs in that we shall argue that these organizations are justified on economic grounds —not solely on the grounds of equity or political power. Their economic raison d'etre is not based, however, on profit maximization. In an important sense, the CDC should be seen as an economic club—one that takes as its objective function the maximization of the community's welfare, rather than private profits, sales, or re-election.

If, as is argued above, the CDC should take as its objective the maximization of community welfare, all of its outputs should be judged relative to that objective. A major evaluation problem, from this point of view, is to devise a method for determining the contribution to community welfare of the various outputs of the CDC. I will discuss this problem below. Prior to doing that, however, a few words are required to relate the methods to be described to standard evaluation techniques.

EVALUATION STANDARDS

The classic evaluation technique is to develop quantitative relationships between resources used in production processes and the outputs of those processes.[8] The normal measures of effectiveness are either ratios of outputs to costs or benefits to costs. The first of these is used in cost-effectiveness analysis and the second in cost-benefit analysis. A number of standard problems with these techniques are well known and their limitations have been discussed extensively in the literature. It will not be necessary to go into them here, since the primary reasons for arguing that standard cost-effectiveness or cost-benefit analysis does not work well in evaluating CDC spring from a different set of concerns than these well-documented limitations.

To indicate the nature of my concern, a few words need to be said about what is involved in moving from a cost-effectiveness analysis to a cost-benefit analysis. Outputs in a cost-effectiveness study are normally stated in physical or other identifiable units. Consequently, in an organization which produces a wide range of different outputs, it is impossible, when using cost-effectiveness analysis, to compare the output costs ratios directly. It is possible, of course, to make statements like, "for the same costs which it takes to produce one housing unit, it is possible to produce two theatrical productions," and the like. The ratios are commensurable, but only in terms of the costs.

A major part of the reason for converting a cost-effectiveness study into a cost-benefit study is to produce commensurability on the benefit side of the ratio as well. This is accomplished by converting physical outputs into value terms, the most customary value being a monetary unit. For those outputs which have market prices, this step is a fairly straightforward exercise. For those outputs which do not have market prices, it is common to develop a proxy or shadow price for the output. If all of the diverse outputs of an organization can be valued in this way, both the benefits and the costs can be stated in commensurable terms.[9] Hence, a benefit-cost ratio greater than one means that the value of the outputs produced exceeds the value of the costs incurred and is, therefore, a desirable thing to do. Relative ratios of benefits and costs can be used to make comparisons across output classes or between activities producing a similar output in terms of their value efficiency.

The use of market prices or proxies for market prices, however, is only one of many possible ways of valuing outputs. It has been argued above that a community development corporation is distinguishable from a straightforward multiproduct firm in that it takes as its objective the improvement of community welfare rather than the maximization of profit. A consequence of this is that the relevant structure of implicit prices which should be applied to the outputs should be reflective of the preference structures and the intensities of these preferences from the point of view of the community. If there are reasons to believe that market prices do not accurately reflect these preferences, then the analysis of benefit streams which use market prices will be distorting. Are there, then, reasons to suspect that these preference structures might reflect different valuations from those reflected in market prices?

The answer to this seems to be clearly yes. It is well known that the existence of positive and negative externalities (spillovers) creates a divergence between market performance and social performance. Negative externalities abound in most communities served by CDCs.

These externalities are created, in part, by the actions of individuals and private market firms. Furthermore, once a neighborhood is suffering from a massive combination of adverse effects (such as abandoned and deteriorating housing, high crime rates, high unemployment rates, low income levels, and inferior schools), any given individual or firm is less likely to undertake the risk of positive action than in a neighborhood which is not so suffering.

The corrective mechanism for this kind of situation is meant to be the government. The government, in theory, taxes away positive externalities and provides transfers to compensate for negative externalities. However, one does not need to be a specialist in government to know that the governments in question do not have the finely tuned measuring instruments to identify all these externalities and, perhaps more importantly, do not have the political incentive to correct the accumulated deterioration of individual neighborhoods.

Some institution, therefore, is needed to perform the combination of private and public activities which address the collective concerns of the community, without, on the one hand, being constrained to maximize profits from its private activities or, on the other hand, being constrained by the political interests of the broader city of which most urban communities served by CDCs are a part. A CDC, I would argue, is such an institution. Its performance, therefore, should be judged on a different basis from a private firm or a broader governmental entity. *That is, CDCs should be concerned with providing a mix of activities whose combined outputs make the greatest contribution to community benefits. Trade-offs which are made among all outputs should be in terms of contributions to community benefits, not between profits and community benefits.*

This criterion for performance (maximization of community benefits) does not mean that any given CDC is necessarily achieving this objective. The criterion is neutral and comes from an assessment of the legitimizing role of a CDC, rather than from the observation of any given CDC's performance. It is logically possible, but unlikely, that the maximization of profits and community benefits would entail the same choices. It is almost certainly true, on the other hand, that the generation of some revenues by the CDC is consistent with maximizing community benefits. These latter two points are somewhat obscured in Harrison's (1974) formulation cited earlier.

How, then, might one develop a means of attaching a community benefit value to the output structure of a CDC? What is required is a method for adjusting either the market or proxy market values used in ordinary cost-benefit analysis to take account of the particular set of adverse circumstances facing the community being served by the CDC.

To help get at what is implied, an example may help.[10] To simplify the exposition, let us imagine a CDC which produces two outputs, x_1 and x_2, which have respective market prices of p_1 and p_2. The situation facing the CDC in making its production choices is shown in Figure 1. The CDC can produce, for a given cost, any combination of x_1 and x_2 along its production frontier, shown as AB. If all resources are devoted to producing x_1, the quantity x_1^m can be produced or, if all resources are devoted to x_2, x_2^m can be produced. Any combination along the curve AB is possible. The output combination that will maximize profit for the CDC is shown as x_1^* and x_2^*. The market price ratio is the slope of the line CD. Any output ratio other than x_1^* and x_2^* will produce less profit to the CDC, at the prevailing prices at which the outputs can be sold. But suppose that the relative value of the two outputs to the community are different from the market valuation. Say that the relative values to the community are shown by EF in Figure 1. If the CDC is to maximize community benefits it will choose output combination x_1^c and x_2^c. This choice will result in a revenue loss of

$$p_1(x_1^* - x_1^c) + p_2(x_2^* - x_2^c)$$

relative to the profit-maximizing solution. What is gained in the process is an increase in community-valued benefits.

The major analytic problem is to find a way of operationalizing the idea of a set of adjustments to market prices which reflect community values better than unweighted market prices do. It must be said that there is no completely satisfactory way to do this. A real advantage of market price valuation is that prices are usually a matter of public knowledge. There is no directly equivalent public repository of community valuations.

Political scientists sometimes argue that community valuations are accurately reflected in voting patterns. Unfortunately, however, voting is seldom done in cases where the issues are as clearly or as finely defined as would be required for current purposes. Furthermore, community issues tend to get mixed with issues of the broader polity in most cases. One of the arguments frequently advanced by proponents of CDCs is that the leadership of the CDC is sufficiently close to the community that the leaders can know what the community values and with what intensity the community values

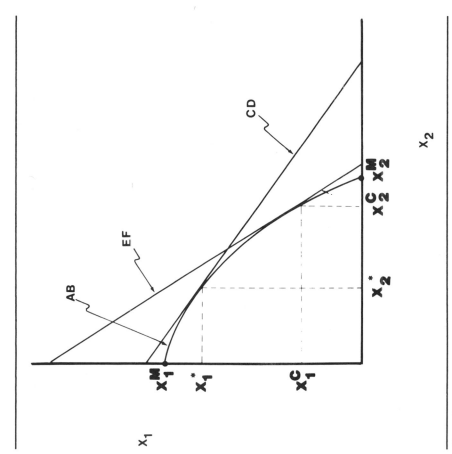

Figure 1.

different outputs. There is, indeed, some merit to this view; but, relationships between the CDC leadership and the community are so varied that it is not possible to use the leadership choices as uniformly reliable guides to community preferences.[11] In addition, it would be somewhat circular to evaluate the merit of the choices made by the leadership on the basis of their own assessments of the value of their own choices.

An obviously possible, if expensive, alternative is to systematically sample the population of the community. Many of the difficulties with attitude surveys are well known. However, there are some recent developments in survey technology which bear watching. Carl Snead (Garn, Snead, and Tevis, 1974) has noted that

Specifically, the work of the Social and Community Planning Research Project in London appears to have promise for use in CDCs. The SCPR is promising because they have been developing policy-oriented survey instruments which enable one to get some feel for desired outcomes and priorities as well as for intensities of preferences. The basic idea of their work is to use a survey instrument to elicit information about the trade-offs those surveyed would make among competing alternative outcomes in a situation where they are constrained by the costs of achieving various outcomes. One of the ways in which the methods employed by SCPR are more useful than traditional surveys, therefore, is that SCPR not only measures the direction of preferences, but the intensity of preferences as well.

Thus, if community residents were asked to rank order what they perceived to be the most severe problems facing the community, it is desirable to know by how much a given problem dominates another in the minds of the residents. If a community saw its most severe problem as lack of control over economic or political resources in the community, lack of meaningful job opportunities, and the undereducation of community residents, it may be that the intensity of preference for solutions to these various problems is about the same, suggesting that a nearly equal emphasis in attacking these problems will be of most benefit to the community. If on the other hand, the rank ordering of the problem remained the same, but the problem of economic and political resources was viewed as a much more severe problem than education, then an unequal response by a CDC in favor of the problem that is felt to be most severe may be more beneficial to the community. The technique developed by SCPR was designed to assess just such preference intensities.

Another advantage of their approach is that it overcomes some of the limitations of traditional "attitude" research which does not force respondents to trade-off some of their preferences against others, i.e., their procedure prevents respondents from opting for ideal outcomes in all areas regardless of the impossibility of achieving such outcomes at any feasible cost.

One of SCPR's main developments in the measurement of preference and social values has been the introduction into survey research of the concept of trade-off preference. When, for example, a family chooses to locate itself in a particular residential area, it has already made a series of trade-offs between a host of factors such as proximity to schools, shops, work, the character of the neighborhood, educational, and recreational facilities. The final choice is an expression of relative values and usually involves the trading-off of one set of priorities for another. It is a complex process that is taken for granted in many aspects of behavior. The challenge for survey research was the unravelling of the preferences involved in a way that would assist planners in their allocations of resources to competing demands. The priority evaluation approach developed by SCPR to assist in this task is now widely used.

SCPR has developed a game technique in their survey research to measure the preference of respondents which they call the *Priority Evaluator*. The Priority Evaluator is a device which permits respondents to indicate priorities between competing and costed alternatives, so that the overall results of their choices are instantly visible to them and bear a close resemblance to real life situations.

A survey technique which reaches solutions about both preferences and their intensities would enable the evaluator to weight market prices or proxies with cardinal values representing the community's preferences and would be usable. It remains an open question whether or not the benefits are justified relative to the costs of gathering data in this way.[12]

Another, more readily available set of adjustment factors can be developed which is consistent with the earlier argument that the raison d'être of a CDC is to address the collective adverse conditions in the community. If one were to imagine a situation in which each community within a city or metropolitan area shared equally in the benefits available within that city and suffered equally from adverse conditions, it would be unlikely that the best means of improving the overall situation would be through community organization and development. Organization at the city or metropolitan level would seem more appropriate. At any rate, it is plausible to argue that there would be far less apparent requirements for community organization.

This line of argument suggests an alternative means of adjusting the market price valuation which uses publicly available data, reflects differences in the degree of severity of problems among communities, and changes over time to show improvement or worsening of the community relative to the city of which it is a part. The adjustment factors[13] consist of ratios of conditions in the community to the corresponding condition in the city. Typical ratios could be the rate of unemployment, median family incomes, violent crime rates, rate of home ownership, to cite some examples. A set of such ratios is constructed for each CDC being evaluated. Individual outputs are weighted by a ratio which the output could plausibly

[574]

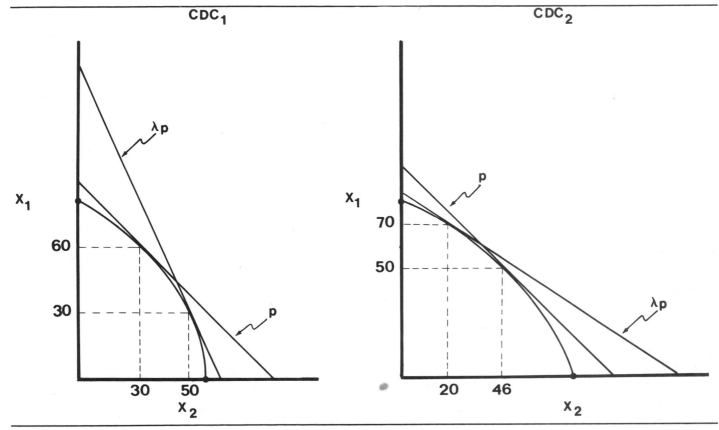

Figure 2.

affect.[14] The ratios are constructed so that the ratio is greater than one if the condition is worse in the CDC community than it is in the city. If, over time, conditions in the CDC community move closer to those in the rest of the city, the ratios would each approach the value of one.

The mechanics of utilizing these factors is analogous to converting outputs into benefits using market prices. The progression is shown below:

Output/Cost	Benefit/Cost	Community Benefit/Cost
O/C	$\dfrac{O \cdot p}{C}$	$\dfrac{O \cdot p \cdot \lambda}{C}$

where O = outputs, p = market or proxy market price, λ = community welfare weight, and C = cost. The effect of using these community welfare weights is easiest to understand as an implicit change in the slopes of the market price lines facing the CDC in the direction of valuing more highly those outputs addressing community conditions which are the more adverse relative to the city. To illustrate this, let us return to our example of a CDC producing two outputs and add a second CDC producing the same two outputs. Figure 2 shows the production possibilities for CDC_1 and CDC_2 respectively for the same cost in the two CDCs. The figure has been drawn to show that CDC_2 could produce more of x_2 than CDC_1 for the same cost. This can arise from the use of difficult technology, cheaper unit costs of inputs, or the existence of more positive spillovers on the cost side for CDC_2.

Lines p show the choices that would have to be made to maximize benefits, if benefits are construed as revenues from output, assuming both CDCs face the same relative prices for the outputs. CDC_1 should produce 60 units of x_1 and 30 units of x_2; while CDC_2 should produce 50 units of x_1 and 46 units of x_2 in order to maximize revenue. The differences stem from the greater cost effectiveness of CDC_2 in producing x_2.

Now let us suppose that the community served by CDC_1 has a relatively greater need for output x_2 and the community served by CDC_2 has a relatively greater need for output x_1 than is reflected in the market price ratios. This is depicted in the figure as lines λp. The line p is the line reflecting market price ratios and the line λp is the line reflecting the adjustment for community conditions. The optimal choice to maximize community benefits for CDC$_1$ would be 30 units of x_1 and 50 units of x_2. The optimal choice of CDC$_2$ would be 70 units of x_1 and 20 units of x_2.

One final figure needs to be examined in order to help assess the significant differences between a solution which maximizes market-valued benefits and the solution which maximizes community benefits.

In this figure, λp reflects the appropriate community benefit line and p represents the appropriate market-valued benefit line as in Figure 2. Line $(\lambda p)^i$ represents the implicit community benefit line *if an output which maximizes market-valued benefits is chosen.* Line p^i represents the implicit market-valued benefit line *if an output which maximizes community benefits is chosen.* These are drawn in to show that the magnitude of the differences between the two criteria can be stated in two different ways. First, to maximize market-valued benefits creates a loss in community benefits of the income equivalent of the difference between budget line λp^i and λp.[15] Obviously, this loss will be greater as the difference between the relative value assigned outputs in the market and that value assigned by the community is greater. Second, to maximize community benefits creates a revenue loss equivalent to the difference between budget lines p^i and p.[16] Again the loss will be greater as the difference between the relative valuations are greater.

Several further points should be made before leaving the discussion of Figure 3. Point A on the figure is the maximum revenue (and, with the given cost, the maximum profit) solution. Point B is the maximum community benefit solution. Point A may or may not represent a solution in which profits are positive. The scale of the activity and the relationship between fixed and variable costs could make point A a solution in which (although revenue is maximized) the CDC is losing money, breaking even, or earning positive profits. That is, it is conceivable to be as efficient in the use of inputs as is possible technologically, and still not make a profit.[17]

If we assume, for the moment, that the CDC would break even at the output combination of point A, the revenue loss implied by the difference between lines p and p^i is the amount of subsidy which would be required from alternative sources for the CDC to keep operating with the output mix which maximizes community benefits. It is not necessary to go into all the permutations possible in this example, but *it is important to stress that it does not follow from the observation that if a CDC is not making a positive profit (in any*

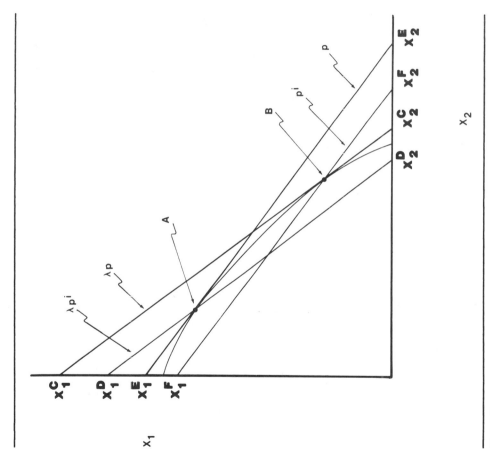

Figure 3.

given activity or overall) that it is (a) inefficient or (b) should not adopt the output mix which it has adopted. Point B is technologically as efficient as point A and, in value terms, more responsive to community requirements than point A—even though the CDC earns less revenue at point B.

In evaluating the performance of a particular CDC, therefore, the use of this analytic approach requires a judgment by the evaluator of the outputs of the CDC relative to three distinct (if not unambiguously identifiable) reference points:

● *First, the production possibility curves.* A particular CDC may be operating an activity in a way that is wasting resources. This means it is operating inside the production possibility curve, not on it. Regardless of the valuation criterion used this would be inefficient, since more of all outputs could be produced with no increase in cost.

- *Second and third, the optimal output in terms of market valuations and in terms of community-valued benefits.* This enables the evaluator to assess the costs in terms of both revenue and community benefit losses of the achieved CDC outputs.

I have gone into this example in considerable detail because the principles apply regardless of the mechanism which is used to establish community valuations. The principles will be unimportant in any evaluation only if the intended beneficiaries of a program can be unambiguously assumed to attach identical value to all outputs or resources used as is reflected by their market values.[18] This assumption is seldom justified. Evaluation studies which overlook this fact—including virtually all known to the author—are unfortunately deficient. *The valuations of outputs by program beneficiaries are at the heart of the evaluation problem and may well be different* from the valuations of those who operate the programs or the more generalized forces or groups who determine market prices.

TWO FURTHER CONSIDERATIONS

This paper would be seriously deficient if notice were not taken of two additional factors which must be built into a thorough assessment of CDC performance. The first comes under the heading of output identification and joint production by the program beneficiaries. The second can be called resource mobilization versus resource allocation. Each of these will be treated briefly.

1. OUTPUT IDENTIFICATION AND PRODUCTION

It is customary in evaluation studies (and in much economic and operations research) to treat the production of outputs and the consumption of outputs as completely independent activities carried out by completely separate sets of actors. One of the negative consequences of this for our current purposes is that it is often assumed or implied that consumers (beneficiaries of CDC programs, in this case) are completely indifferent about who produces the outputs and how. Consumers are thought to be interested only in the output itself and its cost.

It has been my observation that one of the key factors operating in most communities served by CDCs is precisely that they are concerned with both who the producers are and how they go about their activities. Otherwise identical outputs can be and are treated as different outputs depending on the "who" and "how" of production.[19] Such responses, clearly, are relevant to the assessment of the performance of a CDC.

In this same vein, there are particularly important aspects of the production of services which have not been incorporated effectively in evaluation.[20] In most cases of service delivery—an important category of CDC activity—the ostensible producer and beneficiary must interact to produce the output (that is, there is no independent producer-side production function).[21] The beneficiary is concerned, therefore, with how the person delivering the service performs for two distinct reasons: (1) it influences the beneficiary's valuation of the process, and (2) it influences the participation of the beneficiary in the production of the outcome. For example, the way health care or education is provided affects the reaction people have to it and influences the degree of efficacy of the producer's actions. The role of the beneficiary is also partly determined by forces external to the behavior of the producer. An evaluation which takes no notice of these factors misses an important aspect of CDC roles and performance as well as the ability to properly ascribe responsibility for the level of output achieved in the interaction.

2. RESOURCE MOBILIZATION AND RESOURCE ALLOCATION

The earlier discussion has proceeded as if the CDC were vested in some mysterious manner with resources such that its major problem is to make "correct" allocation choices among activities. For a full-scale analysis of CDCs this is insufficient for two major reasons: (1) in most cases, funds available to CDCs are to be used for quite specific activities and no others; there is relatively little discretionary money; and (2) CDCs have to develop a resource mobilization strategy (utilizing already scarce resources for this purpose) which is distinct from their resource allocation strategy.

The first reason is important for CDC evaluation and analysis in a variety of ways. A smooth production possibility curve, such as those described above, only exists over a range of outputs for which resources are potentially transferable. Only to the extent that this is so can the CDC be expected to respond effectively in its allocation choices to information about changes in either community or market valuations. In short, the CDC's range of choices may be sharply constrained by the requirements of those providing funds rather than by their ability to identify more appropriate output mixes and produce them.

A solution, of course, is more discretionary funds. I suspect that this is part of what Harrison (1974) had in mind in his statement about trade-offs between profits and community benefits cited earlier in this paper. An obvious way to try to get discretionary funds is to generate them from one's own revenues, and this may entail abandoning some activities which do not generate revenues. But the strain which is implied as a trade-off between profits and community benefits in Harrison's argument is really between resource mobilization and allocation strategies, not between two different kinds of outputs.

The CDC, as a development institution, operates in a highly complex, mutually interacting system in which the generation of discretionary funds is neither assured nor automatic. A *development institution* such as a CDC is inevitably placed in a situation where it must perform simultaneously a dual role, since it is not self-sufficient. On one hand, the CDC's planning, programming, and program allocations must relate to the needs and interests of its community constituents. On the other hand, the CDC's resource mobilization activity must relate to development support institutions. If the contrast between the views of those providing funds and the community's are sharp, the CDC may not be able to generate sufficient funds to become self-sufficient even if most programs are devoted to generating revenue. While the CDC (or any other development institution) is not unique in having to play this dual role, the problems posed by the dual roles are significant in an overall assessment of CDC progress and prospects.

All institutions necessarily perform both roles if they are to survive. If programs are to be valued, they must be responsive to the needs of clients. If programs are to be maintained, the necessary resources to pay for them must be mobilized. A major problem for organizations like CDCs, therefore, results from the separation of resource suppliers from the "customers" or "clients" for whom programs are maintained and the potential conflicts which result from this separation.

Relatively little analysis has been done which treats resource mobilization strategies and allocation strategies distinctly but simultaneously.[22] Therefore, most of the extant analysis tends to overlook the importance of examining situations in which the appropriate strategies of each type complement or conflict with each other. Some recent analysis of resource mobilization in public service delivery has moved closer to the needed examination. The thrust of an excellent recent example of this literature is that conflicts between resource mobilization and program allocation strategies in the delivery of public services tend to be resolved in favor of the budgetary and tax authorities rather than clients, resulting in a lack of responsiveness to client interests and needs.[23] The analysis is somewhat deficient, however, since the author then argues that the problem of responsiveness would be ameliorated simply by creating private market-like incentives in public service delivery.

This latter argument would tend to appeal to those familiar with the economic analysis of the private firm. The prototype for much of this kind of economic analysis is a small, ongoing business which derives its revenues primarily from its customers (in our terms, its resource mobilization strategy and its program allocation strategy are the same). Pursuing a successful resource mobilization strategy, for this prototypical firm, requires responsiveness to customer interests and needs; conflict between the two is unlikely to arise.

A major deficiency of such an economic analysis (aside from the obvious fact that many businesses are insulated from customer disapproval by diversifying outputs, stock sales, and utilization of retained earnings) for examining CDCs is that CDCs have not been able to generate sufficient revenues to maintain even their current staffs and activity levels (let alone address the complete range of community needs) from their profit-making ventures. Nor is it clear that a CDC *should* do so from the perspective of community valuations. Similarly, a CDC is unlikely to be vested with either the taxing or budget decision authority of a local government required to provide resources to allocate to social programs.

In short, the separation of resource suppliers to and the customer/clients (the community) of the CDC is likely to continue along with the possible conflict, which this implies, between resource mobilization and program allocation strategies. In this circumstance, it is important to those providing resources to CDCs, the CDCs, and the communities served to identify those arrangements in which the CDC *can* pursue complementary resource mobilization and resource allocation strategies or, if that is not possible, arrangements in which the probable conflicts can be ameliorated.

SUMMARY

In this paper, some important requirements for the analysis and evaluation of CDCs have been explored. While it is not possible, in a short paper, to fully explore them, I have attempted to develop two major themes. *First, the activities of CDCs are difficult to evaluate properly.* Evaluation methodology cannot rest with narrowly conceived efficiency studies and claim to get the analysis straight. In particular, more attention must be paid to alternative valuation criteria, output identification and beneficiary roles in output production, and the importance of distinguishing and then understanding the interactions between resource mobilization strategies and resource allocation strategies. To miss these factors (as most evaluation studies do) would be likely to lead to unsatisfactory evaluative conclusions.

Second, policy analysis is required that is more self-conscious about the assumptions and methods employed than one customarily finds. To date policy analysis has been dominated by the perspectives of economists, people in management science, and political scientists. The perspectives of economics lend themselves fairly directly to an emphasis on the transformation of resources into outputs—allocation choices. Within the context of economic theory, the process of valuing outputs is assumed to occur fairly automatically through a responsive market system. In those cases where the market can be expected to operate suboptimally—for example, when there are external economies or diseconomies—the traditional economic view recognizes the necessity for corrections to be developed through governmental processes. Political institutions tend to be treated in economic analysis either as efficient transmitters of information from the populace or as efficient inferrers of the components of social good. The real flavor of political conflict, divergent values, and the muddiness of information (about people's preferences and the social good as well as the efficacy of programs) seldom comes through. All of society's systems end up being treated as if they either are or should be interacting in an equilibrium context or moving relentlessly toward an equilibrium.[24]

Management scientists and organization theorists tend to start with the view that the economist wipes out the most important aspects of policy development, namely, that policies are made in a context characterized better by uncertainty than efficient information flow and in institutional settings characterized by a complicated interplay of individually motivated and bureaucratically motivated behaviors. This view does, indeed, add some dimensions to the analysis of policy which tend to be overlooked in the economic analysis, but this advantage tends to come at a cost. The cost is that much of the analysis treats the institutions in which policy is made as if they were insulated from concern about the outcomes of the processes initiated or assumes that bureaucratic behavior is never primarily motivated by aspirations to produce stated outcomes.

The political scientists tend to focus their attention on sources of conflict and possible interactions which lead to compromise. The picture which frequently emerges from this analysis is that of a besieged government bureaucracy alert to information from the outside, but by virtue of the conflicting signals derived from this information unable to make policy decisions except in an incremental, plodding manner.

The last few paragraphs may appear to be an overly extensive digression in a paper which has talked about community development corporations. The major ingredients of these alternative ways of viewing policy analysis, however, have counterparts in the frameworks which have been designed to study issues associated with community development corporations and in their evaluation. Some observers of the prospects and problems of CDCs think of the community development corporation as yet another example of a private firm whose objective is to maximize profits and whose efficiency is to be judged solely by its skill in carrying out its productive activities. Others focus their attention on the internal organization and characteristics of community development corporations. They notice internal bureaucratic struggles, effect of leadership on relationships with external funding sources, and the nature of interorganizational networks, with relatively little focus on questions other than those associated with whether or not the CDC as an organization can handle its struggles without bursting at the seams. Still others emphasize that the political aspects of the community development corporation are paramount and, therefore, that the critical dimensions of CDCs is the degree to which they are or are not successful in creating political support and public recognition of the problems of the community. Sometimes, indeed, CDCs are treated as if their only possible function is the creation of a political community. Perhaps not surprisingly, the three views sketched in this paragraph correspond roughly with the economic view, the management science view, and the political science view of policy analysis, respectively.

The problem is that each of the views is partially correct, but, taken singly, are misleading when dealing with the range of issues which must be involved in analyzing and evaluating community development corporations. The CDCs do have some similarities with multiproduct firms. They are periodically engulfed in internal bureaucratic struggles. And they do not infrequently attempt to supplant or modify activities which are normally pursued by governmental bodies. But in each case there are significant differences between the CDC and the standard models. Examination of these differences is important in understanding CDCs. Some adaptations, extensions, and integration of these models are required, therefore, for persuasive analysis of policy issues related to CDCs. It has become fairly common recently to claim that policy analysis and evaluation have demonstrated that programs, particularly those addressed to the serious needs of the poor and depressed communities, do not work. Given the requirements for thorough policy analysis and program evaluation discussed in this paper, such judgments are certainly premature, and may well be wrong.

NOTES

1. Sviridoff (1973). Eight CDCs were being supported at that time by the Foundation.

2. The growth which has occurred has been a bit unsteady. Otto Hetzel has captured the flavor of some of the problems encountered in such efforts in his insightful article (Hetzel, 1971: 68-98). In this article he identifies the games, all of which have been played during the development of CDCs. These games are: (1) Come Play with Me, (2) See if You Qualify, (3) Gotta Play with Those Guys, (4) Gotta Bring Your Own Ball, (5) Only One Game at a Time, (6) Change the Rules, (7) No Mistakes Allowed, (8) Rig the Game, (9) Withdraw the Aid, and (10) You Fail. He might have added, had he written the article now, "Take Away the Game." Unless there is some change in current federal policy directions, the growth phase for CDCs appears to be nearly over.

3. A detailed report of the initial phase of the performance evaluation of three CDCs (Bedford Stuyvesant Restoration Corporation, the Woodlawn Organization, and Zion Non-Profit Charitable Trust/Zion Investment Associates) being conducted by the Urban Institute is planned for early 1975.

4. For further discussion of Kotler's view, see Burton and Garn (1972).

5. National Advisory Council on Economic Opportunity (1973: 8).

6. National Advisory Council on Economic Opportunity (1973: 38). The italics are mine.

7. See p. 585 below for a more extended discussion.

8. This discussion will focus on the valuation of outputs. At a later stage in the paper, a further complication in using standard evaluation techniques—namely that such techniques do not adequately treat service delivery systems—will be introduced. For purposes of the present discussion, outputs are taken as given and the problem is to find out how much they are worth.

9. An implicit norm tends to emerge in most cost-benefit studies at this point. The prices which are sought are "perfect market prices." Of course, in the real world, analysts use actual market prices as if they were "perfect market prices." A failure to notice the distinction, however, is crucial, because economists are accustomed to thinking of "perfect market prices" as reflecting social as well as market value. If prices used in the analysis can be assumed to reflect social value, the problem of most concern to us here disappears, since, as I will argue below, there are good reasons to suspect a significant divergence between market and community valuation of outputs.

10. In the example which follows I apologize in advance to non-economist readers. The terms are consistent with those used in basic economic texts, should the meaning of the language not be clear from the text.

11. There are other issues associated with the necessary relationships between the CDC leadership and external development support institutions which further cloud the picture. Some of these are discussed in more detail below.

12. For a discussion of additional, growing concerns about surveys among the scientific survey community, see "Report . . ." (1974).

13. In our current evaluation, we call these factors "community welfare weights" to distinguish them from market prices or proxies for market prices which are normally used to weight outputs in benefit terms. A paper detailing these is being prepared.

14. The language here is carefully guarded for two major reasons. First, although we, among others, are doing research to establish causal linkages between possible CDC outputs and more general community conditions, such linkages have not yet been established in any sophisticated or numerical way. Second, the scale of CDC activities to date is too small to establish unambiguous statistical relationships between, for example, CDC job-creation efforts and changes in unemployment rates in CDC communities. Furthermore, such rates are clearly affected by larger scale systemic factors than the CDCs can control. Although political rhetoric may seem to require it, CDC claims to such improvements as may occur in these ratios are as premature as the claims of those on the other side, who say the CDC has failed if significant overall changes in such factors as income levels have not occurred.

15. This can be expressed as either $p_1(x_1^c - x_1^d)$ or $p_2(x_2^c - x_2^d)$.

16. This can be expressed as either $p_1(x_1^e - x_1^f)$ or $p_2(x_2^e - x_2^f)$.

17. The point being made here is that the CDC may have insufficient resources to operate over any given time period above the break-even point. Efficiency relates to the absence of resource waste while profit is a function of both costs and revenues.

18. I have said relatively little in this paper about costs. The argument can be carried through in terms of costs as well as benefits. See Tevis (1974) for a discussion of cost issues in CDC evaluation.

19. In terms similar to those used by Kelvin Lancaster, the goods and services produced can be assumed to have characteristics associated with them as a function of their origin and

method of production, operation, and methods of providing goods and services affect the perception of the goods and services. See Lancaster (1966 and 1971).

20. A slightly more extensive discussion of these issues can be found in Garn (1973a).

21. The ramifications of this will be explored further in a forthcoming paper by the author.

22. For an interesting, partial exception to this, see Boulding (1973).

23. See Porter (1970 and 1974).

24. This is not the place to go into detail about the notion of equilibrium and its role in the economic analysis of policy. It should be noted, however, that what starts out as a methodological assumption to facilitate analysis often turns out to be vested with substantial normative content.

REFERENCES

BLOCK, C. E. (1971) "Marketing techniques for the community-based enterprise," in Community Economic Development: Part II, Law and Contemporary Problems. Durham, N.C.: Duke University School of Law (Spring).

BOULDING, K. E. (1973) The Economy of Love and Fear, A Preface to Grants Economics. Belmont, Calif.: Wadsworth.

BRIMMER, A. F. and H. HARPER (1970) "Economists' perception of minority economic problems: a view of emerging literature." Journal of Economic Literature 8, 3 (September): 788-806.

BURTON, R. P. and H. A. GARN (1972) "The President's Report on National Growth 1972: a critique and an alternative formulation," in National Growth Policy, Part 2, Selected Papers Submitted to the Subcommittee on Housing of the Committee on Banking and Currency, House of Representatives, Ninety-Second Congress, Second Session. Washington, D.C.: Government Printing Office.

"Community development corporations: a new approach to the poverty problem" (1969) Harvard Law Review 82 (January): 644-667.

CROMWELL, J. and P. MERRILL (1973) "Minority business performance and the community development corporation." Review of Black Political Economy (Spring): 65-81.

CULBERTSON, H. M. (1968) "On the many sides of leadership." Public Relations Quarterly 13 (Fall): 25-30.

DOCTORS, S. I. and S. LOCKWOOD (1971) "New directions for minority enterprise," in Community Economic Development: Part I, Law and Contemporary Problems. Durham, N.C.: Duke University School of Law (Winter).

——— (1971) "Opportunity funding corporation: an analysis," in Community Economic Development: Part II, Law and Contemporary Problems. Durham, N.C.: Duke University School of Law (Spring).

Duke University School of Law (1971) Community Economic Development: Part I (Winter) and Part II (Spring), Law and Contemporary Problems. Durham, N.C.: Duke University.

FAUX, G. (1971a) CDCs: New Hope for the Inner City, Report of the Twentieth Century Fund Task Force on Community Development Corporations. New York: Twentieth Century Fund.

——— (1971b) "Politics and bureaucracy in community-controlled economic development," in Community Economic Development: Part II, Law and Contemporary Problems. Durham, N.C.: Duke University School of Law (Spring).

FERENCE, T. P. (1971) "Managerial assistance: promises and pitfalls," in Community Economic Development: Part II, Law and Contemporary Problems. Durham, N.C.: Duke University School of Law (Spring).

"From private enterprise to public entity: the role of the community development corporation" (1960) Georgetown Law Journal (May): 956-991.

GARN, H. A. (1973a) "Public services on the assembly line." Evaluation 1, 2.

--- H. J. BRYCE, and N. L. TEVIS (1973b) "CDC evaluation: a discussion paper," Urban Institute Working Paper 0719-01-1 (April). Washington, D.C.: Urban Institute.

GARN, H. A., C. E. SNEAD, and N. L. TEVIS (1974) "An approach to issue analysis for CDCs," Urban Institute Working Paper 0719-01-3 (August). Washington, D.C.: Urban Institute.

GARRITY, P. G. (1971) "Community economic development and low-income housing development," in Community Economic Development: Part II, Law and Contemporary Problems. Durham, N.C.: Duke University School of Law (Spring).

HARRISON, B. (1974) Urban Economic Development: Suburbanization, Minority Opportunity, and the Condition of the Central City. Washington, D.C.: Urban Institute, SURI 48000.

HENDERSON, W. L. and L. C. LEDEBUR (1971) "Programs for the economic development of the American Negro community: the moderate approach." American Journal of Economics and Sociology 30, 1 (January).

HETHERINGTON, J.A.C. (1971) "Community participation: a critical view," in Community Economic Development: Part I, Law and Contemporary Problems. Durham, N.C.: Duke University School of Law (Winter).

HETZEL, O. J. (1971) "Games the government plays: federal funding of minority economic development," in Community Economic Development: Part I, Law and Contemporary Problems. Durham, N.C.: Duke University School of Law (Winter).

HOPPS, J. G. (1973) "Ghetto economic corporation theory, reality and policy implications." Review of Black Political Economy (Spring): 43-64.

KOTLER, M. (1971) "The politics of community economic development," in Community Economic Development: Part I, Law and Contemporary Problems. Durham, N.C.: Duke University School of Law (Winter).

LANCASTER, K. J. (1971) Consumer Demand: A New Approach. New York: Columbia University Press.

--- (1966) "A new approach to consumer theory." Journal of Political Economy 74, 2.

LUTTRELL, J. D. (1971) "The effect of the private foundation provisions of the Tax Reform Act of 1969 on community development corporations," in Community Economic Development: Part II, Law and Contemporary Problems. Durham, N.C.: Duke University School of Law (Spring).

National Advisory Council on Economic Opportunity (1973) Sixth Annual Report, June 1973. Washington, D.C.: Government Printing Office.

OLKEN, C. E. (1971) "Economic development in the model cities program," in Community Economic Development: Part II, Law and Contemporary Problems. Durham, N.C.: Duke University School of Law (Spring).

OXENDINE, J. E. and A. N. PURYEAR (1971) "Profit motivation and management assistance in community economic development," in Community Economic Development: Part I, Law and Contemporary Problems. Durham, N.C.: Duke University School of Law (Winter).

PERRY, S. E. (1971) "National policy and the community development corporation," in Community Economic Development: Part II, Law and Contemporary Problems. Durham, N.C.: Duke University School of Law (Spring).

PORTER, D. O. (1974) "Incentives and responsiveness in administering a federal system." Publius (special issue).

--- (1970) Who Slices the Pie: A Study on Mobilizing Resources and the Politics of Budgeting. Detroit: Wayne State University.

"Report on the ASA Conference on Surveys of Human Populations" (1974) American Statistician 28, 1 (February): 30-34.

ROBIN, R. S. (1971) "A taxpayer's choice incentive system: an experimental approach to community economic development tax incentives," in Community Economic Development: Part I, Law and Contemporary Problems. Durham, N.C.: Duke University School of Law (Winter).

ROSENBLOOM, R. S. and R. MARRIS [eds.] (1969) Social Innovation in the City: New Enterprises for Community Development. Cambridge, Mass.: Harvard University Press.

STRANG, W. A. (1971) "Minority economic development: the problems of business failures," in Community Economic Development: Part I, Law and Contemporary Problems. Durham, N.C.: Duke University School of Law (Winter).

STURDIVANT, F. D. (1971) "Community development corporation: the problem of mixed objectives," in Community Economic Development: Part I, Law and Contemporary Problems. Durham, N.C.: Duke University School of Law (Winter).

SVIRIDOFF, M. (1973) "Building from disappointed hopes," Ford Foundation Annual Report 1972 (October 1 to September 30, 1972). New York: Ford Foundation.

TEVIS, N. L. (1974) "Costs, estimation and use in the evaluation of CDCs." Urban Institute Working Paper 0719-01-4 (August). Washington, D.C.: Urban Institute.

VIETORISZ, T. and B. HARRISON (1970) The Economic Development of Harlem. New York: Praeger.

19

An Evaluation of Evaluation Research:
When Professional Dreams Meet Local Needs

MICHAEL BARNDT

☐ THIS STATEMENT WILL EXAMINE evaluation methodology and aspects of its impact in practice. The discussion will be drawn from a limited context—evaluation programs designed to determine funding policy for human service organizations. In this context, the paper will identify limits that evaluation tools encounter which seem to be inherent in the nature of human service programs and the environment within which they operate.

A. MODEL FOR DECISION PROCESS

The critique of evaluation process which follows is grounded in the assumption that the political process (small p) is a primary factor in decisions regarding the future of organizations. As long as resources are limited and programs face competition over the allocation of funds, the decision processes determining that allocation of funds will be primarily political. As the objectives for a program are established or reevaluated, political differences emerge based upon the *value* differences of decision participants. Political differences are enhanced by differences of *interest* as decisions are made regarding the management of a program. These differences may range from conflicts over suitable sponsorship of a program to differences over the particular mechanisms by which management will be held accountable. Additional differences may arise based upon phenomena inherent in much of organizational behavior—the distorted *perception* of participants as a result of the specific roles they play within or outside the organization.

Figure 1 represents an image of the substance of many urban decision processes. Decisions are first a product of history. Decisions tend to sustain established patterns, to reflect prior concensus, institutionalized procedures, or political exchanges. Incremental adjustments in such a process are largely determined by changes in political circumstances. When the inertia of history and the contingencies of the political climate has been met, more refined choices may be subject to rational procedures. In the human services area rational choice is further limited by the state of the art. Programs are frequently experimental tests of procedures not well understood. Frequently serendipity governs the final choice.

Program evaluation is a process which seeks to extend the influence of rational factors in the decision process. To a certain extent it is successful. This critique of evaluation, however, will dwell upon the limits to the process. Those limits will be primarily derived from a discussion of the impact of the political elements of the process upon the implementation of evaluations and the utilization of outcomes. The paper will not attempt to assess the degree to which the limitations discussed are critical. The comments may serve as a framework from which to raise questions of particular evaluation process. But many comments will not be appropriate to a specific case.

Evaluation technology will be critiqued by drawing from illustrations of the experience of local program managers with the process. The discussion will be divided into three parts. The first will identify assumptions about the behavior of organizations which are necessary to much of evaluation technology. The technology is limited as the reality of organizational process is identified. The second part will identify particular types of organizations which are so organized as to be antithetical to assumptions about "normal" organizations. Evaluation procedures are severely limited in these instances. The third part challenges the presumption that evaluation technology in the future will be much more exact and sophisticated

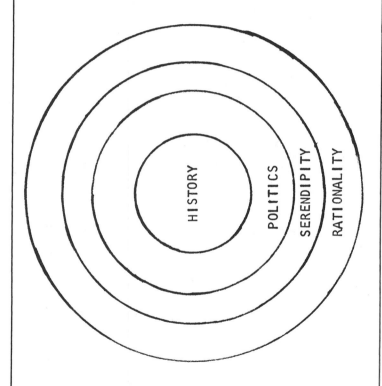

Figure 1: ELEMENTS OF DECISION PROCESS: A SYMBOLIC REPRESENTATION

than it is today. Difficulties with costs, measurement, and modeling suggest barriers to very substantial improvements in evaluation techniques for most human services organizations.

Discussion will not include the full range of evaluative processes. The critique will be limited to those evaluations which assist an organization or its funding sources in decisions regarding the general nature or scope of future programs. Evaluations which are merely exploratory or those which serve as a test of alternative means for possible use in other programs are not of interest here.

B. ASSUMPTIONS INHERENT IN EVALUATION TECHNOLOGY

Many basic forms of evaluation depend upon the accuracy of certain assumptions about organizational process. More sophisticated evaluation methods may accommodate to some degree to a more realistic view of organizations. For some organizations the assumptions are not merely limited but are contrary to actual organizational process. Key assumptions and the limitations encountered are identified and illustrated below. The list is followed by a discussion of organizational forms for which the assumptions may be entirely inappropriate.

Assumption 1: That program objectives are agreed upon by all key actors to the funding contract.

Limitation 1A: Objectives may be left somewhat unspecified because of the political costs of confronting substantial ideological differences between key actors.

Example: A drug counseling program sought to communicate with counter-culture drug users through a street newsletter. The publication included information on drugs that were being misrepresented in drug sales. Staff generally felt that their primary objective was to eliminate the abuse of drugs, not drug use. LEAA (Law Enforcement Assistance Agency) funding officials would have sharply disagreed with that objective. And many officials would have argued with the implication left by the newsletter in its attack on misrepresented drugs only. The funding proposal had been carefully designed and deliberately short to insure approval without resolving these questions.

Limitation 1B: A program may be identified officially as encompassing objectives not intended by participants but specified to satisfy funding resources.

Example: One community leadership training program funded by HEW espoused the objective of serving to develop leadership skills from a broadly based selection of residents within a largely black community. Such an open policy was desired by HEW. The program management had intended the funds to be used basically as a staff training device for a single organization. The resulting emphasis in recruiting and training failed to meet the stated objectives but not the real intentions of program managers.

Assumption 2: The success of a program may be understood by identifying "indicators" which are easily measured and which are expected to parallel the level of excellence promoted by the program.

Limitation: Chosen indicators may be shallow measures which do not distinguish a minimal program from an excellent one.

Example: LEAA-funded agencies are encouraged to use the

degree of recidivism among clients as an indicator of success. A halfway house program sponsored by a local neighborhood organization identifies as its objective the rehabilitation of the "whole man." A staff member complains that recidivism only identifies those who "are stupid enough to get caught." Programs which offer "something extra" are threatened in the long run if they are expected to compete on a cost bases with programs making a less substantial effort that still satisfies minimum expectations as measured by standard indicators.

Assumption 3: A program may be evaluated by comparing its accomplishments to projections made at the start of a program year. Explicit statements of objectives may include performance contracts committing an agency to a specific scale of service and/or rate of success. Failure to meet such goals may jeopardize future funding. Failure to choose sufficiently ambitious goals may jeopardize current funding requests.

Limitation 3A: Accurate predictions are not possible when shifts in the extent of the problem or related problems occur.

Example: Drug problems within counter-culture communities have been substantially different from year to year as the character of the problem has changed. From one year to another a community may experience general shifts in the specific drugs in common use, the level of abuse, the pattern of individual personal problems, and the opportunities open to individuals to change their lifestyles. The scope and the success rate of an agency will be considerably variable under these conditions.

Limitation 3B: When an agency relies upon other systems for referrals its own predictions become less reliable.

Example: A youth services program dependent largely upon the school system for referrals found that the number and character of referrals varied substantially with school personnel. Turnover and the effect of changing relationships with individuals within the school system became important factors in the variation of referrals.

Limitation 3C: Predictions are not reliable when little is known about the effect of a relatively new program.

Example: The first major employment programs to seek to serve the "hard core unemployed" predicted high rates of success. Only after several years did consensus emerge that a 40% rate of success could be considered satisfactory. In the interim, program proposals consistently projected more optimistic expectations for subsequent years, in part because of pressure to substantially improve earlier statistics.

Limitation 3D: Predictions are not reliable when external factors affect the outcome for clients.

Example: Halfway houses for parolees serve to help establish the client in a stable lifestyle. This becomes much more difficult in periods of serious unemployment.

Assumption 4: Programs can be evaluated in a way that allows comparison across programs.

Limitation 4A: A funding agency is concerned with setting priorities for funds. The decision to fund agency A to deliver a service may be made at the expense of additional funding to agency B to serve another clientele or to use different means of service delivery. Can the relative costs be appropriately compared?

Example: The local LEAA funding agency extended the use of recidivism statistics to as many agencies as possible in order to have comparative data for its board. Summaries of such statistics are misleading when agency records are compared without accounting for differences in the difficulty of serving specific clientele or the resources necessary for particular means.

Assumption 5: Programs achieve stability after the initial creation of the organization allowing the charting of organizational progress from one year to the next.

Limitation 5A: Organizations are likely to face significant transformations in operation as the major events such as choosing a director, hiring and training staff, developing organizational management systems, and so on are first experienced. But problems continue to face many small organizations. Each year may include significantly unique events which lend major qualifications to measures of success.

Example: Consider the effect of the following events upon organization "X":

First year—Hired director, 5 staff; designed and publicized program.

Second year—Bought and rehabilitated a larger building for the program; substantial staff time involved in the transition.

Third year—A grant supporting 25% of the program was discontinued; staff was reduced and reallocated; programs were also modified.

Fourth year—The director resigned; the transition was made more difficult by the months he spent as "lame duck" within the organization and the subsequent month with no director when recruitment lagged.

Fifth year—A new agency with a broader mandate siphoned off a substantial number of the lower-risk clientele.

The exceptional nature of each event makes subtle judgments of agency progress meaningless.

Limitation 5B: The life of many programs is very short. Within a few years an organization may be replaced by another or may choose to stress substantially different goals for its programs.

Example: In the past decade federal funds have been substantially fluid. The funding pattern has emphasized experimental programs with only a short-term commitment of federal funds. Small "high-risk" organizations have been granted funds but were not always capable of the programs they took on. Frequent shifts in federal emphasis has led organizations to modify programs substantially as a pragmatic response to survival needs. A substantial number of organizations with such a history can only be evaluated using very open-ended evaluation techniques of limited sophistication.

Assumption 6: Organizational means and objectives are stable during a program year.

Limitation: A program proposal represents a frozen image of an organization in motion. An organization is likely to be continually reassessing its objectives and procedures. It is not reasonable to expect that organizations will delay program modifications until the end of a program year.

Assumption 7: An organization can be evaluated by reviewing its programs as though isolated from the service system of which they are a part.

Limitation: Service systems for criminal offenders, for alcoholics, for the "multiproblem" family, and so on are fragmented among different agencies and frequently lack any authoritative coordination. Individual agencies may have little control over the referrals to them and little influence over the programs to which they refer clients. Failures elsewhere in the system may affect the program's opportunity for success.

Assumption 8: An organization's objectives are compatible with one another.

Limitation: An organization may have some objectives which could be advanced at the expense of others—efficiency versus thorough service; staff advancement and on-the-job training versus fully professional competence; individualized counseling versus continuity and coordinated management. Such objective sets do not contradict each other but under the pressures of costs, time, and constraining events choices are necessary between them.

Example: An organization charged with serving as an advocate for neighborhood youth is expected to develop effective relationships with youth-serving agencies so that it may assist in coordinating the overall effort. The evaluators have chosen good relations with referral agencies as one measure of program success. Another objective for the organization is to pressure other agencies to modify their procedures when necessary in the interests of the client. This objective is much more difficult to assess. The agency is likely to make more effort to measure up to the more specific indicator—to maintain good agency relations—than to threaten such relations when client advocacy seems appropriate.

C. PROGRAMS WHICH CONTRADICT

THE USUAL MODEL FOR

ORGANIZATIONAL EVALUATION

All organizations are in some ways not appropriately subsumed by the assumptions of evaluation process but certain programs stand out as contradictory to the assumptions. The usual model for organizational behavior is taken from experience with organizations with clear objectives, stable structures, easily measured inputs and outputs, and the like. As has been demonstrated in the field of sociological theory it is sometimes appropriate to consider the straightjacket a standard model might represent. Ralf Dahrendorf (1958) challenged the assumptions of a basic sociological framework, structural-functionalism, by suggesting that the assumptions be turned inside out. He attributed the following assumptions to structural-functionalism:

—Every society is a relatively persisting configuration of elements.
—Every society is a well-integrated configuration of elements.
—Every element in a society contributes to its functioning.
—Every society rests on the consensus of its members.

Dahrendorf countered these points with a set diametrically opposed to the first:

—Every society is subjected at every moment to change: social change is ubiquitous.

—Every society experiences at every moment social conflict: social conflict is ubiquitous.

—Every element in a society contributes to its change.

—Every society rests on constraint of some of its members by others.

Dahrendorf suggested that both sets may serve as useful vantage points. The second set appears to be more useful as a vantage point for certain organizations which see themselves (a) as catalysts of change; (b) as "counter-institutional"; (c) as advocates of a particular position; (d) as open to substantial diversity of means and ends; or (e) as intra-institutional brokers in service systems which have not "institutionalized" the interfaces between participating institutions. A few specific forms are worthy of illustration:

(1) the community development corporation,

(2) the organization with an ideological/political agenda in addition to service functions,

(3) the organization with an open-ended treatment program,

(4) the organization for which "process" is as important as standard service delivery,

(5) the coordination/referral agency,

(6) the "temporary" consultation/training agency.

(1) A community development corporation accepts a contract to manage an apartment complex. How does one evaluate the appropriate costs of efficient management? Committed to community development principles, the corporation chooses to offer extended services to tenants, to organize programs with efforts to increase participation, to hire and train local residents, and to contract with other community-sponsored programs for specific maintenance services. "Process" objectives have been identified which are as important to the organization as the "task"—efficient management. The objectives tend to contradict each other. The extra programs have costs. The limited experience of staff increases the risk of program inefficiency. The "process" objectives are especially hard to measure in ways that would allow trade-offs to be closely evaluated. Indicators of community health—increase in local skills,

improved neighborhood communication, a growing commitment by residents to the community, and so on—are likely to remain subject only to inexact measures.

(2) A black community organization seeks to increase the consciousness of political agendas among residents and leaders. Dual "brother" corporations are created—one to operate a service program, the other to act on political matters. The design is intended to free members of the political arm from Hatch Act constraints on political activity. In clear but subtle ways the organization intends the service program to be a mechanism through which the community can be reached, served, educated, and organized. The political interests of the system do drain energy from the operation of an efficient program, however. Evaluative systems tend to evaluate only the nonpolitical aspects and to focus entirely upon the objectives of the service program corporation.

(3) A youth counseling center program claims to have no specific means of client service. Any individual who comes to the center with a personal problem is served. Each client is treated as a unique experience. The mix of problems with a variety of drugs, with parents, with jobs, with school, and with personal trauma challenge any efforts to summarize the "costs" or the rate of "success." Furthermore, the individual staff member is free to determine the client's treatment to avoid imposition of what has been feared to be less appropriately prescribed diagnosis by a centralized procedure. Thus, the organization may be viewed as eight organizations utilizing a larger number of means to satisfy an unlimited number of objectives as contracted with clients. Only very general measures of program success may be possible.

(4) A live-in drug therapy center wishes to understand the development of its program from year to year. This year is unique because nearly everyone has been immersed in the effort to rehabilitate an old apartment building the organization purchased. But the staff admits that the energy ellicited by the experience was an important part of the therapeutic process. Something else unique and as disruptive to a "stable" (routine) operation will need to be found next year—especially to enhance the commitment of new members. In essence, the organization depends upon the "Hawthorne" effect for its continued success. But events may affect the costs, the staff resources, and the predictability of the system.

(5) An organization working with youth services has been created to coordinate the delivery of existing services for youth. Referrals are

made to the new agency by the schools. Youth participate in short term counseling and are then referred to other agencies. The staff follows up on the results of the referrals. But the new agency has no authority over other agencies and cannot expect to have much impact upon the way they operate. The proposal for evaluation of the program suggested using area delinquency rates as an indicator. But can the responsibility for success or failure be located in these figures? The role of an organization with a substantial responsibility but little direct authority is clearly awkward. But such agencies are now being funded in an effort to bring some degree of reform to delivery systems.

(6) To serve the needs of a group of "counter-culture" agencies an organization has been given a grant to provide consultation, coordination, and training to the various staffs. But there is a clear resistance to excessive coordination from the agencies served. Consultation and training functions are made difficult by conflicts among some of the organizations as to appropriate methodologies for dealing with clients. The new consulting agency must develop a role for itself in an uncertain environment. Furthermore, when individual agencies are strong enough, many of the functions of this consulting will be less important. The entire history of the consulting agency will be in flux—appropriate for the context but a challenge to evaluate.

The six illustrations on the last few pages are not intended to be comprehensive. It is likely that additional circumstances not fitting the above patterns would serve to be as challenging as these examples to the assumptions inherent in most evaluation efforts. Evaluation of such systems cannot rely on established techniques but require a more open-ended assessment of the various processes occurring; the relative progress of the organization on each; and the relative values placed upon these processes by participants, clients, and funding sources.

D. LIMITS TO INCREASING THE SOPHISTICATION OF EVALUATION TECHNOLOGY

This section will examine the potential for improving the sophistication of evaluation techniques. A number of directions for change will be considered:

(1) That measurement of organizational activity become more specific and more inclusive.

(2) That processes for valuing or prioritizing objectives reduce the ambiguity concerning the intent of the organization.

(3) That the organization increase its efforts in recording information for evaluation.

(4) That evaluators be granted the responsibility for monitoring a system of agencies whose work is integrally related.

(5) That evaluation become the role of a problem subsystem manager who will monitor and coordinate the overall efforts of agencies within that system.

(6) That evaluation designs develop formulas which will allow the recombination of the more detailed information available so that decision-makers are able to use the data without being overwhelmed by its complexity.

In each of the directions above there is clear opportunity for improvement. But in all cases limits may be identified which seem to be problems inherent in the capacity of the techniques to deal with the very real complexities of the programs under less than ideal conditions.

(1) Evaluation may be improved by more complex measurement techniques. The simpler forms of evaluation tend to depend upon definitions unable to account for important variations within and between organizations. The broad definitions for measurement categories may be set to allow summary comparison of one agency with another or between staff or time periods within an organization. But the comparisons are frequently misleading. Examples:

The output of an agency that assists prereleased offenders chosen for good prison records is compared to an agency that chooses to help parolees most in need of special assistance.

Within a community housing corporation community staff who work with individual client problems report their "contacts" in the same category as the number of "contacts" made when they work with group recreation programs for youth.

A youth counseling service counts the clients with drug problems without distinguishing the type of drug or the extent to which the problem is complicated by other factors.

The quality of measurement may improve by defining more specific categories. But a number of limitations occur. Categories may define ideal types which rarely occur. For example, the youth counseling center may develop a list of client problems from which a client profile may be identified. But the profile may be unique, based upon a combination of symptoms from the specific list. Even then clients may most appropriately fall between categories. For instance, most clients using LSD may be neither completely addicted to it nor independent of it.

Creation of more specific categories may reduce the opportunity for inappropriate comparisons but, as specificity increases, the opportunity for any comparison decreases. Decision-makers who are expected to make choices which favor one agency over another or reward one staff member more than another must face the inevitable limitation that measurements which allow comparisons are susceptible to errors of over-generalization.

The quality of measurement may improve by selecting indicators which are more precisely measured. Any time that a measure is expected to indicate the effect of another variable not being measured, the evaluator must assume that the two variables are related. The use of an indicator may gain in precision what is lost by the appropriateness of the assumption. But frequently the assumption may be less valid. Many times the interrelatedness of variables is not well understood. For example, a community-based program may demonstrate the quality of its relationship to the neighborhood by reporting the number of community residents who attend meetings called by the organization. Active resident participation could be spurred to some extent by a distrust of what the organization would choose to do if not carefully monitored—an explanation counter to the assumption. Furthermore, participation is related to a number of factors—the class of participants, the intensity and novelty of the issue, immediate self interest, and so on—connected very little to the quality of the relationship between the organization and community.

The quality of measurement may improve by broadening the range of measures to include important but less direct objectives. As suggested in the previous section certain organizations have both "task" objectives and "process" objectives. "Process" objectives are generally very difficult to measure. As a result they may be ignored or treated lightly by an evaluator. As indicated in the previous paragraph, certain objectives may resist precise measurement.

Consider, for example, process objectives associated with a community development organization: (a) to identify and develop the skills of local leadership, (b) to train local residents in professional skills the neighborhood will require, (c) to insure that citizens are able to shape the development decisions which affect them, (d) to restore in residents a positive identity and a commitment to their neighborhood, and (e) to increase the political strength of the neighborhood in larger decision systems. Some of these objectives can only be accurately assessed over a substantial period of time. Many will be only partially affected by one organization but it may be difficult to specify an appropriate level of impact. All of the objectives depend upon the effect of other events beyond the control of one organization. Some of the objectives are not well understood.

(2) When organizational objectives are more detailed ambiguity may be further reduced by procedures for valuing or prioritizing objectives. Efforts have been made to measure process variables precisely, to the point of attributing numerical scores to all components of organizational process. The numerical results lead to other difficulties. How are scores for different objectives to be compared? Is a 20% drop on one indicator the equivalent of a 20% drop on another? One very ambitious evaluation design (Mantel et al., 1972; Service et al., 1972) determined weights to be assigned to each criteria. The result was a single "quality" measure:

"Quality" = 9.5 x "effectiveness" + 8.0 x "efficiency" + 8.5 x "capability" + 6.0 x "accessibility" + 5.0 x "interrelationship" + 7.0 x "system contribution."

Such a system may be appropriate for relatively stable organizations, but is a very poor performance on one measure counterbalanced by an excellent performance on another? The formula requires that relationships among variables are linear. Any other assumption would lead to a substantially unwieldy equation.

As more measures are taken a more substantial profile of an organization emerges. The more difficult measures may be less reliable but they are important. However, as the amount of information grows the challenge to summarize or to combine the data grows as well. The limits to recombining data are discussed later in this section.

The use of sophisticated evaluation models suggests a need for more adequate valuing processes. As evaluation is taken more seriously by decision-makers it becomes more important that the objectives of a program are more clearly understood. In many cases this requires that the evaluator be in a position to assist selected participants in the system with a careful analysis of the objectives of the program, the threshold accomplishments anticipated, and the relative importance of each objective. When an evaluation model is designed which is more specific and all-inclusive the work of the evaluator with the system becomes more difficult. Limits to this process are discussed in the next few paragraphs.

When the objectives of a system are specified, the evaluator may play a critical role. He can provide frameworks drawn from conceptual models of programs which have been appropriate to similar organizations. Such models might identify the objectives for comprehensive community development, for effective organizational management, for efficient economic development, for professional diagnosis and treatment of clients, and the like. The individual models are well constructed. But as much as the models may assist participants with the selection of evaluation criteria they may also limit the expression of points of view which seem to be foreclosed by a model. Many models are not sensitive to the contradictions organizations face. For example, it would be difficult to argue with high standards for organizational efficiency unless participants were able to recognize that a commitment of the organization to hiring and training residents may well reduce such efficiency. Participants may also be unaware that many of the basic measures for efficient management would rate an impersonal well-run bureaucratic institution more highly than a more flexible person-centered one.

As objectives are specified, participants must decide upon threshold expectations for the organization. The first section of the paper identified a number of limitations such a process faces. It is not necessary, however, that predetermined expectations be taken as immutable at the end of an evaluation process.

When an appropriate set of objectives has been generated participants must still determine how to value them. A number of problems may occur in the valuing process. One technique for valuing, the Delphi method, will be described and critiqued as an illustration of a number of problems which may occur.

The Delphi method is designed to allow a number of participants to reach a consensus on the assignment of numerical weights reflecting the relative value of a list of objectives. The participants individually assign weights from 1 to 5 to each item on a prepared objectives list. The aggregate assignments are evaluated by an independent monitor. When participants substantially agree on an item, it is removed from further review. Items for which consensus is low are to be revalued by participants. A "discussion" round precedes further valuing. Under the standard procedure, a participant who was most negative and one who was most positive with respect to an item are asked to write brief statements supporting their positions. All statements are shared with other participants by the monitor without identifying the source. The distribution of the ratings is also disclosed. The participants assign values again—presumably more informed about the way items may be interpreted. Consensus is likely to be achieved on a number of additional items. The process is repeated for additional rounds for those items for which disagreement remains. Usually four rounds is enough to complete the process. Closure is hastened to some extent by the willingness of participants to move their rating toward the mean rating given by the group.

The primary dilemma for Delphi and most other valuing procedures is the process of identifying participants. Traditionally, the program manager and evaluator have been free to make evaluation decisions alone. In some recent experiences the process has been opened up to include professional staff, board members, selected clients, community residents, external experts, funding source representatives, and others. The open process is valued for allowing a more democratic process. But when the relative importance of values is in dispute, the mechanism for determining how to combine inputs from participants may be controversial. As illustrated in the first section of the paper, participants aware of differences may seek to avoid opening up the issues, presuming that a program staff may be free to act if it is discreet about its intentions. The Delphi process is designed to address the question of participation by leveling the contribution and influence of participants. The process can accommodate a large number of participants. The anonymity of participants as they rate the objectives and when they contribute to the "discussion" phase eliminates the tendency of participants to be influenced by powerful members of the group. Such a configuration is hardly likely to reflect the nature of an organization even as a desired state. However, no easy consensus is likely regarding the participation question. Participation in the valuing process may be further complicated by the capacity of participants to understand the process. A broadly based selection of participants includes those with less information and competence to judge. An example of the problem was encountered when a housing agency prioritized its tenant services objectives by involving tenants, all staff, and a few board representatives in the process. The numerical weight of the residents and nonprofessional staff (including secretaries in this case) was greater than the professional staff and board. Client participation may require that additional effort go into educating participants to the implications of objectives. In practice, techniques like the Delphi process reduce levels of communication in an effort to eliminate the influence of powerful participants and to balance the presentation of issues. The resulting short formal statements may not even permit participants to learn whether they are perceiving terms to have the same meaning as other participants do. Greater communication, however, would reintroduce the problems of influence and balance.

To some extent the process of assigning numbers to concepts is beyond the competence of all participants. As has been suggested earlier in this section, the relative value of an objective is likely to be a nonlinear function of the achievement of the organization with respect to that objective. Furthermore, the value is likely to be interdependent with other objectives. The limits of the perfect model are less of a problem, however, than the likelihood that participants may not account for or comprehend the trade-offs involved in allocating values among objectives. (Note the discussion of the persuasive character of singular conceptual frameworks for illustrations of this problem.)

(3) Evaluation efforts may be improved by seeking more complete monitoring of programs. A major improvement would result merely from encouraging organizations to take the process seriously. Within limits, organizations can redesign the systems of information they use internally to serve evaluation needs and reduce recording loads. But efforts to employ advanced designs of greater specificity are likely to increase the responsibilities and necessary skills of staff. More sophisticated evaluation implies more resources invested in evaluators. A more sensitive assessment process may include hiring external consultants to evaluate specific program components. Additional costs may be involved in designs which recheck or avoid the potential for staff sabotage of the information process. (Motivation for sabotage has been discussed in part in the first section and will be illustrated again below.) The cost of

evaluation is a very real limitation which is further aggravated by the scarcity of resources in human services programs, the limited impression the evaluation process has made to date upon decision-makers, and the predominance of the political process as a mechanism to set allocation decisions. In the opinion of this writer, even with the reduction of political considerations the cost effectiveness of complex evaluation designs should be seriously questioned.

For certain organizations the recording process may be complicated by the style of the organization. For example, a center for runaway youth wished to be informal, nearly spontaneous in its work. There was even a feeling that "facts" about the previous history of a client prejudice the work of the staff. The need to record information which was not immediately appropriate to the work of staff with the client was perceived to over-formalize the relationship. Such an organization fears the changes that take place in staff as they adapt to bureaucratic procedures or as they are replaced by staff with a greater tolerance for such procedures. A second illustration of perceived indirect costs is that of the free clinic which stressed its unique relationship to a community which was especially wary of the health system. The program created a clinic atmosphere which was open and nonthreatening to community residents. A health planning body asked the clinic to demonstrate that it was in fact attracting new clientele to the health system. In exchange for staff support, the clinic was to give each client a thorough intake interview to trace his past medical history. The act of the interview process would have destroyed the very climate it was to be measuring.

(4) Another mechanism which may allow evaluators to improve the quality of evaluation design is accomplished by monitoring a system of related programs regardless of agency sponsorship. The interrelated character of programs has already been noted. Certain agencies with coordination/referral responsibilities may be substantially dependent upon information which is available from other agencies. Other programs may be heavily affected by changes in their environment—conditions which are barriers to client progress or institutions which are expected to provide some degree of the "treatment" a client requires. The costs of evaluations noted above may be much higher when efforts to evaluate these external factors are entertained. Aside from the complex models involved, the very process of gathering information takes on additional costs. In many instances appeal to a higher level of governance is necessary. For

example, only after the state of Wisconsin began channeling major funds to mental health and drug agencies was it possible to press for coordinated information on the service clients receive from several agencies. Most funding sources have limited influence even within their "jurisdiction," but most problem systems extend beyond the influence of government or private funding sources. The failure of program agencies to coordinate efforts is especially serious from the perspective of the social planner. But the evaluator may be restricted even when many coordinated program systems exist. Inevitably, the full impact of some program effort depends upon factors beyond the "subsystem" which has yielded to coordination. Factors may include the employment market, the current community norms, the neighborhood environment, and so on. "Systems Theory" has been refined to allow consideration of a number of critical factors by identifying the most critical institutions and conditions and treating this set in isolation. The assumption that a social subsystem is a "closed system" is only a matter of convenience. The assumption implies that factors external to the subsystem are either constant or unimportant. Evaluation models will always be limited by the failure of that assumption.

(5) Difficulties of obtaining information from a subsystem would be dramatically reduced if the subsystem were managed by a centralized planning, coordinating, monitoring, allocating authority. The negative consequences of such authority on certain program systems is an additional cost which should be considered. The development of sophisticated evaluation systems with an increased flow of information about the system implies the capacity for more thorough management of the system. An information system can become a vehicle for a substantial increase in the role of program managers in the system. Such a system lends itself to centralized, institutionalized administration. Centralized information suggests the potential to erode efforts to retain decentralized authority, to leave substantial discretion with professionals and/or to allow the community or client to make choices for himself. This may not be an issue unless a manager seeks to challenge these other principles, but he then is given an additional advantage through his control of information.

Individuals or organizations who perceive a threat to the release of certain information may resist or quietly sabotage the effort. Professionals may wish to protect their judgments from critical review. For example, university faculty within a large department resisted for a period of time the use of a standardized student evaluation form. The form implied by design the criteria for good teaching, but was not sensitive to the priorities that certain faculty placed within the teaching process. Faculty wanted the right to determine these priorities for themselves. A second example illustrates the effort of an independent organization to resist possible abuse of information. A youth counseling center allows first-name-only relationships with clients to assure that information will be confidential. The agency has been under pressure from the police department to report cases of drug abuse and to assist in tracing drug pushers by sharing other information received by clients.

Concern for confidentiality has been raised more frequently as proposals are presented for complex computer-based information systems. For example, a state funding source has designed an information system to monitor the path of individuals through any of a number of agencies in the mental health system. The state-supervised computer file would assign a unique code derived from individual names before eliminating client names from the system. Only legitimate agencies would have access to the information. The aggregate data would allow very sophisticated review of the overall effectiveness of the network of organizations in the system. But the proposal has been soundly rejected by many agency representatives because safeguards against the abuse of information cannot be assured. Failsafe procedures are not really possible.

(6) Evaluation designs which are complex challenge the ability of the user of the results to recombine the information so that direct decisions may be made. The difficulty of comparing progress on one objective to progress on another has already been discussed. One solution has been to weigh and recombine the variables before the decision-maker is to use the information. The formula for "quality" specified near the beginning of this section is an illustration of such a process. Aside from the errors associated with the valuing process already discussed, efforts to combine numbers into single variables are limited by the effect of the combination of terms, each containing a degree of error. William Alonzo (1968) has reviewed the ways that errors accumulated in a section of his article headed, "Predicting Best With Imperfect Data." He demonstrates that complex models, even when using more accurate data, tend to accumulate a larger error term than simple models. He demonstrates that in many cases a simple model is more appropriate. He concludes

that perhaps the best strategy is to build several simple models which among themselves use all of the data, and to derive some summary from the set of them. "The strategy is not to build one master model of the real world, but rather a set of weak models as alternative models for the same set of phenomena. Their intersection will produce 'robust theorems'. As complementary models they shed light on different aspects of the same problem" (Alonzo, 1968: 187). An extension of Alonzo's point would suggest that when the perceptions and values of participants differ, the evaluation model may seek to clarify and report the evidence for each of the different perceptions of participants rather than designing a process which assumes early in the design process that consensus has been reached.

The analysis of this section has identified possible *conceptual* limitations to sophisticated evaluation systems which should demonstrate that the evaluation process can hardly be expected to achieve an ideal state. The first section demonstrated contextual limitations to the process from the experience of difficult-to-evaluate programs. Neither section should not be extended and refined. A great many questions were raised in the process of the critique. The questions may be tested in any particular case as they appear to be relevant. The answers to problems identified by such a process are usually not addressed in these pages. The product of any evaluation process must, in the final analysis, improve the relevance of a design when it can and put up with the limitations of the design when little can be done about them. It does seem important, however, that the critical limitations to a design be understood.

REFERENCES

ALLERHAND, M. (1971) "The process outcome research model: an alternative in evaluation research," pp. 131-150 in R. O'Toole (ed.) The Organization, Management, and Tactics of Social Research. Cambridge: Schenkman.

ALONZO, W. (1968) "The quality of data and the choice and design of predictive models," pp. 178-192 in G. Hemmens (ed.) Urban Development Models. Washington, D.C.: Highway Research Board, National Academy of Sciences.

CHOMMIE, P. W. and J. HUDSON (1974) "Evaluation of outcome and process." Social Work 19, 6: 682-687.

DAHRENDORF, R. (1958) "Toward a theory of social conflict." Journal of Conflict Resolution 11, 2: 174-179.

FERMAN, L. (1971) "Some perspectives in evaluating social welfare programs." Annals of the American Academy of Political and Social Science 385: 143-156.

HERZOG, E. (1959) Some Guidelines for Evaluation Research. Washington, D.C.: Government Printing Office.

HOPKINS, T. K. (1967) "Evaluation research," in Research Department, Community Council of Greater New York (comp.) Issues in Community Action Research. New York: Community Council of Greater New York.

LAPATRA, J. W. (1974) Applying the Systems Approach to Urban Development. Stroudsburg, Pa.: Dowden, Hutchinson & Ross.

MANN, J. (1971) "Technical and social difficulties in the conduct of evaluative research," pp. 175-184 in F. G. Caro (ed.) Readings in Evaluation Research. New York: Russell Sage Foundation.

MANTEL, S. et al. (1972) A Measurement Model for Planning and Budgeting for the Jewish Community Federation of Cleveland. Cleveland: Jewish Community Federation.

ROSSI, P. and W. WILLIAMS [eds.] (1972) Evaluating Social Programs: Theory, Practice, and Politics. New York: Seminar Press.

SERVICE, A. L., S. MANTEL, and A. REISMAN (1972) "Systems analysis and social welfare planning: a case study," in M. Mesarovic and A. Reisman (eds.) Systems Approach and the City. Amsterdam: North-Holland.

SHERWOOD, C. (1967) "Issues in measuring the results of action programs." Welfare in Review 5, 7: 13-17.

SHULBERG, H. and F. BAKER (1968) "Program evaluation models and the implementation of research findings." American Journal of Public Health 58, 7: 1248-1255.

SUCHMAN, E. (1971) "Action for what? A critique of evaluation research," in R. O'Toole (ed.) The Organization, Management and Tactics of Social Research. Cambridge: Schenkman.

WEISS, C. H. (1974) "Alternative models of program evaluation." Social Work 19 (November): 675-681.

——— (1973) "Where politics and evaluation meet." Evaluation 1, 3: 37-45.

——— [ed.] (1972a) Evaluating Action Programs: Readings in Social Action and Education. Boston: Allyn & Bacon. [Contains a substantial bibliography.]

——— (1972b) Evaluation Research: Methods of Assessing Program Effectiveness. Englewood Cliffs, N.J.: Prentice-Hall.

WEISS, R. S. and M. REIN (1969) "The evaluation of broad-aim programs: a cautionary case and a moral." Annals of the American Academy of Political and Social Science 385 (September): 118-132.

WHOLEY, J. S. (1972) "What can we actually get from program evaluation?" Policy Sciences 3: 361-369. [Contains a substantial bibliography.]

——— et al. (1970) Federal Evaluation Policy: Analyzing the Effects of Public Programs. Washington, D.C.: Urban Institute.

WILLIAMS, W. and J. EVANS (1969) "The politics of evaluation: the case of Head Start." Annals of the American Academy of Political and Social Science 385 (September): 133-142.

20

AN EPILOGUE ON POLICY

Alternative Agendas for Urban Policy and Research in the Post-Affluent Future

GARY GAPPERT

□ IF ONE DEVELOPS a socioeconomic perspective on the city, one is perforce concerned with policy relating to the control and non-control of the behavior of institutions and individuals. If one perceives the city as a series of socioeconomic systems, it is necessary to lay out policy and research agendas which explicitly concern themselves with questions of social value, purpose, and behavior. These questions may not be perfectly resolved, but they do lead to answers associated with institutional and organizational development, a primary component of urban development policy. The social development of cities has not been the main purpose of American urban policy. Instead, to paraphrase Winston Churchill: First we shape our buildings, then our buildings shape us.

AUTHOR'S NOTE: *Many of the perspectives in this paper have been developed in the urbane, collegial atmosphere of the Milwaukee campus of the University of Wisconsin. I am particularly indebted to Karl Flaming, John Ong, and Warner Bloomberg for their participation in a floating social systems seminar from which some of these ideas were first formed.*

This final chapter represents a simple attempt to lay out and discuss some concerns associated with urban policy questions. The analysis is directed by two major themes. First, a perspective on urban policy and problems must relate to a model of the future of the national social economy. The outline of such a model, sketched in Part I below, considers that American urban society, moving from an industrial period of growth—which began in the nineteenth century—to a postindustrial system in the late twentieth century, will experience a transition period of post-affluence. This post-affluent transition will be characterized by a series of economic and social dislocations which will contribute to a number of discontinuities in socioeconomic behavior.

Second, the analysis of urban policy is also directed around a series of alternative policy perspectives introduced in Part II below. Assuming that urban policy traditionally has been inchoate as well as misdirected by political rhetoric, four policy paradigms are suggested to better organize a discussion of the social ends and economic means of urban development. Part III is an attempt to project some policy and research agendas into the post-affluent decade ahead.

I. POST-AFFLUENCE AND THE COMING OF POST-INDUSTRIAL SOCIETY

It is likely that future historians will characterize the three decades from 1945 to 1975 in U.S. society as a period of rare but foolish affluence. During these decades American society used its unique economic power and prosperity to extend the American dream to its logical but somewhat absurd limits. President Kennedy once said, in the early 1960s, a rising tide floats all boats. In 1974 the energy crisis became the symbol of that tide of affluence disappearing into the political sands of the Middle East. As Russell Baker put it, "The meaning of 70 cents gasoline, $250 electric bills and 12 per cent mortgages is that the good life is dead." Celanese Corporation took out ads in major newspapers to indicate that the belief in the "limitless" nature of American resources had finally been recognized as an "illusion" and the "irrational gospel of more is being replaced by the more pragmatic gospel of less!"

In several ways 1974 represents a symbolic turning point in the evolution of the American social economy. The energy crisis was only one facet. The Congressional election in November reduced the median age of the Congress by almost twenty years, from the low 60s to the low 40s. Also witnessed was the first decline in all measures of real income since the Eisenhower recessions of the 1950s. Double-digit inflation, the so-called European disease, arrived on the shores of the New World along with rising unemployment; this was conceptually impossible according to neo-Keynesian economics. Double-digit mortgage rates sharply curtailed market demand for single-family housing; but overall consumer debt, led by installment purchasing, increased to a national level of over $560 billion by the end of the year. Delinquency ratios and personal bankruptcy cases, reaching new highs, became significant new indicators reported almost as often as the jobless rate.

At one level it was possible for some observers to claim that these manifestations of post-affluence in 1974 would be beneficial to cities. In this view the city itself was an energy-saving device. Edward Hamilton, then Deputy Mayor of New York, put it this way, "In the city you don't need a 300 horsepower car to go to the drugstore for a pack of cigarettes."

At a different level, however, other observers simply view 1974 as the year in which some fundamental contradictions in American society received public attention or recognition as certain evolutionary trends passed over minimum threshold levels, thereby signaling a sense of apparent crisis.

It is the perspective of this analysis that the apparent crisis is not some temporary aberration which can be tinkered away but a permanent condition representing a series of economic dislocations which will lead to a stepping-down of American material prosperity to a lower but perhaps more stable level. This condition of post-affluence is best viewed as a transition period, lasting perhaps for a generation or more, leading to some kind of post-industrial, post-affluent culture in the early twenty-first century.[1]

The characteristics of this post-affluence and the conditions setting it apart must be better understood if the social development of cities is to be the successful object of public policy in the decades ahead. Public policy in general can no longer be projected on the assumption of affluence—on the assumption of a larger share of a growing pie. Public policy specifically for cities, which have been the partial victims of affluence, must be projected on the assumption of a reallocation or better utilization of a zero-growth or declining amount of resources.

CONDITIONS AND CHARACTERISTICS OF POST-AFFLUENCE

To understand post-affluence requires a better perspective on the decades of affluence in American society. During those three decades, from 1945 to 1975, American society, after the years of deprivation experienced in the Depression of the 1930s and the sacrifices of World War II, attempted to buy back lost time by becoming the most lavish consumer society in history. The spending spree was financed in large measure by an explosion of personal credit, by federal subsidies for homeownership and highway construction, and by the floating of local bond issues to build the schools, sewers, hospitals, and other elements of the self-supporting growing suburbs organized to reflect the values of the self-supporting nuclear family (Gappert, 1974a, 1974b).[2]

The ratio of outstanding installment credit to disposable personal income increased from 1.4% in 1945 to about 14% by the mid-sixties, then crept up to about 16% by the end of 1974. That tapering up in the overall ratio masks the fact that American households increased their overall indebtedness from $400 billion in 1970 to over $560 billion by the end of 1974—a 42% increase in consumer credit. During the same period, after-tax personal income rose by only 37%, and government spending rose by less than 28%. This kind of credit-financed prosperity is being abruptly checked by several other conditions:

First, the baby-boom generation, the children born into this splurge of affluence, are now emerging into the workforce and adulthood. From 1964 to 1971, the 17-year-old was the largest single age group (as the colleges well know). From 1974 to 1981, the 27-year-old will be the largest single age group. This age group, this bulge generation, will be expressing all the strong consumer demands of young adulthood. They will have to be housed, employed, and provided with transportation, furnishings, energy for all the new households formed, and so forth. This generation's demand on resources will remain strong for the next three decades. (By 2004, the 57-year-old will be the "bulge" in the national demographic profile; the demands then will be more for maintenance resources.)

A second condition checking the growth of affluence is the change from a goods-producing labor force to a service-producing economy. Daniel Bell, who has chronicled the emergence of a post-industrial workforce since the late 1950s (*The Coming of Post Industrial Society*, 1973), projects that by 1980 over seven in every ten workers will be in service occupations.[3] The goods-producing sector will still be employing about 30 million workers (29 million in 1968), but this will represent only about 22% of the labor force. Econometricians project that only 2% of the labor force will be working in manufacturing by the year 2000. Some rather interesting projections can be made as to what services will be produced by such a post-industrial workforce (massage parlors, Karate schools, daycare centers, law firms specializing in services to professional athletes, fast-food site locators, airport security guards, and so on), and the relative proportions produced by the public and private sectors. (With revenue sharing it is projected that the federal share in all government expenditures will fall to below 40% by the early 1980s from a peak of over 60% in the 1950s. Local and state governments are producing more services in both old and new areas.)

But the real constraint on material prosperity as American society has known it comes from the inflationary bias built into the structure of a service-based economy. Between 1965 and 1970 the consumer price index rose about 30%. The price of automobiles and durable goods rose 15% to 18% while the price of services (medical care, insurance, recreation, and so on) rose over 40%.

Dennis Little (1973: 259-262)[4] describes the underlying condition in these words:

As a result of the existing system, unions have pushed up wages at an average rate of 7 per cent annually during the past four years, though productivity rose at only 3 per cent a year. In the manufacturing economy where the costs of labor are about 30 per cent of the total cost, a 10 per cent wage increase means only a 3 per cent increase in the cost of production and this frequently can be offset by production economies. In a service industry where the costs of labor may run 70 per cent or more of the total costs, a 10 per cent increase in wages adds 7 per cent to the cost of services. Productivity increases in the service sector currently average between 1.2 and 1.9 per cent.

This inflationary bias of a service-producing workforce cannot be easily removed by strictures on productivity. Instead, the society tends to participate in an elaborate game of "money-illusion" where everyone, or almost everyone, receives an increase in money wages. But oligopolistic power is required to ensure that the increase is greater than the decline in purchasing power.

Third, the problems of an evolving social economy organized around a growth in service employment are further compounded by a condition checking the expansion of material prosperity. This is the redistribution of income abroad to those countries possessing sizeable amounts of mineral or agricultural wealth. As more and more countries enter their period of mature industrialization, the cost of these raw materials will continue to rise sharply. Unlike Italy and some other countries, the United States is not facing bankruptcy due to the changing international political economy, but it will continue to experience various dislocations as its national markets respond to the forces involved. U.S. exports, for instance, rose from $8 billion annually to $20 billion annually in less than two years; it can afford to pay more abroad by selling more abroad, but the end results are shortages and higher prices domestically. A crash program of energy independence will have different but similar disruptive effects. The various estimates of the capital investment required range from $300 to $600 billion. A barrel of oil from Colorado shale is expected to cost over $10 as compared to $3 per barrel for North Sea oil. This is quite a change for a society accustomed to negotiating for Mid-Eastern oil, which still costs only about ten cents a barrel to produce.

These conditions then—the limits of credit-financed prosperity, the emergence of the bulge generation into consumerhood, the inflationary bias of a post-industrial workforce, and the shift in international comparative advantage—have combined to signal an end to the affluence of the industrial stage of growth in American society. They have created the circumstances of a post-affluent transition to a post-industrial society. The form of that society is still unclear, but one thing is certain: the precursors of post-industrial society which exist today are not adequate reflections of what post-industrial society will be as it emerges from the post-affluent period of transition. However, a number of consequences of post-affluence can be suggested, and a series of questions may be raised from which an agenda of policy issues can be generated.

CONSEQUENCES OF AFFLUENCE

The behavioral manifestations and consequences which will result as the national social economy evolves in the post-affluent transitional period cannot be precisely projected. On the one hand economics is facing a crisis of analysis; on the other the mixed capitalistic system is experiencing a crisis of control. The analytical and political crises stem from the same phenomenon—a series of market shifts and dislocations which are improperly understood. Three general considerations can be made. First, the economic and social structures of the society will increasingly overlap. Workstyle and lifestyle will become more congruent in the post-industrial society. The service economy demands new kinds of interpersonal relationships among workers, between workers and management, and between workers and consumers. Indeed, in the service economy, production and consumption are uniquely tied together. Some people will find these relationships to be a source of great tension; others will find them a source of great satisfaction.

Second, there will be more of a divergence between the culture and the socioeconomic structure. Cultural style will continue to emphasize self-indulgence and hedonistic manifestations while the economy will continue to require specialization, lengthy training and retraining, and a self-denying rationality.

A third consideration is that the socioeconomic structure, the cultural-media system, the political strata, and the technological institutions will each tend to separate in specialized ways. The divergence of these large, complex, interdependent national systems will contribute both to a proliferation of lifestyles and to a frustrating sense of unconnectedness.

A number of related consequences should be mentioned. First, there will be a diffusion of a kind of post-affluent consciousness. There will be a sizeable number of third and fourth generation Americans who remember affluence as a youthful excess experienced during the 1950s and 1960s. The old saying "from shirt sleeves to shirt sleeves in three generations" will take on new meaning as the grandchildren of the industrial society immigrants discover that they are not "making it" in post-affluent America. The natural decline of economic ambition in this group will be heightened by revulsion for the materialism of middle-class suburbia. Though the sons and daughters of blue-collar America will not necessarily share these reactions, a significant number of young Americans will cast a cynical eye at the conspicuous-consumption syndrome. "Post-affluent consciousness" will infect large numbers of people who may pride themselves on "living poor with style." Social thrift will become a fashion if not a value. The bulge generation, emerging into effective adulthood, will be the carriers of this post-affluent consciousness. A sizeable minority of this generation can be expected to initiate innovative and deviant lifestyles, values, and social demands. The spread of these new phenomena can be viewed as the accelerated or inhibited by various institutional controls and re-actions in different cities and regions.

Second, there will be a decline in the kind of nuclear family loneliness represented by the individualized consumption of sub-urban tract housing. There is no way for the more than 10 million new families which will be formed in the next decade to live in such splendid isolation. The capital to build yet another ring of suburbs just will not be available. The next decade or so will experience a boom in multifamily buildings complete with provisions for social space to be used for recreation, daycare, community meetings, and so on. New forms of collective consumption, voluntary and involuntary, will develop in mass transit, urban recreation, environ-mental protection, street security, and other fields.

Third, two complementary forces in urban development, the frontier of growth and the frontier of deterioration, will be slowed as less new housing is built and less older housing is abandoned. Inner suburbs and outer urban neighborhoods will experience redevelop-ment as they become important points of centrality in the policentric post-industrial city.

Fourth, intrafamily transfers of wealth will become an important phenomenon. The wealth accumulated by middle and upper-middle class Americans in the decades of affluence will be passed through to the next generation and will enable many sons and daughters in the post-affluent bulge generation to temporarily enjoy a living style in excess of their earnings.

Fifth, another consequence of the post-affluent social economy will be some major changes in the style of family life. The two-income family will become more important; the number of women in the workforce will increase. The househusband, sporad-ically employed, will not be unusual; single parent families, second marriages, and urban cooperative "families" will all increase in numbers and in diversity.

Sixth, the affluent population per se will not disappear. It may even increase in sheer numbers, but it will be relatively less well distributed, and new forms of exclusiveness will be developed as the upper-middle class learns a style of "subtle consumption."

The decades of post-affluence will provide an atmosphere of social turbulence, and all the consequences of this cannot now be understood even though different anticipations may be held. It is the nature of futurist types of study to project questions, not to provide precise answers. A series of questions for the post-affluent transition are suggested below. These questions can help chart different agendas for policy debates even though the outcomes of such debates cannot be known:

(1) Which segment of the post-affluent generation will prevail from the standpoint of political and social control? What will this generation do with its inherited wealth? Who will inherit the control of strategic institutions? How are "affirmative action" policies going to affect the distribution of power?

(2) Which deviant behavior and activity present today will diffuse and dominate? What new deviant activity will emerge during the period of post-affluence? What are the retarding and accelerating factors which will affect the forces of diffusions?

(3) How will the post-affluent consciousness change the nature of bureaucratic life? Will bureau-phobic men prevail over bureau-philic man? What public policies can influence the emergence of a post-industrial organizational form which will contribute to personal growth and development?

(4) What will be the nature of the crisis of economic redistribution? How will post-affluent society see to the effective control of capital and its judicious reinvestment? Will the inflationary destruction of affluence end with indicative planning or with a more authoritarian command economy?

(5) With the advent of a more humane and rational welfare system, what will be the nature of the urban underclass? Will poverty be less of a punishing experience? Will voluntary poverty be a social credential in the future? What kind of economic alternatives can be developed and subsidized for those who remain underskilled in a post-industrial economy, or for those who decline to participate in the attendant bureaucratic forms?

(6) According to *Evaluation* magazine, HEW programs could cost well over $200 billion than is currently being spent—nearly 15 million more people could be employed. What will happen to this "human service shortfall?" Will it become the WPA of the advent of the post-industrial society? How should human services evolve and be guided in the post-affluent transition?

(7) What will happen with the soft technology of "human self change?"

(8) Will the existing power centers in U.S. society maintain their power through the control of the multinational systems? Will the New Federalism lead to a flourishing of localism because it will allow the power elites of the national system to attend to the controlling of international systems?

Will entire communities participate in self-change through the techniques of "participatory futuristics?" Can "self-change" and "personal recycling" be effectively marketed?

POST-AFFLUENCE, THE CITY, AND THE SOCIAL DILEMMA

What then of the city in a period of post-affluence? Part of the fundamental problem is that popular expectations of what constitutes the good life have rested upon the assumption of an industrial growth economy with its rising production and distribution of material products. What will the good life come to mean in a service growth economy? A second part of the problem is associated with whether the political culture will choose to become more adept in encouraging a fairer distribution of resources or a more intense competition over discretionary resources. A third part of the problem is that the city, in American industrial society, has seldom been an acceptable symbol of a collective consumption ethic or of an integrative cultural style which links both pauper and prince. Instead the industrial city has been a symbol of individual struggle for private success or of ethnic competition for the control of public resources. The metropolis of the future must fit in a post-industrial society, largely engaged in abstract or indirect transactional activities, hungry for collective rites to offset social fluidity and transience, and increasingly oriented to non-material satisfactions and standards. Or perhaps it will be in the inner suburb (Oak Park, Chevy Chase, Bayonne), or in the outer neighborhood (Rogers Park in Chicago, Northridge in Milwaukee, Hiltonia in Trenton) that the turbulence of post-affluence can be avoided through the redevelopment of locality based associations.

But the social dilemma of a changing social economy within a society oriented to industrial growth expectations is real. The truth of the post-affluent condition must be voiced to the centers of power in American society. Different observers have expressed this dilemma differently. Kenneth Clark[5] (1975) puts it this way:

The paradoxical problem of American society is that it has been too successful.... The crisis of inconsistencies in American life—the American dilemma—is primarily a crisis of moral ambivalence. It is an honesty-dishonesty dilemma that pervades all dimensions of our social, economic, political, educational and, indeed our religious institutions.

This systemic dilemma within American society is complicated and probably made all the more virulent because it is inextricably entangled with status striving, success symbols, moral and ethical pretensions and the anxieties and fear of personal and family failures.

The conflicts in the American character structure and social system—conflicts intensified by the frequently fulfilled promises of upward mobility—must be resolved if the individuals are to continue to pursue the goals of status and success.

Anthony Downs,[6] speaking to the Republican Governors Conference at Hot Springs several years ago, said it this way:

In my opinion, our myths obscure an important hidden community of interest between the most deprived Americans and many of those in the middle class. We may not like to admit it, but most members of both groups need subsidies or government assistance of some kind to cope with some of the key urban problems that face them.

To do his job in the 70's, every elected official will have to face the need for many actions toward urban problems that are unpopular, that change long-established and cherished institutions and traditions, and that even the majority of citizens oppose at first.

II. ALTERNATIVE PERSPECTIVES ON
URBAN ANALYSIS

American society is not very accustomed to attempting to solve problems or to resolve its dilemmas by the use of what might be called "strategies." The idea of "strategy" (partially borrowed from both the military and foreign relations policy fields) implies the development of a single comprehensive, long-term plan to cope with some significant condition. A strategy may on the one hand involve a myriad of complex laws and amendments to laws enacted over several years or decades. On the other hand it may also involve a series of assumptions about behavior and a mixture of rules and rituals by which incentives are provided for desirable changes in behavior and penalties provided as sanctions against undesirable actions.

Alice Rivlin[7] has commented (1973) on the preoccupation of policy analysts with "objectives" rather than with strategies. (See also Downs, 1970: 37-39.) She writes:

One reason there is so little discussion of major strategies in the domestic area is that many people believe that the conditions that would make such a discussion worthwhile simply do not obtain. People concerned about domestic spending tend to be specialists. They worry about child development or health policy or legal services for the poor, but rarely about the choices among them. Anyway, "do-gooders" avoid difficult discussions about budgets and costs, preferring to make their case on moral grounds and taking refuge in the belief that if military spending could only be cut there would be plenty of resources for domestic programs and hard choices would be unnecessary. Moreover, goals of social programs often seem difficult to define and analysts appear to have little to contribute. But above all, no one is in charge. There is no unified social budget in the sense that there is a defense budget. Spending decisions are diffused and usually incremental, and there seems to be little point in arguing about alternative social strategies, because there is no obvious mechanism for translating general principles into budgetary and legislative decisions.

A strategy involves a set of basic assumptions about a style of action judged most appropriate to accomplish some specific aims. Since the strategic aims of American urban policy are so often left unspecified, it is more appropriate to suggest that sets of policy objectives fall into different paradigmatic categories.

THE CONCEPT OF PARADIGM

It may be that "paradigm" will be one of the most overused words of this decade. Ever since Kuhn wrote his *Structure of Scientific Revolutions* (1962), social theorists have been using the concept of alternative paradigms to juxtapose the different assumptions which lead public pursuits in alternative and sometimes opposite directions.

A paradigm consists of particular theories, techniques, beliefs, and values shared by the practitioners of a science. More specifically, it includes a particular set of symbolic generalizations, belief in the efficacy of certain well-defined models of reality, and shared values among practitioners as to the criteria for effective or adequate solutions.

Paradigms can be developed or constructed not to represent an exhaustive classification scheme of the phenomena in question, but rather to provide for the codification of inchoate concepts found in existing approaches to certain problems. The purpose of such paradigm construction is to expose or to make explicit the underlying, somewhat twisted logic which serves as the basis for the way in which practitioners and theoreticians proceed to interpret and to project social realities. (See Kramer, 1974: 78-106.)

Paradigms are more than models. Different models may be contained within a paradigm. Paradigms reflect more of a spirit of inquiry, a kind of world view through which beliefs and reality experiences are generalized into theory. Lacking any acceptable urban theory which is more than an expedient assemblage of propositions borrowed from sociology, political science, and other disciplines, the urbanist, be he scholar or practitioner, has to be more self-conscious of the biases which govern the way in which he formulates the problem and generates and selects "solutions" or "explanations."

There are perhaps four paradigms which can be used to characterize (or caricature) different approaches to understanding and interpreting urban systems for the purposes of strategic policy development. These approaches, shown in Figure 1, include: the technocratic, the utopian, the reconstructionist and the redemptionist paradigms.[8] The use of such paradigmatic categories are seen as a useful alternative to so-called interdisciplinary attempts to generate urban policy analysis. (Other alternatives include the inter-organizational analysis of Roland Warren [1974] and the intermediate systems analysis suggested by Leonard Goodwin [1973].)

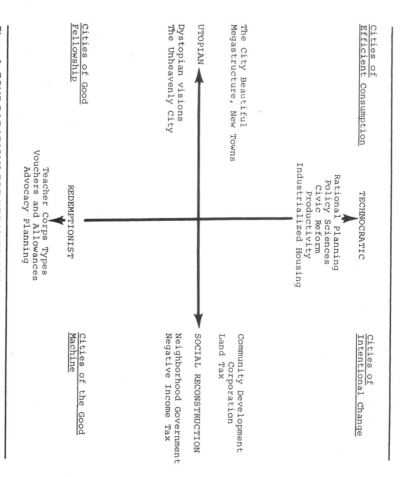

Figure 1: FOUR PARADIGMS FOR URBAN POLICY

THE FOUR PARADIGMS

The problems of paradigm construction are of special significance if the analysis is correct that the American social economy is experiencing a major evolutionary turning point in the mid-seventies. Yankelovich (1974), after completing a series of surveys, wrote:

These first few years of the decade of the 1970's point to vast changes in the complexion and outlook of an entire generation of young people. Indeed, so startling are the shifts in values and beliefs between the late 1960's when our youth studies were first launched and the present time that social historians of the future should have little difficulty in identifying the end of one era and the beginning of a new one. Rarely has a transition between one decade and the next seemed so abrupt and so full of discontinuities.

This kind of shift renders absurd a considerable number of policy objectives and program proposals which are only oriented to a treatment of "symptoms." Many policies are, of course, naturally absurd since they are offered without any regard for the structural context of the problem for which they have been designed as a solution.

President Johnson's assistant for domestic affairs put it succinctly when he (Califano, 1970) said, "The disturbing truth is that the basis of recommendations by an American Cabinet officer of whether to begin, eliminate or expand vast social programs more nearly resembles the intuitive judgment of a benevolent tribal chief."

At the same time however, the wise politician has learned the value of ambiguity in discussing policy objectives. Even the most technically-versed policy maker has cause to be cautious in offering a "clear definition" of the problem to all the actors on the policy systems. Peter McHugh (1968: 8) has discussed the problems of a "common" definition of the problem situation, saying, "They imply . . . that there is no one-to-one correspondence between an objectively real world and peoples' perspective of that world, that instead something intervenes when events and persons come together, an intervention that makes possible the variety of interpretation which Alfred Schwartz calls multiple realities."

Paradigm construction is an attempt to suggest what are some of the "multiple realities" associated with proposals to deal with American urban problems.

The *technocratic* perspective or paradigm is governed by the notion that the urban milieu is best perceived as an aggregation of substantially separable systems, each of which is associated with a different discipline or profession. This perspective seeks to determine which parts of the overall system need to be repaired, replaced, or simply made more efficient, and also generates a concern with seemingly precise planning which identifies the malfunctioning segment of the system and prescribes new techniques or technologies for implementation within the larger system.

The rational planning model and the more elaborate constructions of what is called policy science reflect the technocratic attempt to provide the urban decision maker with a process by which more "efficient" solutions can be reached. Such beliefs led to the development of **PPBS** systems and to the use of operations research techniques to locate fire stations and plan garbage routes. At a different level of the technocratic response to urban problems is an effort such as Operation Breakthrough with its promise of industrialized housing technologies. This effort is a good example of a major difficulty for the proponents of the technocratic paradigm: What do *you* do when the total system rejects the "efficient" new component?

There is little doubt that American cities could benefit from both new techniques and technologies. Besides providing direct benefits, new techniques are also indirectly useful in that their adoption and diffusion provides for access to participation for new kinds (usually younger) of specialists in municipal government. The problem with the technocratic paradigm is twofold. First, it assumes that a goal can be clearly posited and that all information with respect to alternative means is readily available. In open urban systems, meeting this assumption is often impossible or too costly. Second, the partial or incremental nature of technolocratic problem-solving may be appropriate for "muddling through" an immediate crisis but it is less appropriate and sometimes dysfunctional with respect to questions of social goals and values over which there is a substantial degree of conflict. The technocratic set of practices and perspectives does not operate successfully in environments where social purpose itself is changing. This has often led some technocrats to propose a more comprehensive or total systems approach to urban planning which is part of the utopian paradigm.

The *utopian* paradigm is characterized by attempts to affirm a well-defined perfect (or past) future state of the city as the object of efforts to understand and interpret urban realities. As Bloomberg (1972) has put it, "To the extent that we have increasingly come to hold highly sophisticated fantasies as the description of a reality which can be fully achieved once technological problems have been solved and rascals replaced in high office by those who will listen to us, to that extent we are devotees of illusion, romantics, dreamers, if you will: utopians."

In American society, utopian outcomes are not just generated by fanatics and poets but by hard-headed people who can get things done. The utopian myth of "managed progress"—which asserts that with the proper combination of skill, technology, resources, and good intentions, our society can make it all come out right in the end—has been a particular characteristic of urban industrial culture since the late nineteenth century when the "City Beautiful" and the garden city movements of Charles Burnham and Ebenezer Howard began. They find their latter expressions in Robert Moses (recently celebrated in Caro, 1974) and in the idea of a national system of new towns.

Such technocratic urban utopias are seldom suggested by technocrats but rather by those who can command both technique and technology. The "Marshall Plan for the Cities" notion, with its promise of billions of dollars to rebuild urban America, is a somewhat different manifestation of the utopian paradigm. This manifestation echoes the belief of the city planning profession that the built environment can be developed and manipulated sufficiently so that urban problems somehow disappear. The "success" of urban renewal in distributing social and economic conflict to other portions of the city has perhaps weakened that tenet of the utopian paradigm. But the notion that a dynamic, conflict-ridden urban environment can be adequately represented by an abstract model or sets of models combines with the notion that a well-planned and well-designed manbuilt environment can adequately contain urban behavior to preserve the utopian paradigm as a generation of urban policy options. From Oberman's (1972) proposals for the economy of Philadelphia to Newman's (1972) proposals for "defensible space," the utopian belief that society can successfully manage its way into the urban future guided only by a well-designed preconception is still the most important source of proposals dealing with the conditions of American cities.

Disillusionment with the impulses of utopian paradigm, with the belief in well-designed and well-managed progress, has become endemic in the wake of the alleged failure of the Great Society efforts which were aborted by the escalation of the Viet Nam war. This disillusionment is reflected by a *dystopian* paradigm which posits a well-defined future state of urban misery and repression. Sternlieb (1971) and Long (1971) in their projections of the inner city as a "sandbox" and a "reservation," have (along with Banfield [1970]) provided a dystopian component to this paradigmatic category. Current and past disagreeable urban realities are used in a model which assumes a static socioeconomic environment to be an adequate representation of the future. As with the rest of the utopian paradigm, the policy proposals so generated tend to be sterile and ineffective since they reflect an equally sterile representation of reality which is changing in ways unprescribed by the model used.

A more positive and humane response to the dystopian perspective is found in the assumptions of the *redemptionist* paradigm. This paradigm, although it is committed to a belief in the innate human potential for good, rejects or ignores the idea of perfectability, either in the solution of discrete problems or in the development of some ultimate urban system. The paradigm reflects not only the perspective of the French philosopher, Georges Lapassade (1967), who maintains that the individual is engaged by choice in situations without an end, in situations which remain "unachieved," but also Maslow's notion of self-actualization. Rather than developing or perfecting social forms or techniques which institutionalize the "good city," the redemptionist paradigm considers that what is redeeming is the way in which one responds to problems.

The success of one or another of the so-called solutions can neither be predicted nor experienced soon enough to serve as a guide to action. This paradigm does not agree that "maximum feasible participation" is only acceptable to the extent that it is compatible with getting the job done or with producing a measurable end product. Rather, it expresses the concern that people should experience the opportunity to both define and express their own needs and problems and be provided with the responsibility to determine the response which is to be made.

This paradigmatic category tends to generate policy proposals which relate to patterns of participation as well as with the regulation of processes (e.g., reforms in campaign financing). There is also a special concern with the producer-consumer relationship in the service sectors of government.

A central tenet of this paradigm tends to generate a degree of confusion. Does the belief in, and preference for, "better" interpersonal relationships and interactions imply an assumption that the human personality is malleable and easily pliable by behavioral modification? Are the policy proposals generated by this paradigm unduly influenced by the lifestyle preferences of those who seek to serve in the public interest and in the helping professions? Mutuality and responsiveness in relationships, self-actualizing experiences, spontaneity, sharing and a sense of mutual achievement, cooperative rather than competitive effort, social assertiveness and social risk-taking, a toleration of deviance—these are all personality attributes of the professional "making it" in an urbane environment unfettered by interpersonal embarrassments. Both Richard Sennett in *The Uses of Disorder* (1970) and Jane Jacobs in *The Death and Life of Great American Cities* (1961) represent the redemptionist paradigm in its most elaborate urban manifestation. It is in the disorderly processes of the complex city that an authentic personal identity is forged. As Sennett describes it prescriptively (1970: 169):

> Dense, disorderly cities would challenge the capacity of family groups to act as intensive shelters, as shields from diversity. For the whole thrust of these urban places will be to create a feeling of need in the individual that he has to get involved in situations outside the little routines of his daily life in order to survive with the people around him. . . . In a society where men could actually experience constant social change in their own lives, the inevitable dislocations involved in the change of generations could be borne with more grace.

It is unclear whether a greater amount of urban behavior is not oriented to escaping from such disorder rather than to embracing it. It is also unclear whether the "being" and "redeeming" processes of certain occupational lifestyles are not a function of the ability of that occupational group to claim a disproportionate amount of resources, either from the public sector or through intergenerational transfers. But support for "redeeming" life- and workstyles merits a place on the policy agenda.

The fourth paradigmatic category reflects a *reconstructionist* perspective. The social reconstruction paradigm tends to rest upon some general "gestalt" of the whole society and the evolution of the urban community within such a holistic conception. This perspective combines a concern with broad cultural themes with an ad hoc concern to intervene into the basic causal factors which direct the evolution of the overall social system. This paradigm is not particularly concerned with tinkering with discrete malfunctions per se, nor is it concerned with attempting to achieve radical interventions which will drive the system toward some kind of utopian outcome. The emphasis is more on long-range, large-scale programs, or even with the generation of new social movements and partially independent social systems.

On the one hand this paradigm is concerned with the distribution and diffusion of power (and the sources of power) within society. But on the other hand the effective development of policy from this paradigm requires an avoidance of the rather simplistic views of social power associated with the Radical Left. The problem of the reconstructionist paradigm is the problem of social strategy generally: correlated changes between the different components of a complex institutional system can be prescribed, but where does one stand to bring about the coordination of such simultaneous shifts? The proposals of Harrison (1974) for "deepening" urban development and for achieving the redevelopment of the inner city through community development corporations represent well-thought-out policies generated from a reconstructionist perspective.

But such proposals meet what Harrison calls the "failure of policy." He maintains that the institutional rules under which cities operate are rigid and out of date. As he writes:

It is this institutional inflexibility which is ultimately responsible for the increasing fiscal difficulties of central cities. What we face is the inability of constituted governments to design new political institutions and fiscal instruments to meet new conditions created in part by action of those very governments.

The reconstructionist paradigm, like the other three, contains its limits within its tenets. Can the prescriptive analysis of systemic change be viable from a vantage point within the system itself? Unlike technocratic analysis which is conceived with efficient movement toward some agreed-upon goal, reconstructionist analysis challenges the very goals of key sections of the system.

The four paradigmatic categories suggest both the problems and the prospects of analysis linked to the development of different strategies to be generated for the policy agenda for the American city in the post-affluent future.

The utopian paradigm assumes, in its physical or econometric designs, forms of behavior which are either too simple or too unchanging. The technocratic paradigm, with its emphasis on the technical improvement of portions of the overall system, neglects the composite problems of complex change and redirection as new goals evolve to which behavior seeks to conform. The reconstructionist paradigm prescribes complex change but lacks the cultural authority to initiate it without serious conflict, and often lapses into ad hoc, opportunistic reactions when major portions of the systems are identified as being in serious trouble. The redemptionist paradigm, with its heuristic admonitions to reform behavior, lacks a concern with the structure of systems and loses sight of the outcomes toward which processes are directed.

III. THE EVOLUTION OF NEW AGENDAS

American society is not undersupplied with policy proposals. It is perhaps less well supplied with research in support of policy development, and it is certainly short of strategic formulations toward which policy research of several kinds can be directed. It is suggested that paradigm construction is a device by which policies in support of different strategies can be related to a sharper sense of social purpose.

As the society shifts from the material affluence of an industrial stage of growth to a post-affluent, post-industrial condition, it lacks a reliable, predictive evolutionary model of change. Economic goals will be changing and the economic means to achieve some traditional social goals will no longer be easily available. Strategies of inquiry to develop different approaches to urban development are needed so that analytical capability is congruent with changing social belief and behavior.

TWO ADAPTATIONS: FROM PARADIGM TO POLICY

The paradigms suggested in the previous section can be used to illuminate policy development in distinct ways. Different paradigmatic assumptions can be combined to propose different urban outcomes, or a set of national policy proposals can be separated by different paradigmatic components.[9]

In Figure 1, four different city outcomes are suggested. The "City of Efficient Consumption" is an expression of the well-functioning city which reflects both utopian goals and technocratic processes. Some of the New Town proposals reflect such a combination of a well-designed urban form with the latest technocratic processes for everything from solid waste disposal to parental participation in education. Chamber of Commerce groups usually generate proposals which reflect such combinations.

The "City of Good Fellowship" represents a combination of the utopian assumption of a stable and internally consistent socio-economic structure with the redemptionist understanding that there is some finite set of urban relationships which insures a healthy urban culture. Elazar's notion of the medium-sized civic community which he analyzes in *Cities of the Prairie* (1970) reflects this

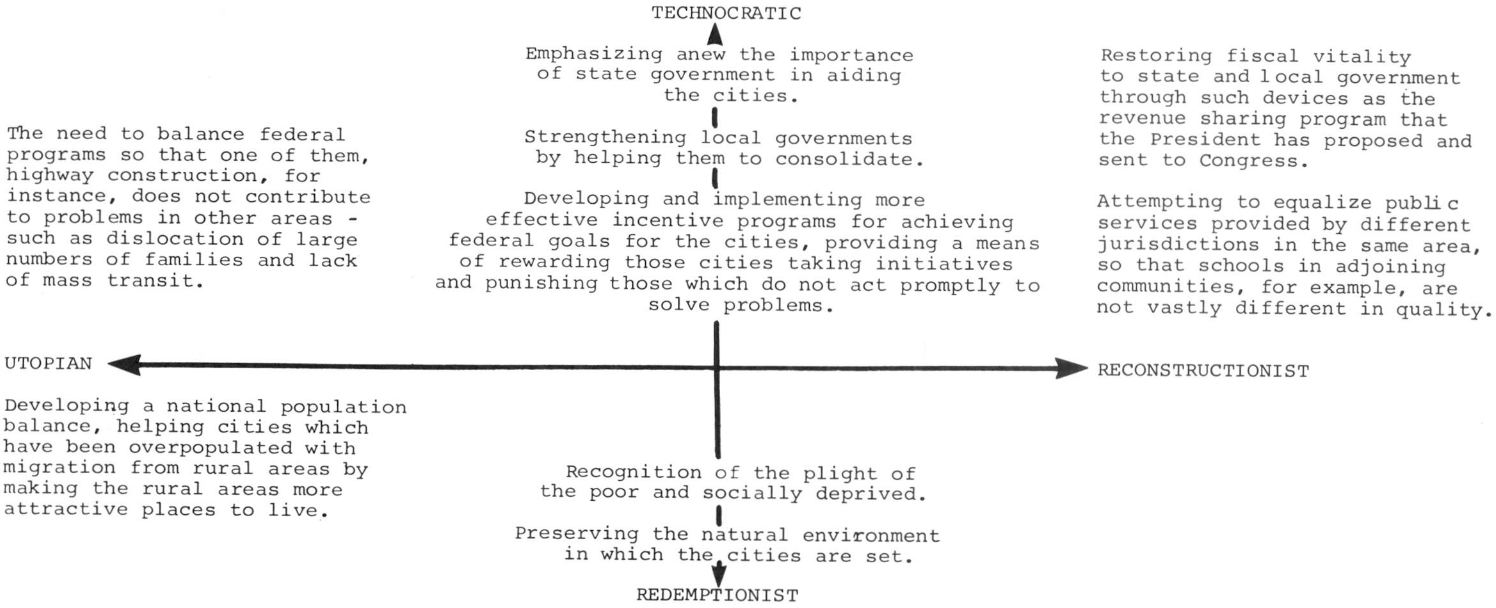

TECHNOCRATIC

Emphasizing anew the importance
of state government in aiding
the cities.

Strengthening local governments
by helping them to consolidate.

Developing and implementing more
effective incentive programs for achieving
federal goals for the cities, providing a means
of rewarding those cities taking initiatives
and punishing those which do not act promptly to
solve problems.

The need to balance federal
programs so that one of them,
highway construction, for
instance, does not contribute
to problems in other areas -
such as dislocation of large
numbers of families and lack
of mass transit.

Restoring fiscal vitality
to state and local government
through such devices as the
revenue sharing program that
the President has proposed and
sent to Congress.

Attempting to equalize public
services provided by different
jurisdictions in the same area,
so that schools in adjoining
communities, for example, are
not vastly different in quality.

UTOPIAN ←———————————————————————————→ RECONSTRUCTIONIST

Developing a national population
balance, helping cities which
have been overpopulated with
migration from rural areas by
making the rural areas more
attractive places to live.

Recognition of the plight of
the poor and socially deprived.

Preserving the natural environment
in which the cities are set.

REDEMPTIONIST

SOURCE: Moynihan (1970).

Figure 2: MOYNIHAN'S ELEMENTS OF URBAN POLICY

combination. The advocates of decentralized neighborhood government are projecting cities of good fellowship in the midst of the megalopolis.

The "City of the Good Machine" combines the reconstructionist notion of a change in the urban governing elite with the redemptionist desire to make delivery systems more responsive to new client populations. Hatcher's Gary and Jordan's Jersey City are often cited as exemplars of this kind of proposed urban outcome. The experiences of Carl Stokes in Cleveland and Mayor Gibson in Newark are more contradictory in their evidence that the traditional urban political machine can be taken over and redirected through an emphasis on redemptionist processes.

The "City of Intentional Change," with its combination of technocratic processes directed to new social goals which eliminate the inequities of the previous socioeconomic order, is an urban outcome more difficult to place. Harrison's (1974) proposals for ghetto redevelopment are illustrative as is Hansen's proposal (1970) to redirect unemployed labor to medium-sized growth centers. Downs' (1973) proposals for "opening up the suburbs" also combine technocratic understandings with a belief in the need to reconstruct urban social purposes.

A different adaptation of the four paradigms can be found in Figure 2 where the ten points of Moynihan's national urban policy are laid out. The policy agenda as announced in 1970 seemed to be humane, comprehensive, and practical. In retrospect it appears to have been primarily technocratic in its implementation. But at the same time the revenue-sharing provisions of the New Federalism can be seen as a reconstructionist ploy which shifted the initiative and resources for urban social policy from the federal government to the local governing elites. At the present time, it is still unclear whether that shift will lead to additional social reconstruction as local groups struggle to understand and participate in the urban budget making process.

THE PARADIGMS AND THE POST-AFFLUENT DILEMMA

It is less clear that the device of paradigm construction can be used to develop a socioeconomic policy strategy which can come to grips with the urban condition in a time of post-affluence.

The utopian paradigm could be used to suggest two design

strategies. First, the national economy sorely needs a comprehensive planning system concerned with the composition and distribution of economic activity. Some set of interurban regional accounts is necessary. This would encompass economic activities, demographic movements, financial transactions, and unemployment and job vacancy rates. The model would not have to be prescriptive but could be indicative. A second design strategy would deal with the prospects of energy and material-saving megastructures such as those proposed by Paolo Soleri and George Danzig. These highly dense and complex fantasies are not of immediate practical value, but utopian proposals have their value in the inspirational qualities of their strategy (e.g., international disarmament).

In a post-industrial economy, the redemptionist paradigm is clearly needed to provide proposals to insure more humane interpersonal relationships in both industrial and bureaucratic workplaces. Proposals for job enrichment, for public accountability, and for community participation all represent a strong redemptionist initiative which will be even more elaborated and sophisticated as the post-affluent condition continues.

In a time of scarce resources, the efficiency imperatives of the technocratic paradigm will not be ignored. More likely, reliance on some kind of "technocratic fix" to get society out of its post-affluent difficulties will obscure the need for a better social assessment of technology. Current attempts to design more efficient and effective welfare measures, tax reforms, and housing allowances are technocratic policy initiatives which will continue to expand. A "distributional impact" device to assess the equity affects of government policy is clearly needed, as is some capacity to develop and field test institutional innovations at the scale of the region and locality.[10]

The further direction of policy strategies generated from the reconstructionist paradigms is less clear. Policy which can facilitate the development of intentional communities is needed. Support for the trying out of deviant lifestyles in the face of new scarcities would be appropriate. Tying revenue-sharing to a few experiments of metropolitan integration (such as those partially attempted in Dayton, Ohio and Hartford, Connecticut) would be useful. It is not clear that the make-work proposals associated with emergency public employment are as desirable in a reconstructionist sense as would be a more ambitious effort to redistribute desirable work through a shorter workweek. This latter kind of proposal, however, raises questions about a synergistic lifestyle which integrates work, leisure, and intimacy to bring the society back to a concern with its utopian visions.

CONCLUSION

Ultimately the development of urban policy in a post-affluent future is the study of who we are and how it is that we should express ourselves as members of an urban post-industrial culture. The commitment to urban analysis is a commitment to an urbane civilization which can support networks of social relationships in an atmosphere of pluralistic tolerance and humane subsistence.

To survive and thrive as members of communities in a metropolitan nation facing a transition to a post-affluent, post-industrial condition will require new unifying myths by which we may organize our economic activities and social affairs. At one level this may be viewed as a naive, existential commitment. But society must, in the words of Robert Kennedy, "dissent from cities which blunt our senses and turn the ordinary acts of daily life into a painful struggle." A longer time ago, Edmund Burke put it this way,

Men are qualified for civil liberties in exact proportion to their disposition to put moral chains on their appetites. Society cannot exist unless a controlling power upon will and appetite be placed somewhere, and the less of it there is within, the more there must be without . . . men of intemperate mind cannot be free. Their passions forge their own fetters.

NOTES

1. This post-industrial, post-affluent culture has been projected by some as "The Greening of America." Two good projections are T. Roszak (1972) and Willis Harman (1974).

2. I am indebted to three graduate students, Joan Grosz, Jack Bayer, and Russ Dixon, for their work with me on the conditions of post-affluence.

3. The December 1973 issue of *The Futurist* has two good critiques of Bell's interpretations of post-industrial society. These are by Michael Marion and Dennis Little,

two outstanding professional futurists. Sociologist Daniel Bell, in analyzing the post-industrial economy since the late 1950s, lists its five principal elements as: (1) the change from a goods-producing labor force to a service-producing economy with an increase of services sold to other services; (2) the growing influence of the professional and technical class throughout society; (3) the centrality of theoretical knowledge as the source of innovation and of policy formulation for society; (4) increased control of, and planning for, technological growth; and (5) the creation of a new intellectual technology to make possible the management of large-scale systems through integrated information manipulation.

4. Some of the "economic" questions which will be persistent in a post-industrial service society are: (1) What types of institutions will restrict or facilitate the working out of the pricing mechanism? (2) What will be the nature of the supply/response mechanism in the various service sub-sectors? (3) What will be the persistent forms of private ownership of land and capital? (4) What will be the nature of the financing of personal service consumption over the life cycle? and (5) How will the public economy operate with regard to public demands for technology, amenity space, and quality of life and environment?

5. There are those who think that the anguish of affluence has significantly warped the American social character and who welcome the realism of post-affluence.

6. December 12, 1969.

7. Rivlin identifies four alternative strategies for federal social policy. These are: (1) individual cash transfers; (2) individual service allowances including vouchers; (3) revenue sharing and bloc grants to local governments; and (4) institution changing attempts including categorical grants.

8. The notion of a need for a paradigm "shift" to more adequately express a changing view of the world is widely held by many social theorists. I am indebted to John Ong, Alexe Christakis, and Dick Wakefield for their leadership in discussing this concern. Warner Bloomberg has had a major influence on me in developing this approach to paradigms for the urban fields.

9. The use of and need for paradigms as a source of policy reflects Daniel Bell's assumption of the centrality of theoretical knowledge in the policy formulation process.

10. Leonard Goodwin (1973) has discussed a form of "experimental social research" which could be adapted to a process of research for urban innovations. Unfortunately, the Urban Observatory network has been too closely tied to the quasi-technocratic concerns of elected officials to achieve much in this area.

REFERENCES

BANFIELD, E. (1970) The Unheavenly City. Boston: Little, Brown.

BELL, D. (1973) The Coming of Post Industrial Society. New York: Basic Books.

BLOOMBERG, W., Jr. (1972) "Urban studies: a personal assessment." Jersey City, N.J.: Council of University Institutes of Urban Affairs, St. Peters College.

CALIFANO, J. A. (1970) Washington Post (January 7).

CARO, R. (1974) The Power Broker. New York: Knopf.

CLARK, K. (1975) New York Times (February 16).

DOWNS, A. (1973) Opening Up the Suburbs. New Haven, Conn.: Yale University Press.

——— (1970) Urban Problems and Prospects. Chicago: Markham.

ELAZAR, D. (1970) Cities of the Prairie. New York: Basic Books.

GAPPERT, G. (1974a) The Future of Work in Post-Affluent America: A Post-Industrial Perspective. Milwaukee: Manpower, Inc.

——— (1974b) "Post affluence: the turbulent transition to a post-industrial society." The Futurist 8, 5 (October).

GOODWIN, L. (1973) "Bridging the gap between social research and public policy." Journal of Applied Behavioral Science 9, 1 (June).

HANSEN, N. (1970) Rural Poverty and the Urban Crisis. Bloomington: Indiana University Press.

HARMAN, W. (1974) "Humanistic capitalism." Journal of Humanistic Psychology (Winter).

HARRISON, B. (1974) Urban Economic Development: Suburbanization, Minority Opportunity and the Central City. Washington, D.C.: Urban Institute.

JACOBS, J. (1961) The Death and Life of Great American Cities. New York: Random House.

KRAMER, K. L. (1974) Policy Analysis in Local Government. Washington, D.C.: International City Management Association.

KUHN, T. S. (1962) The Structure of Scientific Revolutions. Chicago: University of Chicago Press.

LAPASSADE, G. (1967) Groupes, Organisation et Institutions. Paris: Gauthier-Villars.

LITTLE, D. (1973) "Post-industrial society and what it may mean." The Futurist 7, 5 (December).

LONG, N. (1971) "The city as reservation." The Public Interest 75 (Fall).

McHUGH, P. (1968) Defining the Situation: The Organization of Meaning in Social Interaction. New York: Bobbs-Merrill.

MOYNIHAN, D. P. (1970) Towards a National Urban Policy. New York: Basic Books.

NEWMAN, O. (1972) Defensible Space. New York: Macmillan.

OBERMAN, J. (1972) Planning and Managing the Economy of the City. New York: Praeger.

RIVLIN, A. (1973) "Social policy alternative strategies for the federal government." Reprint Series. Washington, D.C.: Brookings Institution.

ROSZAK, T. (1972) Where the Wasteland Ends: Politics and Transcendence in Postindustrial Society. New York: Doubleday.

SENNETT, R. (1970) The Uses of Disorder: Personal Identity and City Life. New York: Knopf.

STERNLIEB, G. (1971) "The city as sandbox." The Public Interest 75 (Fall).

WARREN, R. et al. (1974) The Structure of Urban Reform. Lexington, Mass.: D. C. Heath.

YANKELOVICH, D. (1974) The New Morality. New York: McGraw-Hill.

THE AUTHORS

MICHAEL BARNDT is Assistant Professor in the Department of Urban Affairs, University of Wisconsin—Milwaukee. His principal interests include strategies of neighborhood change, community development, and limitations of urban decision-making technology.

BRIAN J.L. BERRY is Irving B. Harris Professor of Urban Geography at the University of Chicago, where he is also Chairman of the Geography Department and Director of the Center for Urban Studies. His recent books include *The Human Consequences of Urbanization* and *The Geography of Economic Systems*.

JOHN P. BLAIR is Assistant Professor of Urban Affairs and Business Administration, University of Wisconsin—Milwaukee. His research interests lie in the fields of regional economics and public finance.

JESSE BURKHEAD is Maxwell Professor of Economics at Syracuse University. His field is public finance, and his most recent (co-authored) volume is entitled *Public Expenditure*.

CHARLES M. CHRISTIAN is Assistant Professor in the Department of Geography and Institute of Urban Studies, University of Maryland. He is author of *Three Papers on Social and Economic Aspects of the Black Community of Chicago*.

JULES COHN is Professor of Political Science, City University of New York. His most recent book is *The Conscience of the Corporations*.

MICHAEL J. FLAX is Senior Research Associate at the Urban Institute. He has conducted research and published in the areas of computer and communication system applications, management techniques, and social indicators.

GARY GAPPERT, an economist on leave from the Department of Urban Affairs, University of Wisconsin—Milwaukee, is currently serving as Assistant Commissioner of Education in New Jersey in charge of the Division of Research, Planning, and Evaluation. His interest in the social economy of cities derives from previous work on futuristics, economic inequality, and urban social planning.

HARVEY A. GARN is Project Director for Urban Development, Processes, and Indicator Research at the Urban Institute. In addition to his social indicator research, he is currently directing a policy study of large-scale community development corporations.

JEAN-ELLEN GIBLIN is Assistant Professor of Economics and Chairperson of the Social Science Department at the Fashion Institute of Technology, State University of New York. She is also an Associate of the Research Center for Economic Planning, New York.

AMIRAM GONEN is on the faculty of the Department of Geography, Hebrew University of Jerusalem. His main field of research has recently focused on the spatial structure of Israeli cities.

ANN LENNARSON GREER is Assistant Professor of Urban Affairs and Sociology at the University of Wisconsin—Milwaukee. She is currently on leave to the Office of Planning, Evaluation, and Legislation of the Health Resources Administration. Her most recent book is *The Mayor's Mandate: Municipal Statecraft and Political Trust*.

SCOTT GREER, Professor of Sociology and Urban Affairs at the University of Wisconsin—Milwaukee, is currently (1974-1975) Senior Advisor for Social Research, Office of Scientific Affairs, Health Resources Administration. His published works include *Governing the Metropolis*, *The Logic of Social Inquiry*, and *The Urbane View*.

DAVID HARVEY is Professor of Geography at Johns Hopkins University. He is the author of *Social Justice and the City* (1973) and numerous articles on the theme of the political economy of urbanization in advanced capitalist countries.

DAVID LEY is Assistant Professor of Geography at the University of British Columbia. He is the author of *The Black Inner City as Frontier Outpost* and editor of *Community Participation and the Spatial Order of the City*.

MARTIN D. LOWENTHAL is Director of the Social Welfare Regional Research Institute, Co-Director of the Urban Affairs Program, and Assistant Professor of Sociology—all at Boston College.

ROBERT MIER is an Associate and Research Coordinator at the Research Center for Economic Planning, New York. In September 1975, he will join the Planning faculty at the University of Illinois—Chicago Circle.

HUGH O. NOURSE is Professor and Chairman of the Economics Department at the University of Missouri—St. Louis. He is the author of *Regional Economics* and *The Effect of Public Policy on Housing Markets*.

CERI PEACH is Lecturer in Geography and Fellow of St. Catherine's College, Oxford University. He is author of *West Indian Migration to Britain* and editor of *Urban Social Segregation*.

DONALD PHARES is Associate Professor of Economics and Fellow in the Center of Community and Metropolitan Studies at the University of Missouri—St. Louis. He is author of *State-Local Tax Equity: An Empirical Analysis of the Fifty States*.

HAROLD M. ROSE is Chairman of the Department of Urban Affairs, University of Wisconsin—Milwaukee, and Professor of Geography and Urban Affairs. His published works include *The Black Ghetto: A Spatial Behavioral Perspective* and numerous articles dealing with migration patterns of black movers.

SEYMOUR SACKS is Professor of Economics at the Maxwell School of Syracuse University. He is the author of *Metropolitan America* and *City Schools/Suburban Schools*.

MILTON SANTOS is Professor of Geography at the University of Dar-es-Salaam. He is author of *Les Villes du Tiers Monde* and *The Shared Space*.

JEREMY B. TAYLOR is a Research Associate at the Urban Institute. His research and publications have been in the areas of cross-cultural training and social and urban indicators.

MICHAEL SPRINGER is a member of the Senior Research Staff of the Urban Institute. He has served as a consultant to federal and local government agencies in the areas of housing policy, government planning, and social indicators.

THOMAS VIETORISZ is Professor of Economics on the Graduate Faculty of Political and Social Science, New School for Social Research, and President of the Research Center for Economic Planning, New York. He is co-author (with Bennett Harrison) of *The Political Economy of Harlem*.

DAVID C. WARNER is Lecturer in the Department of Epidemiology and Public Health and the School of Organization and Management and Research Associate at the Center for the Study of Health Services in the Institution for Social and Policy Studies—all at Yale University.

STUART WINCHESTER is Research Officer in the Community Studies Section of the British Home Office Research Unit. His current interests are the development of territorial social indicators and testing the validity of "area" approaches to social deprivation.

ROBERT WOODS is Lecturer in Quantitative Social Science at the University of Kent at Canterbury. His research is directed toward urban social analysis and race relations in Britain.

3 5282 00024 3207